Biochemical and Biological Markers of Neoplastic Transformation

NATO Advanced Science Institutes Series

A series of edited volumes comprising multifaceted studies of contemporary scientific issues by some of the best scientific minds in the world, assembled in cooperation with NATO Scientific Affairs Division.

This series is published by an international board of publishers in conjunction with NATO Scientific Affairs Division

A Life Sciences **B Physics**	Plenum Publishing Corporation New York and London
C Mathematical and Physical Sciences	D. Reidel Publishing Company Dordrecht, Boston, and London
D Behavioral and Social Sciences **E Applied Sciences**	Martinus Nijhoff Publishers The Hague, Boston, and London
F Computer and Systems Sciences **G Ecological Sciences**	Springer Verlag Heidelberg, Berlin, and New York

Recent Volumes in Series A: Life Sciences

Volume 50—Somatic Cell Genetics
edited by C. Thomas Caskey and D. Christopher Robbins

Volume 51—Factors in Formation and Regression of the Atherosclerotic Plaque
edited by Gustav R. V. Born, Alberico L. Catapano, and Rodolfo Paoletti

Volume 52—Chemical Carcinogenesis
edited by Claudio Nicolini

Volume 53—Cancer of the Prostate and Kidney
edited by M. Pavone-Macaluso and P. H. Smith

Volume 54—Leukotrienes and Prostacyclin
edited by F. Berti, G. Folco, and G. P. Velo

Volume 55—Durable Resistance in Crops
edited by F. Lamberti, J. M. Waller, and N. A. Van der Graaff

Volume 56—Advances in Vertebrate Neuroethology
edited by Jörg-Peter Ewert, Robert R. Capranica, and David J. Ingle

Volume 57—Biochemical and Biological Markers of Neoplastic Transformation
edited by Prakash Chandra

Volume 58—Arterial Pollution: An Integrated View on Atherosclerosis
edited by H. Peeters, G. A. Gresham, and R. Paoletti

Biochemical and Biological Markers of Neoplastic Transformation

Edited by

Prakash Chandra

University of Frankfurt School of Medicine
Frankfurt am Main, Federal Republic of Germany

Springer Science+Business Media, LLC

Proceedings of a NATO Advanced Study Institute on
Biochemical and Biological Markers of Neoplastic Transformation,
held September 28–October 8, 1981,
in Corfu, Greece

Library of Congress Cataloging in Publication Data

NATO Advanced Study Institute on Biochemical and Biological Markers of Neoplastic Transformation (1981: Corfu, Greece)
Biochemical and biological markers of neoplastic transformation.

(NATO advanced science institutes series. Series A, Life sciences; v. 57)
"Published in cooperation with NATO Scientific Affairs Division."
"Proceedings of a NATO Advanced Study Institute on Biochemical and Biological Markers on Neoplastic Transformation, held September 28–October 8, 1981, in Corfu, Greece"—Verso t.p.
Includes bibliographical references and index.
1. Carcinogenesis—Congresses. 2. Cancer cells—Congresses. 3. Tumor proteins—Congresses. 4. Cell transformation—Congresses. I. Chandra, Prakash, 1936– . II. North Atlantic Treaty Organization. Scientific Affairs Division. III. Title. IV. Series. [DNLM: 1. Neoplasms—Etiology—Congresses. 2. Cell transformation, Neoplastic—Analysis—Congresses. QZ 202 N278b 1981]
RC268.5.N35 1981 616.99′4 82-24548
ISBN 978-1-4684-4456-8 ISBN 978-1-4684-4454-4 (eBook)
DOI 10.1007/978-1-4684-4454-4

Originally published by Plenum Press, New York in 1983
Softcover reprint of the hardcover 1st edition 1983
A Division of Plenum Publishing Corporation
233 Spring Street, New York, N.Y. 10013

PREFACE

This volume is a record of the proceedings of a NATO Advanced Study Institute on "Biochemical and Biological Markers of Neoplastic Transformation" held September 28 - October 8, 1981, at Corfu, Greece.

As early as 1860, Rudolf Virchow provided the first genetic concept of cancer by postulating "*Omnia cellula e cellula ejusdem generis*", a modification of the then exisiting cell theory "*Omnis cellula e cellula*". Thus, the idea that all cells originate from the parent cell was extended to the idea that all cancer cells come from the "*parent*" cancer cell. But how the first cancer cell arose, or in other words, how a normal cell changed to a cancer cell, is, even after 120 years, a mystery.

Experimental studies of the past have convinced us that a number of factors contribute to the neoplastic transformation of a normal cell, but our knowledge on the mechanisms involved in this process is still in an embryonic state. In the last few years, however, this field has witnessed a most remarkable advancement catalyzed by the development of modern technology in the allied fields of immunology, the production of monoclonal antibodies, molecular biology, and sequencing and cloning of *ONC* genes. Presently, it is becoming more and more evident to the wishful mind of those engaged in this research that we are approaching a turning point. Thus, an assessment of the present situation will be most desirable at this time. The purpose of this Advanced Study Institute was to solidify the present state of the art and stimulate new approaches.

It is hoped that the concepts and information presented by the contributors will shed light on the turning point to the right direction.

March, 1982
Frankfurt

Prakash Chandra

ACKNOWLEDGMENTS

I am especially indebted to Robert C. Gallo (Bethesda, U.S.A.), Peter Duesberg (Berkeley, U.S.A.), Michael Feldman (Rehovot, Israel), Fred Rapp (Hershey, U.S.A.), Constantin Sekeris (Athens, Greece) and Harald zur Hausen (Freiburg, Germany), who served as members of the scientific committee of this Advanced Study Institute. Their specific suggestions and general encouragement provided invaluable assistance in designing the concept of this Study Institute. I am grateful to the session chairpersons, N. Mitchison, E.S. Lennox, M. Seilgmann, Fred Bollum, R.J.B. King, H. Kaplan, Eva Klein, and Heinz Bauer, for organizing stimulating discussions.

I wish to express my appreciation to the NATO Scientific Affairs Division, the support of which made this Institute possible. This Institute was also supported by a Conference Grant (Gran No. 1R13 CA 31717-01A1) from National Cancer Institute, U.S.A.

The help of my coworkers, V. Paffenholz, Angelika Vogel, Marilena Rezzonico and Ilhan Demirhan has been invaluable in the organization and management of this Institute.

Last, but not least, I want to thank the contributors and the publishers, especially Dr. Andrews, for making the publication of this volume possible.

March, 1982 Prakash Chandra
Frankfurt

CONTENTS

ANTIGENS AND IMMUNE REACTIVITY

IMMUNOBIOLOGY OF METASTASIS

MHC - ANTIGENS ON TUMOR CELLS

GENETIC DETERMINANTS OF CANCER

BIOCHEMICAL MARKERS AND ASSOCIATED ACTIVITIES IN NEOPLASIA

SOME CONSIDERATIONS ON HLA AND MALIGNANCIES

Jean Dausset

Hôpital Saint-Louis
Paris

At the beginning of this conference, it is certainly worthwhile to survey our present knowledge of the possible association or linkage between human malignancies and the most powerful and polymorphic marker : the HLA system. Indeed a considerable number of studies have been carried out since the first two studies done in France on acute lymphoblastic leukemia[1] and Hodgkin's disease[2]. This research was rather disappointing in comparison with the rich harvest obtained by the study of many other non-malignant diseases[3,4,5]. It is striking that among the diseases undoubtedly associated with HLA, there are practically no infectious diseases due to a well-known agent, and no malignant diseases.

However, the role of the MHC in malignancies is clearly demonstrated in animal experiments. This intervention can take place at several levels.

1. Products of the HLA genes themselves, or very closely linked genes, could act as receptors of the specific agent (? virus) or of the oncogenic chemical.

2. The role of the two classes of MHC products in the immune response is now well-established : class I products are involved in the presentation of the antigen to T lymphocytes; class II formed with the antigen, the target for the T effector lymphocyte. A hierarchy seems to exist in the ability of MHC allelic products for form a complex with the antigen. This hierarchy could explain susceptibility and resistance of some strains of mice,

for example, to leukemia. Sometimes the resistance does not preclude the appearance of the disease but only influences its progression and can be measured in terms of survival time.

3. The MHC also contains the structural genes of some complement factors (C2, C4S, C4F, Bf in humans) which could be important for the immune defence.

4. The metastic phenomenon is analogous to the homing process, which has been shown to be governed by the class I MHC products[6].

5. Altered or alien histocompatibility antigens have been demonstrated in some malignant cells.

6. The HLA complex may contain unknown metabolic genes involved in the malignant transformation.

7. The MHC is also known to genetically control the level of cyclic AMP and of some hormones.

This list is certainly not exhaustive.

In spite of all these possible mechanisms, most of which were demonstrated in animals, data on human beings remains astonishingly poor. This probably reflects the polygenic control of cancer in the outbred human population.

The methodology to overcome this difficulty is limited. Below are given the possible lines of research which have been taken up until now, with limited success.

1. A linkage disequilibrium between a gene of the HLA complex conferring susceptibility or resistance to a given malignant disease (the so-called association) has been looked for in the total population of patients without sub-division and also in various categories (age at onset, sex, response to treatment, metastasis, survival, etc).

2. Familial studies of rare families with several cases are the most promising approach. By genotyping the patient, it is possible to establish a homozygosity for HLA antigens or haplotype. The sharing of antigens or haplotype in parents has the same meaning. Segregation of this disease among children could be very instructive, in looking at the HLA status (HLA identical semi-identical or different) of the pairs of affected

sibs. The number of each category of pairs can be pooled. An excess of HLA identical pairs suggests a recessive mode of transmission. Segregation with the same haplotype would suggest a dominant mode of transmission.

General reviews on this subject were published in 1977,[7] and in 1979.[8] Since this time numerous publications have appeared. We have made a selective choice from among the former and new studies, illustrating the various methodologies mentioned above.

I. POPULATION STUDIES

A. Acute Lymphoblastic Leukemia (ALL)

Since the first study on ALL, several studies have shown a remarkable tendancy in the general patient population towards a slight excess of HLA-A2. The results of the latest edition of the International Registry, (not yet published) kindly communicated to us by Ryder and Svejgaard,[9] underline this association. Although rather weak, this is now well-established (Table I). It has been confirmed by Albert et al.,[10] using a large series of patients and healthy sibling controls. Part of the excess of HLA-A2 disappears in the prospective studies. This is in keeping with the observation of Rogentine et al.,[11] that the HLA-A2 antigen is a marker of long-term survival. Re-examining our series (after 7 years), we also have observed an excess of HLA-A2 among long-term survivers[12]. This finding is a true genetical influence or may reflect transfusion therapy with HLA compatible blood and the frequency of HLA-A2 in the general population (≃50%).

The association with HLA-B12 (Table I) is probably due to the well-established linkage equilibrium with A2.

Another interesting sub-group has been isolated by Tursz et al.,[13] that of ALL children treated with BCG therapy. Among the 13 patients, 4 were HLA-Aw33. This observation deserves confirmation.

B. Hodgkin's disease

The original observation by Amiel, that the public antigen 4C (comprising B5, B15, B18 and Bw35) was increased in Hodgkin's disease, was controversial. Fourteen years later, the finding is still being debated. However a clear excess of A1 is now apparent in the pooled data[9].

Table I

HLA	N° of Studies (N° of Patients)	Range of Frequencies Patients	Controls	R.R.	P <
		HLA and ALL*			
A2	15(1099)	46-68	37-60	1.39	6.10^{-7}
B12	15(1099)	17-75	16-36	1.24	3.10^{-3}
B5	14(1055)	9-26	8-24	1.31	3.10^{-3}
B15	13(988)	4-21	3-20	1.30	9.10^{-3}
		HLA and Hodgkin's Disease*			
A1		29-62	20-49	1.37	10^{-10}
B5	27(2841)	0-25	3-22	1.33	10^{-6}
B8		11-35	12-33	1.24	10^{-5}
B18	22(2478)	3-29	2-16	1.46	10^{-6}
		HLA and Nasopharyngeal Carcinoma**			
A2	1(153)	63	52	1.62	10^{-2}
Bw46	1(153)	39	25	1.84	10^{-2}
A2-Bw46	1(130)	35	18	2.38	10^{-3}

R.R. = Relative Risk

* From the International Registry, 1981 (9)

** From the International Registry, 1979 (4)

HLA-B5, B8 and B18 show weaker associations (Table I). The HLA-A1, B8 linkage disequilibrium is well-known whereas there is no known allelic association between A1 and B5 or A1 and B18, at least in the caucasoid population. Could the data reflect the influence of 2 HLA genes : on the one hand, A1, and on the other hand, 4^{c} (B5, B18)?

When the total patient population is sub-divided into various epidemiological categories, it appears that A1 is mostly increased in males, in the cervical form of the disorder, and in mixed cellularity histological type.

The influence of HLA antigens in survival was studied in particular by Falk and Osoba[14]. They first attributed a favourable pronostic influence to A1 (and B8). Subsequently they demonstrated a shorter survival in patients carrying Aw19 (public antigen comprising : A28, 30, 31, 32, 33)[14].

C. Nasopharyngeal Carincoma (NPC)

NPC is relatively frequent in South China, but is not limited to this geographical area. In 1975, Simons et al[15] claimed an association with A2 and Sin-2 (now Bw46) (Table I). No association, however, was found in Tunisians[16] or in Eskimos[17]. Repeated, but not independent studies in China seem to confirm the association with the A2, Bw46 haplotype (in older patients ⩾ 30 years old), and another association with the Aw19, B17 haplotype (in younger patients < 30 years)[18]. The possible relationship of this tumour with EBV infection should be considered.

D. Various Other Malignancies

Among numerous studies, associations with HLA antigens have been noted : excess of Aw30 or 31 (cross-reacting antigens) in hypernephroma[19]; excess of A9 in the metastatic form of malignant melanoma[20]. In one study of carcinoma of the lung, Aw19 and B5 would seem to favour resistance[21]. These studies should be confirmed in other population groups.

II. FAMILY STUDIES

A. Homozygosity

The rationale for the search for HLA homozygosity

in patients is the hypothesis that the immunological defence of homozygotes would be less than that of heterozygotes because of diminished complex formation with class I or II products. Thus homozygous patients could be considered to carry a pair of recessive genes for susceptibility, these being either the HLA genes themselves or a closely-linked gene (or genes). In 1974, Gerkins et al. claimed that young (< 36 years old) cancer patients are probably more often homozygotes at the HLA-B locus (excess of B Blank in phenotyped patients) than in older (> 75 years old) normal individuals.

In a systematic study, Dausset[7] and Dausset et al.[22] looked for homozygosity in 136 families of genotyped patients, and compared the frequency with that in 168 normal families. A slight but significant excess of HLA-A locus homozygosity was detected in aplastic anaemia and in the Fanconi syndrome (both of these diseases are considered as pre-leukaemic states). No significant differences were observed for acute leukaemias of either lymphoblastic or myeloblastic type. However, in one study of ALL patients, an excess of parent pairs with shared HLA antigen was found, thus establishing the potential for homozygous off-spring.

B. Pairs of affected sibs

1. ALL (a) The literature concerning monozygotic twins, both leukaemic, is rather abundant. The relative risk of leukaemia among twins decreases according to age : the probability of a twin developing leukaemia is 100% if the other twin develops leukaemia before the age of 1 year.

(b) Fifteen per cent if the other twin develops leukaemia between 1-4 years of age, and

(c) Similar to the risk of a sibling of a leukaemic patient if the twin develops leukaemia after 4 years of age.

The relative risk for the sib of a leukaemic patient is four times greater than that for unrelated individuals. We have found only two families reported to have a pair of affected (ALL) sibs. One pair shared one HLA haplotype;[12] the other pair was HLA identical.[23]

2. Hodgkin's disease. The data concerning Hodgkin's disease are more consistent. We were the first to draw attention to multiplex families, describing four families each with multiple cases in a sibship[12]. Since that time, we have been collecting new cases from the literature.[24-30] Characteristics of the 15 families ascertained are presented in Fig. 1. In 9 families, there exists at least one pair of affected HLA identical sibs. When these families are pooled, we note 11 pairs of HLA identical sibs, 5 single haplotype identical pairs and 2 pairs which do not have shared haplotype. This distribution is signifigant at $p < 0.01$. Striking features are seen in some families, such as one reported by Nagel et al.,[27] with four affected sisters, and another by Torres et al.[29] with three brothers, HLA identical, for whom the diagnosis was made during a three month period. It is certainly important to search for new familial cases to confirm or disprove the impression that at least in some families susceptibility to Hodgkin's disease is a recessive character determined by an HLA linked determinant. Such a finding would suggest an HLA linked (dominant) gene for resistance to the disease (R-Hod). The distribution among pairs of siblings with juvenile diabetes mellitus, of HLA haplotype is similar. For this disease, an intermediate genetic model between recessivity and dominance has been proposed. Possibly in Hodgkin's disease, a similar genetic mechanism exists. Although the exact mechanism is unknown, complementation between HLA genes in cis or transphase, analogous to the IA and IE genes in mice, might be the basis for resistance. In this case, two HLA genes would be involved, but possibly only one gene product.

This interpretation of very preliminary data is in agreement with the experimental models, where resistance to viral-induced leukaemia in mice is dominant. One can wonder, in this case, what type of virus could be involved. It is striking that both in Hodgkin's disease and in nasopharyngeal carcinoma, the EBV virus is implicated.

3. Malignant Melanoma. This is another disease which can occur in families. In this respect, two families are striking.

In the first,[31] 12 cases are found in three generations. Nine of these showed the same haplotype. In another branch of the family, a different haplotype is common to three other cases. To explain this observation, a recombination involving these haplotypes can

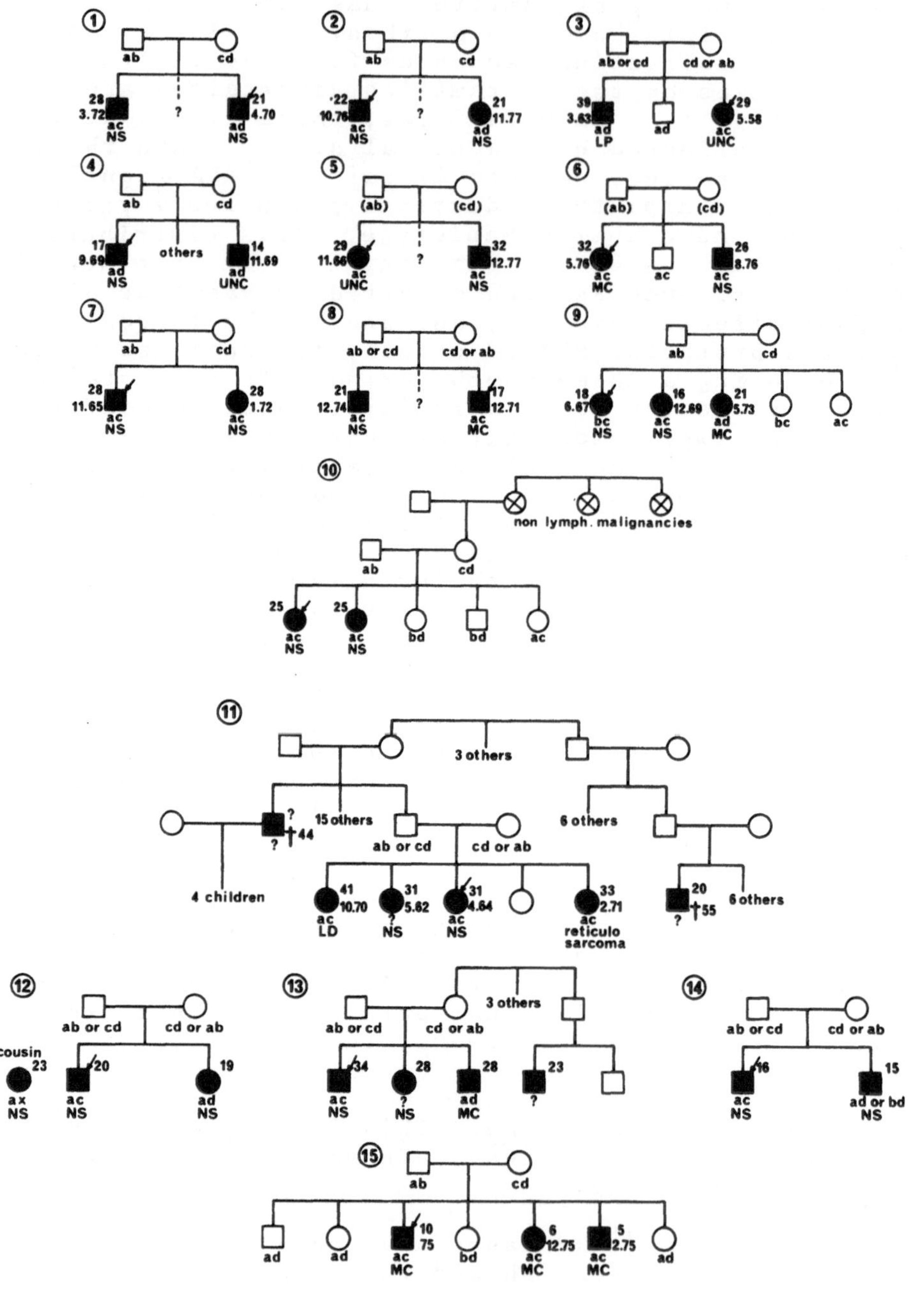

non lymph. malignancies
3 others
15 others
6 others
4 children
reticulo sarcoma
cousin

←

Figure 1.

Fam.N°				
1	(Maldonado et al., 1972)	Shared haplotype		A1, B5
2	(Hors et al., 1980)	"	"	A30, B13
3	(Hors et al., 1980)	"	"	A2, B12
4	(Dausset and Hors, 1975)	"	"	A2, B8/A10, B22
5	(Hors et al., 1980)	"	"	A2, B27/A3, B7
6	(Hors et al., 1980)	"	"	A1, B8/A26, B40
7	(Dausset and Hors, 1975)	"	"	A29, B12/Aw30, B5
8	(Hors et al., 1980)	"	"	A1, B8/A1 or⁻, B21
9	(McBride et al., 1977) a: Aw31, B7	b: A2, B7		c: A1, B8
10	(Bowa, 1977)	Shared haplotype		A2, B7/A2, B15
11	(Nagel et al., 1978)	"	"	A3, B7/A11, B12
12	(Greene et al., 1979)	"	"	Aw30, B8
13	(Greene et al., 1979)	"	"	A1, Bw35
14	(Greene et al., 1979)	"	phenotype	A2, 24 : B12, w35
15	(Torres et al., 1980)	"	haplotype	A1, B5/A26, B18

be postulated. The second family (five patients in two generations) was analysed in our laboratory[32]. Four patients possessed the same haplotype, and the fifth, not typed, could possess this same haplotype.

We were aware of other cases of familial malignant melanoma[33] - six families with two of three patients in each. A single haplotype is shared by 7 patients in three families. The same genotype is shared by 4 patients in two other families. However, in this last family, the siblings were HLA different. Taken together, these data suggest a dominant mode of inheritance.

4. Intestinal polyposis. This is another familial disease. One family has been HLA typed[34]. Eight out of 9 patients observed in three generations were A2, B12, again suggesting a dominant mode of inheritance (Fig. 2).

If confirmed, what could be the mechanism of this dominant susceptibility? The balance between immunity and tolerance of helper and suppressor T cell activity might suggest an explanation. Excess suppressor activity could theoretically lead to susceptibility appearing as a dominant character. Indeed, a dominant Is gene, against tetanus toxoid and membrane streptococcus A antigen, has been reported in humans[35].

5. Breast cancer. This is also familial. We have collected five pairs of affected sisters. Two pairs are HLA semi-identical, two are HLA different and only one is HLA identical. This normal distribution is inconsistent with the families published by Rosner et al.[36]

6. Kidney cancer can also be familial (Fig. 3).

In conclusion, a careful survey of the possible role of the human MHC in resistance (or susceptibility) to various malignancies is not as negative as previously believed. Many analyses are still not statistically significant, but a general trend is perceptible and the hints are sufficiently clear to allow some preliminary conclusions.

Looking first at populations of unrelated patients, it is noticable that the HLA-A locus is apparently more often associated than the HLA-B locus, in contrast with "classical" association claimed for non-malignant disorders (with the exception of idiopathic haemochromatosis). This is particularly evident in haematopoeitic malignancies. This observation concerns the association

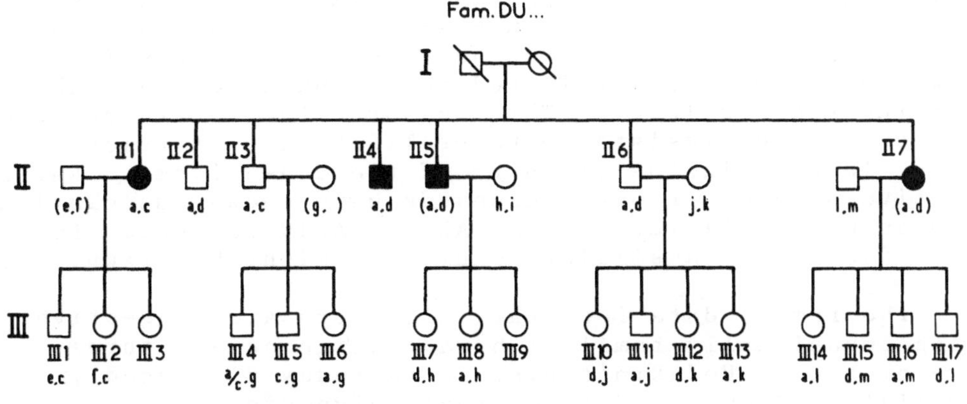

Figure 2. Familial Rectocolic Polyposis : Note that all patients carry the a (A2, B12) haplotype (Dupré et al., 1977, ref. 34).

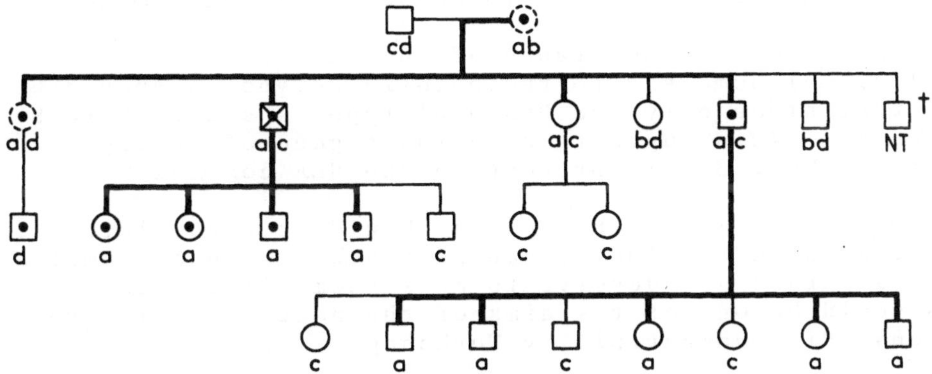

Figure 3. Familial Kidney Cancer : Note that three patients are HLA identical, the fourth being HLA semi-identical. Haplotypes a = A2, B12; c = A3, Bw35; d = Aw30, 31, B13 (Valleteau et al., 1974, ref. Nouv. Presse Med. 24:1539, 1974).

of long-term survival in ALL and HLA-A2, in Hodgkin's disease, A1. Of course, an HLA-B association is also observed in these survivors, but the relationship is weaker and could be attributed to a linkage disequilibrium with the HLA-A allele. This supports the existence of a gene conferring resistance tó proliferation of tumour cells, located in the vicinity of the HLA-A locus. Retrospective studies of other diseases have shown association with an HLA-A allele, such as A2 and Aw19 for NPC. For these disorders, a similar conclusion might be drawn.

The family data for other disorders are suggestive. Among occasionally familial human malignancies, three have emerged : Hodgkin's disease (already discussed), familial melanoma, and intestinal polyposis. The apparence of Hodgkin's disease in sibships is a well-established fact, but the role of a genetics versus environmental factors has always been debated without any clear answer emerging. The data collected since 1975 by J Hors et al.,[30] now reach a p value of 0.05 significance: familial Hodgkin's is mainly present among HLA identical siblings. This suggests a recessive or pseudo-recessive mode of susceptibility in those individuals lacking a gene of resistance as described in mouse viral leukaemias.

The mode of inheritance of the susceptibility to familial melanoma and to recto-colic polyposis should be different since only one HLA haplotype is shared here by patients. This suggests a dominant gene of susceptibility (S-Mel, S-Pol) present on the HLA complex.

Obviously many other cases should be accumulated before a valid conclusion can be reached - these familial cases are rare and deserve to be systematically collected. This could be one of the aims of the next (9th) International Histocompatibility Workshop.

REFERENCES

1. F. M. Kourilsky, J. Dausset, N. Feingold, J. M. Dupuy and J. Bernard, Etude de la répartition des antigènes leucocytaires chez des malades atteints de leucémie aiguë en remission, in: "Advance in Transplantation," Munksgaard, Copenhagen (1967).
2. J. L. Amiel, E. S. Curtoni, P. L. Mattiuz and R. M. Tosi, Study of the leukocyte phenotypes in Hodgkin's disease, in:"Histocompatibility Test.67".

3. J. Dausset and A. Svejgaard (eds), "HLA and Disease," Munksgaard, Copenhagen (1977).
4. L. P. Ryder, E. Andersen and A. Svejgaard (ed), "HLA and Disease Registry," Munksgaard, Copenhagen (1979).
5. "HLA and Malignancy," G.P. Murphy, ed., Alan R. Liss, New York (1977).
6. L. Degos, M. Pla and J. Colombani, H2 restriction for lymphocyte homing into lymph nodes, Eur. J. Immunology 9:808 (1979).
7. J. Dausset, HLA and association with malignancy: A critical view, in: "HLA and Malignancy," G. P. Murphy, ed., Alan R. Liss, New York (1977).
8. W. E. Braun, "HLA and Diseases: A comprehensive review," CRC Press Inc., Boca Raton, Florida (1979).
9. L. P. Ryder, C. Dam-Soerengen and A. Svejgaard, "4th Edition of the HLA and Disease Registry," Copenhagen (in press).
10. E. D. Albert, B. Nisperos and E. D. Thomas, HLA antigens and haplotypes in acute leukemia, Leukemia Res. 1:4:261 (1977).
11. G. N. Rogentine, R. J. Trapani, R. A. Yankee and E. S. Henderson, HLA antigens and acute lymphocytic leukemia: the nature of the HL-A2 association, Tissue Antigens 3:470 (1973).
12. J. Dausset and J. Hors, Some contributions of the HLA complex to the genetics of human disease, Transf. Rev. 22:44 (1975).
13. T. Tursz, J. Hors, M. Lipinski and J. L. Amiel, HLA phenotypes in long-term survivors treated with BCG immunotherapy for childhood ALL, Brit. Med. J. 1:1250 (1978).
14. J. Falk and D. Osoba, The HLA system and survival in malignant disease: Hodgkin's disease and carcinoma of the breast, in: "HLA and Malignancy," (ref. 5).
15. M. J. Simons, G. B. Wee, S. H. Chan, K. Shanmugaratuam, N. E. Day and G. de Thé, Probable identification of an HLA second locus antigen associated with high risk of nasopharyngeal carcinoma, Lancet 1:142 (1975).
16. H. Betuel, M. Camoun, J. Colombani, N. E. Day, R. Ellouz and G. de Thé, The relationship between nasopharyngeal carcinoma and the HL-A system among Tunisians, Int. J. Cancer 16:249 (1975).
17. A. Lanier, T. Bender, M. Talbot, S. Wilmeth, C. Tschopp, W. Henle, G. Henle, D. Ritter and P. Terasaki, Nasopharyngeal cancer in Alaskan Eskimos, Indians and Aleuts: a review of cases and studies of EBV, HLA and environmental risk factors, Cancer 46:2100 (1980).

18. S. H. Chan, N. E. Day, T. H. Khor, N. Kunaratnam and K. B. Chia, HLA markers in the development and prognoses of NPC in disease, in: "Cancer Campaign," vol. 5, Gustar Fischer Verlag, Stuttgart, New York (1981).
19. B. M. E. Kuntz, G. D. Schmidt, S. Scholz and E. D. Albert, HLA antigens and hypernephroma, Tissue Antigens 12:407 (1978).
20. B. Cavelier, M. Daveau, D. Gilbert, M. Fontaine, J. P. Cesarini and B. Delpech, Augmentation de l'antigène HLA-A9 dans les melanomes malins, principalement dans les formes metastiques ou les recidives. A propos de 105 melanomes dont 34 formes graves, CR Acad. Sci. 291:241 (1980).
21. G. N. Rogentine and P. B. Dellon, Prolonged disease-free survival in bronchogenic carcinoma associated with HLA Aw19 and B5: A follow-up, 1er Int. Symp. on HLA and Disease, INSERM, Paris (1976).
22. J. Dausset, E. Gluckman, F. Lemarchand, A. Nunez-Roldan, L. Contu and J. Hors, Excès de HLA-A2 parmi les malades atteints d'aplasie médullaire et maladie de Fanconi, N. Rev. Fr. d'Hémat. 18:315 (1977).
23. W. A. Blattner, J. L. Naiman, D. L. Mann, R. S. Wimer, J. H. Dean and J. F. Fraumeni, Immunogenetic determinants of familial ALL, Ann. Int. Med. 89:173 (1978).
24. J. E. Maldonado, H. F. Taswell and J. M. Kiely, Familial Hodgkin's disease, Lancet 2:1259 (1972).
25. A. McBride and J. J. Fennelly, Immunological depletion contributing to familial Hodgkin's disease, Eur. J. Cancer 13:549 (1977).
26. J. K. Bowers, C. F. Moldow, C. D. Bloomfield and E. J. Yunis, Familial Hodgkin's disease and the major histocompatibility complex, Vox Sanguinis 33:273 (1977).
27. G. A. Nagel, E. Nagel-Studer, W. Seiler and H. O. Hofer, Malignant lymphoma in four of five siblings, Int. J. Cancer 22:675 (1978).
28. M. H. Greene, E. A. McKeen, F. P. Li, W. A. Blattner and J. F. Fraumeni, HLA antigens in familial Hodgkin's disease, Int. J. Cancer 23:777 (1979).
29. A. Torres, F. Martinez, P. Gomez, C. Gomez, J. M. Garcia and A. Nunez-Roldan, Simultaneous Hodgkin's disease in three siblings with identical HLA genotype, Cancer 46:838 (1980).
30. J. Hors, G. Steinberg, J.M. Andrieu, C. Jacquillat, M. Minev, J. Messerschmitt, G. Malinvaud, F. Fumeron and J. Dausset, HLA genotypes in

familial Hodgkin's disease. Excess of HLA identical affected sibs, Eur. J. Cancer 16:809 (1979).

31. B. R. Hawkins, R. L. Dawkins, A. Hockey, J. B. Houliston and R. L. Kirk, Evidence for linkage between HLA and malignant melanoma, Tissue Antigens 17:540 (1981).
32. J. P. Cesarini, M. Daveau, B. Cavelier, N. Feingold, P. Puissant, F. Fumeron and J. Hors, More arguments in favor of a linkage between HLA and susceptibility to malignant melanoma (in preparation).
33. D. Rovini, M. T. Illeni, A. Ghidoni, M. Vaglini, E. Privitera, G. Pellegris and N. Cascinelli, HLA antigens in familial melanoma patients, 6th Int. Cong. Human Genetics, Jerusalem (1981).
34. N. Dupré, E. Nouzha, C. Raffoux, J. C. Morvan and J. P. Gras, Polypose rectocolique familiale et système HLA. A propos d'une observation, Bordeaux Medical 30:2125 (1977).
35. T. Sasazuki, H. Kaneoka, Y. Nishimura, R. Kaneoka, M. Hayama and H. Ohkuni, An HLA-linked immune suppression gene in man, J. Exp. Med. 152:2:297 (1980).
36. D. Rosner, E. Cohen, S. G. Gregory, U. Khurana and C. Cox, Breast cancer and HLA relationship in a high risk family, _in_: "HLA and Malignancy," (ref. 5).

MONOCLONAL ANTIBODIES SPECIFYING ANTIGENS ASSOCIATED WITH RAT TUMORS AND THEIR USE IN DRUG DELIVERY SYSTEMS

R.W. Baldwin, M.J. Embleton, MV. Pimm, R.A. Robins,
J.A. JOnes, C. Holmes and Barbara Gunn

Cancer Research Campaign Laboratories
University of Nottinghom
Nottingham NG7 2RD, U.K.

INTRODUCTION

Neoplastic transformation results in the expression in transformed cells of antigens which are not present in their normal cell counterparts, at least in the adult host. One class of neoantigen expressed upon carcinogen-induced tumours and less frequently on naturally arising tumours was originally defined as a tumour specific transplantation antigen (Baldwin and Price 1981). The outstanding characteristic of these cell membrane associated products is their high degree of polymorphism so that tumours appear to express indivdually-distinct rejection antigens. Cell surface antigens with identical specificities have also been detected by interaction with antibody in sera from animals immunized against syngeneic tumour grafts, although the relationship of these antigens to the tumour rejection antigens has not been elucidated.

Carcinogen-induced and spontaneous animal tumours also express oncofoetal antigens, these being defined as products associated with tumour and foetal cells, but absent from, or expressed at low levels in adult normal tissues (Fishman and Busch 1979). For example, tumour cell membrane associated oncofoetal antigens have been detected upon rat mammary carcinomas by membrane immunofluorescence reaction with serum from multiparous rats (Baldwin et al. 1974).

These investigations indicate that a complex series of changes in cell surface antigen expression may occur following malignant transformation. Consequently typing of cell surface antigens provides an approach for investigating malignant transformation and the subsequent metastic spread ot tumors. Antisera produced following immunization of syngeneic hosts with tumour cells or sub-cellular fractions have provided one source of antibody to tumour cell surface antigens. In many instances, however, it has been difficult to obtain antibody preparations of sufficient specificity and potency. Likewise, antisera obtained following xenogeneic immunization with tumour cells or fractions have proved to have limited value because of difficulties in rendering them 'tumour'specific' following absorption with normal tissue extracts. These difficulties can be overcome by using monoclonal antibodies produced against cell surface antigens on normal and malignant cells. This is illustrated in this paper which presents a series of studies on the development and application of monoclonal antibodies specifying tumour associated antigens on carcinogen-induced and naturally arising rat tumours.

MONOCLONAL ANTIBODY SPECIFYING A TUMOUR SPECIFIC ANTIGEN ASSOCIATED WITH RAT MAMMARY CARCINOMA SP4

Rat mammary carcinoma Sp4 which arose naturally in WAB/Not rats is immunogenic, immunization of syngeneic rats consistently producing immunity against a rechallenge with this tumour, but not other mammary carcinomas (Baldwin and Embleton 1969). Syngeneic immune rats were subsequently found to mount a specific antibody response to this tumour (Baldwin and Embleton 1970). Developing from these studies an interspecies hybridoma has been prepared following fusion of spleen cells from a syngeneic immune rat and P3NS1 mouse myeloma cells (Gunn et el. 1980). Table 1 summarizes the binding of antibody produced by this hybridoma (Sp4/A4) with a range of target cells, both normal and malignant, as detected by a radioisotopic assay. This shows that the antibody reacts specifically with mammary carcinoma Sp4 cells, but not other mammary carcinomas or unrelated tumours including sarcomas

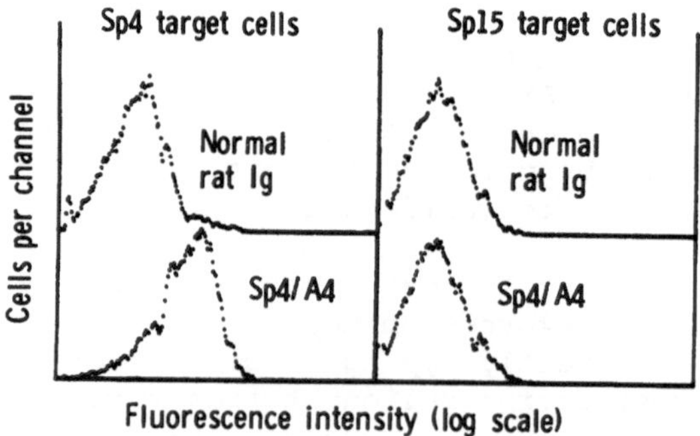

Fig. 1. FACS analysis of Sp4/A4 monoclonal antibody binding to mammary carcinoma Sp4 and Spl5 target cells

TABLE 1

Reactions of monoclonal antibody to spontaneous rat mammary carcinoma Sp4 against various rat target cells

Target cells		Cell binding of antibody		
		Radio Immunoassay [1]	FA [2]	FACS [3]
Mammary carcinoma	Sp4	5.61	0.74	7.53
	Sp15	1.26	0.01	0.98
	Sp22	1.21	0.01	1.11
Hepatoma	D23	1.32	0.32	NT
	D192A	1.28		0.83
Sarcoma	Mc7	1.07		1.08
	Mc57	1.08		NT
	Mc97A	1.21		NT
	Mc98	1.48		NT
	Mc100	1.33		NT
	Mc106B	NT	0.04	
	Mc107B	NT	0.02	
Normal cells	Spleen	0.66		NT
	Lymph node	0.96		NT
	Erythrocytes	0.77		NT
	Peritoneal exudate	0.85		NT
	Areolar tissue	1.39		NT
	Liver	0.96		NT
	Embryo	1.07		NT
	Normal breast	NT		0.90

1. Binding ratio = mean cpm with monoclonal antibody ÷ mean cpm with P3NS1 spent medium
2. Fluorescence index in fluorescent antibody (FA) test = %cells unstained in controls - % cells unstained by monoclonal antibody ÷ % cells unstained in controls (see text)
3. Binding ration = mean fluorescence per cell with monoclonal antibody ÷ mean fluorescence intensity per cell with medium (see Fig. 1)

and hepatocellular carcinomas, or cells derived from normal tissues. Membrane immunofluorescence tests whereby cell bound Sp4/A4 was detected by fluorescein-conjugated sheep anti-rat IgG also indicated that the antibody reacted specifically with mammary carcinoma Sp4 cells (Table 1). In an extension of this approach, binding of Sp4/A4 antibody to tumour cells has been examined using a fluorescence activated cell sorter (FACS). This approach, illustrated in Fig. 1 shows that all the carcinoma cells in the Sp4 tumour cell preparation react with Sp4/A4 antibody, but there is no binding to cells of mammary carcinoma Sp15. For comparative purposes, the mean fluorescence intensity (FACS channel number) is calculated following reaction of tumour cells (2 x 10^5) with standard amounts of antibody. By this criterion it was again shown that cells from Sp4, but not other tumours or normal tissues reacted with Sp4/A4 (Table 1).

SP4/A4 MONOCLONAL ANTIBODY-DRUG CONJUGATES IN THERAPY

One of the objectives in developing monoclonal antibodies to tumour associated antigens is to provide a means for localizing anti-tumour agents in tumour deposits (Baldwin et al. 1981). It is essential for this purpose to demonstrate that Sp4/A4 antibody has *in vivo* localizing properties. In these studies, antibody from hybridoma culture supernatants was isolated by affinity chromatography on Sepharose 4B-linked goat anti-rat IgG and labelled with ^{125}I using the Iodogen reagent (Pimm et al. 1981). As illustrated in Fig. 2 intravenous injection of ^{125}I-labelled antibody resulted in a preferential localization in subcutaneous tumour compared to normal tissues in the tumour-bearing rat. Also antibody localized prefentially in tumour Sp4 when compared with another mammary tumour (Sp15) developing at a contralateral site.

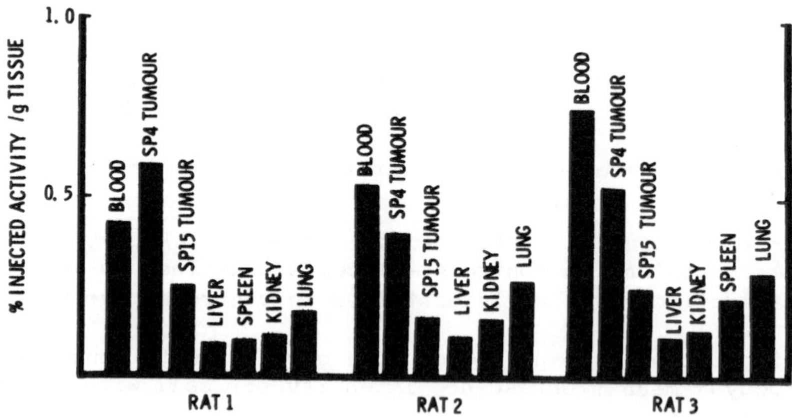

Fig. 2. *In vivo* distribution of ^{125}I labelled anti-Sp4 monoclonal antibody after injection into rats with growths of Sp4 and Sp15

Developing from these studies the use of Sp4/A4 antibody as a carrier for adriamycin in the therapy of Sp4 tumour has been examined (Pimm et al. 1981). In the initial series, adriamycin (Doxorubicin) conjugation to antibody or normal rat IgG was effected through a dextran bridge (Hurwitz et al. 1975), yielding conjugates containing between 7 : 3 and 15:7 molar ratios of adriamycin to protein (Table 2). These conjugates retained antibody activity, this being illustrated by the FACS analysis of binding of free Sp4/A4 and an adriamycin-Sp4/A4 conjugate to Sp4 tumour cells (Fig. 3). Normal rat IgG coupled to adriamycin did not bind to tumour Sp4 cells. Also none of the reagents bound to mammary tumour Sp15 cells. Adriamycin-Sp4/A4 conjugates also retained drug activity as determined by inhibition of DNA synthesis in treated Sp4 tumour cells. Subsequently it was shown that adriamycin-Sp4/A4 conjugates were more effective than free drug or adriamycin-normal IgG (adria-NIgG) conjugates in suppressing growth of Sp4 tumour. This is illustrated in Fig. 4 which compares the therapeutic response of subcutaneous grafts of Sp4 to adriamycin or adriamycin-protein conjugates. In this test adriamycin and an adria-NIgG conjugate did not prolong survival whereas a therapeutic response was obtained with the adria-Sp4/A4 antibody conjugate.

There is not yet sufficient data to conclude that the antibody specifically directs the adriamycin moiety to tumour cells thereby

TABLE 2

Compositional Analysis of Adriamycin Conjugates[1]

Protein	Composition μg/ml[2]		Molar ratio
	Adriamycin	Protein	Adriamycin:Protein
Normal IgG	14.0	400	9.5 : 1
Sp4-Moab	10.5	320	9.0 : 1
Normal IgG	12.0	350	9.3 : 1
Sp4-Moab	13.0	480	7.3 : 1
Normal IgG	5.8	100	15.7 : 1
Sp4-Moab	4.5	80	15.2 : 1

[1]Normal rat IgG or Sp4/A4 monoclonal antibody was coupled to adriamycin through a dextran bridge (Hurwitz et al. 1975)

[2]Adriamycin content was determined from absorbance at 495nM : protein content determined by the Lowry procedure

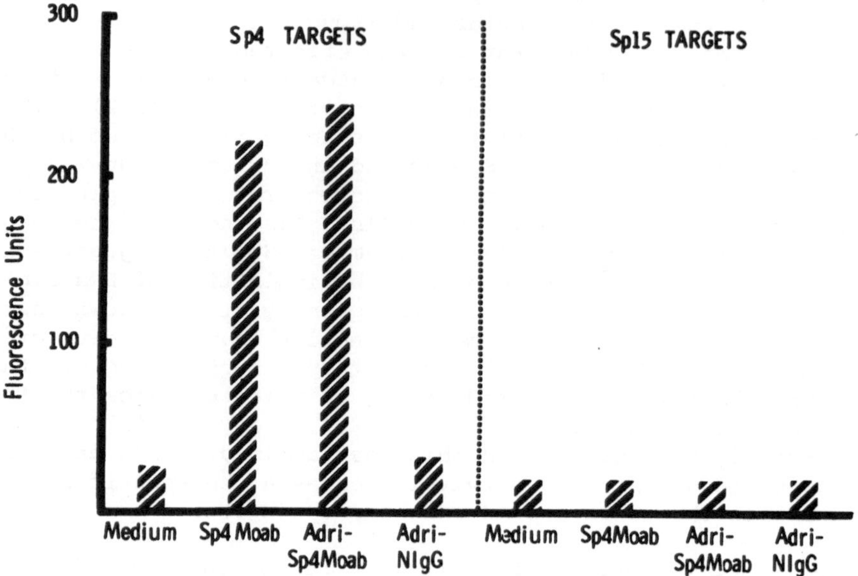

Fig.3. FACS analysis of binding of Sp4/A4 monoclonal antibody and adriamycin conjugates to mammary carcinomas Sp4 and Sp15 target cells

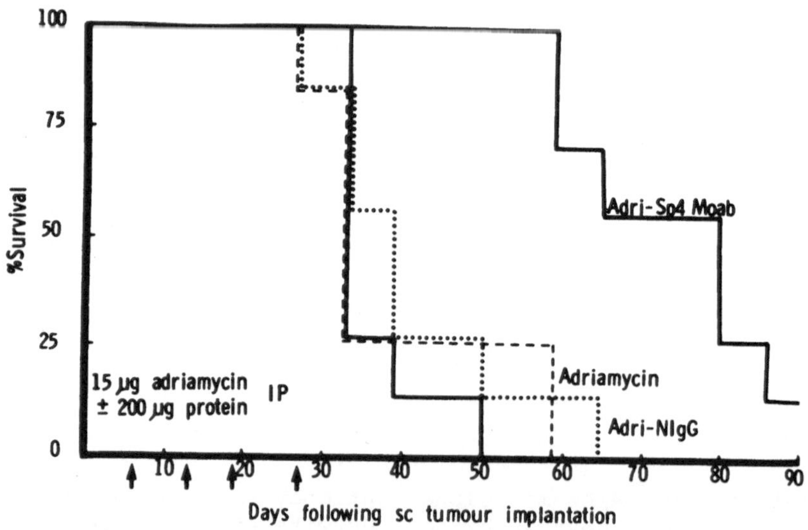

Fig. 4. Treatment of mammary carcinoma Sp4 by adriamycin - Sp4/A4 monoclonal antibody conjugates

enhancing its therapeutic effectiveness. Although the labelled antibody studies indicate a preferential localization of Sp4/A4 antibody in tumour deposits, there is no conclusive proof that this reflects specific binding to tumour cells. It is possible, for example, that adria-Sp4/A4 conjugates bind to extracellular antigen released into the milieu of the tumour. This may be a sufficient response, however, if following tumour localization, as distinct from cell binding, adriamycin is released through the action of tumour-associated proteases. This would then allow intracellular penetration of adriamycin and subsequent interaction with DNA (Myers 1980).

The alternative postulate is that specific binding of adria-Sp4/A4 antibody to tumour cells is an integral step in the anti-tumour action produced by the conjugates. If this is the case, the therapeutic effectiveness will depend upon the capacity of antibody conjugates to extravasate in tumour and also upon the degree of tumour antigen expression upon cells within the tumour mass. This will be influenced by a number of factors including tumour antigen masking/ modulation by host or tumour products. For example, it has been established by FACS analysis that mammary tumour Sp4 cells derived either from an ascites line or from solid tumour disaggregated by collagenase display, at best, low levels of Sp4 antigen. Brief exposure of tumour cells to trypsin results in a rapid and marked increase in tumour antigen expression (Jones et al. 1981). This suggests that the tumour antigens detected by Sp4/A4 may show defective expression in tumours growing *in vivo* and this could be a limiting factor in antibody directed chemotherapy.

MONOCLONAL ANTIBODIES SPECIFYING FOETAL ANTIGENS

Foetal antigens expressed in the normal foetus, often at a particular stage of gestation, but not on normal adult tissue cells have been identified on malignant cells (Baldwin and Price, 1981: Fishman and Busch, 1979). Theyinclude secretory products such as alphafoetoprotein associated with hepatocellular carcinomas. In addition cell surface associated foetal antigens have been identified upon a wide range of animal tumours including naturally arising tumours and tumours induced by chemical carcinogens and oncogenic viruses. For example, foetal antigens have been detected upon naturally arising and carcinogen-induced rat mammary carcinomas by their interaction with antibody in sera from multiparous donors (Baldwin et al. 1974). Although, in the past, multiparous sera have provided a source of antifoetal antibody for membrane immunofluorescence and complement-mediated cytotoxicity studies, these antisera have proved widely variable in their reactivity. This is illustrated by tests summarised in Fig. 5 where cell surface antigens on rat hepatoma D23 have been detected by reaction with a range of antisera produced in syngeneic WAB/Not rats, followed by a second reaction with fluorescein labelled sheep anti-rat IgG. Antibody binding was then detected by analysis in

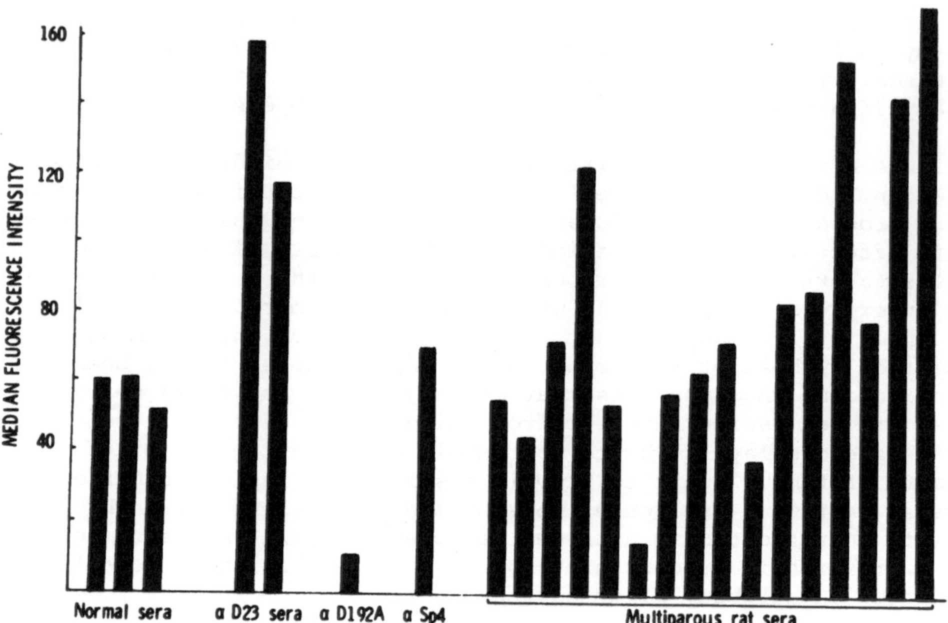

Fig. 5. FACS analysis of antibody binding to rat hepatoma D23 target cells (see text)

a fluorescence activated cell sorter. Compared to background reaction with normal rat serum, the two antisera produced following immunization with hepatoma D23 cells showed positive reactions whilst antisera to another hepatoma D192A and mammary carcinoma Sp4 were non-reactive. In comparison only 4/16 multiparous sera resulted in fluorescence reactions with hepatoma D23 cells which can be considered as significant responses (Baldwin et al. 1981). Likewise, in previous studies it was found that antisera obtained following

TABLE 3

Binding of anti-foetal antibodies to normal and adult rat cells[1]

Target cells	Binding ratio with antibody No.[2] 66	144	186
Rat embryo:-			
10mm (14 days)	5.80	5.59	4.69
11mm (15 days)	4.99	6.51	5.06
23mm (19 days)	4.10	4.55	4.55
Adult cells:-			
Spleen	0.85	0.91	0.85
Kidney	1.42	1.00	1.06
Lung	1.23	1.29	1.14
Testis	1.36	1.21	0.94
Brain	2.51	2.76	2.99
P.E.C.	1.24	1.43	1.53
L.N.C.	NT	1.47	1.64
Thymus	NT	2.60	2.02

[1]Antibody preparations derived from hybridomas obtained following fusion of spleen cells from multiparous WAB/Not rats and mouse p3NS1 myeloma.

[2]Cell binding assayed by uptake of ^{125}I-labelled sheep anti-rat IgG F(ab^{1})$_2$ following treatment of cells with antibody. Binding ratio:

$$\frac{\text{cpm cells treated with antibody}}{\text{cpm cells treated with P3NS1 medium}}$$

TABLE 4

Binding of anti-foetal antibodies to cultured rat tumours

Target cells	Binding ratio with antibody no. 66	144	186
Hepatoma D23	5.69	10.41	8.40
Hepatoma D30	NT	5.21	8.83
Sarcoma Mc7	5.09	9.75	6.33
Mammary carcinoma Sp22	4.56	4.71	5.17

Target cells derived from cultured cell lines established from WAB/Not rat hepatocellular carcinomas (D23, D30) fibrosarcoma (Mc7) and mammary carcinoma (Sp22). See Table 3 for antibody binding test.

immunization of syngeneic WAB/Not rats with embryo cells taken at a stage when foetal antigens were present reacted weakly and inconsistently with embryo cells.

Because of the limitations of these syngeneic antisera, hybridomas have been produced which secrete antibody reacting with foetal antigens. As in the Sp4 tumour studies (Gunn et al. 1980), this has been accomplished following fusion of spleen cells from multiparous WAB/Not rats with the mouse P3NS1 myeloma (Holmes et al. 1981). The reactivity of antibodies in supernatants from four hybridomas with a range of target cells is summarized in Table 3. These tests were carried out using a cell binding assay in which target cells in microtiter plate wells are treated with supernatant fractions and bound antibody detected by a second reaction with ^{125}I-labelled sheep anti-rat IgG $F(ab^1)_2$ (Gunn et al. 1980). All four antibody preparations reacted strongly with embryo cells derived by mechanical dissociation of 14 to 19 day old embryos (binding ratios 2.84 to 5.80). In tests with a range of target cells derived from normal adult tissues, positive reactions were obtained only with brain and thymus cells (binding ratios 2.02 to 2.99). As indicated in Table 4, all three antibody preparations also reacted strongly with cultured cells derived from a range of WAB/Not rat tumours including hepatomas (D23 and D30), a sarcoma, Mc7, and a mammary carcinoma Sp22 (binding ratios 4.56 to 10.41).

The nature of the antigen detected by these antibodies has still to be elucidated, but since they are produced by hybridomas obtained following fusion of spleen cells from multiparous rats, embryo cell-associated antigens would seem to be likely candidates. It is important to recognise that the hybridomas were produced using multiparous rat spleen cells as a source of antibody-producing cells. This approach leads to the production of hybridomas secreting antibodies to antigens which are recognised by syngeneic WAB/Not rats. These 'auto-antigens showing preferential expression upon embryo-derived cells may be related to the multiplicity of regulatory products produced during foetal development (Murgita and Wigzell, 1981) but speculation as to their nature is unwarranted until further characterization has been carried out. The other application of using 'anti-foetal cell' antibodies as carriers for therapeutic agents is attractive since as indicated in Table 4 the antigens are expressed upon cultured tumour cells irrespective of histological type or aetiology. Before this approach can be developed it is necessary, however, to establish that the antigen(s) are expressed at adequate levels upon cells in a developing tumour or are concentrated within the milieu of a tumour. This point has already been emphasized in relation to mammary carcinoma Sp4 antigen. Furthermore, previous studies on the antigen(s) detected upon tumour cells by reaction with multiparous rat serum indicated that it was not a stable cell surface product but rather a secretory product (Price and Baldwin 1977).

REFERENCES

Baldwin, R.W. and Embleton, M.J., 1979, Immunology of spontaneously arising rat mammary adenocarcinomas, Int. J. Cancer, 4:430

Baldwin, R.W. and Embleton, M.J., 1970, Detection and isolation of tumour specific antigens associated with a spontaneously arising rat mammary carcinoma, Int. J. Cancer, 6:373

Baldwin, R.W., Embleton, M.J., Price, M.R. and Vose, B.M., 1974, Embryonic antigen expression on experimental rat tumours, Transplant. Rev., 20:77

Baldwin, R.W., Embleton, M.J. and Price, M.R., 1981, Monoclonal antibodies specifying tumour-associated antigens and their potential for therapy, in: Molecular Aspects of Medicine 4/5:329, Pergamon Press, Oxford

Baldwin, R.W. and Price, M.R., 1981, Neoantigen expression in chemical carcinogenesis, in: Cancer: A Comprehensive Treatise, Vol. 1, Plenum Press, New York, editor, F. Becker.

Fishman, W.H. and Busch, H., 1979, Oncodevelopmental antigens, in: Methods in Cancer Research 18:1, Academic Press, New York

Gunn, B., Embleton, M.J., Middle, J. and Baldwin, R.W., 1980, Monoclonal antibody against a naturally occurring rat mammary carcinoma, Int. J. Cancer, 36:325

Holmes, C.A., Austin, E.B., Embleton, M.J., Gunn, B. and Baldwin, R.W., Oncodevelopmental antigens associated with rat tumours specified by monoclonal antibodies, in preparation

Hurwitz, E., Levy, R., Maron, R., Wilchek, M.S., Arnon, R. and Sela, M., 1975, The covalent binding of daunomycin and adriamycin to antibodies with retention of both drug and antibody activities. Cancer Res. 35:1175

Jones, P.D.E., Robins, R.A. and Baldwin, R.W., Expression of Sp4/A4 monoclonal antibody defined antigen on mammary carcinoma Sp4 cells, in preparation

Murgita, R.A. and Wigzell, H., 1981, Regulation of immune functions in the fetus and newborn, Progress in Allergy, in press

Myers, C.E., 1980, Antitumor antibiotics I: anthracyclines in: Cancer Chemotherapy Annual 2, 66-83, ed. H.M. Pinedo, Excerpta Medica

Pimm, M.V., Tumour localisation of monoclonal antibody against a rat mammary carcinoma and suppression of tumour growth with adriamycin-antibody conjugates, Cancer Immunol. and Immunother., in press

ACKNOWLEGEMENTS

This work was supported by a grant from the Cancer Research Campaign, London, U.K.

IMMUNOLOGICAL MARKERS OF HUMAN LEUKEMIC CELLS OF THE B LINEAGE

M. Seligmann, L.B. Vogler, P. Guglielmi, J.C. Brouet and J.L. Preud'homme

Laboratory of Immunochemistry and Immunopathology (INSERM U 108, Research Institute of Blood Diseases and Laboratory of Oncology and Immunohematology of CNRS), Hôpital Saint-Louis, Paris (France)

INTRODUCTION

The study of immunological cell markers in human leukemias and lymphomas has made it possible to establish the nature and origin of the proliferating cells and to demonstrate the heterogeneity of many conditions which appeared previously rather homogeneous to the hematologist. Since leukemic cells usually appear to be frozen at a given stage along the differentiation pathway, these studies have also provided important insights into the biology of the progenitors of hematopoietic cells which correspond to the normal counterpart of the leukemic cell. The maturation arrest characteristic of many human leukemic cells can sometimes be reverted after in vitro stimulation. Such studies may eventually lead to an understanding of crucial differences between the regulation of normal and leukemic cells.

PRE-B CELL LEUKEMIAS

Approximately 20% of patients (mainly children) with acute lymphoblastic leukemia (ALL) have leukemic blasts with features of pre-B cells[1], i.e. rapidly dividing cells that contain small amounts of cytoplasmic Ig μ heavy chains but lack detectable surface Ig. Our studies of this type of leukemic cells have suggested the possibility of previously unrecognized steps in the early differentiation sequence of the B lineage.

Human leukemic pre-B cells were found to lack intracytoplasmic light chains by immunofluorescence studies[2,3]. This finding was in contrast to prevalent concepts of Ig expression in normal pre-B cells. Such an asynchrony of μ chain and light chain expression in immature cells of the B lineage has now been observed in normal fetal liver murine pre-B cells[4], Abelson virus-infected bone marrow cells[5], pre-B/myeloma cell hybridoma[6] and also some human cell lines[7]. Our initial observations on the absence of light chains in human pre-B leukemic cells have been confirmed by the use of biosynthetic techniques which have also shown that the intracytoplasmic μ' chains are larger than secreted normal μ chains, with an apparent molecular weight of approximately 82,000.

We have encountered several patients with pre-B cell leukemia in whom a rather large proportion of the lymphoblastic cells bore scant amounts of surface μ chains without light chains [2,8]. This phenotype is believed to represent cells at a stage of differentiation intermediate between that of the pre-B cell and B lymphocyte. These cells have persistent expression of the nuclear enzyme terminal-deoxynucleotydil-transferase (TdT)[8]. Since they still lack membrane-bound and cytoplasmic light chains we believe that this observation argues against the suggestion that the synthesis of light chains and their integration with heavy chains facilitates the incorporation of Ig molecules into the cell membrane[9].

Another set of findings in pre-B leukemic cells strongly suggest that mechanisms responsible for the sequential expression of Ig heavy chain genes (isotype switch) are operable at early stages in B cell development and are independent of surface Ig expression[8]. Patients with pre-B cell leukemia may have a very small proportion of leukemic cells that express cytoplasmic γ or α chains in addition to cytoplasmic μ chains[1,2]. In some patients, cells containing γ chains account for more than 10% of the leukemic population and sometimes outnumber μ-containing leukemic cells. In some of these cases, the expression of cytoplasmic κ chains was associated with that of γ heavy chains. Furthermore, one patient with ALL and one with blast crisis of CML had 6 to 8% of lymphoblasts that contained γ heavy chains in the absence of μ chains and of detectable light chains. We have recently studied a patient with ALL in whom the vast majority of blast cells contained γ chains only without detectable immunoglobulin chains on the membrane. The distribution and pattern of immunofluorescence staining was identical to that of μ staining in pre-B cells and double immunofluorescent studies showed the coexpression of cytoplasmic γ chains and nuclear TdT. The γ chains found in these leukemic cells may reflect abnormal regulation of Ig gene expression related to malignant transformation. We would however, like to suggest that such cells reflect early stages in the development of <u>normal</u> B cells. Since this phenotype was observed in a small population of leukemic cells from a minority of patients, class switching may be a rare event at the pre-B stage of differentiation and could represent a pathway alternative to the major sequence of maturation.

In some patients with pre-B cell leukemia, a rather large proportion of leukemic cells without any detectable Ig is found together with the pre-B leukemic cells with intracytoplasmic-μ chains[8]. This finding, the reactivity of many "null" ALL cells with anti-B1 and BA2 monoclonal antibodies[10,11], and mainly the data on immunoglobulin gene rearrangements provided by Korsmeyer et al.[12] strongly suggest that a large proportion of "null" ALL represents the proliferation of precursors of pre-B cells.

The acute lymphoblastic crisis in chronic granulocytic leukemia often involves leukemic cells of the pre-B phenotype (as seen in several laboratories including ours), whereas such blast crises involving precursors of the T lineage are exceedingly rare or do not occur. On the other hand, the study of blood lymphocytes from patients with chronic phase CGL, with either glucose 6-phosphate deshyrogenase isoenzymes[13] or combined chromosomal and SIgM analysis[14], have shown that B-lymphocytes and not T-lymphocytes are involved in the leukemic clone. These findings should be taken into account in the schemes dealing with the early steps of hematopoietic differentiation and pluripotential stem cells in man. They suggest that the T lineage becomes independent at a very early stage preceding the occurence of a stem cell for both the myeloid lines and the B-lymphoid line. This hypothesis is in accordance with the distribution of the Ia-like antigens and with some recent data in the mouse.

ACUTE LEUKEMIA WITH BURKITT CELLS

B cell ALL is exceedingly rare with the exception of those leukemias featured by Burkitt cells which represent a very homogeneous group on clinical, cytological, immunological and cytogenetic grounds[15,16,17]. This subgroup of ALL is featured by a very severe prognosis with a poor and brief response to therapy and a medium survival of 4 months. The blast cells in these patients all have the cytologic, cytochemical and electronmicroscopic features of Burkitt lymphoma cells (L3, according to the FAB classification). The leukemic cells display the membrane markers of B cells. Only one of the 25 patients studied in our laboratory[17] had leukemic cells without surface Ig, and his cells were not studied for intracytoplasmic Ig. This patient may have had a pre-B cell type of ALL since we have recently observed such a phenotype in a case of acute leukemia with Burkitt cells. Surface Ig on the blast cells of the 24 other patients were monoclonal, with a striking and unexplained predominance of cases with λ light chains. They consisted most often of high density IgM. Surface IgD was associated with IgM in only a minority of cases and this is in contrast to the findings on normal B cells and on chronic lymphocytic leukemia (CLL) cells. The membrane-bound Ig was of the IgG class in 3 patients and IgA in another patient. The blast cells lacked detectable receptors for the Fc of IgG in more than half of the cases. Small amounts of serum monoclonal IgM were found in several patients whose leukemic cells bore surface IgM and a Bence-Jones protein was detected in 2 of the 3 cases featuring blasts with surface Ig.

The position within the B cell differentiation pathway which should be assigned to Burkitt cells has been a matter of controversy. We believe that the immunological phenotype of these Burkitt leukemic cells corresponds to activated B cells rather than to immature B lymphocytes. The loss of surface IgD and of receptors for the Fc of IgG is compatible with this hypothesis. The high density of membrane bound Ig molecules, the finding of cases with surface IgM or IgA in the absence of surface IgG and the relatively frequent occurence of serum monoclonal Ig argue in favor of a relatively late stage in the B cell differentiation pathway, as does the usual failure of these leukemic cells to react with the anti-B2 monoclonal antibody[18].

B-DERIVED CHRONIC LYMPHOCYTIC LEUKEMIA

It is well established that the malignant process in the majority of patients with CLL affects one clone of surface Ig bearing B lymphocytes. Ten years ago, in view of the strikingly faint immunofluorescence staining of membrane-bound Ig of CLL lymphocytes, and of its uniformity in a given patient, we suggested that the cells from patients with common CLL were "frozen" at an early maturation stage in the B lymphocyte lineage[19]. In patients with CLL who had a serum monoclonal Ig component, the less uniform staining pattern and the presence of identical Ig isotype on lymphocytes, in plasma cells and the serum monoclonal Ig led us to conclude that this form of CLL corresponds in most instances to the persistence of some degree of differentiation of the leukemic lymphocytes into Ig secreting plasma cells[19], a situation similar to the one documented in Waldenström's macroglobulinemia[20]. Studies performed with anti-idiotypic antibodies confirmed this view[21].

This concept of a maturation arrest of the leukemic B lymphocytes in patients with common CLL at a relatively early stage of differentiation has been supported by the results of recent studies of these cells with various monoclonal antibodies. This hypothesis is, however, difficult to reconcile with the finding that CLL lymphocytes commonly express surface Ig with two different (most often μ and δ) or even three (for instance μ, δ and γ) heavy chain isotypes[22]. This pattern is commonly associated with mature rather than with immature B lymphocytes, which express only surface IgM. A recent study has demonstrated that the lymphocytes from patients with common CLL are able to export monoclonal Ig molecules both *in vivo* and *in vitro*[23]. Indeed, the majority of the serum pentameric IgM molecules in patients with CLL without a detectable monoclonal Ig spike have the same idiotype as that of the leukemic lymphocyte surface Ig. This finding does not necessarily argue in favor of a late level of maturation since careful study of murine B cells, particularly at the messenger RNA level, has shown that they invariably produce both membrane-bound and secretory IgM.

The maturation arrest of leukemic B lymphocytes may be due either to some intrinsic defect in the developmental abilities of the cell or to some extrinsic factor. To explore the latter, several groups have tried to induce *in vitro* further differentiation of CLL lymphocytes. The data in the literature are contradictory, since one group was able to induce *in vitro* differentiation only in those patients with a serum monoclonal Ig[24], whereas in another study[25], it was observed in common CLL also. We have studied the effect of various mitogens on CLL lymphocytes. Our results[26] show that in the majority of patients, CLL lymphocytes may be induced to differentiate into large cytoplasmic Ig-containing blast cells and to a lesser extent to cells with a plasmacytic appearance. The mitogen from Nocardia opaca was effective in 7 of 8 patients and phytohemagglutinin displayed inductive properties in all 5 patients tested. In contrast, pokeweed mitogen was effective in only half of the patients studied and these were patients who had a significant percentage of circulating T cells. It is worth noting that in one case, we observed a heavy chain switch from μ to γ after stimulation by Nocardia. One third of the blast cells with intracytoplasmic μ and λ chains also contained γ chains, as shown by double staining experiments. Strikingly, the differentiation of CLL lymphocytes did not appear to require cell multiplication and the mitotic activity observed in certain cultures was mostly due to the proliferation of residual normal T lymphocytes. This was even more evident when CLL lymphocytes were stimulated by staphylococcal protein A, since this mitogen did not stimulate the leukemic B lymphocytes but induced a very strong proliferation of the residual T cells[27]. In addition to mitogens, the phorbol ester TPA was recently shown to be very effective in inducing differentiation of CLL B lymphocytes[28].

Leukemic cells from patients with CLL may appear to display dual B and T markers. An interesting feature of the lymphocytes from many patients with B-derived CLL is the presence of antigenic determinant(s) shared by normal T cells but not by normal B cells from the blood and various lymphoid organs. These shared antigenic determinants were first recognized by rabbit antisera raised against human T cells[29] and more recently by several monoclonal antibodies such as OKT1[30-34]. Many of these shared antigenic determinants appear to belong to a 65,000 to 70,000 dalton molecule on T cells. Moreover, a 65,000 dalton protein, similar to that found on T cells, was precipitated by T 101 antibody from the surface of B CLL cells[32]. The nature and meaning of these shared antigens, which are not found on normal B cells and other B cell malignancies, remains unknown. A possible relationship between this antigen and the G_{IX} system in the mouse has been suggested[31]. The apparently mixed B+T phenotype of the leukemic cells noted in some CLL cases is quite distinct and may be a false appearance due for instance to an anti sheep erythrocyte antibody activity of the monoclonal surface Ig produced by leukemic B lymphocytes[35] or to the binding to leukemic T lymphocytes of exogeneous monotypic Ig[36] (i.e. antibody with restricted heterogeneity directed against a component of the leukemic cell surface). We have however encountered two CLL patients whose cells appeared to display true dual B and T properties, since they produced monoclonal surface Ig, formed E-rosettes and had T cell specific antigens. These leukemic cells were not studied with monoclonal antibodies to T cell subsets. The finding of dual B and T properties is more common in hairy cell leukemia.

HAIRY CELL LEUKEMIA

Although hairy cell leukemia appears to be a homogeneous and well defined disease on the basis of clinical presentation, light and electron microscopic features and cytochemical characteristics, the study of immunological markers of hairy cells from many patients reveals some degree of apparent heterogeneity. The most common phenotype is that of a B cell with some properties of the monocytic series. Hairy cells usually express monoclonal surface (and in certain cases cytoplasmic) Ig, Ia-like antigens and receptors for the Fc of IgM and IgG and for mouse erythrocytes. Additionally, they are capable of phagocytosis, glass adherence, and lysozyme and peroxidase synthesis. Cases have, however, been reported in which hairy cells failed to express one or more of these properties. In certain cases, they even display a T cell phenotype, while in others, features of both T and B cells are expressed, as observed in this and several other laboratories[37,38,39]. Moreover, in a few recently studied patients a marked fluctuation in the phenotype of the hairy cells was observed in a given patient at different times or in different tissues. These surprising discrepancies led us to hypothesize that hairy cells from the same patient might be able to express different phenotypes following an appropriate stimulus. We therefore studied immunological parameters of hairy cells stimulated by various mitogens[40].

In the study of 6 patients, we found differences in the phenotype of hairy cells from the same patient when the cells were studied twice (during an interval of several months), and spontaneous variations in control cultures without mitogens, such as an acquirement of cytoplasmic Ig or a switch from a B+T to a B phenotype. The changes induced by mitogens were much more striking.

Phytohemagglutinin induced a switch from a B to a T or a B+T (T cell features with production of intracytoplasmic monoclonal Ig) phenotype in hairy cells from every patient studied. The T cell markers acquired by PHA-stimulated hairy cells were the ability to form E rosettes and reactivity with heterologous anti-T sera. The latter did not interfere with E-rosette formation and did not react with B CLL cells. Several control experiments showed that the acquired T cell markers did not result from adsorption of membrane molecules released in the culture medium by PHA-stimulated normal T cells and that the cells which expressed these markers were actually derived from the hairy cells themselves and not from residual T lymphocytes. The kinetics of the phenotypic switch varied from patient to patient. The newly acquired T phenotype was quite stable in certain patients while in others, and especially with spleen cells, the T cell phenotype was acquired only by a subpopulation of the hairy cells and was transient. In a case recently studied with anti-T monoclonal antibodies, fresh splenic hairy cells expressed monoclonal Ig and were positive with OKT4; a moderate proportion also reacted with OKT8 and OKT10. Since these fresh cells did not form E-rosettes and did not react with Fab'2 fragments of polyclonal pan-T antisera, this reactivity with some monoclonals may be due to non-specific attachment to the cell membrane. After mitogen stimulation, SIg and OKT4 antigen were no longer detectable; whereas a subpopulation of hairy cells became able to form E-rosettes and acquired reactivity with the pan-T antisera and OKT3.

In certain patients, stimulated hairy cells retained their typical morphology and enzymatic properties. In other cases, morphologic changes (mostly towards large basophilic blastic cells) paralleled the phenotypic modifications. The effect of Concanavalin A was studied in a few cases and the findings were similar to those with PHA. Pokeweed mitogen had only a marginal effect, whereas Nocardia induced phenotypic changes which varied from patient to patient and from experiment to experiment. In two cases, we observed a plasmacytic differentiation, but in one of these cases the cells with a plasmacytic appearance and large amounts of cytoplasmic Ig were also able to form E-rosettes.

Hairy cells may thus be considered as very unusual and remarkable leukemic cells capable of expressing different phenotypic properties after (and sometimes even before) appropriate *in vitro* stimulation. The meaning of these findings remains unknown at present. The hypothesis of an aberrant gene expression in these leukemic cells cannot be ruled out. A minor subpopulation of normal lymphoid cells sharing these unusual phenotypic and functional properties has not yet been identified; this normal counterpart should probably be sought in tissues other than blood.

Acknowledgements

We thank Mrs. Sylvaine Labaume and Miss Annette Chevalier for expert technical assistance.

This work was aided in part by a grant from INSERM (CRL 811048).

REFERENCES

1. L.B. Vogler, W.M. Crist, D.E. Bockman, E.R. Pearl, A.R. Lawton, and M.D. Cooper. Pre-B cell leukemia. A new phenotype of childhood lymphoblastic leukemia. New Engl. J. Med, 298:872 (1978).
2. J.C. Brouet, J.L. Preud'homme, C. Penit, F. Valensi, P. Rouget, and M. Seligmann. Acute lymphoblastic leukemia with pre-B cell characteristics. Blood, 54: 269 (1979).
3. M.F. Greaves, W. Verbi, L.B. Vogler, M.D. Cooper, M. Ellis, K. Ganeshaguru, V. Hoffbrand, G. Janossy, F.J. Bollum. Antigenic and enzymatic phenotypes of the pre-B subclass of acute lymphoblastic leuke mia. Leuk. Res., 3:353 (1979).
4. D. Levitt, and M.D. Cooper. Mouse pre-B cells synthesize and secrete mu heavy chains but not light chains. Cell, 19:617 (1980).
5. E.J. Siden, and D. Baltimore. Immunoglobulin synthesis by lymphoid cells transformed in vitro by Abelson murine leukemia virus. Cell, 16:389 (1979).
6. P. Burrows, M. Lejeune and J.F. Kearney. Evidence that murine pre-B cells synthesize mu heavy chains but no light chains. Nature, 280:838 (1979).
7. P. Guglielmi and J.L. Preud'homme. Immunoglobulin expression in human lymphoblastoid cell lines with early B cell features. Scand. J. Immunol., 13:303 (1981).

8. L.B. Vogler, J.L. Preud'homme, M. Seligmann, W.E. Gathings, W.M. Crist, M.D. Cooper and J.F. Bollum. Diversity of immunoglobulin expression in leukemic cells resembling B lymphocyte precursors. Nature, 290:339 (1981).
9. N. Sakaguchi, T. Kishimoto, H. Kikutami, T. Watanabe, N. Yoshida, A. Shimizu, Y. Yamawaki-Kataoka, T. Honjo and Y. Yamamura. Induction and regulation of Immunoglobulin expression in a murine pre-B cell line, 70Z/3. 1. Cell cycle. Associated induction of sIgM expression and k-chain synthesis in 70Z/3 cells by LPS-stimulation. J. Immunol., 125:2654 (1980).
10. L.M. Nadler, J. Ritz, R. Hardy, J.M. Pesando and S.F. Schlossman. A unique cell surface antigen identifying lymphoid malignancies of B cell origin. J. Clin. Invest., 67:134 (1981).
11. J.H. Kersey, T.W. LeBien, C.A. Abramson, R. Newman, R. Sutherland and M.F. Greaves. p24: A human leukemia-associated and lymphohemopoietic progenitor cell surface structure identified with monoclonal antibody. J. Exp. Med., 153:726 (1981)
12. S.J. Korsmeyer, P.A. Hieter, J.V. Ravetch, D.G. Poplack, P. Leder and T.A. Waldmann, Patterns of immunoglobulin gene rearrangement in human lymphocytic leukemias, in: "Leukemia Markers", W. Knapp, ed. Academic Press, London (1981) pp. 85-97.
13. P. J. Fialkow, A.M. Denman, R.J. Jacobson and M.N. Lowenthal. Origin of some lymphocytes from leukemic stem cells. J. Clin. Invest., 62:815 (1978).
14. A. Bernheim, R. Berger, J.L. Preud'homme, S. Labaume, A. Bussel and R. Barot-Ciorbaru. Philadelphia chromosome positive blood B. lymphocytes in chronic myelocytic leukemia. Leuk. Res. 5:331, (1981).
15. G. Flandrin, J.C. Brouet, M.T. Daniel and J.L. Preud'-homme. Acute leukemia with Burkitt's tumor cells. A study of six cases with special reference to lymphocyte surface markers. Blood, 45:183 (1975).
16. R. Berger, A. Bernheim, J.C. Brouet, M.T. Daniel and G. Flandrin. t(8, 14) translocation in Burkitt's type of lymphoblastic leukemia (L3). Br. J. Haematol. 43:87 (1979).
17. J.L. Preud'homme, J.C. Brouet, F. Danon, G. Flandrin and G. Schaison. Acute leukemia with Burkitt cells: Membrane marker and serum immunoglobulin studies. J. Natl. Canc. Inst., 66:261 (1981).
18. L.M. Nadler, P. Stashenko, R. Hardy, A. van Agthoven, C. Terhorst and S.F. Schlossman. Characterization of a human B-cell specific antigen (B2) distinct from B1. J. Immunol., 126:1941 (1981).

19. J.L. Preud'homme and M. Seligmann. Surface-bound immunoglobulins as a cell marker in human lymphoproliferative diseases. Blood, 40:777 (1972).
20. J.L. Preud'homme and M. Seligmann. Immunoglobulins on the surface of lymphoid cells in Waldenström's macroglobulinemia. J. Clin. Invest. 51:701 (1972).
21. S.M. Fu, R.J. Winchester, T. Feizi, P.D. Walzer and H.G. Kunkel. Idiotypic specificity of surface immunoglobulin and the maturation of leukemic bone-marrowderived lymphocytes. Proc. Natl. Acad. Sci. USA, 71:4487 (1974).
22. J.L. Preud'homme, J.C. Brouet and M. Seligmann. Membrane-bound IgD on human lymphoid cells, with special reference to immunodeficiency and immunoproliferative diseases. Immunological Reviews, 37:127 (1977).
23. F.K. Stevenson, T.J. Hamblin, G.T. Stevenson and A.L. Tutt. Extracellular idiotypic immunoglobulin arising from human leukemic B lymphocytes. J. Exp. Med., 152:1484 (1980)
24. S.M. Fu, N. Chiorazzi and H.G. Kunkel. Differentiation capacity and other properties of the leukemic cells of chronic lymphocytic leukemia. Immunol. Rev., 48:23 (1979).
25. K.H. Robert. Induction of monoclonal antibody synthesis in malignant human B cells by polyclonal B cell activators. Immunol. Rev., 48:123 (1979).
26. P. Guglielmi, J.L. Preud'homme, R. Ciorbaru-Barot and M. Seligmann. Mitogen-induced maturation of chronic lymphocytic leukemia B lymphocytes. J. Clin. Immunol. in press (1981).
27. P. Guglielmi and J.L. Preud'homme. Stimulation of T lymphocytes by protein A from Staphylococcus Aureus in B derived chronic lymphocytic leukemia. Clin. Exp. Immunol., 41:136 (1980).
28. T.H. Tötterman, K. Nilsson and C. Sundström. Phorbol ester-induced differentiation of chronic lymphocytic leukemia cells. Nature, 288:176 (1980).
29. L. Boumsell, A. Bernard, V. Lepage, L. Degos, J. Lemerle and J. Dausset. Some chronic lymphocytic leukemia cells bearing surface immunoglobulins share determinants with T cells. Eur. J. Immunol., 8:900 (1978).
30. L. Boumsell, H. Coppin, D. Pham, B. Raynal, J. Lemerle, J. Dausset and A. Bernard. An antigen shared by a human T cell subset and B cell chronic lumphocytic leukemic cells. Distribution on normal and malignant lymphoid cells. J. Exp. Med., 152:229 (1980).

31. C.Y. Wang, R.A. Good, P. Ammirati, G. Dymbort and R.L. Evans. Identification of a p69,71 complex expressed on human T cells sharing determinants with B-type chronic lymphatic leukemic cells. J. Exp. Med., 151:1539 (1980).

32. I. Royston, J.A. Majda, S.M. Baird, B.L. Meserve and J.C. Griffiths. Human T cell antigens defined by monoclonal antibodies: the 65,000 dalton antigen of T cells (T65) is also found on chronic lymphocytic leukemia cells bearing surface immunoglobulin. J. Immunol., 125:725 (1980).

33. P.J. Martin, J.A. Hansen, R.C. Nowinski and M.A. Brown A new human T cell differentiation antigen: unexpected expression on chronic lymphocytic leukemia cells. Immunogenetics, 11:429 (1980).

34. E.G. Engleman, R. Warnke, R.I. Fox, J. Dilley, C.J. Benike and R. Levy. Studies of a human T lymphocyte antigen recognized by a monoclonal antibody. Proc. Natl. Acad. Sci. USA, 78:1791 (1981).

35. J.C. Brouet and A.M. Prieur. Membrane markers on chronic lymphocytic leukemia cells: a B cell leukemia with rosettes due to anti-sheep erythrocytes antibody activity of the membrane-bound IgM and a T cell leukemia with surface Ig. Clin. Immunol. Immunopath., 2:481 (1974).

36. J.C. Brouet, G. Flandrin, M. Sasportes, J.L. Preud'homme and M. Seligmann. Chronic lymphocytic leukemia of T cell origin. An immunological and clinical evaluation in eleven patients. Lancet, 2:890 (1975).

37. G.F. Burns, C.P. Worman, and J.C. Cawley. Fluctuation in T and B characteristics of two cases of T-cell hairy cell leukemia. Clin. Exp. Immunol., 39:76 (1980).

38. H.J. Cohen, E.R. George and W.B. Kremer. Hairy cell leukemia: cellular characteristics including surface immunoglobulin dynamics and biosynthesis. Blood, 53:764 (1979).

39. J. Jansen, H.R.E. Schuit, G.M.T. Schreuder, H.P. Muller and C.J.L.M. Meijer. Distinct subtype within the spectrum of hairy cell leukemia. Blood, 54:459 (1979).

40. P. Guglielmi, J.L. Preud'homme and G. Flandrin. Phenotypic changes of phytohaemagglutinin-stimulated hairy cells. Nature, 286:116 (1980).

EXPRESSION OF NORMAL DIFFERENTIATION ANTIGENS ON HUMAN LEUKEMIAS AND LYMPHOMAS

Lee M. Nadler, Jerome Ritz, Ellis L. Reinherz, and Stuart F. Schlossman

Div. of Tumor Immunology, Sidney Farber Cancer Institute
Harvard Medical School
Boston, Massachussetts, USA

INTRODUCTION

Leukemias and lymphomas are presently classified according to morphologic and histochemical criteria. (1-7). Although attempts to catalogue tumors according to these criteria have provided important insights into their presentation, clinical course, and response to therapy, a single universally accepted classification system is currently lacking. Moreover, within each histologically defined subgroup there is considerably greater clinical heterogeneity than has been defineable by either morphology or histochemistry. In the last decade, cell surface markers have provided an alternative means to identify normal lymphoid and myeloid cells possessing unique biologic properties. The application of these markers to the classification of lymphoid and myeloid tumors promises to complement presently existing classification schemes.

The development of immunologic techniques identifying human B, T and Null lymphocytes have provided the tools to delineate the cellular origin of malignant lymphocytes (8-12). Using these markers, leukemias and lymphomas of T, B and Null cell lineage have been identified. Human B lymphocytes express either cytoplasmic immunoglobulin (cIg) (13) or surface immunoglobulin (sIg)(14-15) receptors for the Fc portion of human immunoglobulin (16-18), receptors for the third component of the complement system (C3)(19), and the HLA-D related Ia like antigens (Ia)(20,21). Normal human T lymphocytes form rosettes with sheep erythrocytes (E-rosette) (22) and react with t cell specific heteroantisera (23). Null cells lack surface immunoglobulin and T lymphocyte markers but are heterogeneous with respect to Ia-like antigens, Fc and C3 receptors (24). Monocytes express Ia-like antigens and Fc and C3 receptors but lack surface Ig or T cell markers (24). There is considerable cellular overlap in the expression of many of these cell surface markers. Moreover, cells with Fc receptors may bind Ig and give a spurious positive result for cell surface immunoglobulin (25). Some cells, such as T cells will develop both Ia-like antigens and Fc receptors after activation (26-28). The lack of specificity and lineage restrictions of these cell surface markers have limited their diagnostic utility.

In recent years attention has been given to the definition of serologically and functionally distinct subsets of human immune cells. Heteroantisera have been developed which are capable of defining cell surface antigens expressed on populations of normal lymphoid and myeloid cells. These heteroantisera have proven to be extraordinarily useful in defining subpopulations of lymphoid and myeloid cells, but significant technical difficulties were encountered in their preparation. To render these antisera specific, it was necessary to extensively absorb them with cells of different cellular lineages. Since these antisera underwent many absorptions, the resulting titers were frequently low, thereby limiting their utility for biologic or clinical investigations. Finally, given the diversity of the immune response, it has been difficult to reproduce antibodies with identical specificity.

With the development of hybridoma derived monoclonal antibodies against cell surface antigens, many of the difficulties experienced with heteroantisera have been overcome (29). In our laboratory, we have recently developed and characterized a series of monoclonal antibodies which define cell surface differentiation antigens expressed on lymphocytes of T and B cell origins. In addition, unique leukemia and lymphoma associated antigens have also been described. In this report we will review some of the work from our laboratory utilizing monoclonal antibodies against normal differentiation and leukemia associated antigens. We hope to demonstrate that the lymphoid malignancies reflect the same degree of heterogeneity and maturation as is seen in normal T and B cell ontogeny.

RESULTS AND DISCUSSION

T-Cell Antigens on Human Leukemias and Lymphomas

Considerable evidence now exists in man to support the notion of T cell heterogeneity as defined by cell surface markers including E-rosettes, complement receptors, and Fc receptors (30-35). More definitive studies of T cell heterogeneity were based upon heteroantisera prepared against human peripheral blood T cells, thymocytes, and malignant T cells (21, 36-40). These heteroantisera provided the first reagents which defined functionally distinct subpopulations of human T cells (40, 41), and also provided important insight into the cellular origins of T cell malignancies. Given the difficulties of working with heteroantibodies, many laboratories have developed monoclonal anti-T cell antibodies. Several of these antibodies have proben to be useful in studying thymic differentiation (42), the heterogeneity of mature T lymphocytes, in clinical studies directed at an understanding of immunologic disorders, and for the classification of leukemias and lymphomas.

In man, the earliest lymphoid cells within the thymus bear antigens shared by some bone marrow cells but lack antigens expressed on mature T cells (42). This prothymocyte population accounts for approximately 10% of thymic lymphocytes and is reactive with two monoclonal antibodies which were prepared against human thymocytes (anti-T9 and anti-T10) (43). These antigens are extremely useful for studying T cell differentiation but are not T cell lineage restricted and have also been found on some normal bone marrow cells, fetal tissues, transformed cells and some B cell malignancies (43,44, unpublished observations, Reinherz, Nadler). With maturation thymocytes lose T9, retain T10, and acquire a thymocyte-distinct antigen (T6) which is the TL (thymic leukemia) equivalent (45). Concurrently, these cells express antigens defined by monoclonal antibodies anti-T4 and anti-T5/8 (46-48). Thymocytes which express T4, T5/T8, T6 and T10 account for approximately 70% of the total thymic population and are primarily cortical in location (49-50). With further maturation, thymocytes lose T6, acquire T1 and T3 antigens found on mature T-cells, and segregate into either T4+ or T5+/T8+ subsets (42). These thymocytes are found primarily in the medullary region of the thymus and account for approximately 10% of the thymic population (49-50). Immunologic competence is acquired at this stage of differentiation but is not fully developed until the cells are exported. With exportation, the cells become $T10^-$ and the circulating T cell is either $T1^+$, $T3^+$, $T4^+$, or $T1^+$. $T3^+$, $T5^+/T8^+$. The former population represents circulating inducer (helper) cells whereas the latter defines the cytotoxic/suppressor population.

The ability to define discrete stages of normal T cell differentiation has permitted a better understanding of the heterogeneity of the T cell malignancies. Earlier studies employing the T cell subset specific heteroantiserum TH2 allowed one of the first demonstrations of heterogeneity within the T cell lymphoblastic malignancies by serologic methods (51). Tumor cells isolated from the majority of patients with T cell acute lymphoblastic leukemia (T-ALL) were unreactive with the anti-TH2 antiserum, whereas tumor cells isolated from patients with lymphoblastic lymphoma (LL) were largely anti-TH2 reactive (52). The differential expression of the TH2 antigen in these diseases suggested that the malignant lymphoblasts were derived from different T cell populations. Moreover, the TH2 phenotype correlated with both the clinical presentation and the course of the disease. Preliminary studies suggested that TH2 phenotype also correlated with a favorable response to therapy.

A more precise dissection of the T cell malignancies result from studies utilizing monoclonal anti-T cell antibodies. The T cell malignancies do, in fact, reflect the same degree of heterogeneity and maturation as seen in normal T cell ontogeny (44-53). All T cell leukemias tested are reactive with the pan anti-human T cell heteroantisera (A99)(23), have a variable ability to form E-rosettes and lacked both surface immunoglobulins and Ia-like antigens. We have recently investigated a large group of patients with T-ALL and LL (54). As shown in Table I, most T cell acute leukemias possessed antigens found on early thymocytes or prothymocytes (Stage 1). Approximately 20% of the T-ALLs expressed antigens found on Stage II thymocytes, and only rarely on Stage III thymocytes.

T cell lymphoblastic lymphomas share many of the clinical features of T-ALL in that they arise predominantly in adolescent males who often present with a mediastinal mass (55-57). By defination, patients with T cell malignant lymphoma (LL) lack marrow and blood involvement. As LL progresses, however, bone marrow and peripheral blood tumor infiltration occurs and this disease appears indistinguishable from T-ALL. In fact, they have often been considered as stages of the same disease. Analysis of cell surface phenotypes of tumor cells from patients with LL would suggest that a majority of patients with LL arise from Stage II or III thymocytes (54). A smaller proportion of patients with LL have tumor cells which like T-ALL, have the phenotype of Stage I thymocytes. These results suggest that T-ALL and LL are not different clinical stages of a single neoplastic process. The differences noted in clinical presentation, survival, and response to therapy may well result from the specific thymic pool, the differentiative stage, or the drug susceptibility of a distinct malignant T lymphoblast.

In addition to T-ALL and LL, several other T cell malignancies were studied to correlate these diseases with T cell differentiation (58,59). Patients with Sezary's syndrome and T cell CLL are quite distinct and had a phenotype identical to a mature T cell subset; i.e., either a T1, T3, T4 inducer cell or a T1, T3, T5/T8 cytotoxic/suppressor cell. As shown in Table 1, patients with Sezary's syndrome were T1, T3, T4, whereas patients with T-CLL were either T1, T3, T4 or T1, T3, T5/T8. It should be noted that these cells did not express or coexpress antigens associated with earlier stages of thymic maturation. In addition, some of these patients' cells also expressed Ia antigens (58). Although Ia is commonly found on B cells, monocytes and a fraction of Null cells, its presence on activated peripheral T cells has now been well documented (28).

The studies described above support the conclusion that heterogeneity of T cell malignancies, for the most part, reflects stages

TABLE I

CELL SURFACE ANTIGENS EXPRESSED ON T CELL

LEUKEMIAS AND LYMPHOMAS

Stage	T Cell Antigen by Stage	T Cell Malignancies
I	T9, T10 or T10 (prothymocyte)	Majority of T-ALL
II	T6, T4, T5/T8, T10 (thymocyte)	Majority of T-LL, Minority of T-ALL
III	T4,T5/T8, T10, T1, T3 (thymocyte)	Minority of T-LL, Rare T-All
Mature	T1, T3, T4 (inducer)	All Sezary and Mycosis Fungoides Majority of T-CLL
Mature	T1, T3, T5/T8 (cytoxic/suppressor)	Rare T-CLL

of normal T cell differentiation. Anomalous expression of T cell antigens on T-ALL and LL are also seen. Some cases bear both early and late antigens but these represent a minority of the patients that were studied to date. Whether these cases reflect a true aberration of the malignant cell or actually have a normal counterpart in a minority of cells in the thymus is still not clear.

Expression of B cell antigens on Leukemias and Lymphomas

Human B lymphocytes possess a distinct cell surface phenotype that distinguishes them from other lymphoid cell populations. These markers include integral membrane immunoglobulin (14,15), receptors for C3 and the Fc portion of IgG (16,19), and the presence of antigens encoded by the HLA-D locus, the so-called DR antigens (20-21). Because many of these phenotypic markers are not restricted to B cells, their utility in cell enumeration and fractionation has been somewhat limited.

In the past year several laboratories have reported monoclonal antibodies reactive with human B cell antigens distinct from conventional human immunoglobulins, Ia-like antigens, and Fc or C3 receptors (60-64). We have two B cell specific monoclonal antibodies designated anti-B1 and anti-B2 (61,62). The B1 antigen is present on greater than 95% of B cells in peripheral blood and lymphoid organs. The B1 antigen is absent from resting and activated T cells, monocytes, Null cells and granulocytes. Functional studies demonstrated that the removal of the B1 reactive cells by cell sorting or by complement mediated lysis eliminated all cells from peripheral blood or spleen capable of immunoglobulin synthesis.

A second human B lymphocyte specific antigen, B2 has also been identified (62). By indirect immunofluorescence and quantitative absorption, B2, like B1, has been shown to be expressed exclusively on Ig^+ B cells isolated from peripheral blood and lymphoid tissues. In contrast to the B1 antigen, the B2 antigen is weakly expressed on peripheral blood B cells but strongly expressed on B cells isolated from lymph node, tonsil and spleen. Chemical characterization indicated that the B1 and B2 antigens were different. The B2 antigen is a single band with a molecular weight of approximately 140,000 daltons whereas the B1 antigen is approximately 30,000 daltons. Like B1, only anti-B2 reactive splenocytes could be induced to differentiate by PWM into plasma cells.

Recent experiments have provided evidence that B1 and B2 are B cell differentiation antigens (65). This evidence has been derived from studies employing a pokeweed mitogen (PWM) driven model of B cell differentiation. Pokeweed mitogen induces differentiation of B cells to antibody secreting cells, a function which is T cell dependent in man (66). Either peripheral blood or splenic B cells were stimulated with PWM and were followed over time for the disappearance of B1 and B2. It was found that the B2 antigen was lost from the cell surface by four days and B1 was lost by day six or seven. When B2 was lost from the cell surface, the cells had transformed into lymphoblasts, many of which contained IgM. Moreover, when the B1 antigen was lost, cytoplasmic IgG was noted.

With the *in vitro* evidence that the B1 and B2 antigens are expressed on distinct stages of B cell differentiation, we then investigated the expression of these antigens on tumor cells. It has been hypothesized that B cell non-Hodgkins lymphomas represent clonal proliferations of distinct stages of B cell differentiation (67). Indeed, these tumor cells express either κ or λ light chains but not both. Tumor cells from all patients with B cell chronic lymphocytic leukemia and all patients with B cell lymphomas except myelomas were reactive with the anti-B1 antibody (Table II) (68). Anti-B1 was unreactive with tumor cells from patients with T cell lymphomas and leukemias, acute myeloblastic leukemia, and the stable phase of chronic myelocytic leukemia.

Although all B cell lymphomas except myelomas express the B1 antigen, the expression of B2 appears more restricted (62). Tumor cells from patients with Waldenstrom's macroglobulinemia, Burkitt's lymphoma, nodular mixed lymphocytic lymphoma, and diffuse histiocytic lymphoma (large cell transformed lymphoma) were almost all B2 negative (Table LL). In contrast, approximately half of all diffuse, poorly differentiated lymphocytic (D-PDL) and nodular poorly differentiated lymphocytic (N-PDL) lymphomas were reactive with anti-B2 (Table LL). Tumor cells from patients with undifferentiated lymphomas and B cell CLL expressed B2. These observations suggest that the B cell tumors corresponding to the later stages of differentiation lack the B2 antigen. Specifically transformed or large cell lymphomas, Waldenstrom's macroglobulinemia, and plasma cell myeloma are B2 negative. These tumors correspond to either transformed B lymphoblast (11) or to the cells of the secretory phase of B cell differentiation (12).

Previous studies have shown that acute lymphoblastic leukemia cells can be divided into two major subgroups: T and non-T (9,10). T cell ALL cells express T cell surface antigens but are Ia negative (21). The non-T ALL cells are predominantly Ia$^+$ but lack surface Ig and were considered to be "Null" cell tumors (21,69). Most of the non-T ALL cells express the common ALL antigen (CALLA) and these patients could be divided into two groups: Calla$^+$, Ia$^+$ which accounted for 80% of the total, and CALLA$^-$, Ia$^+$ which accounted for the remaining 20% (21,70-74). While the presence of Ia antigens, and the failure to express either mature T cell or thymic antigens on the non-T cell leukemias suggested their B cell derivation, this was not conclusive evidence of B cell lineage. Recently 30% of these patients were found to express cytoplasmic IgM (cIgM)(75-77). We have found that leukemic cells from 50% of patients with non-T cell ALL are reactive with anti-B1 (68). Moreover, the tumor cells from almost all patients with non-T ALL and CML in blast crisis which express the B1 antigen also express the CALLA antigen. Taken together, these studies

TABLE II

CELL SURFACE ANTIGENS EXPRESSED ON B CELL DERIVED HUMAN LEUKEMIAS AND LYMPHOMAS

Tumor	B Cell Antigens					Hypothetical B Cell Differentiative Stage
	Ia	sIg	cIg	B1	B2	
Myeloma	-[7]	-	+++	-	-	Plasma Cell
Waldenstrom's	+/-	++	+/-	+	-	Late Secretory Cell
Histiocytic (N[1] or D[2])	+++	+++	-	+	-	Transformed Lymphoblast
PDL[3] (N or D)	+++	+++	-	+++	+/-	Virgin B Cell
CLL[4]	++	+	-	+/++	+/++	Young B Cell
NULL-ALL[5] CALLA+[6]	+++	-	30%+	75%+/++	10%+	Pre-B Cell
CALLA-	+++	-	0%	0%	0%	Unknown

1 Nodular
2 Diffuse
3 Poorly differentiated lymphocytic lymphona
4 Chronic lymphocytic leukemia
5 Acute lymphoblastic leukemia
6 Common acute lymphoblastic leukemia antigen
7 - no reactivity
+/- marginal reactivity
+ weak reactivity
++ moderate reactivity
+++ strong reactivity

suggest that a significant fraction of the CALLA positive ALLs are of B cell lineage.

SUMMARY

In this report, we have reviewed primarily work from our own laboratory which demonstrates that antibodies can be utilized as probes to identify unique cell surface antigens on normal and malignant lymphoid cells. These antibodies provide a panel of reagents to study normal differentiation. By understanding normal cellular differentiation it is now possible to relate the malignant cell to its normal cellular counterpart. Acute lymphoblastic leukemias can now be readily distinguished by virtue of cell surface markers which define lineage and state of differentiation. An understanding of intrathymic differentiation and mature T cell subsets has permitted the demonstration that most T-ALL, T-LL, T-CLL and Sezary syndrome are derived from distinct T cell differentiative stages. Similarly, it is now possible to assign a B cellular origin to the tumor cells from many patients with common or "Null" cell ALL. These studies, at the very least, demonstrate the inadequacy of our present morphologically based classification schemes. It is hoped that the approach outlined in this review will eventually assist in the classification of leukemias and lymphomas and may define groups of patients who share similar disease presentations and/or therapeutic responses.

REFERENCES

1. Hayhoe, F.G.J. and Flemans, R.J. (1970). *In* "An Atlas of Haematological Cytology", pp. 110-265. Wiley-Interscience, New York.
2. Hayhoe, F.G.J., Quagliano, M. and Doll, R. (1964). "The Cytology and Cytochemistry of Acute Leukemias: A Study of 140 Cases". Her Majesty's Stationery Office, London.
3. Rappaport, H. (1966). *In* "Atlas of Tumor Pathology". Section 3, Fascicle 9, p. 13. Armed Forces Institute of Pathology, Washington, D.C.
4. Braylan, R.C., Jaffe, E.S. and Berard, C.W. (1975). *Pathol. Annu.* 10, 213-270.
5. Bennett, J.M., Catovsky, D., Daniel, M.T. et al. (1976). *Br. J. Haematol.* 33, 451-458.
6. Gralnick, H.R., Galton, D.A.G., Catovsky, D., Sultan C. and Bennett, J.M. (1977). *Ann. Int. Med.* 87, 740-753.
7. Dorfman, R.F. (1977). *Cancer Treat. Rep.* 61, 945-952.
8. Aisenberg, A.C. and Bloch, K.J. (1972). *N. Engl. J. Med.* 287, 272-276.
9. Borella, L. and Sen, L. (1973). *J. Immunol.* 111, 1257-1261.
10. Brouet, J.C. and Seligmann, M. (1978). *Cancer* 42, 817-827.
11. Siegal, F.P. (1978). "The Immunopathology of Lymphoreticular Neoplasms" (Ed. J.J. Twommey and R.A. Good). pp. 281-324. Plenum, New York.

12. Mann, R.B., Jaffe, E.S. and Berard, C.W. (1979). *Am. J. Pathol.* 94,104-191.
13. Gathings, W.E., Lawton, A.R. and Cooper, M.D. (1977). *Eur. J. Immunol.* 7, 804-810.
14. Froland, S.S. and Natvig, J.B. (1970). Int. Arch. Allerg. & Appl. Immunol. 39, 121-132.
15. Froland, S.S., Natvig, J.B. and Berdal, P. (1971). Nature 234, 251-252.
16. Huber, H., Douglas, S.D. and Fidenberg, H.H. (1969). Immunol. 17, 7-21.
17. Bianco, C., Patrick, R. and Nussenzweig, V. (1970). J. Exp. Med. 132, 702-720.
18. Dickler, H.B. and Kunkel, H.G. (1972). J. Exp. Med. 136, 191-196.
19. Bianco, C., Patrick, R. and Nussenzweig, V. (1970). J. Exp. Med. 132, 702-720.
20. Winchester, R.J., Fu, S.M., Wernet, P., Kunkel, H.G., Dupont, B. and Jerslid, C. (1975). J. Exp. Med. 141, 924-929.
21. Schlossman, S.F., Chess, L., Humphreys, R.E. and Strominger, J.L. (1976). Proc. Natl. Acad. Sci. USA 73, 1288-1292.
22. Jondal, M., Holm, G. and Wigzell, H. (1972). J. Exp. Med. 136, 207-215.
23. Pratt, D.M., Schlossman, S.F. and Strominger, J.L. (1980). J. Immunol. 124, 1449-1461.
24. Chess, L. and Schlossman, S.F. (1977). In "Advances in Immunology" (Ed. F.J. Dixon and H.G. Kunkel). pp. 213-247. Academic Press, Inc., New York.
25. Winchester, R.J., Fu, S.M. and Kunkel, H.G. (1975). J. Immunol. 114, 1210-1212.
26. Mendes, N.F., Tolnai, M.C.A., Silveira, N.P.A., Gilbertsen, R.B. and Metzgar, R.S. (1973). J. Immunol. 111, 860-867.
27. Evans, R.L., Faldetta, T.J., Humphreys, R.E., Pratt, D.M., Yunis, E.J. and Schlossman, S.F. (1978). J. Exp. Med. 148, 1440-1445.
28. Reinherz, E.L., Kung, P.C., Pesando, J.M., Goldstein, G. and Schlossman, S.F. (1979). J. Ext. Med. 150, 1472-1482.
29. Kohler, G. and Milstein, C. (1975). Nature 256, 495-497.
30. Ross, G.D., Rabellino, E.M., Polley, M.D. and Grey, H.M. (1973). J. Clin. Invest. 52, 377-384.
31. Borella, L. and Sen, L. (1975). J. Immunol. 114, 187-190.
32. Jaffe, E.S., Shevach, E.M., Sussman, E.H., Frank, M., Green, I. and Berard, C.W. (1975). Br. J. Cancer 31, (suppl. 2), 107-120.
33. Moretta, L., Webb, S.R., Grossi, C.E., Lydyard, P.M. and Cooper M.D. (1977). J. Exp. Med. 146, 184-200.
34. Moretta, L., Ferrarini, M., Mingari, M.C., Moretta, A. and Webb, S.R. (1977). J. Immunol. 117, 2171-2174.
35. Moretta, L., Mingari, M.C., Moretta, A. and Lydyard, P.M. (1977) Clin. Immunol. Immunopathol. 7, 405-409.

36. Chechik, B.E., Pyke, K.W. and Gelfand, E.W. (1976). Int. J. Cancer 18, 551-556.
37. Borella, L., Sen, L., Dow, L.W. and Casper, J.T. (1977). In "Hematology and Blood Transfusion", p. 77. Springer, Berlin.
38. Boumsell, L., Bernard, A., Coppin H., et al. (1979). J. Immunol. 123, 2063-2067.
39. Evans, R.L., Breard, J.M., Lazarus, H., Schlossman, S.F. and Chess, L. (1977). J. Exp. Med. 145, 221-233.
40. Evans, R.L., Lazarus, H., Penta, A. C. and Schlossman, S.F. (1978). J. Immunol. 129, 1423-1428.
41. Reinherz, E.L. and Schlossman, S.F. (1979). J. Immunol. 122, 1335-1341.
42. Reinherz, E.L. and Schlossman, S.F. (1980). Cell 19, 821-827.
43. Reinherz, E.L., Kung, P.C., Goldstein, G., Levey, R.H. and Schlossman, S.F. (1980). Proc. Natl. Acad. Sci. USA 77, 1588-1592.
44. Reinherz, E.L. and Schlossman, S.F. (in press). Cancer Res.
45. Boyse, E.A., Stockert, E., and Old, L.J. (1967). Proc. Natl. Acad. Sci. USA 58, 954-957.
46. Reinherz, E.L., Kung, P.C., Goldstein, G. and Schlossman, S.F. (1979). Proc. Nat. Acad. Sci. USA 76, 4061-4065.
47. Reinherz, E.L., Kung, P.C., Goldstein, G. and Schlossman, S.F. (1979). J. Immunol. 123, 2894-2896.
48. Reinherz, E.L., Kung, P.C., Goldstein, G. and Schlossman, S.F. (1980). J. Immunol. 124, 1301-1307.
49. Bhan, A.K., Reinherz, E.L., Poppema, S., McCluskey, R.T. and Schlossman, S.F. (1980). J. Exp. Med. 152, 771-782.
50. Janossy, G., Tidman, N., Selby, W.S., et al. (1980). Nature 288, 81-84.
51. Reinherz, E.L., Nadler, L.M., Sallan, S.E. and Schlossman, S.F. J. Clin. Invest. 64, 392-397.
52. Nadler, L.M., Reinherz, E.L., Weinstein, H.J., D'Orsi, C.J. and Schlossman, S.F. (1980). Blood 55, 806-810.
53. Nadler, L.M., Reinherz, E.L. and Schlossman, S.F. (1980). Cancer Chemother. Pharmocol. 4, 11-15.
54. Bernard, A., Boumsell, L., Reinherz, E.L., et al. (1981) Blood, 57, 1105-1110.
55. Nathwani, B.N., Kim, H. and Rappaport, H. (1976). Cancer 38, 984-985.
56. Jaffe, E.S. and Berard, C.W. (1978). Ann. Int. Med. 89, 415-416.
57. Rosen, P.J., Feinstein, D.I., Pattengale, P.K., et al. (1978) Ann. Int. Med. 89, 319-324.
58. Reinherz, E.L., Nadler, L.M., Rosenthal, D.S., Moloney, W.C. and Schlossman, S.F. (1979). Blood 53, 1066-1075.
59. Boumsell, L., Bernard, A., Reinherz, E.L., et al. (1981) Blood 57, 526-530.
60. Brooks, D.A., Beckman, I., Bradley, J., McNamara, P.J., Thomas, M.E. and Zola, H. (1980). Clin. Exp. Immunol. 39, 477-485.

61. Stashenko, P., Nadler, L.M., Hardy, R. and Schlossman, S.F. (1980). J. Immunol. 125, 1678-1685.
62. Nadler, L.M., Stashenko, P., Hardy, R., van Agthoven, A., Terhorst, C. and Schlossman, S.F. (1981) J. Immunol. 126, 1941-1947.
63. Greaves, M.F., Verbi, W., Kemshead, J. and Kennett, R. (1980). Blood 56, 1141-1144.
64. Abramson, C., Kersey, J. and LeBien, T. (1981). J. Immunol. 126, 83-88.
65. Stashenko, P., Nadler, L.M., Hardy, R. and Schlossman, S.F. (1981), Proc. Nat. Acad. Sci. 78, 3848-3852.
66. Fauci, A.S., Pratt, K.R.K. and Whalen, G. (1976). J. Immunol. 117, 2100-2104.
67. Salmon, S.E. and Seligmann, M. (1974). Lancet II, 1230-1233.
68. Nadler, L.M., Stashenko, P., Ritz, J., Hardy, R., Pesando, J.M. and Schlossman, S.F. (1981). J. Clin. Invest. 67, 134-140.
69. Fu, S.F., Winchester, R.J. and Kunkel, H.G. (1975). J. Exp. Med. 142, 1334-1339.
70. Chessells, J.M., Hardisty, R.M., Rapson, N.T. and Greaves, M.F. (1977). Lancet 2, 1307-1309.
71. Sutherland, R., Smart, J., Niaudet, P. and Greaves, M. (1978) Leukemia Research 2, 115-126.
70a. Greaves, M.F., Brown, G., Rapson, N.T. and Lister, T.A. (1975) Clin. Immunol. Immunopathol. 4, 67-84.
71a. Borella, L., Sen, L. and Casper, J.T. (1979). Leuk. Res. 3, 353-362.
72. Billing, R., Monowada, J., Cline, M., Clark, B. and Lee, K. (1978). J. Natl. Cancer Inst. 61, 423-429.
73. Pesando, J.M., Ritz, J., Lazarus, H., Costello, S.B., Sallan, S.E. and Schlossman, S.F. (1979). Blood 54, 1240-1248.
74. Ritz, J., Pesando, J.M., Notis-McConarty, J., Lazarus, H. and Schlossman, S.F. (1980). Nature (Lond) 283, 583-585.
75. Vogler, L.V., Crist, W.M., Bockman, D.E., Pearl, E.R., Lawton, A.R. and Cooper, M.D. (1978). N.Engl. J. Med. 298, 872-878.
76. Greaves, M.F., Verbi, W., Vogler, L.B., et al. (1979). Leuk. Res. 3, 353-362.
77. Brouet, J.C., Preud'homme, J.L., Penit, C., Valensi, F.. Rouget, P. and Seligmann, M. (1979). Blood 54, 269-273.

HUMAN-HUMAN HYBRIDOMA MONOCLONAL ANTIBODIES IN DIAGNOSIS AND TREATMENT OF NEOPLASTIC DISEASE

Henry S. Kaplan, and
Lennart Olsson*

Cancer Biology Research Laboratory
Department of Radiology
Stanford University School of Medicine
Stanford, CA 94305

INTRODUCTION

A new era in immunology was introduced by Köhler and Milstein (1975, 1976) when they devised a method for the production of mouse "hybridomas" capable of secreting monoclonal antibodies of predefined antigenic specificity. Immunologists were liberated from the constraints and difficulties previously associated with the preparation and use of heteroantisera, since the clonal selection and immortality of hybridoma cell lines assure the monoclonality, monospecificity, and permanent availability of their antibody products. However, although monoclonal antibodies of murine or other rodent origin have been extraordinarily powerful new reagents in laboratory investigations, their clinical use is likely to be severely limited by the fact that they are foreign proteins. Human immunoglobulin-producing cells have been fused to mouse myeloma cells to generate chimeric hybridomas (Schwaber, 1975; Levy and Dilley, 1978). Although there have been rare instances in which long-term, stable production of human antibody has thus been achieved (Schlom et al., 1980), most such hybrids have tended to be highly unstable due to the selective loss of human chromosomes. An alternative approach to the generation of antibody-producing human lymphocyte cell lines has involved the transformation of human B-lymphocytes with Epstein-Barr virus (EBV) (Steinitz et al., 1977; Zurawski et al., 1978; Koskimies, 1979) but this method appears to be of limited practical usefulness because the cultures tend to cease antibody production after a variable period (Zurawski et al., 1978). We now have been able to obtain human-human hybridomas secreting monoclonal antibodies of predefined antigenic

*Present address: State University Hospital, Copenhagen, Denmark

specificity by fusing primed human lymphoid cells with HAT-sensitive human myeloma cells (Olsson and Kaplan, 1980). This should open the way to the generation of human monoclonal antibodies against a broad spectrum of prefefined antigens.

TECHNICAL CONSIDERATIONS

Selection and Establishment of a HAT Sensitive Human Myeloma Cell Line

Aminopterin blocks the main pathway of DNA synthesis. The rescue pathway used by cells cultured in a medium (HAT) also containing hypoxanthine and thymidine requires the enzymes hypoxanthine-guanine phosphoribosyl transferase (HGPRT) and thymidine kinase (TK). Myeloma cell lines lacking one or both of these enzymes are therefore unable to grow in HAT medium. HGPRT-deficient cells can be selected for resistance to the purine analog 8-azaguanine, and TK-deficient cells can similarly be selected for resistance to the pyrimidine analog 5' bromodeoxyuridine (BUdR).

Murine myeloma cell lines have usually been selected for 8-azaguanine resistance in a single step procedure by cloning the myeloma cells in soft agar in the presence of 8-azaguanine (usual concentration 20 μg/ml). With mouse myeloma cell lines, this procedure yields as many as 20-30 resistant clones per 10^6 seeded myeloma cells. Thus, the more tedious selection procedures in liquid culture have not been necessary.

The human U-266 myeloma cell line was established and characterized by Nilsson et al., 1970. This IgE(λ)-secreting myeloma cell line was initially reported to grow slowly, to require a feeder layer, and to have a very poor cloning efficiency. In 1977, one of us (HSK) had requested the U-266 line primarily to test its permissiveness for the replication of a C-type virus produced by the SU-DHL-1 human histiocytic lymphoma cell line (Epstein and Kaplan, 1974; Kaplan et al., 1977, 1979). When the U-266 cells first arrived, their viability and growth rate were relatively low, and it required four to five months of painstaking selection of the healthiest miniwell cultures at each passage to obtain a cell line with markedly improved viability and growth rate, which no longer required the use of feeder layers. The myeloma cells thus obtained could also be cloned in semisolid agarose at a cloning efficiency of slightly less than 1%.

When the present human hybridoma experiments were initiated, the U-266 cells were recovered from storage in liquid nitrogen and re-established in culture. We tried initially to clone the cells in soft agar or agarose in the presence of 10, 15, or 20 μg/ml of 8-azaguanine. Under none of these culture conditions could we observe the growth of resistant colonies. Several months later, it was discovered that the original cell line was contaminated with mycoplasma, despite earlier negative tests; the parental line, as well as the HAT-sensitive mutant described below, have now been cured of mycoplasma infection by treatment with heat and/or antibiotics and subcloning, and both now express a much greater cloning efficiency in agarose.

Meanwhile, we turned to the use of liquid cultures, and developed the following selection procedure. A large inoculum of U-266 cells was grown in RPMI-1640 medium supplemented with 15% fetal calf serum (FCS) and 20 μg/ml of 8-azaguanine. After five days of cultivation, at least 99% of the cells were dead. The dead cells were then separated from the remaining viable cells on a Ficoll-Hypaque gradient (Böyum, 1968). The viable cells were then cultivated for an additional two to three days in medium containing 20 μg/ml of 8-azaguanine. Viable cells remaining in the culture at this time were again isolated on a Ficoll-Hypaque gradient, washed, and seeded in a single microwell of a 96-well microtiter plate containing RPMI-1640 with 15% FCS and 5 μg/ml 8-azaguanine. Visible cell growth could be seen in this medium within about one week. As soon as growth was observed, the 8-azaguanine concentration was increased to 10 μg/ml. After approximately one additional week, as growth again resumed, the 8-azaguanine concentration was further increased to 15 μg/ml, without any further inhibition of cell growth.

The cell culture was then expanded in RPMI-1640 with 15% FCS and 20 μg/ml 8-azaguanine, and cloned in microtiter plates by the limiting dilution procedure. Growth curves were determined for five actively growing clones. The fastest growing clone, which had a cell population doubling time of about 18 hours in exponential growth phase, was selected for use in human hybridoma production. This clone was initially designated U-266AR_1, and later identified as SKO-007 in the Stanford University Biological Organism Registry. These cells were found to be HAT-sensitive, dying at aminopterin concentrations as low as 4×10^{-8}M. This was the lowest concentration that assured the killing of all myeloma cells. We consider it important to use a relatively low aminopterin concentration in order to obtain a maximum yield of hybridomas. However, variations in culture conditions may influence the HAT-sensitivity of the line, and periodic titration of the aminopterin concentration is therefore recommended. The SKO-007 cell line was found to fuse readily with human lymphoid cells in the presence of polyethylene glycol (PEG). Its HL-A phenotype, kindly determined by Dr. Rose Payne (Department of Medicine, Stanford University School of Medicine), is: HLA-A2, A3, B7, Bw60, Bw6, Cw3. Its karotype, kindly determined by Drs. Barbara Kaiser-McCaw and Frederick Hecht (Southwest Biomedical Research Institute, Tempe, AZ) is: 44, X, -8, -17, -18, +2mar, +2dmin, t(1;1), t(2;?), t(11;?).

Sensitization of Human Lymphoid Cells

In our initial experiments, we considered it important to mimic the Köhler-Milstein procedure as closely as possible by carrying out the immunization *in vivo*. Patients with previously untreated Hodgkin's disease were selected for these studies, since they are routinely submitted to a battery of immunologic tests (Eltringham and Kaplan, 1973), including sensitization and later challenge with dinitrochlorobenzine (DNCB). About two weeks later, such patients are submitted to staging laparotomy with splenectomy as an integral part of their diagnostic investigation (Glatstein et al., 1969). Spleens which, on pathological investigation, appeared devoid of involvement by the disease, were used for hybridoma production. Single cell suspensions were prepared, freed of red blood cells and granulocytes by Ficoll-Hypaque gradient centrifugation and enriched for lymphocytes by repeated brief incubation in plastic dishes and transfer of non-adherent cells. The lymphocyte-enriched mononuclear cell suspensions thus derived were then fused with the SKO-007 human myeloma cell line and incubated in HAT medium as described below. The spleen cells from two of three patients with Hodgkin's disease yielded a total of five HAT-resistant hybridoma clones producing IgG (κ) antibody with specific reactivity in radio-immunoassays for DNP-BSA (Olsson and Kaplan, 1980).

However, the spectrum of antigens to which human subjects can appropriately be immunized *in vivo* is severely limited. It was thus clear that the development of this procedure into a practical approach for the production of human monoclonal antibodies would require the development of techniques for eliciting high titer antibody responses by human lymphoid cells to a spectrum of antigens presented *in vitro*. Many investigators working with the murine hybridoma system have observed that the specific immunization schedule and the time of harvest of spleen cells after the last antigen exposure strongly influence the yield of specific hybridomas, suggesting that the state of differentiation of antigen-primed lymphocytes determines not only the intensity of their antibody response but also their susceptibility to fusion. The limited duration of viability of human lymphoid cells *in vitro* severely restricts the immunization schedules that can be utilized in culture.

For our initial *in vitro* sensitization experiments, we again used lymphoid cells from the uninvolved spleens of patients with Hodgkin's disease, but incubated them with sheep red blood cells (SRBC) *in vitro*. Human spleen lymphoid cell suspensions, cleared of erythrocytes and granulocytes on a Ficoll-Hypaque gradient (Böyum, 1968), were seeded in 30 mm Petri dishes in RPMI-1640 with 15% fetal calf serum (FCS) and 2 x 10^{-5}M mercaptoethanol; SRBC were then added to a final concentration of 0.5%. The cells were incubated at 37°C in air + 5% CO_2 for five days. Viable mononuclear cells were separated from dead cells on a Ficoll-Hypaque gradient, washed three times, and fused as above with SKO-007 cells. Initial screening with ^{125}I-labeled protein A revealed several IgG-producing hybrids, but tests of these hybrids for anti-SRBC antibody production in a Cunningham plaque-forming assay (Cunningham and Szenberg, 1968) revealed that none of these IgG-products were cytotoxic, in the presence of complement, for SRBC. The remaining hybrids derived from this fusion procedure were then tested for anti-SRBC production in the plaque-forming assay, and three hybrid cultures were found to be positive. These cultures were incubated in 0.33% agarose in HAT medium with 15% FCS for one week, and the emerging clones again tested for anti-SRBC production in a plaque-forming assay (Jerne and Nordin, 1963). Two clones with significant IgM antibody production were obtained. Thus, these experiments indicate that it is possible to generate human hybridomas producing antibody against an antigen to which lymphoid cells were primed _in vitro_. However, the antibodies produced in these early experiments have all been IgM molecules of relatively low affinity.

Further optimization of the _in vitro_ immunization procedure will therefore be essential to the success of human hybridoma antibody production. The experimental difficulties are considerable, since complex interactions of multiple variables must be elucidated. We are currently evaluating the influence of culture conditions; specifically, the relative merits of Iscove's medium (Iscove and Melchers 1978) vs. RPMI-1640, and of the polyclonal activation system described by Hoffman (1980) vs. the standard PWM stimulation procedure. Different methods of antigen presentation will require careful evaluation, and it is not unlikely that different classes of antigens will need to be presented in different ways. Presentation by macrophages is likely to be important, but the ratio of antigen-fed macrophages to lymphoid cells may be a significant variable. Antigens may also be presented effectively when incorporated into liposomes, or when coupled to SRBC. In the case of human cell membrane antigens, careful studies will be required to determine whether co-cultivation of lymphoid cells with whole, intact, human target cells (lethally irradiated, if necessary) yields responses as good as those provided by detergent-isolated cell membrane fractions presented by macrophages or in liposomes. We are also undertaking an evaluation of two broad approaches to the sensitization of human lymphoid cells _in vitro_ to human cancer cell membrane antigens (Table 1). Preliminary experiments with autologous lymphocytes from the heavily involved spleen of a patient with Hodgkin's disease yielded a low affinity IgM antibody which gave positive immunofluorescence reactions with cultured neoplastic giant cells from the same patient and from several other patients, but was negative with normal fibroblasts, lymphocytes, and spleen macrophages.

Table 1. Priming of human lymphocytes for production of human monoclonal hybridoma antibodies against human malignant cells

I. Autologous:

a) Peripheral lymphocytes, spleen cells, or cells from lymph nodes draining a given tumor may be fused directly with SKO-007 without *in vitro* priming step

b) Fusion may be preceded by *in vitro* priming of lymphoid cells with autologous tumor cells

II. Allogeneic:
Lymphoid cells may be primed with:

a) Fresh allogeneic tumor material

b) Established malignant cell lines

c) Normal cells from the tumor donor are essential to assay for antibodies reactive with tumor cell antigens but not with HL-A or other normal allogeneic antigens

Fusion Procedure

It remains to be ascertained whether the fusion procedures which have proven most successful in the production of mouse hybridomas will also be optimal for human hybridoma production. In our initial experiments, we used PEG of MW1500 successfully. However, it has recently been reported that the yield of mouse hybrids was optimal when a PEG at a MW of 4000 was used. Other investigators have reported that the use of a slightly alkaline pH (approximately 8.0 - 8.1 (Sharon et al., 1980) and/or of Ca^{2+} free medium during and after fusion (Schneiderman et al., 1979) can significantly augment the yield of viable hybrids.

In the antigen-directed fusion procedure recently described by Bankert et al. (1980), antigen is coupled to the myeloma cell line to increase the probability that antigen-primed B-lymphocytes bearing specific antibody on their cell surfaces will bind and fuse preferentially and selectively to the antigen-coated myeloma cells. The procedure can be carried out successfully in the presence or absence of PEG. In one of the experiments reported by these investigators, the yield of specific antibody producing clones was as high as 90%.

Parks et al (1980) described a highly efficient procedure for the selection of fused hybrids using the fluorescence-activated cell sorter and antigen coupled to fluorescent microspheres. Suitable variations of this procedure, adapted to a variety of antigens, may well also prove helpful for the selection of human hybridomas.

Cultivation and Cloning of Human Hybridomas

In our initial experiments, it was observed that human hybrids require a relatively long time to grow out in HAT medium, as compared with mouse hybrids. The first viable HAT-resistant human hybrids were observed at 8 to 14 days, and additional hybrids emerged during continued incubation for as long as three to four weeks. A variable which remains to be explored is the optimal time of addition of HAT medium to the fused hybrid cultures. At present, we usually delay the addition of HAT medium for 24 hours after fusion to permit the cells to recover from the toxic effects of PEG. However, it is likely that the antigen-directed fusion procedure, carried out in the absence of PEG, would permit the immediate introduction of HAT medium. Experiments are now underway to compare the yield of viable hybrids in different culture media, and in particular to test the relative merits of Iscove's medium vs. RPMI-1640. Earlier experiments demonstrated convincingly that optimal cell density is an extremely important variable in determining both the viability of hybrids and the yield of antibody-producing cultures (Olsson and Kaplan, 1980). Finally, we are evaluating the influence of feeder layers derived from human foreskin fibroblasts, human thymocytes, and human peripheral blood monocytes or spleen macrophages.

It is well established that the level of antibody production by mouse hybridomas can be augmented 100 to 1,000 fold by *in vivo* passage in the peritoneal cavity and harvesting of ascites fluid. If the successful heterotransplantation of human hybridomas into mice were feasible, similar gains in antibody production levels might be achieved, albeit at the price of admixture with murine serum proteins. Our initial attempts to transplant human hybridomas into the peritoneal cavities of Pristane-primed nude mice proved unsuccessful. Additional experiments utilizing variously tolerized or immunosuppressed mice are in progress.

HUMAN vs. MURINE MONOCLONAL HYBRIDOMA ANTIBODIES

Monoclonal antibodies derived from murine or other rodent hybridomas have already proven their value as powerful analytic reagents in laboratory investigations of many problems in biology and medicine. Moreover, murine monoclonal antibodies of high titer and high affinity are already available in good and ever-increasing supply for a broad spectrum of antigenic systems. In contrast, the production of monoclonal antibodies by human-human hybridomas is still in its infancy, and must await the development of optimized techniques for antigen priming of human lymphocytes before a similar array of high titer human monoclonal antibodies can be made available.

It is thus clearly premature to attempt any detailed comparison of the potential of human vs. murine monoclonal antibodies at this time. Nonetheless, a few general comments may be offered.

In general, it seems likely that murine and human monoclonal antibodies will be equally useful for *in vitro* diagnostic and analytic applications involving antigens which are not human gene products. In this circumstance, the murine antibodies are likely to be preferable for practical and economic reasons. However, immunogeneticists have already observed that murine monoclonal antibodies have thus far failed to detect certain human major histocompatibility antigens which are readily detected by conventional anti-HLA antibodies. It has thus been suggested that intra-species lymphocyte responses to major histocompatibility antigens may yield human monoclonal antibodies with a broader range of antigenic specificity than the murine antibodies. If this prediction is fulfilled, human monoclonal antibodies will undoubtedly achieve an important *in vitro* role in HL-A histocompatibility testing. Still another potential application of monoclonal antibodies *in vitro* is the characterization of the membrane antigens which are phenotypically distinctive for human neoplasms of different types.

However, it is clear that the major importance of human monoclonal antibodies stems from the expectation that they can be administered safely and repeatedly to human beings, and thus may well have significant applications in the prophylaxis, diagnosis, and/or therapy of a broad spectrum of human diseases. This is not to suggest that murine monoclonal antibodies will have no role whatever in *in vivo* administration in man. It is quite conceivable that one or a few injections of murine antibodies may be well tolerated in human subjects. However, the highly significant differences in antigenicity of murine and human immunoglobulins make it highly probable that the repeated administration of a given murine antibody in therapeutic quantities will result in severe hypersensitivity reactions. Yet, many types of treatment with monoclonal antibodies are likely to require repeated injections. It is thus clear that human monoclonal antibodies will be preferable in such situations. The fact that many thousands of humans have been safely and repeatedly injected with relatively large amounts of human immunoglobulin supports the expectation that human monoclonal antibodies will prove to be safe in clinical use. However, the fact that each such monoclonal antibody bears a single idiotype raises at least the theoretical possibility of a risk that human subjects may generate anti-idiotype antibodies in response to the administration of a human monoclonal antibody. In the murine disease models in which monoclonal antibodies have been used for therapeutic experiments to date, this appears not to have been a problem. Nevertheless, the possibility of complications stemming from anti-idiotype reactions must be carefully borne in mind in phase I clinical trials with human monoclonal antibodies. The generation of multiple antibodies directed against the same antigen, though perhaps against different antigenic determinants, is one way of circumventing this difficulty if it should prove to be real. Barring these and other, as yet unforseen difficulties, it seems entirely possible that human monoclonal antibodies will achieve a major in vivo role in clinical medicine, and especially in the diagnosis and treatment of neoplastic disease, during the next decade.

REFERENCES

Bankert, R.B., DesSoye, D. and Powers, L., 1980, Antigen-promoted cell fusion: antigen-coated myeloma cells fuse with antigen-reactive spleen cells, Transpl. Proc., 12:443.

Böyum, A., 1968, Separation of leucocytes from blood and bone marrow, Scand. J. Clin. Lab. Invest. 21, Suppl. 97:77.

Cunningham, A.J., and Szenberg, A, 1968, Further improvements in the plaque technique for detecting single antibody-producing cells, Immunology, 14:599.

Eltringham, J.R., and Kaplan, H.S., 1973, Impaired delayed hypersensitivity responses in 154 patients with Hodgkin's disease, Natl. Cancer Inst. Monogr. 36:107.

Epstein, A.L., and Kaplan, H.S., 1974, Biology of the human malignant lymphomas. I. Establishment in continuous cell culture and heterotransplantation of diffuse histiocytic lymphomas, Cancer 34:1851.

Glatstein, E., Guernsey, J.M., Rosenberg, S.A., and Kaplan, H.S., 1969, The value of laparotomy and splenectomy in the staging of Hodgkin's disease, Cancer, 24:709.

Hoffman, M.K., 1980, Antigen-specific induction and regulation of antibody synthesis in cultures of human peripheral blood mononuclear cells, Proc. Nat. Acad. Sci. USA, 77:1139.

Iscove, N.N., and Melchers, F., 1978, Complete replacement of serum by albumin, transferrin and soybean lipid in cultures of lipopolysaccharide-reactive B lymphocytes, J. Exp. Med., 147:923.

Jerne, N.K., and Nordin, A.A., 1963, Plaque formation in agar by single antibody-producing cells, Science, 140:405.

Kaplan, H.S., Gartner, S., Goodenow, R.S., and Bieber, M.M., 1979, Biology and virology of the human malignant lymphomas; 1st Milford D. Schultz lecture, Cancer, 43:1.

Kaplan, H.S., Goodenow, R.S., Epstein, A.L., Gartner, S., Decleve, A., and Rosenthal, P.N., 1977, Isolation of a C-type RNA virus from an established human histiocytic lymphoma cell line, Proc. Nat. Acad. Sci. USA 74:2564.

Köhler, G., and Milstein, C., 1975, Continuous cultures of fused cells secreting antibody of predefined specificity, Nature, 256:495.

Köhler, G., and Milstein, C., 1976, Derivation of specific antibody-producing tissue culture and tumor lines by cell fusion, Eur. J. Immunol., 6:511.

Koskimies, S., 1979, Human lymphoblastoid cell line producing specific antibody against Rh-antigen, Scand. J. Immunol., 10: 371 (abst.).

Levy, R., and Dilley, J., 1978, Rescue of immunoglobulin secretion from human neoplastic lymphoid cells by somatic cell hybridization, Proc. Nat. Acad. Sci. USA, 75:2411.

Nilsson, K., Bennich, H., Johansson, S.G.O., and Pontén, J., 1970, Established immunoglobulin producing myeloma (IgE) and lymphoblastoid (IgG) cell lines from an IgE myeloma patient, Clin. Exp. Immunol., 7:477.

Olsson, L, and Kaplan, H.S., 1980, Human-human hybridomas producing monoclonal antibodies of predefined antigenic specificity, Proc. Nat. Acad. Sci. USA, 77:5429.

Parks, D.R., Bryan, V.M., Oi, V.T., and Herzenberf, L.A., 1979, Antigen-specific identification and cloning of hybridomas with a fluorescence-activated cell sorter, Proc. Natl. Acad. Sci. USA, 76:1962.

Schlom, J., Wunderlich, D., and Teramoto, Y.A., 1980, Generation of human monoclonal antibodies reactive with human mammary carcinoma cells, Proc. Nat. Acad. Sci. USA, 77:6841.

Schneiderman, S., Farber, J.L., and Baserga, R., 1979, A simple method for decreasing the toxicity of polyethylene glycol in mammalian cell hybridization, Somatic Cell Genetics, 5:263.

Schwaber, J., 1975, Immunoglobulin production by a human-mouse somatic cell hybrid, Exper. Cell Res., 93:343.

Sharon, J., Morrison, S.L., and Kabat, E.A., 1980, Formation of hybridoma clones in soft agarose: effect of pH and of medium, Somatic Cell Genet., 6:435.

Steinitz, M., Klein, G., Koskimies, S., and Mäkelä, O., 1977, EB virus-induced B lymphocyte cell lines producing specific antibody, Nature, 269:420.

Zurawski, V.R., Jr., Haber, E., and Black, P.H., 1978, Production of antibody to tetanus toxoid by continuous human lymphoblastoid cell lines, Science, 199:1439.

SURFACE ANTIGENS ON NORMAL AND MALIGNANT LYMPHOID CELLS

Laurence Boumsell° and Alain Bernard§

° INSERM U93, Hôpital Saint-Louis, Paris, France

§ Laboratoire d'Immunologie des Tumeurs de l'Enfant
Institut Gustave Roussy, 94800 Villejuif, France

We shall first briefly review the reactivity on normal lymphoid cells of various monoclonal antibodies we have used to study malignant tumor cell populations. We then emphasize the most important technical requirements for assessing the reactivity of MA with cell surfaces. Finally we report our results from the investigation of 66 cases of T-cell malignancy and of 111 cases of B and non-B, non T cell malignancies.

1. MONOCLONAL ANTIBODIES RECOGNISING ANTIGENS ON HUMAN LYMPHOID CELLS

In table 1, we have categorised these MA according to their pattern of lymphohaemopoietic cell recognition. Clearly, 2 major classes of MA exist ; those which are lineage specific and those which are common to several ineages but, as they are not found on every cell but only on discrete subpopulations, may be reflecting differentiation, maturation or activation within lineages.

1.A. T cell surface antigens

1.A.1. Group I. MA recognising thymocytes.

D66 [1] recognises an antigen present primarily on all thymocytes and in much lesser amounts on peripheral T cells. By contrast, several MA react exclusively with cortical thymocytes, anti-T6 [3] and NAI [2] probably recognise the same molecule of MW 49,000. Yet we obtained one MA, D47 [4] reactive with another epitope discrete of the cortical thymocyte surface that seems to be different from the T6 and NA1 epitopes and that is not associated with them. We were thus

TABLE 1

Monoclonal antibodies recognising leucocyte differentiation antigens.

	Lymphohaemopoietic cells	Designation/ reference		MW§ ($x10^3$)
1A1	All thymocytes	D66	(1)	- Ŧ
	Subset of thymocytes	NA1/34	(2)	49
		T6	(3)	49
		D47	(4)	-
1A2	All E+ cells°	9.6	(5)	50
		U4	(6)	-
	All E+PBL+subset thymocytes	T3	(7)	19
1A3	Subset of E+PBL+subset thymocytes	9.3	(8)	44
		T4	(9)	62
	inducers	3A1	(10)	40
		Leu3a,3b	(11,12)	62
		T5	13	30.33
	cytotoxic/suppressor	T8	3	-
		Leu2a,2b	(11,12)	32-43
1B	SIg+ cells	B1	(14)	30
		B2	(15)	140
1C1	B-CLL cells and E+ cells	10.2	(16)	65-67
		T.65	(17)	65
	all E+	L17	(18)	67
		Leu1	(19)	69-71
	all E+PBL,subset thymoc.	A50	(20)	-
		T1	(21)	70
1C2	c.ALL cells and E+ cells	D44	(1)	-
	and thymocytes	T10	(3)	45
		12E7	(22)	28
1C3	Rapidly dividing cells	T9	(3)	90
		4F2	(23)	40-80
1C4	Primarily or exclusively on non T-cells	Ia	(8,24)	29-33
		J5	(25)	95
		H27	(26)	-

Legend

§ M.W. molecular weight

Ŧ not available

° E-rosette forming cells

able to show a polymorphism of human cortical thymocyte antigens. D47 is moreover associated with β_2-microglobulin, this resembling the mouse thymus leukemia antigenic system.

1.A.2. Group II. MA recognising thymocytes and blood T cells

Among these antibodies, 9.6[5] seems to have the broadest pattern of reactivity with human T cells and to recognise a component involved in E-rosette formation. U4 [6] also has a very broad pattern of reactivity with human T cells, since in children and adults it recognises all E-rosette + (E+) cells. However, in the fetal thymus, below the age of 28 weeks, a proportion of cells are not reactive with U4. U4 does not block E-rosette formation.D66[1] has also a very broad pattern on normal human T cells and blocks E-rosette formation. However it seems unique and different from 9.6.

Anti-T3 [7] recognises a component primarily present on the cells from the medulla of the thymus and most peripheral T cells. This component (19,000 MW) appears to be involved in T cell activation and functions [27-28]

1.4.3. MA defining subsets of peripheral blood T cells

Certainly, one of the most important contributions of anti T cell techniques has been the discovery in humans of antigenic systems that might be the equivalent of the murine Lyt system[29].

We have thus obtained a monoclonal antibody E10[30] which recognizes an antigenic determinant present on subpopulations of thymus cells and peripheral blood T cells, different from the T4/Leu 3 and T8/Leu 2 subsets. Efforts towards obtaining other reagents of this type should be continued, for these antigenically defined subsets are still heterogenous, and in the near future, it will certainly be possible to obtain MA reactive with structures which will permit the recognition of regulatory T cell circuits.

I. B. B cell surface antigens

Human lymphocytes are ascribed to B cell lineage when they express either cytoplasmic immunoglobulin (cIg) or surface immunoglobulins (sIg). However the study of Ig gene configurations has recently provided new insights into the definition of B cell lineage.

B cell specific surface antigens.

Recent experiments have provided evidence that αB1[14] and α B2[15] recognize B cell differentiation antigens

1. C. Antigens shared between several lineages

1.C.1. MA recognising both T cells and B CLL cells.

We have previously shown, by cross-absorption experiments using heteroantisera, that T cells and B CLL cells carry several common surface antigens[31]. It was therefore not surprising to find that a good proportion of MA raised against T cells whould also recognize these determinants shared with B CLL cells, and that several different components, as defined both by their molecular weight and cellular distribution, have been defined with these MA[16-21]

1. C.2. MA recognising both T cells and c.ALL cells

The circumstances of discovery, and the questions raised by these MA defined antigens are similar to the preceding ones, since these antigens were previously discovered using heteroantisera, by Brouet et al.[32]. We have obtained a monoclonal antibody D44[1] which recognizes an antigenic determinant present on all T cells, c.ALL cells, and AML cells but lacking on mature B cells. In addition, treatment by D44 and complement of normal bone marrow cells eliminates all CFU.c and BFU.E.

1. C.3. MA recognizing antigens present on dividing cells

Although not T cell specific, these reagents have proved to be useful in the definition of discrete T cell differentiation stages[3] or functionally specialised T cell subsets[33].

1. C.4. MA recognition structures primarily or exclusively present on non T cells

Molecular heterogeneity of Ia-like antigens in humans has been demonstrated by the use of these MA[34]. This approach will probably lead to the discovery that functionally specialised mononuclear cells in man carry different sorts of Ia-like antigens, as has been revealed in the mouse[35].

J5 recognises an antigen peculiar to non-T non-B ALL cells (CALLA)[25]. Further production of MA recognising antigenic determinants belonging to the CALLA family but different from the determinant recognized by J5, would be extremely useful for a complete analysis of these glycoproteins. Such MA would help to determine the exact cellular distribution of these molecules and would resolve the question as to whether particular antigenic determinants might be displayed by malignant lymphoid cells only.

In addition a monoclonal antibody H2[26] recognizes and antigenic determinant common to monocytes and a subset of B cell from the tonsils, adenoids, spleen, lymph nodes. This subset accounts for only a very small proportion of normal blood B cells, but is already present in fetal spleen as early as 16 weeks of gestation.

2. TECHNIQUES FOR DETECTING CELL RECOGNITION BY MONOCLONAL ANTIBODIES

The availability of MA has not entirely solved the difficulties of the serological analysis of cell surfaces. In addition, the concomitant development of quantitative fluorometry technology has greatly increased the sensitivity of immunofluorescence detection methods[36]. Thus, it follows that when assessing the reactivity of a given monoclonal antibody with cell surfaces, it is mandatory to rely on an extensive knowledge of the patterns of reactivity of the specific MA and to use identical reactive conditions. Moreover, extensive comparisons should be made of the reactivity pattern of a given monoclonal antibody according to the main techniques used.

2. A. Comparison of immunofluorescence and cytotoxicity methods for monoclonal antibodies recognizing subpopulations within the human thymus

As previoulsy reported[20], and as indicated in Table 2, A50 kills in a micro (or macro) cytotoxicity assay 25-60 % of cells within a thymus from a child, depending upon the individual. In an indirect immunofluorescence assay, examined by immunofluorescent microscopy, A50 recognizes the same proportion of positive cells as the cytotoxicity assay. Figure 1 depicts the histogram obtained on a flow microfluorometer after labelling thymocytes with A50 and a fluoresceinated goat anti-mouse antiserum.

Clearly, there are two subpopulations of thymocytes : one is brightly fluorescent, the other dimly fluorescent but nevertheless positive. After killing thymocytes with A50 plus complement, it is obvious from Figure 1 that the bright population has disappeared while no significant decrease in the number of dimly labelled cells has occurred. When examined by immunofluorescent microscopy, the residual cells did not appear to be fluorescent.

One advantage of the quantitative fluorometry method can also be seen in Figure 2. It shows the pattern of reactivity on the three MA we have listed as belonging to Group I, with the malignant cells from a lymphoblastic lymphoma (LL). In figure 2, it can be seen that cells from this particular LL give a superimposable reaction for both NA1 and T6 which is more intense than the reaction with D47. This is typical of the patterns of reaction we observed in all the T-cell malignancies recognized by the MA belonging to this group. Regarding the observation that the intensity of D47 reaction with normal thymocytes varies between different donors, we think this is related to the ability of the D47 antigen to "disappear" from thymus cell surface and again is very reminiscent of the mouse thymus leukemia antigenic system (L. Boumsell and A. Bernard, in preparation).

Is this indirect immunofluorescence labelling method, measured by flow microfluorometry, always the most sensitive technique for detecting the reactivity of a MA with a cell surface ?

2.B. Comparison of immunofluorescence and cytotoxicity methods for a monoclonal antibody recognizing all thymocytes

Table 2 shows that, using D66, another MA we obtained, 100 % of the cells from child thymi were killed, in the peripheral blood

TABLE 2

Distribution of the antigenic determinants recognized by various monoclonal antibodies

		D66	A50	D47	NA1
Children Thymocytes	C.dep.°°° cyto.	90(13)°	25-59§(13)°	80(8) 40(2) 15(3)	75-90(9)
	Indir.Ŧ I.F.	90(6)§§	23-38(6)°°	59-80(11)§§	73-80(3)§§
E^+PBL	C. dep. cyto	80-95(10)	70-90(16)	5(10)	5(10)
	Indir. I.F.	10(5)	75-95(5)	5(5)	5(5)

Legend

° : number of individual tested
§ : value range of cytotoxic index
°° : %age fluorescent cells examined by fluorescent microscopy(FM)
§§ : %age fluorescent cells examined by flow microfluorometry and by FM
°°° : C dependent cytotoxicity
Ŧ : IF indirect immunofluorescence

All thymo.
Thymo after
A50+C

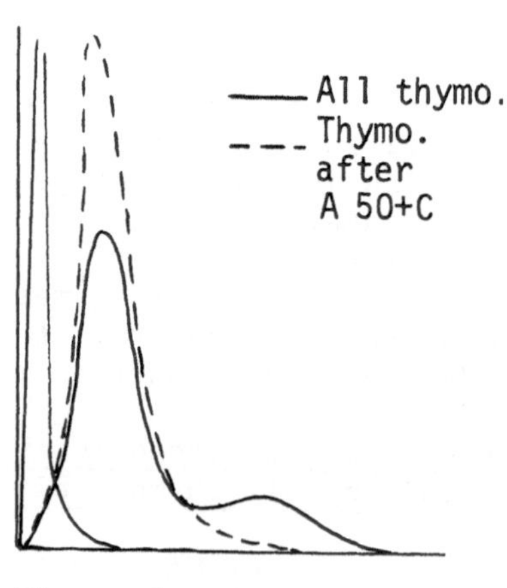

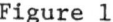

Figure 1

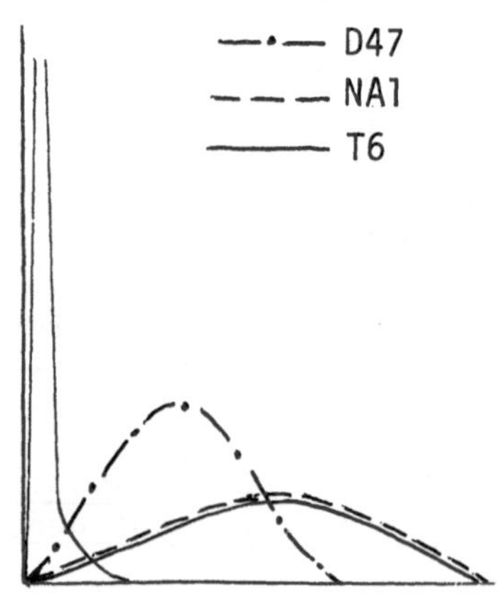

Figure 2

most E-rosette forming cells were killed. However, when we looked at the reactivity of D66 in an indirect immunofluorescence assay, the thymocytes appeared to be positive on the flow microfluorometer histogram, but the E+ PBL gave no significant positivity as compared with the negative control, even after amplification of the cells illumination or the fluorescent signal detected. These results were obtained with numerous anti-mouse immunoglobulins and we are currently preparing an antiserum against D66.

Thus it seems probable that for some MA, the cytotoxicity assay will remain by far the most sensitive technique for detecting reactivity with cell surfaces. Numerous explanations can be offered for this phenomenon.

3. SURFACE ANTIGENS ON MALIGNANT T-CELLS

All the data presented here were obtained with tumor cell populations collected before any treatment.

3. A. Categorisation of children with T-cell malignancies. Differences in the cell phenotype of children with T-cell ALL and T-cell LL

In children (below the age of 20 years), T-cell malignancies are encountered in the context of two clinically defined entities; acute lymphoblastic leukaemias (ALL) and LL[37-38]. By definition, patients with LL have no or minimal bone marrow invasion, whereas patients with ALL present with massive bone marrow invasion[39]. In fact, the clinical similarities between T-cell ALL and T-cell LL have raised the question of their relationship. Are patients with T-cell ALL diagnosed at a more advanced stage of the disease than patients with T-cell LL, or are T-cell ALL and T-cell LL two different entities ?

Table 3 depicts the results we obtained in analysing with MA from the anti-T series, the tumor cell populations from children with T-cell LL[40]. Although a proportion of patients (about one third) had tumour cells not exactly conforming with the cells normally detected in the thymus, it is clear that cells from each of these patients could be related to one of the three major thymocyte populations defined with the anti-T series of MA [3]. About 1/5 of the children with LL could be categorised as having early, 1/2 as common, (cortical) and 1/3 as late (medullary), thymocyte-type tumour cells. This classification pattern of children with T-cell LL contrasts with the pattern seen with tumour cells from children with ALL. Table 3 shows that 2 out of 19 children with T-cell ALL were T3 positive, which accords with another series of T-ALL patients investigated using the same reagents [3].

Table 4 shows that we were able to see further phenotypic differences between malignant T-cells from children with ALL or with LL. Among malignant T-cells related to early thymocytes, only children

TABLE 3

Childhood T-cell lymphoid malignancies

T8	–	+°°	–	+	–	+	–	+	–	+	+	+	–	–
T4	–	–	+	+	–	–	+	+	–	+	–	+	+	–
T6	–	–	–	–	+	+	+	+	+	+	+	–	–	–
T3	–§	–	–	–	–	–	–	–	+	+	+	+	+	+
ALL (19)°	5°	1	1	2	1	1	2	4	1	0	0	0	0	1
LL (25)	4	2	1	2	0	1	1	6	0	2	1	2	3	0

TABLE 5

Adult T-cell lymphoid malignancies

SEZ (9)°	T-CLL (4)	Lympho Lympho. (5)	ALL (4)	T3	T6	T4	T8
–	–	–	2	–§	–	–	–
–	–	2	–	–	+	+	+°°
–	–	1	–	+	+	–	–
–	–	1	–	+	+	+	–
1	2	–	1	+	–	–	+
8	2	–	–	+	–	+	–
–	–	1	1	+	–	–	–

Legends for Tables 3 and 5

° number of patients

§ no detectable amount of T3 antigen found by FMF (cytofluorograph system 30L) in indirect IF using goat anti-mouse IgG/FITC

°° detectable amount of T8 antigen on ≫ 50 % of the cells as assayed by FMF.

TABLE 4

Childhood T-cell malignancies

			D66°°	D47§	U4°°	10.2§	A50°°	4F2°°	J5§
9.6+	ALL°	(5)	5	0	3	3∓	0	2	0
T−	LL	(4)	0	0	0	4	0	2	0
T8+	ALL°	(4)	4	0	3	3	0	3	2
or 4+	LL	(5)	3	0	2	5	2	1	1
T6+	ALL	(8)	5	8	5	8	0	5	2
	LL	(8)	7	8	6	8	8	5	3
T6,3+	ALL	(1)	1	0	1	1	0	1	0
	LL	(3)	2	2	2	3	1	1	3
T3+	ALL	(1)	0	0	1	1	1	1	0
	LL	(5)	4	0	4	5	5	2	0

Legend for Tables 4 and 6

° number of patients with a phenotype defined by the anti-T series

∓ number of positive patients

°° presence of D66, U4, A50, 4F2 was tested in a C-dependent micrototoxity test. A patient is considered positive when ≥30 % cells were killed.

§ presence of D47, 10.2, J5, Ia was tested by flow microfluorometry in indirect IF

with ALL had cells recognized by both D66 and U4. Among malignant T-cells related to common thymocytes, only children with LL had cells carrying substantial amounts of A50 antigen.

3.B. Classification of adult T-cell malignancies

Although thus far our series of adult patients with T-cell ALL is small (4 cases), it is striking that, as seen in table 5, two of them had T3 positive cells. By contrast, in patients with T-cell ALL and LL, table 5 shows that all patients with the Sézary syndrome and T-cell chronic lymphocytic leukemia (T.CLL) or lymphosarcoma cell leukaemia, had tumour cells with a phenotype indistinguishable from the phenotype of mature, functionally specialised normal T-cells as defined by the anti-T series[41]. As shown in table 6, many of these patients had cells which were also recognized by the MA 4F2, which also recognised a subset of T peripheral blood lymphocytes only after activation by mitogen or alloantigen. This suggests that these malignant T-cells are in a state of activation [33]. However we found only one patient with substantial amounts of Ia-like antigens. We consistently observed immunological disorders in some of these patients which could have been mediated by their malignant cells. For example, one patient affected by T-cell CLL with a suppressor/cytotoxic phenotype had no in vitro lymphocyte proliferative response and suffered repeated infections which were the eventual cause of his death.

TABLE 6

Adult T-cell malignancies

		D66°°	D47§	U4°°	10.2§	A50°°	4F2°°	J5§	Ia§
$3^-,4^-,6^-,8^-$									
ALL	(2)°	0	0	1	1	0	1	0	0
$3^-,4^+,6^+,8^+$									
LL	(2)	2	2	2	2	1	1	1	0
$3^+,6^+$									
LL	(1)	0	1	1	1	1	1	1	0
$3^+,4^+,6^+$									
LL	(1)	1	0	1	1	1	1	1	0
$3^+,4^+$									
SEZ	(8)	2	0	8	8	8	3	0	0
T-CLL	(2)	2	0	2	2	2	1	0	1
$3^+,8^+$									
SEZ	(1)	0	0	0	1	1	1	0	0
T-CLL	(2)	2	0	2	2	2	1	0	0
3^+									
LL	(1)	1	0	1	1	1	1	0	0
ALL	(1)	0	0	1	1	1	1	0	1

For legend, please see Table 4

Another patient with T-cell CLL of an inducer phenotype had a monoclonal IgG λ in his serum. These findings are also consistent with the previous observation that Sézary cells can exert in vitro helper activity [42], or delayed type hypersensitivity activity [43].

3.C. Presence of CALLA on malignant T-cells

Tables 4 and 5 show that we found malignant T-cells, recognised by J5, in both adults and children. surprisingly, these malignant T-cells did not have the phenotype of early T-cells, since some of them also displayed T3 and most of the others T6, T8 or T4. On the other hand, we never found malignant T-cells with the phenotype of peripheral T-cells reactive with J5. Thus CALLA is not specific to malignant lymphoid cells differentiated within a particular lymphoid cell lineage, nor is it indicative that malignant cells are arrested at very immature stages of lymphoid differentiation. It may clearly be displayed on cells deeply committed within a lymphoid cell differentiation lineage.

3. D. Lack of correlation between E-rosette forming ability and differentiative status indicated by surface antigens on malignant T-cells

Table 7 shows that we found T malignant cell populations forming insignificant numbers of E-rosettes, although their pattern of surface antigens was indicative of cells of advanced differentiation within the T-cell lineage. The E-rosette assay was performed at 4° C with sheep erythrocytes pretreated with AET, a procedure which has been shown to be the most efficient in detecting malignant T-cells forming E-rosettes. These results show that it is premature to conclude that when a lymphoid malignancy has cells which form no E-rosettes but which are reactive with anti-T-cell antisera, this represents the clonal expansion of pre-T-cells.

3. E. The heterogeneity of T-cells malignancies

In our study, the numerous MA used disclosed a disconcerting degree of heterogeneity from one patient to another, although the tumour cell populations from a single patient appeared extremely homogenous, particularly when fluorometer histograms were considered. For instance, why did MA 10.2, U4 and especially D66 fail to recognise T malignant cells from many patients ? Why did many, if not most patients, with ALL or LL, have cells without a phenotypic counterpart detectable within a normal thymus ? It is remarkable that the degree of heterogeneity appears to be maximal for patients with ALL or LL, most of them having an early or intermediate differentiative status, whereas Sézary diseases and T-cell CLL appear to have a surface antigen pattern very similar to that of normal, mature, and in some cases, activated T-cells.

4. SURFACE ANTIGENS ON MALIGNANT NON-T NON-B ALL AND MALIGNANT B CELLS

We have studied 64 cases of c.ALL, 31 cases of B-CLL, 16 cases of Burkitt lymphoma with a series of monoclonal antibodies as shown in table 8. From these results, it is possible to assign a B cellular origin to the tumour cells from many patients with common or non-T, non-B cell ALL. Interestingly a unique B cell subset seems to be the target of Burkitt like lymphomas, while B-CLL and c.ALL seems to develop with equal frequency from $H27^+$ and $H27^-$ B cell subsets[26].

TABLE 7

Comparison of 9.6 reactivity and E-rosette forming ability of malignant T cells from children

	%E°	9.6 (C.I.)§
$T3^-,T4^-,T6^-,T8^-$	20	21
	01	95
	11	71
$T3^-,T4^+,T6^-,T8^-$	73	94
$T3^-,T4^-,T6^-,T8^+$	100	94
	86	93
	30	90
	01	31
	10	5
$T3^+,T4^+,T6^+,T8^+$	40	82
	80	52
$T3^+,T4^+,T6^-,T8^+$	15	94
	60	71

Legend
° percent E-rosette forming cells
§ value of the cytotoxic index for each patient

SUMMARY

In this report, we have reviewed recent work which demonstrates that monoclonal antibodies can be used as probes of cell surface differentiation antigens on normal and malignant lymphoid cells. We have also stressed technical and fundamental problems related to the use of monoclonal antibodies in assessing cell surface differentiation antigens. We hope that the general approach outlined in this review will eventually assist in the classification of leukemias and lymphomas and may define groups of patients who share similar disease presentations and/or therapeutic responses.

The development of a battery of well characterized antibodies has already permitted to show a remarkable degree of heterogeneity particularly among patients with T cell malignancies. These facts may certainly reflect fundamental events in the process of T-cell differentiation and/or malignant transformation.

In the following several years, an increasing number of monoclonal antibodies with similar tissue distributions and/or chemical characteristics will be obtained. It is thus crucial to compare these antibodies in systematic studies according to phenotypic, functional and biochemical criteria. Only with the outcome of a standardized nomenclature will these antibodies be equally useful to researchers, biologists and clinicians to whom unlimited supplies of antibody could offer new therapeutic means.

TABLE 8

Phenotypic heterogeneity among 111 cases of "non-B non-T" and B-cell malignancies

Origin of cells tested	Monoclonal antibodies tested					
	H27°	B1°	J5°	DR°	A50°	D44°
c.ALL(64)°°	51§	50	80	100	0	96
B.CLL(31)	50	90	0	100	50	03
BLS(16) (Burkitt lymogina)	100	100	NT	100	0	0
B.LCL(4)°°°	100	100	NT	100	0	0

Legend

§ % tumor cell populations positive among tumor cell populations tested

°° number in ( ) represent number of patients tested

°°° B.LCL : B-lymphoid cell lines in culture studied were Raji, Daudi, Ramos, 8866.

ACKNOWLEDGEMENTS

We thank Mrs F. Bayle for typing our manuscript. We are most indebted to Drs J. Hansen, B. Haynes, R. Lévy, A. McMichael for exchanged reagents. Part of this work was done in collaboration with Drs H. Coppin, J. Dausset, J. Lemerle, L. Nadler, D. Pham, B. Raynal, E. Reinherz, Y. Richard, J. Ritz, S. Schlossman & F. Valensi.

REFERENCES

1. Bernard A., & Boumsell L. (1981) in preparation
2. McMichael A. J., Pilch J.R., Galfre G., Mason D.Y., Fabre J.W. & Milstein C. (1979) Eur. J. Immunol. 9, 205-210
3. Reinherz E.L., Kung P.C., Goldstein G., Levy R.H. & Schlossman S.F. (1980) Proc. Natl. Acad. Sci. USA 77 : 1588-1592.

4. Boumsell L. & Bernard A. (1981) in preparation
5. Kamoun M., Martin P.J., Hansen J.A., Brown M.A., Siadak A.W. & Nowinski R.C.J. (1981) J. EXp. Med. 153, 207-212
6. Bernard A., Coppin H., Clausse M., Bayle C., Degos L., Valensi F. Flandrin G., Dausset J., Lemerle J. & Boumsell L. 4th Int. Congr. Immunology 1980, Paris. Abstract Book (eds. J.L. Preud'homme & V.A.L. Hawken) Abs. No. 18.4.07
7. Kung P.C., Goldstein G., Reinherz E.L. & Schlossman S.F. (1979) Science 206, 397-399.
8. Hansen J.A., Martin P.J. & Nowinski R.C. (1980) Immunogenetics, 10, 247-260
9. Reinherz E.L., Kung P.C., Goldstein G. & Schlossman S.F. (1979) Proc. Natl. Acad. Sci. USA 76, 4061-4065
10. Haynes B.F., Eisenbarth G.S. & Fauci A.S. (1979) Proc. Natl. Acad. Sci. USA 76, 5829-5833
11. Ledbetter J.A., Evans R.L., Lipinski M., Rundles C., Good R.A. & Herzenberg L.A. (1981) J. Exp. Med. 153, 310-323
12. Evans R.L., Wall D.W., Platsoucas C.D., Siegal F.P., Fibrig F.M., Testa C.M. & Good R.A. (1981) Proc. Natl. Acad. Sci. 78 : 544-548
13. Reinherz E.L., Kung P.C., Goldstein G. & Schlossman S.F. (1980) J. Immunol. 124, 1301-1307
14. Stashenko P., Nadler L.M., Hardy R. & Schlossman S.F. (1980) J. Immunol. 125, 1678-1685
15. Nadler L.M., Stashenko P., Hardy R., Van Agthoven A., Terhorst C. & Schlossman S.F. (1981) J. Immunol. (in press)
16. Martin P.J., Hansen J.A., Nowinski R.C. & Brown M.A. (1981) Immunogenetics 11, 429-439
17. Royston E., Majda J.A., Baird S.M., Meserve B.L. & Griffiths J.C. (1979) Blood 54 (suppl. 1) 106 A
18. Engleman E.G., Warnke R., Fox R.I. & Levy R. (1981) in press
19. Wang C.Y., Good R.A., Ammirati P., Dimbort G. & Evans R.L. (1980) J. Exp. Med. 151, 1539-1544
20. Boumsell L., Coppin H., Pham D., Raynal B., Lemerle J., Dausset J. & Bernard A. (1980) J. Exp. Med. 152, 229-234
21. Reinherz E.L., Kung P.C., Goldstein G. & Schlossman S.F. (1979) J. Immunol. 123, 1312-1317
22. Levy R., Dilley J., Fox R.I. & Warnke R. (1979) Proc. Natl. Acad. Sci. USA 76, 6552-6556
23. Eisenbarth G.S., Haynes B.F., Schoer J.A. & Fauci A.S. (1980) J. Immunol. 124, 1237
24. Reinherz E.L., Kung P.C., Pesando J.M., Ritz J., Goldstein G. & Schlossman S.F. (1979) J. Exp. Med. 150, 1472-1482
25. Ritz J., Pesando J.M., McConarty J.N., Lazarus H. & Schlossman S.F. (1980) Nature 283, 583-585
26. Coppin, H., Boumsell L., Pham D. & Bernard A. (1981) in preparation
27. Van Wauke J.P., Demey J.R., Goossens J.G. (1980) J. Immunol. 124, 2708-2713
28. Reinherz E.L., Hussey R.E. & Schlossman S.F. (1980) Europ. J. Immunol. 10, 758-762

29. Cantor H. & Boyse E.A. (1975) J. Exp. Med. 141, 1390-1398
30. Boumsell L., Richard Y., Bernard A. in preparation
31. Boumsell L., Bernard A., Lepage V., Degos L., Lemerle J. & Dausset J. (1978) Europ. J. Immunol. 8, 900-904
32. Brouet J.C., Valensi F., Daniel M.T., Flandrin G., Preud'homme J.L. & Seligmann M. (1976) Brit. J. Hameat. 33, 319-328
33. Haynes B.F., Hemler M.E., Mann D.L., Eisenbarth G.S., Shelhamer J., Mostowski H.S., Thomas C.A., Strominger J.L. & Fauci A.S. (1981) J. immunol. in press
34. Charron D.J., McDevitt H.O. (1979) Proc. Natl. Acad. Sci. USA 76, 6567-6571
35. Murphy D.B., Herzenberg L.A., Okumura K., Herzenberg L.A., McDevitt H.O. (1976) J. Exp. Med. 144, 699-712
36. Bonner W.A., Hulett H.R., Sweet R.G. & Herzenberg L.A. (1972) Rev. Scientific Instr. 43, 404-414
37. Lebien T.W. & Kersey J.H. (1980) J. Immunol. 125, 2208-2214
38. Bernard A., Boumsell L., Bayle C., Richard Y., Coppin H., Penit C., Rouget P., Micheau C., Clausse B., Gerard-Marchant R., Dausset J. & Lemerle J. (1979) Blood 54, 1058-1068
39. Murphy S.B. (1978) N. Egl. J. Med. 299, 1446-1448
40. Bernard A., Boumsell L., Reinherz E.L., Nadler L.M., Ritz J., Coppin H., Richard Y., Valensi F., Flandrin G., Dausset J. Lemerle J. & Schlossman S.F. (1981) Blood (in press)
41. Boumsell L., Bernard A., Reinherz E.L., Nadler L.M., Ritz J., Coppin H., Richard Y., Dubertret L., Degos L., Lemerle J. Flandrin G., Dausset J. & Schlossman S.F. (1981) Blood, 57, 526-531
42. Broder S., Edelson R.L., Lutzner M.A., Nelson D.L., MacDermott R.P., Durm M.E., Goldman C.K., Meade B.D. & Waldmann T.A. (1976) J. Clin. Invest. 58, 1297-1306
43. Yoshida T., EDelson R., Cohen S. & Green I. (1975) J. Immunol. 114, 915-918

IMMUNE RECOGNITION OF TUMOR CELLS DETECTED BY LYMPHOCYTE MEDIATED CYTOTOXICITY

Eva Klein and Farkas Vanky

Department of Tumor Biology
Karolinska Institutet
S 104 01 Stockholm, Sweden

INTRODUCTION

The goal of human tumor immunology studies is to demonstrate the recognition of autologous tumor cells. Though the recognition may not ensure that the immune mechanisms can control tumor growth, increasing knowledge of the details in the immune events may lead to therapeutical exploitations. There is considerable interest in the establishment of tumor cell reactive T lymphocyte cultures. They would provide a tool for characterization of tumor-related antigens and may also be useful for therapeutical strategies.

Cell mediated cytotoxicity is the method of choice in many studies concerned with tumor immunology. The technical ease of the short term in vitro assays and the demonstration of direct anti tumor effects give a justified satisfaction for the researchers. The results of such tests have a strong impact on the view on tumor immunology and surveillance.

Due to the complexity of the immune system, the results obtained in the in vitro conditions may not reflect the in vivo events. Absence of lytic effects in a particular system, does not rule out the existence of immune recognition since this may be manifested in a less dramatic way than an efficient and prompt direct lytic effect. Conversely, the demonstration of cytotoxic lymphocytes in vitro does not necessary ensure the manifestation of such lytic effects in vivo.

Auto-Tumor Cytotoxicity Exerted by Fresh and Activated T Cell Populations

In search of evidence for the immunological recognition of tumor cells in patients with sarcomas and carcinomas we have performed blastogenesis (ATS) and 4 h cytotoxicity tests (ALC) with freshly separated blood lymphocytes and tumor cells derived from surgical specimens[1,2,3,4]. In 25% of our 77 experiments cytotoxicity was detectable. In the auto-tumor stimulation assay (ATS) which involves cocultivation of lymphocytes and tumor cells the proportion of positive cases was considerably higher, blastogenesis was recorded in 73% of 200 cases. In the majority of ATS cultures auto-tumor cytotoxicity was generated[3,5,6]. In addition to the ATS cultures, when lymphocytes of the tumor patients were cocultivated with allogeneic lymphocytes (mixed lymphocyte culture, MLC), cytotoxicity against the autologous tumor cells was also generated in some but not all cases[7]. The MLC activated cells often damaged allogeneic tumors derived from unrelated individuals. Our and recently published similar findings with solid tumors[8] confirm the results of Zarling et al.[9] with leukemia patients.

We investigated the specificity of the cytotoxicity exerted towards autologous tumor biopsy cells in the different systems.

The nature of recognition of autologous tumor biopsy cells which results in their lysis may be different in the various systems. The direct effects may reveal the presence of lymphocytes which carry receptors for antigens expressed on tumor cells. Cocultivation with the autologous tumor cells may induce the proliferation and expansion of these clones. On the other hand a large proportion of the PHA or MLC activated lymphocytes may not recognize tumor-related antigens but act through a yet unknown mechanism which causes the lysis of certain cultured cell lines. In contrast to the specific recognition this latter does not depend on antigen recognition (discussed later) and therefore is not exerted by clones with specific receptors.

Presentation of antigen to lymphocyte populations in vitro triggers the receptor carrying T cells to divide. After a few days such mixed cultures develop cytotoxic potential. On the population level the following lytic effects can be generated in antigen containing lymphocyte cultures: 1) Enlargement of the specific clone will result cytotoxicity against a) cells which carry the stimulating antigen; b) cells which carry cross reactive antigens. 2) Transactivation induced by soluble factors produced in the mixed cultures will recruit lymphocytes with other specificities. Thus targets unrelated to the stimulus may be killed if the proportion of lymphocytes with receptors against their antigens is high. 3) Activated lymphocytes kill certain type of targets (these are among cultured lines) independently of antigen recognition.

Two techniques allow the investigation of the specificity of the functioning subsets in one effector population acting on various targets."Cold target competition" is the simpler way which, provided that it is adequately controlled, can give reliable information about the simultaneous presence of various functional subsets with differing specificity in a given lymphocyte population[10]. This assay can be used when the lytic effect of lymphocytes is measured by the release of isotope as a consequence of target damage. Admixture of non-labelled targets compete for the effectors and thus the isotope release is reduced. The other method is the characterization of cloned populations. This is more laborious but it reveals specificities on the single cell level[11,12].

We performed experiments with the cold target competition assay[7,13]. Primarily these were designed to elucidate whether the auto-tumor killer effector sets within the ATS, MLC or PHA activated lymphocyte populations can react with other type of targets such as Con A blasts or allogeneic tumor biopsy cells. The cold target competition assay is widely used and the results obtained with it indicate that the evaluations concerning the demonstration of distinct or overlapping effector populations which react with two different targets, are correct. Representative experiments are shown in Tables I-III.

Table I. Autologous freshly isolated lymphocytes act on tumor cells, donor designated A, at 50:1 effector target ratio. The figures represent the percentage cytotoxicity related to the value without competitor - 25%.

Competitor	Donor	Ratio labeled : Cold targets 5:1	1:1	1:5
Tumor	A	72	48	12
PHA blast	A	100	n.d.	100
Tumor	B	100	n.d.	80
PHA blast	B	100	n.d.	92

Table II. Autologous lymphocytes cultivated with tumor cells, donor designated A, for 6 days. Tested at 50:1 effector target ratio against target A. The figures represent the percentage cytotoxicity related to the value without competitor - 27%

Competitor	Donor	Ratio labeled : Cold targets 5:1	1:1	1:5
Tumor	A	59	26	0
Con A blast	A	85	67	74
Tumor	B	100	96	67
Con A blast	B	74	85	67

Table III. Lymphocytes of tumor patient A cultivated in mixed lymphocyte culture, stimulator S, for 6 days. Tested at 50:1 effector target ratio. The figures represent the percentage cytotoxicity related to the value without competitor - 30%

Competitor	Donor	Ratio labeled : Cold targets 5:1	1:1	1:5
Target A				
Tumor	A	93	53	27
Con A blast	A	93	90	87
Tumor	B	83	80	80
Con A blast	S	100	97	100
Target B				
Tumor	A	100	100	82
Con A blast	A	93	100	79
Tumor	B	22	36	36
Con A blast	B	36	11	7

Auto-tumor cytotoxicity, independently of the mode of its generation, was not inhibited by autologous lymphocytes (blasts)[7]. This suggests that the active set of lymphocytes recognizes tumor or perhaps organ-specific antigens. Allogeneic tumor cells did not inhibit either.

Auto-tumor cytotoxicity of the lymphocytes activated in allogeneic MLC was not inhibited by the stimulator lymphocytes (Con A blasts) and by allogeneic tumor cells. Thus the damage of autologous tumor cells was not the consequence of cross reactivity with the stimulator allogeneic lymphocytes.

Allo-Tumor Cytotoxicity Exerted by Fresh and Activated T Cell Populations

In a high proportion of experiments interferon treated blood lymphocytes of tumor patients were cytotoxic for cells derived from allogeneic tumors[4,7]. Cold target competition tests showed that different allogeneic tumor biopsy cells were lysed by distinct sets of the effector population[7,13]. Cytotoxicity against a given allogeneic tumor target was inhibited only by identical cells or lymphocytes (Con A blasts) from the same individual but not by unrelated tumors or blasts[7]. The competition by the lymphocytes of the individual whose tumor cells were used as targets suggests that the cytotoxic effectors recognize alloantigens on the tumor targets (Table III). Such results were also obtained when lymphocytes of healthy donors were used as effectors. With tumor patients lymphocytes, the allo-tumor cytotoxicity was not inhibited by admixture of the patients own tumor.

The lack of cross reactivity between the autologous and allogeneic tumors (even with similar histological types) may be due to a real antigenic difference. Alternatively antigenic cross reactivities may exist but in the short term T cell cytotoxicity assays they may not be detectable due to the MHC restriction[14]. Cytolytic T cells have been shown to recognize cell surface antigens on the target in combination with major histocompatibility antigens (MHC). It has been demonstrated that well defined antigens (chemical and viral) are not recognized as cross reactive when expressed on cells with different MHC make-up.

Lymphocytes activated in MLC lysed the allogeneic stimulator lymphocytes, tumor cells from the stimulator and also third party tumor cells[5]. Similar to the IFN activated fresh effectors these populations acted on different allotargets with different sets also and their effect was inhibited by blasts from the target cell donor[7]. Allo-killing generated in MLC of tumor patient's was not inhibited by admixture of tumor cells from the lymphocyte donor which responded in the MLC.

The selective cytotoxicity for the autologous tumor cells generated in the ATS cultures and the wider target panel of the MLC-s was maintained when the lymphocytes were propagated with the help of TCGF. Due to the exclusive reactivity of activated T-cells with the growth promoting factor IL-2, the initial events in the cultures determined thus the characteristics of the TCGF-cultures[15,16]. The early committment determining specific reactivity has been directly shown with T-cell colonies raised following brief exposure to allogeneic lymphocytes and grown with TCGF[11].

<u>Cytotoxicity of Blood Lymphocytes Against Certain Cultured Cell Lines</u>

In recent years considerable attention has been directed towards the cytotoxic potential of fresh blood lymphocytes affecting certain cell lines. The phenomenon was detected in the course of the studies aimed to detect recognition of tumor cells in patients and experimental animals. Cell lines were used in these studies because it was assumed that interpretation of the results with standard well defined targets would be easier since variation in target sensitivity would not influence the results. It was found that lymphocytes of unprimed donors exerted also lytic effects. Since tumor derived lines were sensitive to this effect and the activity occurred without the need of sensitization it was proposed that the natural killing (or spontaneous cytotoxicity, as designated by various investigators) is an important surveillance mechanism.

Consequently the nature of the recognition, the surface moiety on the target responsible for the recognition by lymphocytes, the phenotypic characteristics of the reactive lymphocytes have been the subject of many investigations.

The lymphocytes subsets responsible for the effect were found to be heterogeneous with regard to cell surface characteristics[17,18,19,21,22]. The majority carry T cell markers, such as receptors for sheep erythrocytes, SRBC. However, the mature T cells with high avidity to these erythrocytes did not show the killing or were relatively inefficient. The majority of active cells were found to carry Fc receptors[17,20,22]. In contrast to the antibody dependent cytotoxicity, the Fc receptor on the cells does not have a function in the direct lytic effect[20].

The characteristics that determine the degree of a given cell line's sensitivity to natural killing are unknown. Cell lines can be divided into three categories: 1) sensitive in short term assays; 2) sensitive only in long term assays; 3) insensitive. The difference between the first and second categories is probably quantiative. In the long term assays the interaction between effectors and targets induces the production of interferon which is a potent activator of the lytic function.

We assume that with the cell lines as targets, membrane properties and not the expression of certain surface antigens may define the consequences of the majority of target-lymphocyte encounter. Experiments with mouse somatic hybrid cell lines support this assumption[23]. In hybrids between high and low NK sensitive cells alloantigens, virally determined cell surface antigens and other defined membrane antigens such as TSTA were expressed codominantly but the NK sensitivity was low.

Short term exposure of lymphocytes in vitro to IFN enhances their cytotoxic activity, "interferon activated killing" (IAK)[21,22,24,25,26]. Importantly, IAK was found to be mediated by the same lymphocyte subsets (defined by surface characteristics) which function in NK[21,22,25]. The ranking order of the lymphocyte subsets was similar, but the IAK effects were stronger. In all likelihood, IFN elevates the function of lymphocytes with lytic potential[21,22,26]. As a result, IAK can damage also such targets that resist the majority of unmanipulated (NK) killer cells[22].

It is important to note that short term NK and IAK assays differ only operationally, in that IAK involves IFN pretreatment of the effector cells. The results with lymphocytes of individuals with high NK-activity are similar to IFN-activated lymphocytes from donors which function at a lower level of natural (NK) activity[21]. Thus there is an individual difference in the inherent activation profile of the lymphocyte population. The similarities of NK and IAK suggest that the rules emerging from IAK experiments are likely to be valid for the NK as well.

The cytotoxic activity against NK sensitive target lines gradually disappears when the lymphocytes are cultivated in vitro in human serum[27,28]. However, when lymphocytes are activated in culture by specific stimuli, e.g. in MLC, cytotoxicity against the NK sensitive cell lines are generated[5,28,29,30,31]. This activated killing, AK, can also affect such cell lines that are NK resistant in short term tests[27,28]. In addition to allo-antigens, calf serum[27,32], some modification that occurs on the surface of autologous cultured tumor cells[33], autologous B cells[34], EBV transformed autologous and allogeneic lymphoid lines[35], interleukin-2, IL-2[27] also triggers this type of cytotoxicity.

There are some phenotypic differences between the natural and the cultured activated killer cell populations. Relatively higher proportion of the activated killer cells adheres to nylon wool[28]. The proportion of Fc receptor carrying killer cells is lower in the AK than in the NK system[28] and the lymphocytes do not react with the monoclonal OKM1 reagent, a marker for a large part of the NK active lymphocytes[36]. They are similar in one aspect, inasmuch T cells with high affinity receptors to sheep erythrocytes are the least active both in the fresh (NK) and cultured (AK) populations[28,29].

The AK of a lymphocyte culture is not brought about by surviving NK cells, but is triggered de novo. Lymphocyte populations depleted of the NK active subsets i.e. FcR negative T-cells with high avidity sheep erythrocyte receptors became cytotoxic on exposure to appropriate activating stimuli[30,37] and lymphocytes kept in autologous plasma in vitro for several days which have lost NK activity became cytotoxic when cultured further with K562 cells[28].

Allogeneic MLC provides conditions where specific CTL (against the stimulator blast cell as targets, or a corresponding EBV-transformed B-cell line of recent origin which is practically non-sensitive to the NK effect) and AK (e.g. against autologous B cell line or the HLA negative, K562 and Daudi cell lines) could be studied in parallel. The properties of the two cytotoxic systems can be contrasted as follows:

Generation of AK in mixed lymphocyte culture does not require cell division[38]. This implies that AK is not dependent on the enlargement of the specific alloreactive clone(s). It seems that precursor cells, present at the initiation of the culture are triggered to AK, as a sequel to the specific recognition step. This is further illustrated by the fact that AK which accompanies a specific secondary MLC response, generated by a restimulation with the specific target, does not show a similarly strong secondary peak[29].

General Consideration

We consider the lysis of K562 cells (the prototype of the human NK sensitive line) to be a sign of the activated state of the lymphocytes. The main difference between this activity and the lysis of freshly harvested tumor cells is that the anti K562 effect is polyclonal[39,40]. This was proven by the killing of K562 by antigen specific clonal lines and by the general competing potential of the K562 cells both when fresh IFN treated effectors and in vitro activated effectors were used against various targets[13,30]. Consequently in the judgement of the specificity of the reactivity exerted by lymphocyte populations against freshly collected tumor cells we do not take into account the anti K562 or anti Daudi effects, but only the results against targets of similar type.

In several series of experiments the natural killing of blood lymphocytes were monitored in various categories of patients using K562 as targets. The impetus was given by the proposition that the natural killer cells contribute to tumor surveillance. Since the anti K562 effect is a measure of the activation profile of the lymphocyte population with regard of lytic function it provides indeed an information about the potentialities of these cells. The lytic potential is relevant also against targets which are damaged on the basis of antigen recognition and are insensitive to the indiscriminative cytotoxicity. For the in vivo antitumor response the relevant information is whether the tumor cells express antigens that are recognized by members of the T cell repertoire[40]. If such lymphocytes with specific receptors exist they are likely to be represented also in subsets with varying functional characteristics. The blood lymphocyte population of individuals with high proportion of lytic subsets may thus exert antigen specific cytotoxicity.

The experiments with MLC-activated lymphocytes show that cells which react with the autologous tumor can be activated by other means than exposure to the specific antigen. Our experiments in which parallel samples of lymphocytes were cultured in MLC and ATS indicate however that the latter is more efficient in generation of auto tumor cytotoxicity[5,7].

The emergence of auto tumor cytotoxicity in the MLC is probably due to production of IL-1 and IL-2 as a corollary of antigen recognition which leads to activation and proliferation of cells which do not bear receptors for the stimulating antigens[41].

The development of auto-tumor reactivity as a consequence of alloimmunization has been shown in vivo also. Allosensitization was shown to confer protection of mice and rats against the outgrowth of syngeneic tumor grafts[42].

While cocultivation with allogeneic cells often activated the auto-tumor reactive lymphocytes, in the ATS cultures alloreactivity was not generated[5,7]. This may be due to the relatively lower proportion of auto-tumor reactive lymphocytes compared to those that react with the MHC antigens. This is indicated by the fact that the blastogeneic response is considerably lower in the ATS cultures compared to the MLC-s. The proportion of antigen reactive cells in the culture probably determines the quantity of the produced soluble amplifying factors. The comparison of the auto-tumor killing effectors generated either by single or by "pool" MLC stimulation shows that this may be the case because more efficient activation occurred under the latter condition[8].

Auto-tumor killers were not activated by short exposure of the fresh lymphocyte population to IFN[4]. In view of the high proportion of cases (73%) in which the ATS cultures showed blastogenesis this would have been expected. The number of tumor recognizing lymphocytes in a stage of differentiation which can be enhanced for lytic function by IFN may be low in the blood of the patients. It is known that not all lymphocyte subsets can be triggered for cytotoxicity by IFN. In the majority of individuals Fc receptor negative T-cells are not cytotoxic even after IFN treatment for highly susceptible targets[21,22]. This subset represents a high proportion of the blood lymphocyte population and contains those lymphocytes which are triggered for proliferation by confrontation with antigens[37]. Thus in the blastogenesis assay the auto-tumor reactive cells may be more readily detected.

Another case of the lack of IFN effect may be that the specific receptor bearing lymphocytes with cytotoxic potential had left the blood streem for the tumor site. However, this possibility seems to be unlikely because experiments performed with lymphocytes of tumor patients collected after surgery gave similar results.

ACKNOWLEDGEMENTS

This project has been funded with Federal Funds from the Department of Health, Education and Welfare under Contract Number NO1-CM-74144, grant No. 1RO1-CA-25184-01A1 awarded by the National Cancer Institute, DHEW and by the Swedish Cancer Society.

REFERENCES

1. B.M. Vose, F. Vánky, and E. Klein, Lymphocyte cytotoxicity against autologous tumor biopsy cells in humans. Int. J. Cancer 20: 512 (1977).
2. F. Vánky, J. Stjernswärd, and U. Nilsonne, Cellular immunity to human sarcoma. J. Natl. Cancer Inst. 46: 1145 (1971).
3. F. Vánky, B.M. Vose, M. Fopp, and E. Klein, Human tumor-lymphocyte interaction in vitro. VI. Specificity of primary and secondary autologous lymphocyte-mediated cytotoxicity. J. Natl. Cancer Inst. 62: 1407 (1979).

4. F. Vánky, S. Argov, S. Einhorn, and E. Klein, Role of alloantigens in natural killing. Allogeneic but not autologous tumor biopsy cells are sensitive for interferon-induced cytotoxicity of human blood lymphocytes. J. Exp. Med. 151: 1151 (1980).
5. F. Vánky, S. Argov, and E. Klein, Tumor biopsy cells participating in systems in which cytotoxicity of lymphocytes is generated. Autologous and allogeneic studies. Int. J. Cancer 27: 273 (1981).
6. B.M. Vose, F. Vánky, M. Fopp, and E. Klein, In vitro generation of a secondary cytotoxic response against autologous human tumor biopsy cells. Int. J. Cancer 21: 288 (1978).
7. F. Vánky, T. Gorsky, Y. Gorsky, M.G. Masucci, and E. Klein, Lysis of tumor biopsy cells by autologous T-lymphocytes activated in mixed cultures and propagated with TCGF. J. Exp. Med., in press.
8. J.L. Strausser, A. Mazumder, E.A. Grimm, M.T. Lotze, and S.A. Rosenberg, Lysis of human solid tumors by autologous cells sensitized in vitro to alloantigens. J. Immunol. 127: 266 (1981).
9. J.M. Zarling, P.C. Raich, M. McKeough, and F.H. Bach, Generation of cytotoxic lymphocytes in vitro against autologous human leukemia cells. Nature 262: 691 (1976).
10. O.M. Landazuri, and R.B. Herberman, Specificity of cellular immune reactivity to virus induced tumors. Nature 238: 18 (1972).
11. J. Kornbluth, and B. Dupond, Cloning and functional characterization of primary alloreactive human T lymphocytes. J. Exp. Med. 152: 164 (1980).
12. C. Taswell, H.R. MacDonald, and J.-C. Cerottini, Clonal analysis of cytotoxic T lymphocyte specificity. I. Phenotypically distinct sets of clones as the cellular basis of cross-reactivity to alloantigens. J. Exp. Med. 151: 1372 (1980).
13. F. Vánky, and E. Klein, Alloreactive cytotoxicity of interferon triggered human lymphocytes detected with tumor biopsy targets. Immunogenetics, in press.
14. R.M. Zinkernagel, and P.C. Doherty, MHC-restricted cytotoxic T cells: studies on the biological role of polymorphic major transplantation antigens determining T cell restriction - specificity, function, and responsiveness. Adv. Immunol. 27: 51 (1979).
15. D. Gillis, and K.A. Smith, Long-term culture of tumor-specific cytotoxic T cells. Nature (Lond.) 268: 154 (1977).
16. J.L. Strausser, and S.A. Rosenberg, In vitro growth of cytotoxic human lymphocytes. I. Growth of cells sensitized in vitro to alloantigens. J. Immunol. 121: 1491 (1978).

17. T. Bakács, P. Gergely, and E. Klein, Characterization of cytotoxic human lymphocyte subpopulations. The role of Fc-receptor carrying cells. Cell. Immunol. 32: 317 (1977).
18. T. Bakács, E. Klein, E. Yefenof, P. Gergely, and M. Steinitz, Human blood lymphocyte fractionation with special attention to their cytotoxic potential. Immunobiol. 154: 121 (1978).
19. M.A. Kall, and H.S. Koren, Heterogeneity of human natural killer populations. Cell. Immunol. 40: 58 (1978).
20. H.D. Kay, G.D. Bonnard, W.H. West, and R.B. Herberman, A functional comparison of human Fc-receptor bearing lymphocytes active in natural cytotoxicity and antibody-dependent cellular cytotoxicity. J. Immunol. 118: 2058 (1977).
21. G. Masucci, M.G. Masucci, and E. Klein, Activation of human blood lymphocyte subsets for cytotoxic potential. To be published.
22. M.G. Masucci, G. Masucci, E. Klein, and W. Berthold, Target selectivity of interferon induced human killer lymphocytes related to their Fc receptor expression. Proc. Natl. Acad. Sci. 77: 3620 (1980).
23. L. Åhrlund-Richter, E. Klein, and G. Masucci, Somatic hybrids between a high NK-sensitive lymphoid (YAC-IR) and several low sensitive sarcoma or L-cell derived mouse lines exhibit low sensitivity. Som. Cell Gen. 6: 89 (1980).
24. E. Saksela, T. Timonen, and K. Cantell, Human natural killer activity is augmented by interferon via recruitment of "pre-NK" cells. Scand. J. Immunol. 10: 257 (1979).
25. D. Santoli, and H. Koprowski, Mechanism of activation of human natural killer cells against tumor and virus infected cells. Immunol. Rev. 44: 125 (1979).
26. S. Targan, and F. Dorey, Interferon activation of pre-spontaneous killer (pre-SK) cells and alteration in kinetics of lysis of both "pre-SK" and active SK cells. J. Immunol. 124: 2157 (1980).
27. M.G. Masucci, E. Klein, and S. Argov, Non-specific cytotoxic potential of human lymphocyte cultures is different from NK and its strength correlates to the level of blastogenesis. J. Immunol. 124: 2458 (1980).
28. A. Poros, and E. Klein, Cultivation with K562 cells leads to blastogenesis and increased cytotoxicity with changed properties of the active cells when compared to fresh lymphocytes. Cell. Immunol. 41: 240 (1978).
29. J.K. Seeley, G. Masucci, A. Poros, E. Klein, and S.H. Golub, Studies on cytotoxicity generated in human mixed lymphocyte cultures. J. Immunol. 123: 1303 (1979).

30. R.I.H. Bolhuis, and C.P.M. Ronteltalp, Generation of natural killer (NK) cell activity after mixed lymphocyte culture (MLC). Activation of effector cells in NK depleted populations. Immunology Letters 1: 191 (1980).
31. D.M. Callewaert, J.J. Lightbody, J. Kaplan, J. Joroszewski, W.D. Peterson, and J.C. Rosenberg, Cytotoxicity of human peripheral lymphocytes in cell-mediated lympholysis; antibody dependent cell-mediated lympholysis and natural cytotoxicity assays after mixed lymphocyte culture. J. Immunol. 121: 81 (1978).
32. J.V. Zielska, and S.H. Golub, Fetal calf serum induced blastogenic and cytotoxic response of human lymphocytes. Cancer Res. 36: 3842 (1976).
33. M.R. Martin Chandon, F. Vanky, C. Carnaud, and E. Klein, In vitro education on autologous human sarcoma generates non-specific killer cells. Int. J. Cancer 15: 342 (1975).
34. K. Tomonari, Cytotoxic T cells generated in the autologous mixed lymphocyte reaction. I. Primary autologous mixed lymphocyte reaction. J. Immunol. 124: 1111 (1980).
35. E.A. Svedmyr, F. Deinhart, and G. Klein, Sensitivity of different target cells to the killing action of peripheral lymphocytes stimulated by autologous lymphoblastoid cell lines. Int. J. Cancer 13: 891 (1974).
36. J.M. Zarling, and P.C. Kung, Monoclonal antibodies which distinguish between human NK cells and cytotoxic T lymphocytes. Nature 288: 394 (1980).
37. G. Masucci, A. Poros, J.K. Seeley, and E. Klein, In vitro generation of K562 killers in human T-lymphocyte subsets. Cell. Immunol. 52: 247 (1980).
38. D.M. Callewaert, J. Kaplan, D.F. Johnson, and W.D.Jr. Peterson, Spontaneous cytotoxicity of cultured human cell lines mediated by normal peripheral lymphocytes. II. Specificity for target antigens. Cell. Immunol. 42: 103 (1979).
39. E. Klein, Natural and activated cytotoxic T lymphocytes. Immunology Today 1: IV (1980).
40. E. Klein, and F. Vánky, Natural and activated lymphocytes which act on autologous and allogeneic tumor cells. Cancer Immunol. Immunother. 11: 183 (1981).
41. H. Wagner, and M. Röllinghof, T-T cell interactions during in vitro cytotoxic allograft responses. I. Soluble products from activated Lyl+ T cells triggered autonomously by antigen primed Ly2,3+ T cells to cell proliferation and cytotoxic activity. J. Exp. Med. 148: 1523 (1978).
42. H. Kobayashi, M. Hoshokova, and T. Oikawa, Transplantation immunity to syngeneic tumors in WKR rats immunized with allogeneic cells. Transpl. Proc. 12: 156 (1980).

SOME CHARACTERISTICS OF NON-IMMUNOGLOBULIN COMPONENTS IN LOW pH EXTRACTS OF A NON-LYMPHOID TUMOR

Maya Ran, Margot Lakonishuk, Esther Bachar and Isaac P. Witz*

Department of Microbiology, The George S. Wise Faculty of Life Sciences, Tel Aviv University, Tel Aviv 69978 Israel
*Incumbent: The David Furman Chair in Immunobiology of Cancer

INTRODUCTION

In situ humoral immunity as an indispensable component of host-tumor relationship is being studied in our laboratory using various tumor systems (1-3). It is now well established that immunoglobulins localize in vivo within the tumor tissue of man and animals (3). Some of these tumor-associated Ig molecules may be anti tumor antibodies induced in tumor-bearers (e.g.2,4).

Tumor associated immunoglobulins and antibodies were dissociated form the tumor tissue by employing a procedure offering a maximal recovery of antibody and a minimal damage to the coated cells (5). Although the dissociation procedure using 0.1 M citrate buffer at pH 3.5 seemed to be mild (5) the possibility was not excluded that the low pH buffer treatment extracted loosely bound cell surface components from in vivo growing tumor cells in addition to Ig molecules coating such cells. Since molecules present in low pH extracts of various tumors may exert important biological functions (3), it is of importance to characterise such molecules.

In this study we analysed non-immunoglobulin components extracted from mouse tumors by low pH citrate buffer.

THE TUMOR SYSTEM

Most of the study was carried out using SEYF-a tumors. This

tumor is a polyoma-virus induced sarcoma passaged as ascites in syngeneic A.BY mice. The antigenic specificities on the membrane of the cells were analysed using artificially raised syngeneic antisera causing complement dependent lysis (CdL) of the cells (6). In addition to antibodies directed against antigens expressed on a wide spectrum of mouse tumors the cells expressed also a polyoma-virus induced membrane antigen (7).

A.BY mice bearing the tumor developed antibodies causing CdL of SEYF-a cells in vitro. Such antibodies localised on the tumor cells in vivo (2,4). The evidence that CdL mediating antibodies localised in vivo on the tumor cells was derived from experiments showing that the addition of exogenous complement to freshly harvested tumor cells caused their lysis. Furthermore, other experiments showed that the addition of low pH eluates and complement to non-coated SEYF-a target cells caused lysis of such cells (2). Using an isopycnic and velocity sedimentation gradients of Percoll as a method of separating tumor seeking lymphocytes and tumor cells, it was established that the potentially cytotoxic antibodies localised on the tumor cells per se, and not on infiltrating host-derived immunocytes (4). The SEYF-a bound anti tumor antibodies belonged to the IgG2a subclass (8) and their pattern of specificity was more restricted than that of the sera of tumor bearing mice (i.e. they reacted with fewer specificities, Ran et al., in preparation). Eluates of SEYF-a cells also contained lymphocytotoxic antibodies, reacting with lymphocyte populations enriched for T cells (9,10). These lymphocytotoxic antibodies inhibited the formation of EA rosettes by activated T cells (10).

RESULTS

The Elution Conditions of Anti Tumor Antibodies from SEYF-a Cells

In order to answer the question as to whether or not non-immunoglobulin components are extracted from SEYF-a cells by low pH citrate buffer, we applied stepwise elution procedures aimed at peeling off the maximal amount of tumor-bound antibodies. We then analysed the various eluates for the presence of residual CdL-mediating antibodies as well as for the presence of non-antibody Ig molecules and non Ig molecules. The elution procedures were as follows: SEYF-a cells were divided into two portions. The one was exposed to two consecutive acid citrate treatments. The second portion was incubated in culture conditions for 60 min. Such incubations were found to cause the shedding of antibodies from viable cells (2). After removing the spent medium the cultured cells were divided. One portion was exposed to 2 consecutive treatments with low pH citrate buffer and the other portion was incubated for an additional 60 min at 37°C. After removal of the spent medium of the second incubation period the cells were exposed to 2 consecutive acid citrate treatments.

Table 1 shows that: 1)Cytotoxic antibodies were recovered both in the medium of short-term cultures as well as in the low pH extracts. 2)It was possible to achieve a complete elution of CdL mediating antibodies especially by using 2 consecutive treatments with the low pH citrate buffer.

Table 2 demonstrates that eluates which hardly contain CdL mediating antibodies still contain appreciable amounts of protein. The non antibody protein fractions were shown to contain IgG2a. This was established by utilising the method of inhibition of hemagglutination (HAI) of sheep erythrocytes (SRBC) conjugated with mouse IgG by rabbit antibodies directed against mouse IgG2a. Experiment II of Table 2 indicates that additional non Ig protein are still detectable in certain tumor eluates (e.g. No.8) which are essentially devoid of CdL mediating antibodies and of HAI mediating IgG2a.

The Presence in Low pH Extracts from Tumor Cells of a Factor Causing Hemagglutination of Antigen Conjugated SRBC Sensitised with Subagglutinating Concentrations of Antiserum

Table 3 shows that low pH extracts from SEYF-a tumors inhibited the agglutination of mouse IgG (MIgG) conjugated SRBC sensitised with agglutinating amounts of rabbit anti-bodies directed against mouse IgG (RaMIgG diluted 1:16.000). The inhibition,as indicated in the previous section and in Table 2 was caused by the IgG usually present in low pH eluates of in vivo propagated mouse tumor cells. However, the same extracts agglutinated MIgG-conjugated SRBC if these cells were sensitised with subagglutinating amounts of the rabbit anti mouse IgG antibodies (diluted 1:64.000). The agglutination of antigen-conjugated SRBC sensitised with subagglutinating amounts of the corresponding antibody by tumor extracts was designated hemagglutination enhancement (HAE) to distinguish this reaction from the direct agglutination of antigen-conjugated SRBC by the corresponding antibody. As seen in Table 3 the sensitivity of HAE could be increased when such a high dilution of sensitising antiserum was used that no HAI could be observed (compare the 1:64,000 dilution to 1:16,000 dilution).

The specificity of the HAE of MIgG conjugated SRBC mediated by SEYF-a eluates was determined as follows: 1)Extracts did not cause HAE of unconjugated SRBC or of MIgG conjugated SRBC in the absence of RaMIgG (Table 3). 2)Extracts from normal splenocytes or from S-19 plasmacytoma did not mediate HAE (Table 3). 3)HAE mediated by low pH tumor eluates could be inhibited by the antigen corresponding to

Table 1. The Elution of Anti-tumor Antibodies from SEYF-a Cells

Elution[1)]	Cytotoxicity Index[2)]		
	1:2	1:4	1:8
1st low pH eluate	0.66	0.57	0.41
2nd " " "	0.49	0.22	0.08
60 min at 37°C[3)]	0.34	0.22	0.12
" " " " + 1st low pH eluate	0.52	0.31	0.17
" " " " + 2nd " " "	0.32	0.06	0.00
60 + 60" " [3)]	0.40	0.12	0.05
" " "" " + 1st low pH eluate	0.51	0.31	0.10
" " "" " + 2nd " " "	0.26	0.03	0.00

1) The various elution procedures are described in the text.
2) The cytotoxicity index values to SEYF-a cells mediated by the indicated dilution of the various eluates.
3) The proteins in the spent medium were precipitated by 50% ammonium sulphate. These were then assayed for CdL activity.

Table 2. Antibody and IgG in SEYF-a Extracts

Elution[1)]	Exp. 1			Exp. 2		
	Protein (mg/ml)	C I[2)]	HAI[3)]	Protein (mg ml)	C I[2)]	HAI[3)]
1st low pH eluate	0.70	0.41	1:32	0.55	0.14	1:8
2nd " " "	0.83	0.08	1:16	0.59	0.11	1:4
60 min at 37°C[4)]	0.65	0.12	1:128	0.65	0.06	1:32
" " " "+ 1st low pH eluate	0.35	0.17	1:16	0.52	0.12	1:2
" " " "+ 2nd " " "	1.12	0.00	1:16	0.83	0.14	<1:2
60+60 " " " 4)	0.70	0.05	1:128	0.75	0.13	1:32
" " " " "+ 1st low pH eluate	0.38	0.10	1:16	0.55	0.13	1:2
" " " " "+ 2nd " " "	0.98	0.00	1:16	0.66	0.04	<1:2

1) The various elution procedures are described in the text.
2) The cytotoxicity index values to SEYF-a cells mediated by the various eluates at a 1:8 dilution.
3) The highest dilution of tumor eluates mediating hemagglutination inhibition (HAI) of IgG-conjugates SRBC sensitised with rabbit antibodies against mouse IgG2a.
4) see footnote 3 of Table 1.

Table 3. Hemagglutination of MIgG-SRBC by Anti MIgG2a and SEYF-a Eluates

Eluate of	Dilution of Anti MIgG2a 1:16,000		1:64,000	
	Dilution of tumor eluate			
	HAI[1]	HAE[2]	HAI[1]	HAE[2]
SEYF-a #1[3]	1:32	1:2	<1:2	1:8
#2	1:16	1:4	<1:2	1:8
#3	1:16	1:2	<1:2	1:4
#4	1:16	1:2	<1:2	1:16
spleen		<1:2		<1:2
S-19 (plasmacytoma)		<1:2		<1:2

1) See footnote 3 of Table 2.
2) The highest dilution of tumor eluates mediating hemagglutination of IgG-conjugated SRBC sensitised with rabbit antibodies against mouse IgG2a at a 1:16,000 or 1:64,000 dilution.
3) SEYF-a extracts did not hemagglutinate non-sensitised MIgG-SRBC or unconjugated SRBC. IgG2a but not IgG2b abrogated HAE.

Table 4. Hemagglutination of DNP.BSA-SRBC by Anti DNP.HSA and Low pH Tumor Extracts

Extract of	Highest dilution of extract mediating HAE[1]
SEYF-a #1	1:8
" #2	1:8
#3	1:16
#4	1:8
L5178Y #1	1:4
#2	1:12
CCRF-CEM #1	1:2
#2	1:8
S-19 (plasmacytoma)	1:2
spleen	1:2

1) No hemagglutination inhibition was obtained in this system.

the sensitising antiserum. For example, IgG2a inhibited HAE of IgG-conjugated SRBC tumor extracts sensitised with antibodies against IgG2a more efficiently than IgG2b and vice versa (results not shown). This indicated that specific binding of the sensitising antibody to the corresponding antigen conjugated to SRBC was a prerequisite for the HAE.

We confirmed the presence of an HAE mediating factor in tumor eluates by using SRBC conjugated with DNP-BSA and sensitised with anti DNP-HSA antibodies (Table 4). These results indicated that the HAE mediating factor in low pH tumor extracts could function as an immunoglobulin binding factor,possibly a Fc receptor (FcR). This assumption was further supported by the following experiments. 1)Low pH extracts of FcR positive T cell lines of mouse (L5178Y) and human (CCRF-CEM) origin also contained a HAE mediating factor (Table 4). 2)The presence of the HAE mediating factor in SEYF-a eluates could only be revealed with undegraded IgG fractions of the sensitising antibody but not when F(ab)2 fragments prepared from the sensitising antiserum were used for sensitisation (Table 5).

The HAE mediating factor was present in extracts prepared both from SEYF-a cells propagated in vivo for short (7-9 days) or long (13-34 days) periods (Table 6). It seemed that older tumors had a higher concentration of the HAE mediating factor than the younger tumors. Table 7 shows that low pH extracts from several mouse tumors in addition to those prepared from the SEYF-a tumor also contain the HAE mediating factor.

An enrichment for the HAE mediating factor was achieved by sequential precipitation of low pH SEYF-a extracts with increasing concentrations of ammonium sulphate. The HAE mediating factor did not precipitate at 40% or 50% saturated ammonium sulphate. The further precipitation of proteins in the supernatants of 40% or 50% ammonium sulphate by increasing the ammonium sulphate concentration to 70% saturation caused a partial precipitation of the factor (Table 8).

The Extraction and Enrichment of a 33K Cell Surface Protein from Low pH Extracts of SEYF-a Tumors

HAE-mediating low pH SEYF-a extracts were analysed by 15% SDS-PAGE. Twofold dilutions of first and second low pH extracts (see above) were applied to the gel. Figure 1 demonstrates the presence of a protein with a mw of 33.000 daltons (33K) in both extracts. The 33K fraction was detectable up to a 1:16 - 1:32 dilution of the extracts. The number of contaminating bands in the second low pH extracts was smaller than those present in the first extracts. The former extracts served therefore for further enrichment of the 33K proteins. It was found that the 33K protein, similarly to the HAE-mediating factor, partially precipitated at 70% saturated ammonium

Table 5. The Requirement of Intact Antibody Molecules for the Hemagglutination of DNP.BSA-conjugated SRBC by Anti DNP.HSA and Tumor Extracts

DNP.BSA-SRBC sensitised with	Highest dilution of SEYF-a extract causing HAE	
	Exp. 1	Exp. 2
Unfractionated NRS	<1:2	<1:2
" anti DNP.HSA	1:128	1:64
IgG of " " " "	1:8	1:4
$F(ab)_2$ of " " " "	<1:2	<1:2

Table 6. Hemagglutinating Factor in SEYF-a Extracts

Days following implantation	No. extracts tested	Protein (mg/ml)	Highest dilution of extract causing HAE (range)
7 - 9	9	0.30 ± 0.13	<1:2 - 1:64
13 - 34	8	0.31 ± 0.14	1:4 - 1:512

sulphate (Fig. 1). It is therefore evident that the ammonium sulphate precipitation procedure which enriched for the HAE-mediating factor also enriched for the 33K protein.

No conspicuous bands were observed in the 33K region when low pH extracts of normal mouse kidney, liver or spleen were analysed by 15% SDS-PAGE.

The question as to whether or not the 33K protein present in low pH extracts of SEYF-a cells originated from tumor cells per se or from infiltrating host cells, was answered by performing synthesis experiments utilising cultured SEYF-a cells. Low pH extracts from such cells similar to those prepared from in vivo grown cells also contained a protein which did not precipitate at 50% ammonium sulphate, but which precipitated partially at 70% saturation. This protein produced a conspicuous 33K band in SDS-PAGE. The 33K protein of low pH extracts of cultured SEYF-a cells could be labelled de novo with (^{75}Se) L-selenomethionine. Exposing labelled cultured cells to low pH buffer resulted in the extraction of a labelled 33K protein detected by autoradiography (Fig. 2).

Table 7. Hemagglutinating Factor in Low pH Extracts of Various Murine Tumors

Tumor	Strain	Days following implantation	Highest dilution of extract causing HAE
mammary carcinoma (1)	RIII	Primary	1:128
210 MC (2) Sarcoma	C57B1/6	20 days	1:8
211 MC (2) Sarcoma	C57B1/6	20 days	1:4
Sa601 MC (2) Sarcoma	NZW	35 days	1:16
S-14 Plasmačytoma	BALB/C	8 days	< 1:2
141 Lymphoma (3)	NZB/NZW	9 days	< 1:2
S-19 Plasmacytoma (4)	BALB/C	8 days	< 1:2

1) A kind gift of Dr. A. Frensdorff
2) Methylcholanthrene
3) A spontaneous B lymphoma
4) A kind gift of Dr. M. Cohn, Salk Institute, San Diego, Ca., USA. This extract served as negative control in the HAE tests performed in this study.

Table 8. Enrichment of Hemagglutinating Factor from SEYF-a Extracts

$(NH_4)_2SO_4$ treatment		Extract		
% saturation	Material tested	O.D(280nm)	Highest dilution causing HAE	Enrichment
None	-	0.64	1:8	-
40%	precipitate	0.22	< 1:2	0
50%	"	0.23	< 1:2	0
70%	precipitate of 40% supernatant	0.18	1:8	x 3.5
70%	precipitate of 50% supernatant	0.17	1:32	x 15.0
70%	supernatant	0.06	1:32	x 42.6

No 33K bands were formed by solubilised membrane preparations from SEYF-a cells after these have been pretreated by low pH citrate buffer.

33K proteins were also not detected in low pH extracts or in solubilised membrane fractions of YAC-1 cells (Fig. 2).

Fc-binding by Low pH Extracts of SEYF-a Tumors

The possibility was raised (see above) that low pH extracts of in vivo grown or of cultured SEYF-a tumors contain an Fc binding component. This possibility was further supported by the results of the following experiment. Low pH extracts of SEYF-a tumors enriched for the HAE-mediating factor and for the 33K protein (by precipitation at 50% ammonium sulphate) were conjugated to sepharose columns. These columns served for affinity chromatography of purified ^{125}I labelled Fcγ fragments. The ^{125}I-Fcγ was eluted from the column either with glycine buffer at pH 2.5 in 0.5 N NaCl or with purified staphylococcal protein A. The elution profile is shown in Fig. 3. ^{125}I labelled F(ab)2 fragments served as controls in these experiments. The results indicated that Fcγ but not F(ab)2 fragments could bind to the tumor extract columns and be eluted from these columns either by a low pH buffer or by protein A.

DISCUSSION

Using a procedure aimed at dissociating tumor-bound immunoglobulins from the tumor cells we found that in addition to Ig molecules other components were extracted from the cells. These non-immunoglobulin molecules agglutinated antigen-conjugated SRBC sensitised with subagglutinating doses of the corresponding antibodies. Enrichment for the agglutinating (HAE-mediating) factor enriched also for a 33,000 dalton (33K) protein present in tumor extracts. However, since we have not yet purified the 33K protein we are unable, at this point, to state with certainty that the HAE-mediating factor and the 33K protein are identical.

Is the HAE mediating factor an Fc binding protein? While no conclusive evidence is yet available to answer this question, indirect evidence suggests that this is indeed the case. 1) HAE could be inhibited by the addition of soluble antigen corresponding to the antibody sensitising the antigen-conjugated erythrocytes.

2) HAE occured only when the antibody molecule sensitising the antigen-conjugated SRBC was an intact IgG. Antigen-conjugated SRBC coated with F(ab)2 fragments of the antibody did not agglutinate in the presence of tumor extracts.

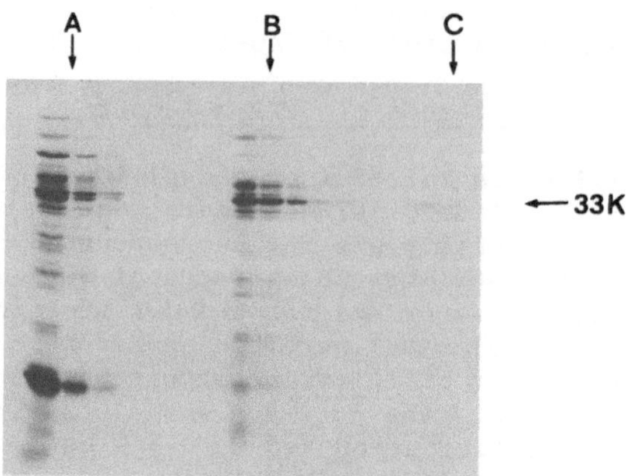

Fig. 1. Analysis of reduction products of low pH extracts of SEYF-a ascites, by 15% SDS-PAGE with 5% 2-mercaptoethanol. The samples from left to right are: Twofold dilutions - undiluted and diluted up to 1:64 of 1st (A) and 2nd (B) low pH extracts. The final supernatant (C) left after sequential precipitation of a 2nd low pH extract at 50% and 70% ammonium sulphate was also analysed. The 33K protein was localised using external markers.

3) Sepharose columns conjugated with extracts enriched for the HAE mediating factor retained ^{125}I Fc γ but not F(ab)2 fragments. The retained ^{125}I Fc γ could be eluted from these columns by protein A - indicating the presence of a Fc γ -binding molecule in the tumor extracts with the ability to compete against protein A for binding sites on Fc γ fragments.

4) Low pH extract of the YAC-1 lymphoma known to express no FcR (11) contained neither the HAE-mediated factor nor the 33K protein. These results, combined with previous ones demonstrating that SEYF-a tumors express FcR (12,13), thus suggest that the low pH treatment of these cells caused the release of a molecule functioning as a soluble FcR.

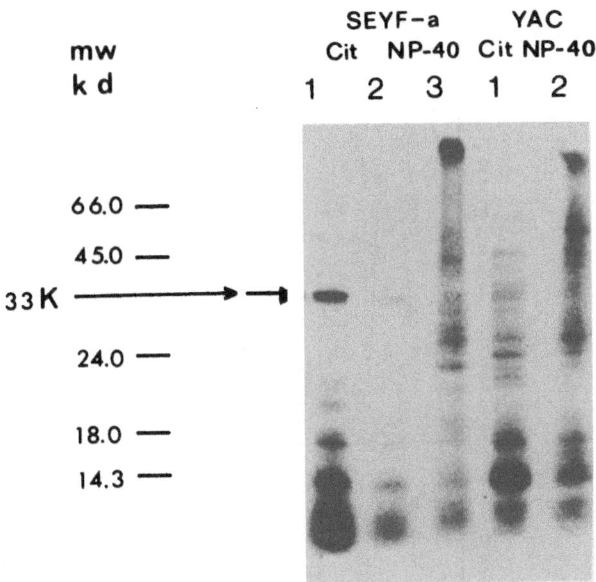

Fig. 2. Analysis of reduction products of low pH extracts of cultured SEYF-a cells pre-labeled with (^{75}Se)-L-Selenomethionine by 15% SDS-PAGE. The 33K protein was localised using coomassie brilliant blue staining of an extract of in vivo grown SEYF-a cells that served as a positive reference marker. The samples from left to right are: 1st low pH SEYF-a extract (1); 2nd low pH SEYF-a extract (2); NP-40 extract from SEYF-a cells after low pH extraction (3); 1st low pH of YAC-1 cells (1); 2nd low pH YAC-1 extract combined with a NP-40 extract from cells after low pH extraction (2).

Results obtained by others show that a protein of a molecular weight of 33.000 daltons could represent at least a part of an Fc receptor (14,15). A protein of similar M.W. has also been detected by James et al (16) in citrate extracts of methylcholanthrene induced mouse tumors. These quthors reported however, that their protein binds in addition to IgG, also BSA.

An FcR-like structure expressed on tumor cells might aid those cells in their survival by modulating immune responses associated with functional FcR on immunocytes (17).

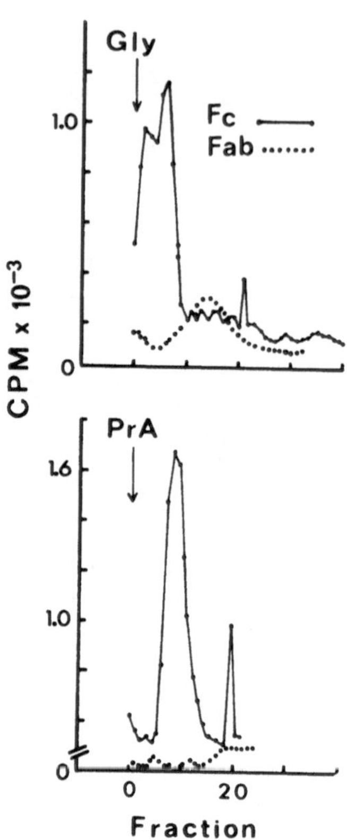

Fig. 3. The elution of ^{125}I-labeled Fcγ fragment of monoclonal human IgG from a sepharose solumn to which a fraction enriched for the 33K protein was conjugated. The protein was extracted from in vivo grown SEYF-a cells. ^{125}I-labeled Fcγ fragment (•——•) was bound to the column and eluted either by glycine pH 2.5 in 0.5 M NaCl or by protein A. ^{125}I-F(ab)2 fragment (·····) served as control.

REFERENCES

1. M. Ran and I.P. Witz, Tumor-associated immunoglobulins. The elution of IgG2 from mouse tumors, Int. J. Cancer 6:361 (1970).
2. M. Ran, G. Klein and I.P. Witz, Tumor-bound immunoglobulins. Evidence for the in vivo coating of tumor cells by potentially cytotoxic antitumor antibodies, Int. J. Cancer 4:90 (1976).
3. I.P. Witz, Tumor-bound immunoglobulins, in situ expressions of humoral immunity, Advances in Cancer Research 15:95 (1977).
4. T. Nethanel, R. Kinsky, N. Moav, R. Brown, M. Ran and I.P. Witz,

Separation of tumor seeking small lymphocytes and tumor cells using percoll velocity gradients, J. Immunol. Methods 41:43 (1981).

5. R. Ehrlich and I.P. Witz, The elution of antibodies from viable tumor cells, J. Immunol. Meth. 26:345 (1979).
6. I.P. Witz, N. Lee and G. Klein, Serologically detectable specific and cross-reactive antigens on the membrane of a polyoma virus-induced murine tumor, Int. J. Cancer 18:243 (1976).
7. G. Klein, B. Ehlin and I.P. Witz, Serological detection of a polyoma-tumor-associated membrane antigen, Int. J. Cancer 23: 683 (1979).
8. N. Moav and I.P. Witz, Characterisation of immunoglobulin eluted from murine tumor cells: Binding patterns of cytotoxic anti-tumor IgG, J. Immunol. Meth. 22:51 (1978).
9. M. Ran, M. Yaakubowicz and I.P. Witz, Lymphocytotoxic auto-antibodies eluted from in vivo propagating sarcoma cells of mice, J. Natl. Cancer Inst. 60:1509 (1978).
10. M. Ran, M. Yaakubowicz, O. Amitai and I.P. Witz, Tumor-localising lymphocytotoxic antibodies, in:"Contemporary topics in immunobiology", I.P. Witz and M.G. Hanna Jr., eds., Plenum Publishing Corp., p.191 (1980).
11. R.S. Kerbel. Increased sensitivity of rosetting assay for Fc receptors obtained by using non-hemagglutinating monoclonal antibodies, J. Immunol. Meth. 34:1 (1980).
12. G.R. Braslawsky, D. Serban and I.P. Witz, Receptors for immune complexes on cells within a polyoma-virus-induced murine sarcoma, Eur. J. Immunol. 6:579 (1976).
13. G.R. Braslawsky, M. Ran and I.P. Witz, Tumor-bound immunoglobulins: The relationship between the in vivo coating of tumor cells by potentially cytotoxic anti-tumor antibodies and the expression of immune complex receptors, Int. J. Cancer 18: 116 (1976).
14. A. Kulczycki Jr., R. Schneider, V. Krause, C. Chew Killian, L. Solanki and J.P. Atkinson, Purification of biologically-active Fcγ receptor from macrophages, Fed. Proc. 39:799 (1980).
15. T. Suzuki, K. Hachimine and R. Sadasivan, Biochemical properties of biologically active Fcγ receptors isolated from human B lymphocytes, Fed. Proc. 39:463 (1980).
16. K. James, S. Davis and J. Merriman, Immunological and biochemical characteristics of acid citrate eluates from tumor cells: a major non-immunoglobulin component, Brit. J. Cancer 43:294 (1981).
17. R.S. Kerbel and A.J.S. Davies, The possible biological significance of Fc receptors on mammalian lymphocytes and tumor cells. Cell 3:105 (1974).

ACKNOWLEDGEMENT

This investigation was supported by Grant No. R01 CA 20078 awarded by the National Cancer Institute, DHHS.

IMMUNOGENETIC DETERMINANTS OF METASTATIC CELLS[1]

S. Katzav, P. De Baetselier[2], B. Tartakovsky,
E. Gorelik[3], S. Segal and M. Feldman

Department of Cell Biology, The Weizmann Institute of Science, Rehovot 76100, Israel

The progression of metastases is controlled by a series of cellular interactions between host and tumor cells. The generation of blood-bourne metastases is initiated by tumor cells capable of adhering to capillary basement membranes, followed by penetration, apparently via enzymatic degradation, of its proteins (1-3). Their survival in the circulation might be associated with their capacity to resist host immune responses and nonspecific defense mechanisms such as those mediated by NK cells. Indeed, we demonstrated that cells from lung metastases of the 3LL Lewis lung carcinoma are significantly less susceptible to NK cells than cells of the local tumor (4), and in vivo selection for NK resistance concomitantly selects tumor cells of a higher metastatic potency (5). The "selection" of organ sites for colonization by metastatic cells might be determined by the capacity of the tumor cells to recognize and thus specifically adhere to cells of a target organ (6, 7), although the progressive growth of such cells might necessitate vascularization induced in response to angiogenic factors secreted by the metastatic cells (8). Such a sequence of cellular interactions is most probably controlled by the cell surface properties of metastatic cells. Approaching the question of whether metastatic cells differ in the expression of cell-surface antigenic determinants from cells of the local tumor, we initially studied the 3LL Lewis lung carcinoma, a tumor which originated spontaneously in a C57BL (H-2^b) mouse. We found that

---

[1]Supported by PHS Grant No. CA 28139 awarded by the National Cancer Institute, DHHS, USA and by a grant from the Schilling Foundation, Essen, FRG.

[2]Dienst Algemene Biologie, Vrije Universitait Brussels, Belgium.

[3]Division of Surgery, National Cancer Institute, NIH, Bethesda, MD 20205, USA.

in vitro sensitization of syngeneic spleen lymphocytes against tumor cells from lung metastases resulted in effector lymphocytes which lysed metastatic target cells significantly more than target cells from the local tumor, whereas sensitization against the local tumor resulted in effector cells directed mainly against the local tumor cells (9, 10). These observations demonstrated that antigenic differences exist between the two populations of the 3LL cells. We found that such differences are recognizable by the tumor-bearing mice (10). Hence, metastatic spread might involve immunoselection of antigenic variants which could "escape" immune responses evoked by and directed against determinants of the local cells.

H-2 HAPLOTYPE EXPRESSION ON CELLS OF TUMOR METASTASES DIFFER FROM ITS EXPRESSION ON CELLS OF THE LOCAL TUMOR

Membrane associated antigenic determinants encoded by the major histocompatibility complex (MHC) play a crucial role in evoking T-cell mediated immune responses against cell surface antigens of syngeneic viral infected or malignant cells. The association, at the tumor cell surface, between MHC gene products and adjacent determinants, might constitute the entities recognized by T lymphocytes. Hence, the expression of MHC components on neoplastic cells could determine host-tumor immune relationships. Whether tumor cells would be subjected to immune rejection or whether they would evade immune destruction and disseminate to generate metastases might therefore depend on the expression of MHC gene products. In fact, Haywood and McKhann (11) demonstrated a correlation between expression of MHC encoded antigens on tumor cells and their capacity to generate metastases.

In view of the determining role of the MHC system in controlling the function of immune effector cells, it seemed of interest to investigate the expression of H-2 antigens on cells from the local tumor, as distinct from their expression on cells isolated from its metastases. For this purpose we chose the T10 sarcoma which was induced by methylcholanthrene in a (C3H/eb x C57BL/6) F_1 mouse (12). (Courtesy of Dr. J. Gordon, Montreal.) Intramuscular, intrafoot-pad and subcutaneous inoculation of T10 cells into syngeneic F_1 mice usually results in local growth, followed by the appearance of pulmonary metastases.

We first analyzed the expression of H-2 antigens of both H-2 parental haplotypes (i.e., H-2^k and H-2^b) on cells isolated from the local tumor (L-T10) versus cells isolated from metastatic nodules in the lungs (M-T10) (13, 14). This was carried out using fluorescent staining, and the relative numbers of brightly stained cells were determined by the fluorescence-activated cell sorter (FACS-II). The results (Table 1) demonstrate that the L-T10 tumor

TABLE 1. Parental H-2 haplotype expression on L-T10 and M-T10 tumor cells

Tumor cells	Relative number of cells stained with:	
	anti $H-2^b$	anti $H-2^k$
L-T20	85**	0
	93	24
	84	16
$M-T10_1$*	44	71
	61	86
	47	90
$M-T10_2$*	45	78
	37	75
$M-T10_3$*	40	71
	40	71

*$M-T10_1$, $M-T10_2$, $M-T10_3$ refer to cells of different metastatic nodules.

**The two or three different numbers in each column represents three successive serial transfers of L-T10 and M-T10 tumor cells.

cells express predominantly antigens encoded by the $H-2^b$ haplotype, whereas M-T10 tumor cells express both parental H-2 haplotypes, namely the $H-2^b$ and $H-2^k$ encoded antigens. The expression of the two parental H-2 haplotypes on M-T10 cells was found to be a stable phenomenon, since the expression of the two haplotypes was retained following three successive subcutaneous transplantations in syngeneic recipients.

The question then arose whether the differences in the expression of H-2 determinants on L-T10 versus M-T10 cells is due to modulation of H-2 expression during metastasis growth or whether the metastatic cells constitute a preexisting subpopulation within the local tumor (15, 16), which express differently the H-2 encoded antigens.

To approach this question the L-T10 tumor was cloned on semi-solid agar. Randomly chosen clones were transplanted into syngeneic F_1 recipients. Ten colonies were analyzed for the expression of $H-2^k$ and $H-2^b$ antigens. Two clones (IE7 and IB9) were found to express both $H-2^b$ and $H-2^k$, whereas the other 8 clones expressed predominantly the $H-2^b$ haplotype and were essentially $H-2^k$ negative (Table 2; 13, 14).

TABLE 2. Parental H-2 haplotype expression on L-T10 and T10 cloned tumor cells and their ability to produce experimental pulmonary metastases

Tumor cells	Percentage of cells brightly stained with Anti H-2^b	Anti H-2^k	Experimental lung metastases after i.v. inoculation
			(Lung weight, mg ± SE)
L-T10	84	23	300 ± 25
Clone IC9	66	14	212 ± 20
IG2	67	16	214 ± 25
IIF3	73	8	241 ± 50
IG3	81	11	243 ± 15
IB9	73	75	607 ± 75
IE7	91	93	947 ± 78
IF7	93	17	271 ± 27
IID6	77	14	214 ± 5
IID9	82	15	Not done
IB7	85	14	Not done

We then tested whether there is a correlation between the expression of the H-2 parental haplotypes and the capacity to generate experimental pulmonary metastases. For this purpose, cells of individual clones were injected intravenously (1 x 10^6 cells per recipient, 10 recipients in each experimental group), and two weeks later the animals were tested for lung metastases.

The results (Table 2) demonstrated that the two clones which were H-2^b and H-2^k positive (IE7 and IB9) were the only clones which produced lung metastases. Cells of the H-2^k negative clones did not generate experimental metastases in immune intact mice. They did however produce metastases when injected into animals which had been immunologically suppressed by total body irradiation (14). These metastases, however, did not express the H-2^k haplotype (Table 3), thus indicating that their formation, should be attributed to the immunosuppressed state of the recipient.

Thus, our experiments revealed that a) while metastatic cells (M-T10) express both parental haplotypes (i.e., H-2^k and H-2^b), the local tumor cells express mainly the H-2^b haplotype; b) the local T10 tumor consists of a heterogeneous population with regards to the expression of H-2. A predominant subpopulation is H-2^k negative and H-2^b postive whereas a smaller fraction is H-2^k and H-2^b positive. c) Only cells of the H-2^k and H-2^b positive clones produce

TABLE 3. Parental H-2 haplotype expression on T10 clones and their pulmonary metastatic cells isolated from irradiated mice

T10 clones	Treatment	Relative No. of cells stained with: Anti $H\text{-}2^b$	Anti $H\text{-}2^k$
IC9	Local tumor cells from nonirradiated mice	6.2	1.1
	Pulmonary metastatic cells from irradiated mice (450 R Co)	10	2.5
IID6	Local tumor cells from nonirradiated mice	57	8.3
	Pulmonary metastatic cells from nonirradiated mice	45	11

metastases when injected intravenously. This suggests that the metastatogenic cells form a preexisting subpopulation of cells within the local tumor cell population (15, 16), characterized by its H-2 phenotype.

TUMOR PROGRESSION: METASTATOGENIC $H\text{-}2^k$ POSITIVE CELLS APPEAR IN VIVO DURING TRANSFERS OF $H\text{-}2^k$ NEGATIVE CLONES

Having a tumor model in which related tumor clones differ in two properties, namely in H-2 expression and in the ability to produce experimental metastases, we turned to testing whether progression from low to high malignancy does take place and whether it is associated with alterations in the H-2 phenotype. To approach this we serially transplanted T10 cloned tumor cells in syngeneic F_1 mice and tested, at various transplant generations, the expression of H-2 antigens and their capacity to produce experimental pulmonary metastases (17). The results (Table 4) indicated that the IB9 and IE7 clones were stable with regards to both expression of the two parental haplotypes and the ability to generate experimental metastases. Changes did take place in the other initially nonmetastatic clones: IC9, IID6, IG3 and IF7 and in the uncloned L-T10 population. After the fifteenth in vivo passage, the $H\text{-}2^b$ positive IID6 clone cells expressed the $H\text{-}2^k$ encoded antigens and concomitantly acquired the capacity to generate metastases.

TABLE 4. Serial transfers of H-2^k negative clones resulted in the appearance of H-2^k positive cells concomitantly with the acquisition of metastatic capacity

T10 tumor cells	Generation*	Percentage of cells brightly stained with:		Experimental metastases after i.v. inoculation	
		Anti H-2^b	Anti H-2^k	Nodules in lungs	Lung weight (mg ± S.E.)
L-10	1	57	6	±	235 ± 12
	5	46	9	±	228 ± 12
	25	62	58	+++	599 ± 81
Clone IC9	1	66	14	-	212 ± 20
	5	21	0	-	223 ± 16
	15	8	14	N.d.	Not done
	25	6	8	-	249 ± 14
	35	58	45	+++	398 ± 80
Clone IID6	1	77	14	N.d.	Not done
	5	77	10	-	214 ± 5
	15	58	39	+++	559 ± 55
	25	86	90	++	409 ± 42
Clone IE7	1	97	73	+++	759 ± 59
	5	91	93	+++	947 ± 78
	15	87	95	+++	1031 ±123
	25	82	90	+++	1202 ± 70
Clone IB9	1	93	65	+++	704 ± 72
	5	73	75	+++	607 ± 75
	15	82	95	+++	800 ± 90
	25	81	93	N.d.	Not done
Clone IG3	1	81	11	-	243 ± 15
	27	79	94	N.d.	Not done
Clone IF7	1	93	17	-	271 ± 27
	27	79	94	++	436 ± 65

*Generation refers to number of in vivo passages in syngeneic F_1 mice.

- = no metastases; ++ = moderate involvement (multiple nodules), +++ = extensive involvement (numerous, confluent nodules).

A similar phenomenon was observed in the IG3 and IF7 clones and in the uncloned L-T10. The IC9 behaved in a different way. This clone lost the H-2^k haplotype on the third passage yet remained non-metastatic. Only at the 35th passage in vivo did these cells start to express both H-2 parental haplotypes and concomitantly acquired the ability to metastasize to the lungs when injected intravenously.

These results indicated that the progeny of non-metastatic tumor cells can acquire metastatogenic properties. Tumor progression towards increased malignancy has taken place in every clone tested. The capacity to generate metastases was shown to be intimately associated with the expression of H-2^k and H-2^b haplotypes. Since these changes took place within the local tumor, it appears that metastatogenic H-2^k variants have a selective advantage within the local subcutaneous tumor growth. Thus, tumor progression towards increased malignancy seems to take place within the local population.

The fact that cells of the IC9 clone, which lost the expression of H-2^b and were H-2^k negative, did not acquire metastatogenic properties until it expressed H-2 encoded antigens (H-2^k and H-2^b) suggests that the expression of the H-2 antigens is required for the metastatogenic process. Since in every case the expression of the H-2^k on the T10 tumor cells was associated with the co-expression of the H-2^b haplotype, it is impossible to attribute the metastatogenic capacity per se just to the H-2^k haplotype. It seems, however, that the metastatogenic property is associated with an active function of gene products of the H-2^k , since the loss of H-2 expression by cells of the IC9 clone on its 3rd transfer was not associated with the acquisition of metastatogenic capacity. Yet, the precise effect of such gene products which enables metastatic growth remains an open question. An inviting possibility is a suppressing effect which the H-2^k products expressed on metastatic cells and which might act on the host's immune system. This possibility gains support from the results of experiments we made on the growth of M-T10 cells in recipients of parental strains. We found (13) that H-2^b positive L-T10 cells grew progressively in C57BL (H-2^b) mice, but were rejected by C3H (H-2^k) mice. On the other hand, M-T10 cells expressing both the H-2^k and the H-2^b grew progressively in mice of both parental haplotypes. It appears, therefore, that the H-2^k expressed on the M-T10 tumor might exert a suppressing effect blocking the response against the tumor's alloantigens. Whether or not such a suppressing effect in semiallogeneic recipients is relevant to the effect exerted by the tumor's H-2^k haplotype in syngeneic recipients, enabling the cells to generate metastases, remains an open question.

HOW DO H-2^k NEGATIVE CELLS TURN INTO METASTATOGENIC H-2^k POSITIVE CELLS?

How are non-metastatic H-2^k negative cells turn into metastatogenic H-2^k positive cells? Two categories of mechanisms for the in vivo shift of the H-2 phenotypes could be considered: (a) A repressed H-2 locus within the H-2^k negative T10 cells is derepressed. This is associated with the acquisition of malignancy and a selective advantage within the original tumor cell population. (b) A T10 H-2^k negative cell acquires an expressable H-2^k chromosome via somatic hybridization in vivo with normal H-2^k positive host cells.

At present, there is no experimental evidence to support any one of these proposed mechanisms. The notion that somatic hybridization in vivo between tumor and normal cells might result in a shift from non-metastatic to metastatic cell was expressed by Goldenberg et al. (18), studying xenografts of human tumors. Such grafts rarely produced metastasis in the recipient animal. Yet, following transplantation of cells from human astrocytic glioma to cheek pouches of hamsters, tumor growth followed by the generation of metastases was observed. Karyotype analysis indicated that in the cases in which metastases were formed, the aneuploid tumor cells contained both human and hamster chromosomes. The authors claimed that somatic cell hybridization between the human tumor and normal hamster cells had taken place and that this cell fusion had conferred metastatogenic properties on the tumor cells. In fact, they proposed (18) that in vivo tumor progression in autochthonous organisms might involve such somatic hybridization. Indeed, there had been previous indications that somatic hybridization between neoplastic and normal cells can take place in vivo within the mouse system, although here no changes in the malignant properties, as a function of cell fusion, were reported (19).

In our laboratory, we carried out studies with the plasmacytoma NSI which, upon transplantation to syngeneic BALB/c recipients, grows locally but does not produce spontaneous metastases. When, however, such plasmacytoma cells were hybridized with spleen B cells of C57BL/6 origin, we obtained hybridomas which produced spontaneous metastases of distinct organ specificities (20). Some hybridomas produced spontaneous metastases in the spleen and in the liver, following trnasplantation to syngeneic recipients. Others produced only liver metastases. Tumor cells from liver metastases of hybridomas which generated both spleen and liver metastases when transplanted to normal animals, grew locally and again showed metastasis formation in the spleen and liver. Metastatic cells from livers of hybridomas which generated only liver metastases metastasized only to the liver when grafted to normal recipients (20). Thus, somatic hybridization in vitro resulted in the acquisition of metastatogenic properties.

Although these studies indicate that somatic hybridization between tumor and normal cells can confer metastatogenic properties on non-metastatogenic plasmacytoma cells, the question of whether the shift in vivo within non-metastatic H-2^k negative T10 clones to H-2^k positive malignant cells involves somatic hybridization with recipient cells must await further study.

REFERENCES

1. Klebe, H.K., Isolation of a collagen-dependent cell attachment factor. Nature 250:248 (1974).
2. Liotta, L.A., K. Tryggvason, S. Garbisa, I. Hart, C.M. Foltz and S. Shafie, Metastatic potential correlates with enzymatic degradation of basement membrane collagen. Nature 284: 67 (1980)
3. Murry, J.C., S. Garbisa and L. Liotta, The role of tumor cell basement interactions in the metastatic process. In: "Metastasis," p. 169, editors: Hellman, K., P. Hilgard and S. Eccles, Martinus Nijhoff, Publishers, The Hague, 1980.
4. Gorelik, E., M. Fogel, M. Feldman and S. Segal, Differences in resistance of metastatic tumor cells and cells from local tumor growth to cytotoxicity of natural killer cells. J. Nat. Cancer Inst. 63:1397 (1979).
5. Gorelik, E., M. Feldman and S. Segal, Selection of 3LL concomitantly selected for increased metastatic potency. Cancer Immunol. Immunother., in press.
6. Fidler, I.J. and G.L. Nicolson, Organ selectivity for implantation, survival and growth of B16 melanoma variant tumor lines. J. Natl. Cancer Inst. 57:1199 (1976).
7. Brunson, K.W., G. Beattie and G.L. Nicolson, Selection and altered tumour cell properties of brain colonizing metastatic melanoma. Nature 272:543.
8. Folkman, J. and C. Handeschild, C., Angiogenesis in vitro. Nature 288:551 (1981)
9. Fogel, M., E. Gorelik, S. Segal and M. Feldman, Differences in cell surface antigens of tumor metastases and those of the local tumor. J. Natl. Cancer Inst. 62:585 (1979)
10. Gorelik, E., M. Fogel, S. Segal and M. Feldman, Tumor associated antigenic differences between the primary and the descendant metastatic tumor cell population. J. Supramol. Struc. 12:385 (1980).
11. Haywood, G.R. and C.F. McKhann, Antigenic specificities on murine sarcoma cells: Reciprocal relationship between normal transplantation antigens (H-2) and tumor-specific immunogenicity. J. Exp. Med. 133:1171 (1971).
12. Brodt, P. and J. Gordon, Anti-tumor immunity in B lymphocyte deprived mice. I. Immunity to chemically induced tumor. J. Immunol. 121:359 (1978).

13. DeBaetselier, P., S. Katzav, E. Gorelik, M. Feldman and S. Segal, Differential expression of H-2 gene products in tumor cells is associated with their metastatogenic properties. Nature 288:179 (1980).

14. Katzav, S., P. DeBaetselier, E. Gorelik, M. Feldman and S. Segal, Immunogenetic control of metastasis formation by a methylcholanthrene-induced tumor (T10) in mice. Transpl. Proc. 13:742 (1981).

15. Fidler, I.J. and M.L. Kripke, Metastatis results from pre-existing variant cells within a malignant tumor. Science 197:893 (1977).

16. Kripke, M.L., E. Gruys and I.J. Fidler, Metastatic heterogeneity of cells from an ultraviolet light induced murine fibrosarcoma of recent origin. Cancer Research 38:2962 (1978).

17. Katzav, S., P. DeBaetselier, B. Tartakovsky, E. Gorelik, S. Segal and M. Feldman, Immunogenetic determinants controlling the metastatic properties of tumor cells. In: "Immunoregulation, 1981," editor N. Fabris, in press.

18. Goldenberg, D.N., R.A. Pavia and M.C. Tsao, In vivo hybridization of human tumor and normal hamster cells. Nature 250:649 (1974).

19. Wiener, F., E.M. Fenyo, G. Klein and H. Harris, Fusion of tumor cells with host cells. Nature New Biol. 238:155 (1972).

20. DeBaetselier, P., E. Gorelik, Z. Eshhar, Y. Ron, S. Katzav, M. Feldman and S. Segal, Hybridization between plasmacytoma cells and B lymphocytes confers metastatic properties on a nonmetastatic tumor. J. Natl. Cancer Inst. in press.

SPONTANEOUS PHENOTYPIC SHIFTS FROM LOW TO HIGH METASTATIC CAPACITY

Volker Schirrmacher, Peter Altevogt and Klaus Bosslet

Institut für Immunologie und Genetik am Deutschen Krebsforschungszentrum
D-6900 Heidelberg, FRG

INTRODUCTION

Tumor progression from low to high malignancy is believed to occur in multiple steps (1). Analogous to mutation/selection the process of tumor progression is thought to have its basis in the continuous emergence of successive clones of tumor cell variants, one gradually replacing another, through the intervention of natural or artificial selection pressures (2). Highly malignant metastasizing tumor cells have been shown to differ from low malignant non-metastasizing ones in a number of properties such as plasma membrane enzyme activities (3), cell surface antigen shedding (4), tumor antigenicity and immunogenicity (5). A critical question is whether all these specific properties have been accumulated in the metastatic cell in a stepwise fashion. Is each new property the result of a process of random variation and host selection? How can a random process result in the generation of cells endowed with not just one but a number of very specific properties that enable them to cross the various biological barriers of the host, to survive and grow in different microenvironments?

Recent results obtained in our own as well as in some other laboratories point out the importance of environmental factors in the evolution of cancer metastasis. The experiments show that cloned homogeneous tumor cell populations can adapt to changes in their microenvironment by changing their whole phenotype within a short time interval. Thus, Diamandopoulos reported that a transplantable SV40 induced hamster lymphocytic neoplasm expressing a single class ($7S_{\gamma 2}$) immunoglobulin, could be adapted to grow either as a leukemia or as a lymphoma. The former was characterized by systemic manifestations and poor prognosis while the latter was characterized by localization and favourable prognosis (6). Brouty-Boyé and Gresser recently demonstrated phenotypic reversion of transformed cells to the parental phenotype by subcultivation at low cell density (7) or by treatment with interferon (8). Minz and Illmensee reported that malignant teratocarcinoma cells depending on the microenvironment could either grow out as tumors or grow and differentiate like a normal cell (9).

We here intend to report on our observations of phenotypic changes observed in a particular tumor model system. It will be shown that shifts from low to high malignancy of a lymphoid tumor are associated with changes in the expression of lymphoid differentiation antigens. This might suggest similarities between the process of tumor progression and processes of normal cell differentiation or dedifferentiation. Finally, a model system will be presented which could explain some of the observed phenomena.

<u>Etiology of the tumor lines studied</u>

A tumor system under intensive investigation in our laboratory is the Eb/ESb system, a model set up to study mechanisms of metastasis formation and the role that cell-mediated anti-tumor immune reactions might play in this process. The etiology of the tumor lines is delineated in Table 1.

Table 1. Etiology of the Eb/ESb tumor system

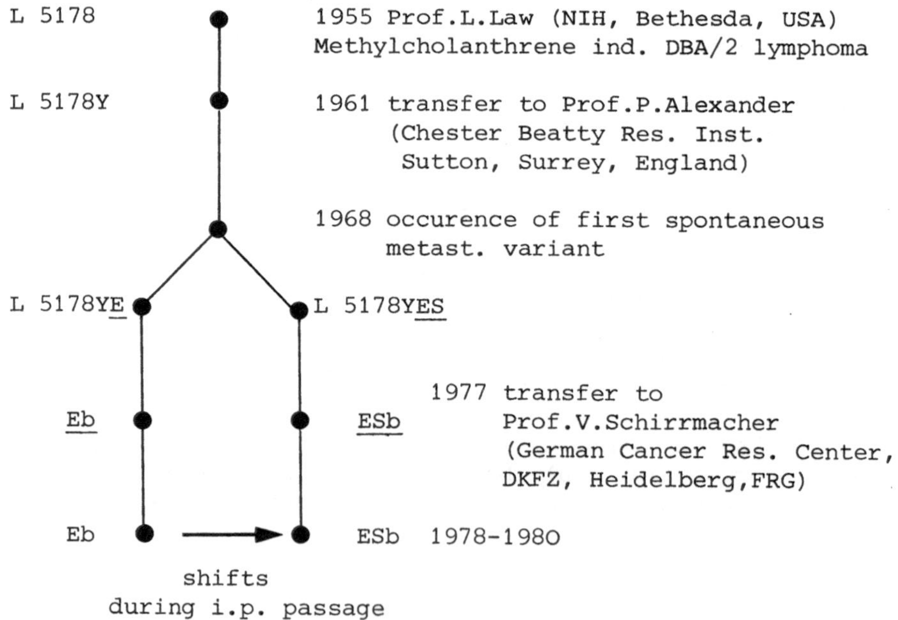

The tumor was induced in 1955 by Prof. L. Law by methylcholanthrene in a female DBA/2 mouse and was established as the well known lymphoma line L 5178. After transfer of the line to England, a spontaneous variant eventually arose in Prof. Alexander's laboratory during routine i.p. transplantation in syngeneic DBA/2 mice. The variant was noticed by its greatly increased ability to produce widespread internal metastases (liver, lung, spleen, bone marrow, brain etc.) and was kept separately as the subline L 5178YES (10). We obtained the low and high metastatic paired tumor lines in 1977 from Prof. Alexander and designated the Heidelberg sublines as Eb and ESb.

Characteristics by which the metastatic variant differs from the parental line Eb

While the parental line Eb morphologically and histologically resembles the type of a locally growing extra-nodal lymphoma with a prominent host cellular infiltrate, the ESb variant is more similar to a "reticularsarcoma" with a lower host cellular infiltrate. Electronmicroscopic studies revealed morphological differences between Eb and ESb with regard to the structure of the nucleus and that of the plasma membrane (11). Both tumor lines express differentiation antigens of T lymphocytes (Thy 1, Ly 1,T 200, T 145, T 130) showing that they belong to the T lymphoid series. With regard to differentiation antigens the Eb cells typed as Thy 1^{+}, Ly 1^{-}, Ly 2.3^{+}, Ly 6^{+}, while the ESb cells typed as Thy 1^{-}, Ly 1^{+}, Ly 2.3^{-}, Ly 6^{-} (12). The conversion from the Eb to the ESb phenotype was thus connected with losses and gains in the expression of defined lymphoid differentiation antigens.

ESb cells were found to differ from Eb cells not only in morphology and differentiation antigens but also in function as tested in two newly established in vitro assays. One demonstrates increased tumor invasiveness (13), the other the ability of ESb but not Eb cells to selectively bind to cultured syngeneic liver cells (hepatocyte-tumor cell rosette assay) (14). The metastatic variant showed also greatly increased expression and shedding of receptors for the Fc portion of IgG immunoglobulin molecules (15).

Most pertinent for our identification of Eb and ESb tumor phenotypes were our studies on the expression of tumor-associated transplantation antigens (TATA's). Both Eb and ESb were found to express TATAs, but these were distinct for each line and noncrossreactive (16). Highly specific cytolytic T lymphocytes (CTL) could be raised in syngeneic hosts against the TATA's (17) and these were used for routine type of Eb and ESb tumor lines for TATA expression.

Spontaneous shift from low to high malignancy during successive i.p. transplantation of Eb tumor cells

Ten years after the first appearance of the ESb variant in Prof. Alexander's laboratory (then designated as L5178YES) (10) we repeatedly observed a similar tumor variant which grew out from the parental tumor line Eb. We realized that upon successive i.p. transplantation the Eb line eventually went through a "crisis" state and became highly metastatic. Such Eb crisis lines were frozen in liquid nitrogen and later characterized for cell surface markers. Table 2 contains a list of Eb tumor lines which became highly metastatic upon i.p. transplantation. In an effort to get more stable parental type Eb cells we started to use carefully cloned cell lines. We soon observed, however, that the clones were as unstable as the uncloned Eb population. The passage numbers at which the uncloned or the once or twice cloned Eb lines shifted are indicated in Table 2.

Table 2. Summary of Eb tumor lines which became highly metastatic after successive i.p. transplantation.

Batch number	Origin	Precloned	Passage number in vivo (PA)	Year
L 5178YES[1)]	L5178YE	-	unknown	1968
no. 568	Eb-288	-	23	1978
no. 598	Eb-PA3	-	13	1979
no. 632	Eb-PA4	-	5	1979
no. 749	Eb-Cl4	once	13	1980
no. 661	Eb-Cl 34	once	12	"
no. 697	Eb-Cl 34.2	twice	6	"

[1)] Original spontaneous ESb variant, arisen in Prof. P. Alexander's laboratory from the L5178YE line.

All shifted Eb lines were found to produce metastases in various internal organs after s.c. inoculation and showed highly increased malignancy (18). The frequency of organ metastases raised from 2% (parental Eb line) to 77% in the shifted lines (Table 3).

Table 3. Metastasizing capacity of shifted Eb tumor cell lines.

Tumor line	Animals group	Metastases[1)] liver	lung	spleen	kidney	Frequency of metastases	%
Eb standard	10	1	0	0	0	1/40	2
Eb shifted	11	9	8	9	7	34/44	77
ESb standard	10	10	9	9	7	35/40	88

[1)] Metastases 12 days after s.c. inoculation of 10^5 tumor cells into syngeneic DBA/2 mice.

Cell surface marker changes during the penotypic shift

The original ESb line and the new Eb tumor variants were similar by several criteria: (i) formation of metastases, (ii) dependency of growth in tissue culture on 2-mercaptoethanol (19) and (iii) high expression of Fcγ receptors (15). However, the most convincing evidence, pointing to a similarity or even identity of the tumor variants was the expression of the individually distinct tumor antigen (TATA).

Table 4. Shifts of tumor antigen during i.p. passage of a twice cloned Eb cell line.

Tumor line	Passage number in vivo	in vitro	% specific cytotoxicity[1)] of CTL anti Eb	anti ESb	anti $H-2^d$
Eb-Cl 34.2	PA 1	PT 5	35	5	79
(736)	PA 3	PT 1	38	2	71
	PA 4	PT 3	25	3	75
	PA 5	PT 2	1	70	88
	PA 6	PT 2	3	67	77
Eb control			36	7	66
ESb control			4	77	80

[1)] % specific ^{51}Cr release in a 4 h assay at a 40:1 ratio of cytotoxic T lymphocytes (CTL) to target cells.

Table 4 shows the shift of the tumor antigen during i.p. passage of a twice cloned Eb cell line. For typing of the tumor antigen we used highly specific cytotoxic T lymphocytes (CTL) which were raised either against the parental line Eb or against the original ESb variant from 1968. The shift occured from the fourth to the fifth in vivo passage. Before that passage the cells were lysed only by anti-Eb CTL, while after the fifth passage the cells were lysed only by anti-ESb CTL (70% specific ^{51}Cr release). Similar results were obtained with the other shifted Eb lines. We previously showed that the tumor antigen recognized by anti-ESb CTL was unique for ESb and did not crossreact with antigens on a number of other syngeneic or allogeneic tumor lines (17).

Apart from the tumor antigen we investigated the tumor variants for expression of differentiation antigens and cell surface receptors. The results are summarized in Table 5.

Table 5. Cell surface marker changes on a shifted Eb cell line.

Tumor line	TATA[1)] Eb	TATA[1)] ESb	Thy1.2[2)]	T130	T145[3)]	$Fc_{\gamma}R$[4)]	SFV-[5)] R	Hep.-[6)] R
Eb control	+	-	+	+	-	-	+	-
ESb control	-	+	-	-	+	+	-	+
Eb-Cl 34.2 preshift	+	-	+	+	-	-	+	-
Eb-Cl 34.2 postshift	-	+	-	-	+	+	-	+
due to shift:	loss	gain	loss	loss	gain	gain	loss	gain

1) tumor-associated transplantation antigen (TATA) (17)
2) T-lymphoid differentiation antigen (12)
3) T-lymphoid differentiation glycoprotein (12)
4) receptor for immunoglobulin (15)
5) receptor for Semliki-Forest Virus (20)
6) receptor for syngeneic hepatocytes (14)

Compared to the Eb line before the shift, the shifted Eb variant was found to have gained some new markers (the $TATA_{ESb}$, the differentiation antigen T145, the Fc_{γ} receptor and a receptor for hepatocyte binding) and lost others (the $TATA_{Eb}$, the Thy 1 and T130 differentiation antigen and a receptor for Semliki-Forest Virus).

Interestingly, all of these phenotypic changes have been previously observed with the original ESb variant from 1968. The new emerging variants thus seem to be identical with the original one.

It is difficult to judge how often the Eb line had been transplanted between 1968 and 1978, how often it had been frozen (using DMSO) and how often it had been stored in liquid nitrogen. The fact that the high malignant variant could coexist with the parental line over such a long period without getting lost or growing out is remarkable. Eb/ESb tumor mixing experiments (18) indicated that as few as 10^1 ESb cells when mixed with 10^6 Eb cells and inoculated s.c. would grow out, metastasize and kill the animal. The possible contamination of the uncloned Eb population with ESb cells was therefore probably less than 10^{-5}. With such a low frequency it appears unlikely that the cloned Eb cell lines could still carry ESb type cells within them.

A model for inducible shifts in tumor cell phenotypes

The long-term coexistence of the variant with the parental line and the short-term shifts observed even with cloned Eb cell lines could perhaps be explained by the assumption that the ESb variant exists within the Eb population in a less malignant, hidden form which needs tumor promotion to develop into the metastatic ESb phenotype. Perhaps every cell of the Eb tumor has the potential to shift towards the ESb phenotype. This could mean that there is a preformed genetic program for the transition from Eb to ESb. According to this interpretation Eb and ESb would represent two different phenotypes of one common genotype.

A preliminary caryotype analysis was kindly performed by Dr. Andreas Radbruch (Institut für Genetik der Universität zu Köln, Köln, FRG). No significant differences were observed in the overall chromosome number of Eb and ESb cells, nor between Eb cells before and after the shift. When 10-15 metaphase plates were analyzed for each cell line, the average chromosome number was 40. The tumor lines Eb and ESb can both be considered as being pseudodiploid with several chromosomal translocations.

Based on our experience with the Eb/ESb tumor system we propose a model to explain shifts of tumor cell phenotypes. More detailes about the theoretical beckground can be found elsewhere (21). A generalized version is illustrated in Figure 1.

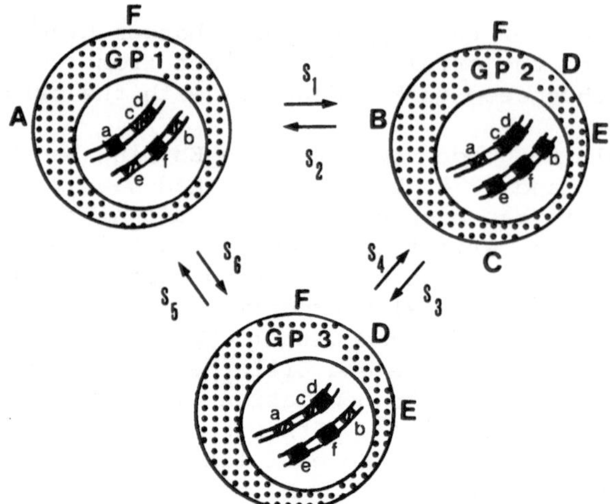

Figure 1. Model showing how microenvironmental signals can cause shifts in tumor cell phenotypes by activating preformed genetic programs.
a-f genes (▨ inactive or repressed; ■ active or derepressed).
A-F phenotypic markers, i.e. products of the corresponding genes.
GP1-3 genetic programs.
S_2-S_6 signals from the microenvironment which induce a shift.

The model states that the apparently coordinated change in a number of cell properties which occurs during a shift might follow a preformed genetic program (GP) which becomes activated by certain inductive signals (S) from the microenvironment. In this model, GP is a gene regulation program which not only dictates the pattern of genes which should be active, it also controls the amount of gene products to be made, their topographical arrangement on the cell surface and the time sequence in which genes are being switched on or off. The three phenotypes shown are just an example of the most stable forms in which a tumor, such as the Eb/ESb tumor, can express its genetic information. GP1 would code for the Eb phenotype, GP2 for the ESb phenotype and GP3 for an immunoadapted TATA negative variant (22). The signals from the microenvironment (S1-S6) are mostly hypothetical. They could be mediated through soluble factors (growth factors, hormones, cell aggregating factors, interferons etc.), through insoluble components of the intercellular matrices (fibrous proteins such as collagen, proteoglycans etc.)

or through direct cell-cell contact (with other tumor cells, cells of the surrounding tissue or cells of the host defense system). If A stands for $TATA_{Eb}$ and B for $TATA_{ESb}$, anti-A immune responses could mediate signals S_1 or S_6 and anti-B immune responses signals S_2 or S_3. C could represent the structure involved in hepatocyte binding, D Fc_{γ} receptors, E a membrane enzyme etc.

CONCLUSIONS

Metastasis has been described recently as a highly selective process (23-25). We suggest here that the influence of the host is not only selective but also inductive. The heterogeneity of tumor cells, their flexibility and the specialization observed in metastasizing tumor cells may be the results of both random and non-random processes: chemical carcinogens, irradiation etc. may affect the genome of their target cells more or less randomly, both before and after malignant transformation, and thus contribute to the generation of variants. Some of these variants may be gene-regulatory variants which have an increased ability to respond to inductive signals from the microenvironment. It is suggested that the ability of a tumor cell to metastasize and to survive depends on their capacity to re-use preformed genetic information which has been repressed during previous cellular differentiation. Although it seems that nothing can be generalized about tumors, because of their enormously variable potentials, a model taking into account inductive microenvironmental factors and the ability of cells to shift in their phenotype, can possibly explain more adequately the behaviour of tumor cells during the metastatic process: "Metastases do not result from random survival of cells released from the primary tumor, but from the selective growth of specialized subpopulations of highly metastatic cells endowed with specific properties that befit them to complete each step of the metastatic cascade" (23). Our model seems to fit with the impression of human cancer pathologists that highly malignant cells appear less differentiated than the tumor cells characterized by localization and favourable prognosis. The model also has an impact on understanding tumor cell heterogeneity in suggesting that part of it may be due to multiple phenotypes becoming expressed by one common genotype.

Observations of reversible shifts in tumor cell phenotypes, of losses and gains of cellular properties (e.g. tumor antigens, enzymes), are relevant for the biology and immunobiology of cancer metastasis in several respects: (i) they suggest that metastasizing tumor cells are not fixed, constant and static, but rather flexible, variable and dynamic entities able to react to and adapt to changing micro-environmental circumstances; (ii) they offer a new explanation for tumor heterogeneity and tumor progression; (iii) they allow to understand failures of more conventional approaches to the therapy of metastases; and (iv) they focus the attention on the importance of the microenvironment. The analysis of microenvironmental factors which might either progress or regress tumor development and their mechanism of action may become of paramount importance in future cancer research.

REFERENCES

1. P. C. Nowell, The clonal evolution of tumor cell populations, Science 194:23 (1976).
2. R. S. Kerbel, Implications of immunological heterogeneity of tumours, Nature 280:358 (1979).
3. S. K. Chatterjee, U. Kim and K. Bielot, Plasma membrane associated enzymes of mammary tumors as the biochemical indicators of metastasizing capacity. Analyses of enriched membrane fragments, Brit.J.Cancer 33:15 (1976).
4. P. Alexander, Escape from immune destruction by the host through shedding of membrane antigens: is this a characteristic shared by malignant and embryonic cells? Cancer Res. 34:2077 (1974).
5. E. V. Sugarbaker and A. M. Cohen, Altered antigenicity in spontaneous pulmonary metastases from an antigenic murine sarcoma, Surgery 72:155 (1972).
6. G. T. Diamandopoulos, Microenvironmental influences on the in vivo behavior of neoplastic lymphocytes, Proc.Nat.Acad.Sci. 76:6456 (1979).
7. D. Brouty-Boyé, I. Gresser and C. Baldwin, Reversion of the transformed phenotype to the parental phenotype by subcultivation of x-ray transformed C3H/10 T 1/2 cells at low cell density, Int.J.Cancer 24:253 (1979).
8. D. Brouty-Boyé and I. Gresser, Reversibility of the transformed and neoplastic phenotype. I. Progressive reversion of the phenotype of X-ray transformed C3H/10 T 1/2 cells under prolonged treatment with interferon, Int.J.Cancer 28:165 (1981).
9. B. Minz and K. Illmensee, Normal genetically mosaic mice produced from malignant teratocarcinoma cells, Proc.Nat.Acad. Sci. 72:3585 (1975).
10. I. Parr, Response of syngeneic murine lymphomata to immunotherapy in relation to the antigenicity of the tumor, Brit.J.Cancer 26:174 (1972).
11. V. Schirrmacher, G. Shantz, K. Clauer, D. Komitowski, H.-P. Zimmermann, M. L. Lohmann-Matthes, Tumor metastases and cell-mediated immunity in a model system in DBA/2 mice. I. Tumor invasiveness in vitro and metastases formation in vivo, Int.J.Cancer 23:233 (1979).
12. P. Altevogt, J. T. Kurnick, A. K. Kimura, K. Bosslet and V. Schirrmacher, Different expression of Lyt differentiation antigens and cell surface glycoproteins by a murine T lymphoma line and its high metastatic variant, Eur.J.Immunol. (in press 1981).

13. M.-L. Lohmann-Matthes, A. Schleich, G. Shantz and V. Schirrmacher, Tumor metastases and cell-mediated immunity in a model system in DBA/2 mice. VII. Interaction of metastasizing and non-metastasizing tumors with normal tissue in vitro, J.Natl. Cancer Inst. 64:1413 (1980)

14. V. Schirrmacher, R. Cheingsong-Popov, and H. Arnheiter, Hepatocyte-tumor cell interaction in vitro. I. Conditions for rosette formation and inhibition by anti H-2 antibody, J.Exp.Med. 151:984 (1980).

15. V. Schirrmacher and W. Jacobs, Tumor metastases and cell-mediated immunity in a model system in DBA/2 mice. VIII. Expression and shedding of Fcγ receptors on metastatic tumor cell variants, J.Supramol.Struct. 11:105 (1979).

16. V. Schirrmacher, K. Bosslet, G. Shantz, K. Clauer and D. Hübsch, Tumor metastases and cell-mediated immunity in a model system in DBA/2 mice. IV. Antigenic differences between the parental tumor line and its metastasizing variant, Int.J.Cancer 23:245 (1979).

17. K. Bosslet, V. Schirrmacher, and G. Shantz, Tumor metastases and cell-mediated immunity in a model system in DBA/2 mice. VI. Similar specificity patterns of protective anti-tumor immunity in vivo and of cytolytic T cells in vitro, Int.J.Cancer 24:303 (1979).

18. V. Schirrmacher and K. Bosslet, Tumor metastases and cell-mediated immunity in a model system in DBA/2 mice. X. Immunoselection of tumor variants differing in tumor antigen expression and metastatic capacity, Int.J.Cancer 25:781 (1980).

19. V. Schirrmacher, D. Hübsch and K. Clauer, Tumor metastases and cell-mediated immunity in a model system in DBA/2 mice. IX. Radioassay analysis of tumor cell spread from a local site to the blood and liver, in: "Metastatic Tumor Growth, Cancer Campaign," E. Grundmann, ed., Gustav Fischer Verlag 4:147 (1980).

20. D. Barz, K. Bosslet and V. Schirrmacher, Metastatic tumor cell variants with increased resistance to infection by Semliki Forest Virus, J.Immunology 127:951 (1981).

21. V. Schirrmacher, Shifts in tumor cell phenotypes induced by signals from the microenvironment: Relevance for the immunobiology of cancer metastasis, Immunobiology 157:89 (1980).

22. K. Bosslet and V. Schirrmacher, Escape of metastasizing clonal tumor cell variants from tumor-specific cytolytic T lymphocytes, J.Exp.Med. 154:557 (1981).

23. G. Poste and I. J. Fidler, The pathogenesis of cancer metastasis, Nature 283:139 (1980).

24. I. J. Fidler and M. L. Kripke, Metastasis results from pre-existing variant cells within a malignant tumor, Science 197:893 (1977).

25. G. I. Nicolson, Cancer Metastasis, Scientific America 240:50 (1979).

HISTOCOMPATIBILITY ANTIGENS ON THE CELL SURFACE OF TUMORS

Giorgio Parmiani[1] and Giuseppe Della Porta[2]

Divisions of Experimental Oncology B[1] and A[2]
Istituto Nazionale Tumori
Via Venezian, 1 - 20133 Milan, Italy

1. HISTOCOMPATIBILITY ANTIGENS ON NORMAL CELLS: TISSUE DISTRIBUTION AND VARIATION OF THEIR EXPRESSION

The histocompatibility antigens (HA) encoded by the MHC are a high polymorphic family of cell surface molecules whose function seems to have evolved to regulate the immune response of T lymphocytes to foreign antigens (Snell et al.,1976). The expression of the class I antigens of the MHC (i.e.: K, D or L gene products for the mouse and HLA-A, B and C for man) is widespread and virtually all the adult tissues studied display these antigens with the possible exception of placenta and B-cells of pancreas (Parr et al.,1979). A more selected tissue distribution is that of class II antigens, i.e. Ia for mice and DR for men.

It should be noted, however, that considerable variations have been reported in the quantitative expression of class I antigens in different tissues.Spleen and other lymphoid organs have the highest expression of class I HA, whereas brain, testis and skeletal muscles have the lowest expression of such HA (Klein J., 1975). In lymphoid cell populations other variations have been described, with B cells expressing more D code HA than T cells (Tartakovsky et al., 1980). Differences in HA have been reported also in the same tissues of different individuals, thus indicating the existance of a genetic control over this antigenic expression (O'Neill, 1980). This variability may then lead to differences in the effectiveness of the T cell-mediated immune control of several external pathogens (O'Neill and Blanden, 1979) eventually resulting in a different susceptibility of individuals and, within tha same individual, of different body tissue to diseases.

2. EXPRESSION OF HA ON NEOPLASTIC CELLS OF MURINE AND HUMAN ORIGIN

Tumors tend to display the HA profile of the normal cells from which they derive. The growth rate or the growth fraction, however, may be considerably different in a tumor tissue as compared to a normal one, and this may influence the expression of several cell-surface structures, including HA (Cikes et al., 1973; Curry et al., 1979). Genetically derived abnormalities of HA expression on cancer cells, however, have been also reported and will be discussed below.

2.1 Quantitative alterations of HA expression on murine and human tumors

Information on the quantitative expression of MHC products on tumor cells is sparse. This is likely to reflect the inherent difficulty of methodology for measuring the presence of HA molecules on tumor cells and the frequent lack of proper normal control cells (i.e.: cells of the same individual and tissue from which the tumor derives), especially in the study of human neoplasms. Several reports, however, can be found in the literature indicating either a reduction and, more rarely, an increase in the expression of MHC products on experimental tumors (*see* Parmiani et al., 1979a), sometimes inversely correlated with the presence of tumor-associated antigens. The use of monoclonal reagents has now open the possibility of a more detailed analysis of the various epitopes of the HA molecules. Variant phenotypes of HA have been reported on tumor cells that may be defined by a loss of a given antigenic determinant (Holtkamp et al., 1981).

Quantitative alterations in the expression of some HLA antigens have been described also on human neoplasms (see Callahan et al., 1978). Partial loss of HLA was reported in cultured human tumors of various types (Takasugy and Terasaki, 1972; Pollack et al., 1980) and a complete loss of HLA-A,-B,-C antigens was found in two choriocarcinoma lines (Trowsdale et al., 1980); other investigations described a normal presence of HLA on melanoma cells when compared to fibroblasts or lymphoid cells of the same patient, but a variable situation (either an increase or a reduction) in sarcoma or carcinoma cells (Callahan and Ferrone, 1978). It is, therefore, difficult to draw clear conclusions, but it is possible to infer from the available evidence that regulation of expression of MHC gene products is often disturbed on cancer cells.

Quantitative alterations of expression of MHC products on cancer cells may have several important biological consequences, including that of an impairment in the host's recognition of tumor-antigens by lymphoid T cells. The lack of CTL immunogenicity of teratocarcinoma cells devoid of H-2 antigens is an example of the relevant role of the MHC products in the recognition of tumor antigens (Golstein et al., 1976). It has also been shown that differential expression of H-2 antigens in variants of a tumor cell population affects the metastatic behaviour of cancer cells (De Baetselier et al., 1980).

2.2 Qualitative alterations in HA expression on tumor cells

Qualitative modifications in the expression of *K*, *D* or *I* region products on tumor cells of mice have been described and reviewed (*see* Parmiani et al., 1979b). Expression of K, D and Ia antigens of foreign haplotypes have been reported on several types of neoplastic cells. Some of the serological data, however, were subsequently interpreted as indicating the presence of previously undetected public antigens on tumor cells, i.e. antigens which could be found by proper techniques also on normal counterparts (Flaherty and Rinchik, 1978, Robinson and Schirrmacher, 1979). The controversial aspects of this area of investigation have been recently discussed (*see* Parmiani et al., 1981), and an unequivocal evidence for the presence of extra HA either of the MHC or of the non-H-2 system was found to be limited to a few tumors, such as chemically induced sarcomas of BALB/c strain and spontaneous reticulum cell sarcomas (RCS) of SJL strain (Table 1).

Table 1.
Tumor systems reported to express extra MHC or non-MHC alloantigens.

Species	Tumor system	Strain of origin	Normal haplotype	Extra antigens	References
MOUSE	RCS	SJL	$H-2^s$	$H-2^{d,b}$, Ia^d	Wilbur & Bonavida, '81
	RCS	SJL	$H-2^s$	$H-2^{d,b}$	Finke et al., '80
	RCS	SJL	$H-2^s$	$H-2^{d,b,p}$	Beisel, '81
	Lung tumors	C3H/HeN C57BL/6	$H-2^k$ $H-2^b$	kv1	Callahan & Martin, '81
	K36	AKR	$H-2^k$	D^d	Schmidt et al.,'80
	LBN	BN/a,b	$H-2^{b,p}$	H-2d	Czarnomska & Capkowa, '80
	MCG4	BALB/c	$H-2^d$	$H-2^{k,b}$	Schirrmacher et al., '80
	C-1	BALB/c	$H-2^d$	K^k,D^k	Parmiani et al., '79
	MM2	C3H/He	$H-2^k$	D^b	Kubata & Manson, '80
	MM102	C3H/He	$H-2^k$	non-H-2	Fujiwara et al., '78
	P815	DBA/2	$H-2^d$	non-H-2	Russel et al.,'79
	P388	DBA/2	$H-2^d$	non-H-2	
	YC8	BALB/c	$H-2^d$	non-H-2	Parmiani & Sensi, '81
CHICKEN	ROUS	White Leghorn	B^2B^2	B^5B^5	Heinzelman et al., '81
MAN	BrCa		HLA-A3,-B5	HLA-A9	Callahan & Ferrone, '78
	TCC		?	HLA-B14	O'Toole, '81
	AdCa colon		?	HLA-A9, -B7,BW16 BW22, BW35	Hirai et al., '80

Two chemically induced BALB/c sarcomas have been described which are an example of unexpected expression of alien H-2K^k and D^k molecules on a H-2^d cell as defined by serology, cell-mediated cytotoxicity and biochemistry (Parmiani et al., 1979b; Schirrmacher et al., 1980), whereas the RCS of SJL strain provide the interesting case of neoplastic cells of H-2^s haplotype displaying Ia^d-like molecules absent from the normal lymphocytes of that strain which are unable to synthetize Eα gene products (Wilbur and Bonavida, 1981).

More recently, the presence of an apparently alien gene product of the B locus (the MHC of chicken) has been reported on Rous virus-induced sarcomas of chicken (Heinzelmann et al.,1981). It is worth to note that in this system the expression of this additional B antigen can be reproducibly obtained by *in vitro* infection of fibroblasts with the virus, whereas in the mouse system such a reproducibility has never been achieved.

As for human tumors, evidence in favor of the expression of genetically inappropriate products of *HLA-A*,-*B* and -*C* loci on neoplastic cells, has been provided in isolated examples of breast and colon carcinoma lines (Callahan et al., 1978; Hirai et al., 1980) and in 6 out of 8 transitional cell carcinomas of the bladder (O'Toole, 1981), but in no case a biochemical analysis of these extra HLA antigens has been reported.

An indirect evidence for the presence of undefined extra HLA antigens on human leukemias was provided by the work or Zarling et al. (1978) who showed that *in vitro* sensitization of peripheral blood lymphocytes with a pool of 20 HLA different normal allogenic lymphocytes, induced a killing activity on autologous tumors. The serological study of HLA profile of other tissue cultures lines of human tumors, however, failed to reveal the presence of alien HLA-A, -B or -C antigens, although abnormal reactions were often noted in the direct cytotoxicity tests which are not confirmed by absorption experiments (Pollack et al., 1980; Curry et al., 1979, McAlack, 1980).

More consistent is the finding of the presence of DR antigens on human melanoma cell lines deriving from normal cells that should not express these class II antigens. Extensive serological and biochemical analysis of this phenomenon has been carried out (Winchester et al., 1978; Wilson et al.,1979; Natali et al., 1981) using both conventional antisera a monoclonal antibodies (Table 2).

It has been shown that also other solid human tumors, beside melanomas, expose DR antigens and that their presence seems to correlate with the degree of differentiation of tumor cells (Natali et al., 1981).

Table 2. Expression of DR-like antigens on cancer cells[a].

Malignant tumor of	Expression of DR antigens as compared to normal cells	Reference
OVARY	Unchanged	Natali et al., '81
STOMACH		
COLON		
UTERUS		
THYROID		
SKIN		
URINARY BLADDER		
CERVIX		
ESOPHAGUS		
BREAST	Reduced or abolished	
MELANOMAS	Increased	Winchester et al., '78,Wilson et al.,79
LUNG		Natali et al., '81
RECTUM		
LIVER		
BRAIN		

[a]Modified from Natali et al. (1981)

Since DR antigens are the human counterparts of Ia which, in the mouse, have been show to play an essential role in antigen recognition by T lymphocytes (Thomas et al., 1977), the presence of DR on neoplastic cells may drastically modify the host-tumor immune relationship (Forni et al., 1975). The putative role of unexpected DR molecules on human cancer cells, however, has still to be worked out by experimental means.

3. HISTOCOMPATIBILITY ANTIGENS AND TUMOR ANTIGENS

It is known that most experimental tumors of rodents

induced either by chemicals or by viruses express tumor--associated cell surface antigens (TSA) which can be revealed by transplantation or serological techniques. Whether tha same situation applies to human tumors is matter of debate but, at least in a fraction of them,TSA could be defined by serology and partially characterized biochemically (Carey et al., 1979; Imai et al., 1981; Schnegg et al., 1981). Then the question rises of the possible relationship between a novel surface structure (TSA) and the normal antigens of the cancer cells, namely the HA.

Three types of ralationships can be considered to occur between tumor antigens and HA: genetic, structural and functional (see Table 3).

Table 3. Possible structural relationship between HA and tumor-associated cell surface antigens (TSA)

- TSA as alien, genetically inappropriate MHC products
- TSA as alien, genetically inappropriate MiHA
- TSA as modified MHC products
- TSA as complex antigen formed by MHC product + X
- TSA as complex antigen formed by β_2-M + X
- TSA as modified normal, non-HA cell surface structure

The possibility that TSA are products of the MHC has been disproved both in experimental systems (Parmiani et al., 1979a; Law et al., 1980) and in human tumors (Carey et al., 1979; Curry et al., 1979; Imai et al., 1981). The genetic relationship between tumor antigens and minor HA, however, have yet to be examined in view of the fact that some authors have suggested that TSA may be an altered minor HA (Fujiwara et al., 1978; Parmiani et al., 1979; Russel et al., 1979).

As for the structural relationships between tumor antigens and HA (see a possible list in table 3), reports suggesting that TSA determinants are borne by the heavy chain of H-2 or HLA molecules or are associated with β_2-microglobulin (Callahan et al., 1978; Thomson et al., 1978) have been contradicted by findings on the lack of such structural associations (Bowen and Baldwin, 1979; Carey et al., 1979; Law et al., 1980).

Since the discovery that MHC regulates the T cell response to extrinsic antigens (Doherty and Zinkernagel, 1975) many efforts have been devoted to explore the possibility that also tumor antigens can be seen by T cells in association with the products of MHC. This functional link between TSA and MHC has been found for antigens of virally and chemically induced tumors (Gomard et al., 1976; Trinchieri et al., 1976; Lannin et al., 1981) but it appears to be still elusive for human neoplasms.

CONCLUSIONS

Malignant cells of different species, including man, often appears to have abnormalities in the qualitative and quantitative expression of products of MHC, including class I and class II antigens. The mechanism of this alteration has not been elucidated, with the exception of a few studies of experimental tumors indicating a repression or a derepression of genes coding for H-2 antigens (Parmiani et al., 1979a). Taking into account the many biological functions attributed to the MHC system (Klein J., 1975) which include not only the regulation of the immune response by lymphocytes, but also the mutual recognition among non-lymphoid cells, the receptor function for viruses or other pathogens, to mention a few of them, one can easily understand how an alteration in the expression and function of MHC-encoded molecules may have profound effects on the relationship between the emerging cancer cells and its natural environment. The frequency of such alterations, however, has not been established and, therefore, the biological significance of these observations remains purely speculative.

ACKNOWLEDGMENTS

The authors work is supported by C.N.R. Grant No. 80.01607.96 of the Finalized Project "Control of Tumor Growth".

REFERENCES

Beisel,K.,1981, Transplant. Proc. (in press).

Bowen,J.G., and Baldwin, R.W.,1979, Tumour antigens and alloantigens.II.Lack of association of rat hepatoma--D23-specific antigen with β_2-microglobulin, Int.J. Cancer, 23:833.

Callahan, G.N., Pellegrino, M.A., McCabe, R.P., Frugis, L., Allison, J.P., and Ferrone, S., 1978, Histocompatibility antigens on tumor cells: spatial and structural relationship with tumor associated antigens, Bëhring Inst. Mitt., 62:115.

Callahan, G.N., Walker, L.E., and Martin, W.J., 1981, Biochemical comparison of H-2K antigen isolated from C3HfB/HeN and C3H/HeN mice, Immunogenetics, 12:561.

Carey, T.E., Ko, L., Takahashi, T., Travassos, L.R., and Old, L.J., 1979, AU cell surface antigen of human malignant melanoma. Solubization and partial characterization, Proc.Natl.Acad.Sci.USA, 76:2898.

Cikes,M., Friberg, S.Jr., and Klein, G., 1973,Progressive loss of H-2 antigens with concomitant increase of cell-surface antigen(s) determined by Moloney leukemia virus in cultured murine lymphomas, J.Nat.Cancer Inst., 50:347.

Curry, R.A., Quaranta, V., Wilson, B.S., McCabe, R.P., Natali, P.G., Pellegrino, M.A., and Ferrone, S., 1979, Expression of HLA antigens on cultured human melanoma cells: lack of association with melanoma-associated antigens, in: "Current trends in tumor immunology", S.Ferrone, S.Gorini, R.B.Herberman and R.A.Reisfeld, eds., Garland Press, New York.

Czarnomska, A., Capkova, J., and Démant, P., 1980, Change of H-2 antigens expression on mouse leukemia LBN/a-2 and LBN/b-3 cells in the course of serial transplantation, J.Immunogenetics, 7:39.

De Baetselier, P., Katzav, S., Gorelik, E., Feldman, M., and Segal, S., 1980, Differential expression of H-2 gene products in tumor cells is associated with metastogenic properties, Nature, 288:179.

Doherty, P.C., and Zinkernagel, R.M., 1975, H-2 compatibility is required for T cell mediated lysis of target cells infected with lymphocytic choriomeningitis virus, J.Exp.Med., 141:502.

Finke, J.H., Fyfe, D.A., Del Villano, B.C., Butler, G.H., and Ponzio, N.M., 1980, Characterization of "foreign" alloantigen-like specificities on a murine lymphoma cell line. Transpl.Proc., 12:53.

Flaherty, L., and Rinchik, E., 1978, No evidence for foreign H-2 specificities on the EL 4 mouse lymphoma, Nature, 273:52.

Forni, G., Sheach, E., and Green I., 1975, Mutant lines of guinea pig L2C leukemia .I. Deletion of Ia alloantigens is associated with a loss in immunogenicity of tumor-associated transplantation antigens, J.Exp.Med., 143:1067.

Fujiwara, H., Aoki, H., Tsuchida, T., and Hamaoka, T., 1978, Immunologic characterization of tumor-associated transplantation antigens on MM102 mammary tumor eliciting preferentially helper T cell activity, J. Immunol., 121:1591.

Golstein, P., Kelly, F., Arner, P., and Gachelin, G., 1976, Sensitivity of H-2-less target cells and role of H-2 in T cell mediated cytolysis, Nature, 262:693.

Gomard, E., Duprez, V., Henin, Y., and Levy; J.P., 1976, H-2 region product determinant in immune cytolysis of syngeneic tumor cells by anti-MSV T lymphocytes, Nature, 250:707.

Heinzelmann, E.W., Zsigray, R.M., and Collins W.M., 1981, Cross-reactivity between RSV-induced tumor antigen and B_5 MHC alloantigen in the chicken, Immunogenetics, 13:29.

Hirai, T., Yamando, H., and Hamaoka, T., 1980, Relationship between carcinoembryonic antigen and major histocompatibility antigen on cultured human carcinoma cells, J.Immunol., 124: 2765.

Holtkamp, B., Cramer, M., Lemke, H., and Rajewsky, K., 1981, Isolation of a cloned cell line expressing variant $H\text{-}2K^k$ using fluorescente-activated cell sorting, Nature, 289:66.

Imai, K., Galloway, D.R., and Ferrone, S., 1981, Serological and immunochemical analysis of the specificity of xenoantiserum 8986 elicited with hybrids between human melanoma cells and murine fibroblasts, Cancer Res., 41:1028.

Klein, J., 1975,"Biology of the mouse histocompatibility--2 complex",Springer-Verlag, Berlin-New York.

Kubota, K., and Manson, L.A., 1981, Characterization of H-2K, D-like structures on MM2 x mouse L cell hybrids. I. H-2K, D-like alloantigen encoded by the D region and/or to the right of the D region of the H-2 gene complex, Int.J.Cancer, 27:537.

Lannin, D.R., Yu,S., and McKahn, C.F., 1981, T cells must recognize tumor antigen in association with self-MHC antigen. Transpl.Proc., 13:739.

Law,L.W., Rogers,M.J., and Appella, E., 1980, Tumor antigens on neoplasms induced by chemical carcinogens and by DNA- and RNA- containing viruses: properties of the solubilized antigens. Adv.Cancer Res.,32:201.

McAlack, R.F., 1980, Normal HLA phenotypes and neo-HLA--like antigens on cultured human neuroblastomas. Transpl.Proc., 12:107.

Natali, P.G., De Martino, C., Quaranta, V., Bigotti, A., Pellegrino, M.A., and Ferrone, S., 1981, Changes in Ia-like antigen expression on malignant human cells, Immunogenetics, 12:409.

O'Neill, H.C., and Blanden, R.V., 1979, Quantitative differences in the expression of parentally-derived H-2 antigens in F1 hibrid mice affect T-cell responses, J.Exp.Med., 149: 724.

O'Neill,H.C., 1980, Quantitative variation in H-2 antigen expression.II.Evidence for a dominance pattern in H-2K and H-2D expression in F1 hybrid mice. Immunogenetics,11:241.

O'Toole, C., 1981, HLA antigen on human transitional cell carcinomas, Immunobiol., 159:189.

Parmiani, G., Carbone, G., Invernizzi, G.,Meschini, A., and Della Porta, G., 1979, Expression of genetically inappropriate histocompatibility antigens on the cell-surface of experimental tumors and their relationship to tumor-associated transplantation antigens, in: "Tumor associated antigens and their specific immune response", F.Spreafico and R.Arnon, eds., Academic Press, London.

Parmiani, G., Carbone, G., Invernizzi, G., Pierotti, M.A., Sensi, M.L., Rogers, M.J., and Appella, E.,1979, Alien histocompatibility antigens on tumor cells, Immunogenetics, 9:1.

Parmiani, G., Ballinari, D., Carbone, G., Cattaneo,M.,

Pierotti, M.A., Sensi, M.L., and Rogers, M.J., 1981, Extra H-2 specificities on tumor cells, Proc.XIV Int. Leukocyte Cult.Conf. (in press).

Parmiani, G., and Sensi,M.L., 1981, Inhibition of syngeneic lymphoma growth by alloimmunization with normal lymphoid cells, Transpl.Proc. (in press).

Parr,E.L., 1979, The absence of H-2 antigens from mouse pancreatic β-cells demonstrated by immunoferritin labeling, J.Exp.Med., 150:1.

Pollack,M.S., Heagney, S., and Fogh, J., 1980, HLA typing of cultured human tumor cell lines: the detection of genetically appropriate HLA-A,B,C and DR alloantigens, Transpl.Proc., 12:134.

Robinson, P.J., and Schirrmacher,V.,1979, Differences in the expression of histocompatibility antigens on mouse lymphocytes and tumor cells: immunochemical studies, Eur.J.Immunol., 9:61.

Russell,J.H.,Ginns,L.C., Terres,G., and Eisen,H.N., 1979, Tumor antigens as inappropriately expressed normal alloantigens, J.Immunol., 122:912.

Schirrmacher,V., Garrido,F., Garcia-Olivares,E., and Koszinowski,U., 1980, Alien H-2-like molecules on a murine tumor (HCG4): target antigens for alloreactive cytolytic T lymphocytes (CTL) and restricting elements for virus specific CTL, Transpl.Proc., 12:32.

Schmidt,W., and Festenstein,H., 1980, Serological and immunochemical studies of H-2 allospecificities on K36, a syngeneic tumor of AKR, J.Immunogen.,7:17.

Schnegg,J.F., Diserens,A.C., Carrel,S., Accolla, R.S., and De Tribolet,N., 1981, Human glioma-associated antigens detected by monoclonal antibodies, Cancer Res., 41:1209.

Snell, G.D., Dausset, J., and Nathenson,S., 1976, "Histocompatibility", Academic Press, New York.

Takasugy,M., and Terasaki, P.I., 1972, Detection of HL-A and other cell-surface antigens on cultured cells by a cytotoxic plating inhibition test, J.Natl.Cancer Inst., 49:1229.

Tartakovsky,B., De Baetselier,P., and Segal, S., 1980, Serological detection of H-2K and H-2D- gene products. I. Principal difference between T and B lymphocytes in expression of H-2D-encoded alloantigens.Immunogenetics, 11:6.

Thomas,D.W., Yamashita,U., and Shevach,E.M., 1977, The role of Ia antigens in T cell activation. Immunol. Rev., 35:97.

Thomson,D.M.P., Rauch,J.E., Weatherhead,J.C.,Friedlander, P., O'Connor,R., Grosser,N.,Shuster,J., and Gold,P., 1978, Isolation of human tumor-specific antigens associated with β_2-microglobulin, Br.J.Cancer, 37:753.

Trinchieri,G., Aden,D.P., and Knowles,B.B., 1976, Cell--mediated cytotoxicity to SV40-specific tumour--associated antigens, Nature, 261:312.

Trowsdale,J.,Travers,P., Bodmer,W.F., and Patillo, R., 1980, Expression of HLA-A,-B, and -C and β_2-microglobulin antigens in human choriocarcinoma cell lines, J.Exp.Med., 152: 11s.

Wilbur,S.M., and Bonavida, B., 1981,Expression of hybrid Ia molecules on the cell-surface of reticulum cell sarcomas that are undectable on host SJL/J lymphocytes, J.Exp.Med., 153:501.

Wilson,B.S., Indiveri,F., Pellegrino, M.A., and Ferrone, S., 1979, DR (Ia-like) antigens on human melanoma cells.Serological detection and immunochemical characterization, J.Exp.Med., 149:658.

Winchester, R.J., Wang,C.Y., Gibofsky,A., Kunkal,H.G., Lloyd,K.O., and Old,L.J., 1978, Expression of Ia-like antigens on cultured human malignant melanoma cell lines, Proc.Nat.Acad.Sci.,USA, 75:6235.

Zarling,J.M., and Bach,F.H., 1978, Sensitization of lymphocytes against pooled allogeneic cells.I.Generation of cytotoxicity against autologous human lymphoblastoid cell lines, J.Exp.Med., 147:1334.

THE IMMUNOGENETICS OF MHC CONTROLLED ANTIGENIC DETERMINANTS ON LEWIS-LUNG-CARCINOMA (3LL) CELLS[1]

N. Isakov[2], S. Katzav, P. De Baetselier[3], B. Tartakovsky, M. Feldman and S. Segal

Department of Cell Biology, The Weizmann Institute of Science, Rehovot 76100, Israel

The transplantation antigens controlled by the major histocompatibility complex (MHC) of mammals play a cardinal role in the control of immune processes. These alloantigens determine the ability to evoke effector cells directed against antigens expressed on malignant cells transformed by either chemical or viral carcinogens (1-4). MHC-encoded components may also control the qualitative outcome of an immune process directed against a given antigen on a certain malignant target cell. Thus, the preferential elicitation of antibody-producing cells, suppressor cells, killer cells, or lymphokines producing T cells against a cell-surface antigen may depend on the nature of the neighboring MHC components that form a complex with tumor antigens. Therefore, to elucidate immune tumor—host relations, one has to analyze both quantitatively and qualitatively whether MHC components expressed on tumor cells are different in their immunogenic properties from those expressed on normal somatic cells. Such differences might be relevant to the capacity of tumor cells to attack the immune barriers of the host and disseminate from its original site of growth to distant anatomical locations.

---

[1]Supported by PHS Grant No. CA 28139 awarded by the National Cancer Institute, DHHS, USA and by a grant from the Schilling Foundation, Essen, FRG.

[2]Present address: Immunobiology Center, Box 724, Mayo Memorial Building, Minneapolis, MN 55455, USA.

[3]Present address: Dienst Algemene Biologie, Insitute for Molecular Biology, Free University of Brussels, Belgium.

Table 1. Mouse strains used in this study and their H-2 haplotypes

<table>
<tr><th rowspan="2">Mouse strain</th><th rowspan="2">H-2 haplo-type</th><th colspan="9">H - 2 r e g i o n s [a]</th></tr>
<tr><th>K</th><th>I-A</th><th>I-B</th><th>I-J</th><th>I-E</th><th>I-C</th><th>S</th><th>G</th><th>D</th></tr>
<tr><td>C57BL/6, B10</td><td>b</td><td>b</td><td>b</td><td>b</td><td>b</td><td>b</td><td>b</td><td>b</td><td>b</td><td>b</td></tr>
<tr><td>B10.BR</td><td>k</td><td>k</td><td>k</td><td>k</td><td>k</td><td>k</td><td>k</td><td>k</td><td>k</td><td>k</td></tr>
<tr><td>B10.D2</td><td>d</td><td>d</td><td>d</td><td>d</td><td>d</td><td>d</td><td>d</td><td>d</td><td>d</td><td>d</td></tr>
<tr><td>B10.HTG</td><td>g</td><td>d</td><td>d</td><td>d</td><td>d</td><td>d</td><td>d</td><td>d |</td><td>• |</td><td>b</td></tr>
<tr><td>B10.HTI</td><td>i</td><td>b</td><td>b</td><td>b</td><td>b</td><td>b</td><td>b</td><td>b |</td><td>• |</td><td>d</td></tr>
<tr><td>B10.A(2R)</td><td>h2</td><td>k</td><td>k</td><td>k</td><td>k</td><td>k</td><td>| d</td><td>d</td><td>• |</td><td>b</td></tr>
<tr><td>B10.A(3R)</td><td>i3</td><td>b</td><td>b</td><td>b</td><td>b</td><td>| k</td><td>d</td><td>d</td><td>d</td><td>d</td></tr>
<tr><td>B10.A(4R)</td><td>h4</td><td>k</td><td>k</td><td>| b</td><td>b</td><td>b</td><td>b</td><td>b</td><td>b</td><td>b</td></tr>
<tr><td>B10.A(5R)</td><td>i5</td><td>b</td><td>b</td><td>b</td><td>| k</td><td>k</td><td>d</td><td>d</td><td>d</td><td>d</td></tr>
</table>

[a]Vertical lines indicate presumed position of genetic exchange leading to a given recombinant haplotype.

Centered dots indicate that the origin of allele or region is not known.

To investigate this question we performed an antigenic and immunogenic analysis of the Lewis Lung Carcinoma (3LL) which developed spontaneously in a C57BL/6J (H-2^b) male and was kept by serial transplantations in syngeneic hosts (5). This tumor is highly malignant and produces a large number of pulmonary metastases after inoculation in the footpad (i.f.p.) or subcutaneously (s.c.). The occurrence and number of metastases in syngeneic mice was found to be enhanced following amputation of the local growing tumor, when the latter reached a certain mass (6). This tumor, unlike most other murine tumors, is capable of lethal growth in allogeneic recipients, irrespective of either the composition of their H-2 haplotype or their non H-2 "background" (7). The progressive growth of the tumor across genetic barriers could be attributed to a number of possible mechanisms: a) The tumor cells do not express MHC components, b) the tumor cells do express such components but these are masked and hence inaccessible to antibodies or cytotoxic cells, c) the tumor cells express MHC components on their cell surface, yet they secrete highly suppressive inhibitory molecules which paralyze the inflammatory responses of the host, d) the tumor possesses modified MHC-encoded antigens which are incapable of eliciting potent

Table 2. Staining of 3LL tumor cells with anti H-2^b, anti H-$2D^b$ and anti H-$2K^b$ antisera

Stained cells	Percentage of brightly positive cells stained with				
	Anti H-2^b (1)	Anti H-$2D^b$(2)	Anti H-$2K^b$(3)	Anti H-$2D^b$(4)	Anti H-$2K^b$(5)
3LL tumor cells	88	56	0	71	0
Spleen cells of:					
C57BL (H-2^b)	85	43		54.1	58
B10A.3R(H-$2K^b$)	50	0			
B10A.4R(H-$2D^b$)	50		13		
B10A.5R(H-$2K^b$)	55		32		

The antisera were obtained as follows:

(1) Immunization of C3H.Disn (H-2^k) against C3H.SW (H-2^b)
(2) Immunization of B10.HTI (H-$2K^b$,D^d) against B10(H-2^b)
(3) Absorption of anti H-2^b antiserum on B10A.4R(H-$2D^b$) spleen cells
(4) Monoclonal B22/249R1, obtained from G. Hammerling; specificity H-2.m2 (private)
(5) Monoclonal 29-13-3S, obtained from D. Sachs; specificity H-239 (public)

effector mechanisms and may in fact lead to the initiation of tumor protective immunospecific mechanisms. To explore these questions, we first analyzed the possible loss of H-2 components on 3LL cells. The membrane expression of H-2 antigens was analyzed using fluorescence serology carried out with the fluorescence-activated cell sorter (FACS II). The mouse strains and their detailed H-2 haplotypes used in this study are given in Table 1.

The results presented in Table 2 show that 3LL tumor cells express H-2 antigens, since they are brightly stained with anti H-2^b antiserum. These cells were further subjected to staining with specific antisera directed against alloantigens encoded by either the H-$2K^b$ region or the H-$2D^b$ region. These experiments, performed in the presence of appropriate controls (spleen cells from B10A.2R (H-$2D^b$), B10A.3R(H-$2K^b$), B10A.5R(H-$2K^b$) or C57BL(H-2^b)) demonstrated clearly that the 3LL tumor cells do not express serologically detectable H-$2K^b$ controlled alloantigens; they do express a high density of alloantigens controlled by the H-$2D^b$ region. Because of the misleading results that might be obtained using antisera prepared in recombinant resistant mice, we repeated these experiments using monoclonal antibodies. Similar results were obtained when we used monoclonal anti H-2K^b and anti H-2D^b antibodies. As presented in Table 2, they support the staining data obtained with the antisera elicited in congenic resistant mice. The staining experiments thus indicate that the H-2K^b is serologically undetected on 3LL tumor cells. Because of the possibility that serologically defined entities are not always the ones that are defined by cytotoxic cells, we performed experiments aimed at analyzing the cellularly accessible H-2K^b and H-2D^b determinants on 3LL cells.

For this purpose we immunized recombinant resistant mice with spleen cells. The cytotoxicity results presented in Table 3 revealed that we obtained a positive cytotoxic reaction using cytotoxic T cells against H-2D^b region, yet we could not observe cytotoxicity when using cells immunized against H-2K^b. These results are in accordance with those obtained by serological stainings.

The experiments performed thus far have clearly shown that the growth of the 3LL tumor cells across allogeneic barriers is not due to the inability of 3LL cells to express MHC components. These cells do express MHC determinants, although not all of the MHC sub-

Table 3. Cytotoxicity of 3LL and EL4 tumor cells by B10.BR and B10.D2 effector cells

Effector cells	H-2 subregion "identity" between target and immunizing cells	% target cell cytotoxicity	
		3LL	EL4
B10.BR anti B10.D2	-	- 4.0	- 2.1
B10.BR anti B10	K,I,D	15.5	26.3
B10.BR anti B10.A(2R)	D	20.7	19.1
B10.BR anti B10.A(3R)	K,I	5.7	19.4
B10.D2 anti B10.D2	-	4.9	0.5
B10.D2 anti B10	K,I,D	44.9	41.4
B10.D2 anti B10.HTI	K,I	3.4	53.9
B10.D2 anti B10.HTG	D	56.2	45.8

regions. These cells are accessible to specific antibodies and they show no resistance to specific cytotoxic T cells, once elicited under appropriate conditions. Since these conclusions were based on in vitro tests, the question still remained as to the immunogenic behavior of these cells in the tumor-bearing organism. For this purpose we studied the immune reactions that 3LL cells elicit in syngeneic and allogeneic recipients.

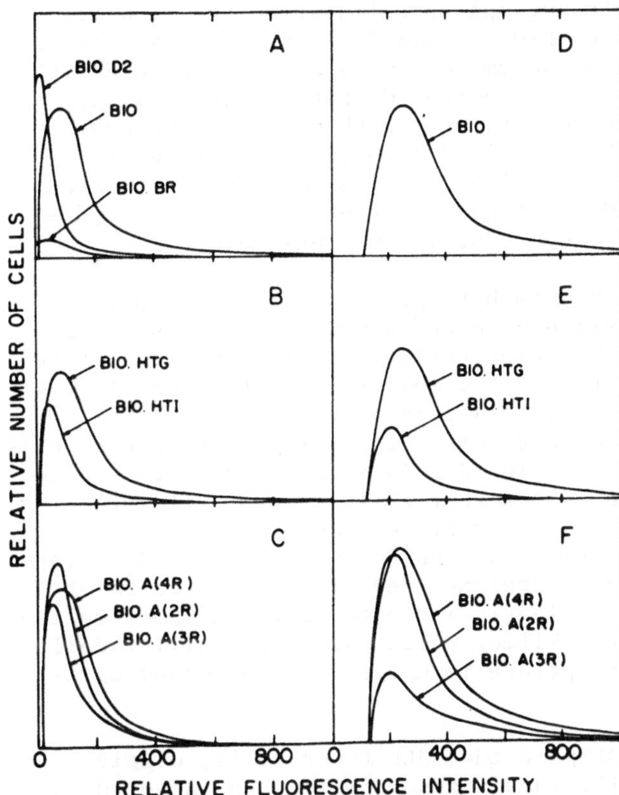

FIGURE 1. Fluorescence profile of spleen cells from congenic mouse strains stained with B10.D2 anti-3LL antiserum and FITC-labelled rabbit anti-mouse IgG_1 and IgG_2.

B10.D2 anti-3LL antiserum was prepared in female mice and reacted with the different sources of spleen cells. The photomultiplier voltage of the FACS was 650 in A-C and 700 in D-F. Forty thousand viable cells were analyzed for intensity of fluorescence. Background staining of spleen cells with FITC-labeled rabbit anti-mouse IgG_1 and IgG_2 antiserum was subtracted from the specific staining (A-C). In D-F, the spleen cells were reacted with B10.D2 anti-3LL antiserum which was previously absorbed once on B10.D2 spleen cells (0.2 ml antiserum on 10^8 spleen cells for 20 min at 4^oC). The background staining with FITC-labelled rabbit antimouse IgG_1 and IgG_2 and the residual staining of the B10.D2 spleen cells was then subtracted from the specific staining.

We analyzed antisera obtained from different recombinant-resistant mice which were inoculated with 3LL cells. The tumors were excized at early stages of growth to ensure the survival of the inoculated host. These antisera were used for fluorescence stainings of spleen cells or hydrocortisone-resistant thymocytes: B10.D2(H-2^d), B10.HTG(H-$2D^b$), B10.HTI (H-$2K^b$), B10(H-2^b), B10A.2R (H-$2D^b$), B10A.4R(H-$2D^b$) and B10A.3R(H-$2K^b$).

Figure 1 presents results of spleen cells stainings from different recombinant mice, with B10.D2 anti-3LL antiserum (A-C on the figure) and with the same antiserum following absorption on B10.D2 spleen cells (D-F). The results indicated that antisera raised against 3LL tumor cells reacted strongly with H-$2D^b$ region gene products (B10.HTG, B10.A(2R), B10.A(4R)) and weakly with H-$2K^b$ region gene products (B10.HTI, B10.A(3R)). The results thus further supported the presence of H-$2D^b$ region on the 3LL cells and indicated its ability to evoke the production of antibodies.

In an attempt to find out whether there is a relation between the imbalance in the expression of H-$2D^b$ and H-$2K^b$ regions on 3LL cells and the capacity of the tumor cells to resist specific effector mechanisms in vivo, we tested the ability of intravenously (i.v.) injected 3LL tumor cells to form pulmonary tumors in B10.D2 (H-2^d) and B10.BR(H-2^k) mice that were preimmunized intraperitoneally (i.p.) with spleen cells originating in congenic resistant mice. The data presented in Table 4 show that while spleen cells from B10(H-2^b) and B10.HTG(H-$2D^b$) lead to protection from metastasis formation after i.v. inoculation, no inhibition is observed in the lungs of those mice which were preimmunized with spleen cells of B10.HTI(H-$2K^b$). The results clearly indicate that 3LL tumor cells are not resistant to effector killer cells once these killer cells are elicited in vivo by appropriate immunogenic MHC-encoded antigenic determinants.

Previous studies demonstrated that 3LL tumor cells, despite their ability to grow locally in allogeneic mice when injected i.f.p. or s.c., are unable to produce spontaneous pulmonary metastases in such allogeneic recipients unless these recipients are immunologially impaired or rendered specifically tolerant (by neonatal injection of cells) to H-2^b alloantigens (7, 8). Yet, when injected i.v., the 3LL cells can grow locally in the lungs of allogeneic recipients, irrespective of their H-2 haplotype or non H-2 "background" (7, 8). The inability to form spontaneous pulmonary metastases by dissemination from the locally growing tumor is probably a result of peripheral immunization induced by the locally growing tumor. I.v. injection of 3LL tumor cells into allogeneic mice whose locally growing tumor had been removed by amputation of the tumor-bearing limb did not lead to formation of pulmonary metastases (7, 8). This might suggest that local and peripheral immunity may require different mechanisms. Thus, immunity against the locally growing 3LL, although not sufficient to block its growth, may prevent dissemination and the development of a secondary lethal disease.

To test whether the ability of 3LL tumor cells to confer resistance against i.v. inoculated 3LL cells following amputation of the local tumor depends on incompatibility with the MHC of the host, we performed the following experiment: B10.HTG(H-2D^b) and B10.HTI(H-2K^b) mice were inoculated i.f.p. with 3LL tumor cells. The tumors were surgically removed and then the mice were inoculated i.v. with increasing doses of 3LL cells. The results obtained indicated that while a strong protection from pulmonary growth was obtained in B10.HTI mice, no protection was observed in B10.HTG mice (Table 5). This confirms our earlier findings concerning the expression of H-2D^b gene products and not of H-2K^b gene products on 3LL tumor cells.

Table 4. Effective immunization of mice with H-2D^b alloantigens but not with H-2K^b alloantigens against primary 3LL Lewis lung tumors

Immunized mice	Origin of spleen cells used for preimmuniation of mice	H-2 region differences	Experimental metastases after i.v. inoculation	
			Lung weight	CPM
			(mg)	
B10.D2	-		130	600
	B10.D2	-	310	5200
	B10	K,I,D	150	1000
	B10.HTI	K,I	380	5000
	B10.HTG	D	200	1000
B10.BR	-		150	500
	B10.BR	-	20-	3950
	B10	K,I,D	155	650
	B10.HTI	K,I	295	3840
	B10.HTG	D	150	850

Table 5. Effect of a primary 3LL tumor in the footpad of B10.HTI and B10.HTG mice on the subsequent progression of secondary i.v.-inoculated 3LL tumor cells

Mouse strain	Treatment of mice	No. of i.v.-injected 3LL cells	Pulmonary metastases	
			Lung weight	cpm
			(mg)	
B10.HTI	-	-	130	400
	Amputation	-	125	355
	Amputation + secondary i.v. inoculation with 3LL cells	0.5×10^6	145	400
		1.0×10^6	140	395
		1.5×10^6	130	400
	Control i.v. inoculation with 3LL cells	0.5×10^6	165	1000
		1.0×10^6	240	2300
		1.5×10^6	375	4300
B10.HTG	-	-	300	350
	Amputation	-	300	4100
	Amputation + secondary i.v. inoculation with 3LL cells	0.5×10^6	375	4200
		1.0×10^6	330	4000
		1.5×10^6	360	4300
	Control i.v. inoculation with 3LL cells	0.5×10^6	235	3100
		1.0×10^6	300	4700
		1.5×10^6	385	5200

We have thus shown that 3LL tumor cells express H-2D^b gene products, that these cells are susceptible to anti H-2D^b killer cells and that they can elicit production of antibodies specific to the H-2D^b region. Yet, these cells grow locally across allogeneic barriers and therefore seem to be incapable of inducing sufficient local immunity in allogeneic hosts.

To further analyze this phenomenon we tested the ability of 3LL tumor cells to generate locally-growing tumors in B10.D2 mice when injected i.f.p. in a mixture with B10.D2, B10, B10.HTI and B10.HTG spleen cells (N. Isakov et al., submitted for publication). 3LL cells were found to form an established tumor which when injected with either B10.D2 or B10.HTI spleen cells eventually killed the host, whereas complete rejection and protection were obtained when 3LL tumor cells were injected in the presence of B10 or B10.HTG spleen cells. The inability of the 3LL tumor cells to induce local immunity could therefore be attributed to a weak immunogenicity of those cells and not to active paralysis of the inflammatory response or to local resistance to effector mechanisms of 3LL tumor cells, when grown as solid tumor masses. Once immunity is induced by strongly immunogenic splenocytes, it is sufficient to completely prevent the growth of the malignant cells.

We cannot yet provide a simple explanation for the escape of this tumor from alloimmune destruction. The loss of H-2K determinants per se is insufficient to explain this, since it has been shown that grafts of various cells can be rejected when there are H-2D region differences only (9). The possibility exists that the H-2D determinants expressed on the 3LL tumor cells are deviant in their immunogenic properties. Indeed, results accumulated in our laboratory support this assumption: a) When this tumor is injected into different animals, we obtain IgG antibodies, but these are only of the IgG_2 isotype. b) Stainings performed with specific anti $H-2^b$ antisera which contain immunoglobulins belonging to the IgG_1 and IgG_2 isotypes reveal that both these isotypes stain equally well various somatic cells, yet 3LL tumor cells are stained predominantly by IgG_2 antibodies (B. Tartakovsky et al., submitted for publication). The ability to elicit only IgG antibodies indicates deviant regulatory characteristics of those MHC determinants, since the requirements for the induction of IgG_1 and IgG_2 seem to be different (10, 11).

The deviant nature of the H-2D expressed on these cells is supported by additional observations made in our laboratory. It could be expected that the more MHC-controlled determinants are expressed on tumor cells, the more immunogenic and the more restrained from rapid growth should the tumor be. Yet, using H-2-rich and -poor cloned populations of 3LL tumor cells, we found that tumor cells which expressed a high density of H-2 alloantigens grew more rapidly in syngeneic and allogeneic hosts and produced a higher number of pulmonary metastases in syngeneic hosts.

These observations suggest that 3LL tumor cells which are highly malignant express MHC determinants of a deviant immunogenic nature. The possible deviant nature of MHC-controlled determinants on tumor cells is not restricted to the 3LL tumor. Different lines of chemically-induced or virus-induced tumors were found to express antigens which were alien to the original host but cross reacted with H-2 associated alloantigens of different H-2 haplotypes (12-17). Absence of H-2K region antigenic specificities from the cell surface of tumor cells was found both in spontaneous tumors and in chemically-induced tumors (18-20; E. Lennox, personal communication).

It is possible that the preferential loss of H-2K controlled antigens and possible modified determinants of H-2D are phenomena relevant to the malignant behavior of the tumor. H-2K-encoded antigens may be mostly involved in efficient immunogenic presentationtion of antigens, as has been reported for the induction of cytotoxic effector cells (21) and of DTH effector cells (22), while the H-2D subregion may under certain conditions favor the induction of antigen-specific suppressor cells (23-27). Similarly, it was found that while normal T cells do not express a high density of H-2D controlled determinants (28), neoplastic T cells do express a high density of H-2D determinants (29).

Malignant transformation then might be associated with molecular events which lead to the formation of structurally "modified" MHC components of both the H-2K and the H-2D controlled determinants. The H-2K determinants might evoke selective pressures which lead to the suppression of their synthesis and to a preferential propagation of those cells expressing a deviant H-2D. The H-2D controlled determinants, being preferentially suppressive in function, will lead to suppression of immune effector cells directed against tumor cells. We do not claim that a deviant and imbalanced expression of MHC is the sole explanation for the escape of tumor cells from immune destruction. It may, however, play an important role in the ability of tumor cells to generate secondary tumors. If the same situation is found in human tumors, then analysis of the MHC structure of the primary tumor may enable the prediction of its disseminative capacity.

REFERENCES

1. Doherty, P.C. and Zinkernagel, R.M. 1975. A biological role of the major histocompatibility antigens. Lancet 1.1406.
2. Gomard, E., Duprez, V., Reme, T., Colombani, M.J. and Levy J.P. 1977. Exclusive involvement of H-2D^b or H-2K^d product in the interaction between T killer lymphocytes and syngeneic H-2^b or H-2^d viral lymphomas. J. Exp. Med. 146:909.
3. Blank, K. J., Freedman, H.A. and Lilly, F. 1976. T lymphocyte response to Friend virus-induced tumor cell lines of mice congenic at H-2. Nature (Lond.) 260:250.
4. Wagner, H., Pfinzenmaier, K. and Rollinghoff, M. 1980. The role of the major histocompatibility gene complex in murine cytotoxic responses. Adv. Immunol. 31:78.
5. Sugiura, K. and Stock, C.C. 1955. Studies in a tumor spectrum. III. The effects of phosphoramides on the growth of a variety of mouse and rat tumors. Cancer Res. 15:38.

6. Gorelik, E., Segal, S. and Feldman, M. 1978. Growth of a local tumor exerts a specific inhibitory effect on progression of lung metastases. Int. J. Cancer 21:617.
7. Isakov, N., Feldman, M. and Segal, S. 1981. Control of progression of local tumor and pulmonary metastases of the 3LL lung carcinoma by different histocompatibility requirements in mice. J. Natl. Cancer Inst., in press.
8. Isakov, N., Feldman, M. and Segal, S. 1981. An immune response against the alloantigens of the 3LL Lewis lung carcinoma prevents the growth of lung metastases, but not of local allografts. J. Inv. Metastases, in press.
9. Klein, J., Chiang, C.L., Lofgreen, J. and Steinmuller, D. 1976. Participation of H-2 regions in heart transplant rejection. Transplantation 22:384.
10. Siskind, G.W., Paul, W. and Benacerraf, B. 1966. Studies on the effect of the carrier molecule on anti-hapten antibody synthesis. I. Effect of carrier on the nature of the antibody synthesized. J.Exp.Med. 123: 673
11. Torrigiani, G. 1972. Quantitative estimation of antibody in the immunoglobulin classes of the mouse. II. Thymic dependence of the different classes. J. Immunol. 108:161.
12. Bonavida, B. and Roman, J.M. 1979. Inappropriate alloantigen-like specificities detected on reticulum cell sarcoma of SJL/J mice. Immunogenetics 9:318.
13. Invernizzi, G. and Parmiani, G. 1975. Tumor-associated transplantation antigens of chemically-induced sarcomata cross reacting with allogeneic histocompatibility antigens. Nature (Lond.) 254:713.
14. Martin, W.J., Gipson, T.G., Martin, S.E. and Rice, J.M. 1976. Derepressed alloantigen or transplacentally-induced lung tumor coded for by H-2 linked gene. Science 194:532.
15. Garrido, F., Schirrmacher, V. and Festenstein, H. 1976. H-2 like specificities of foreign haplotypes appearing on a murine sarcoma after vaccinia virus infection. Nature (Lond.) 261:705.
16. Garrido, F. and Festenstein, H. 1976. Further evidence for depression of H-2 and Ia-like specificities of foreign haplotypes in mouse tumor cell lines. Nature (Lond.) 261:705
17. Meschini, A., Invernizzi, G. and Parmiani, G. 1977. Expression of alien H-2 specificites on a chemically-induced BALB/c fibrosarcoma. Int. J. Cancer 20:271.
18. Holtkamp, B., Fischer Lindahl, L., Segall, M. and Rajewsky, K. 1979. Spontaneous loss and subsequent stimulation of H-2 expression in clones of a heterozygous lymphoma cell line. Immunogenetics 9:405.
19. Martin, W.J. and Imamura, M. 1979. Variation in expression and immunogenicity of an H-2K coded alloantigen on murine tumors. Immunogenetics 9:313.

20. Schmidt, W., Atfield, G. and Festenstein, H. 1979. Loss of $H\text{-}2K^k$ gene products from an AKR spontaneous lymphoma. Immunogenetics 8:311.
21. Levy, R.B. and Shearer, G.M. 1980. Regulation of T cell mediated lympholysis by the murine major histocompatibility complex. II. Control of cytotoxic responses to trinitrophenyl K and D self products by H-2K and H-2D region genes. J. Exp. Med. 151:252.
22. Smith, F.I. and Miller, J.F.A.P. 1979. Delayed type hypersensitivity to allogeneic cells in mice. III. Sensitivity to cell surface antigens coded by the major histocompatibility complex and by other genes. J. Exp. Med. 150:905.
23. Miller, S.D., Sy, M.S. and Claman, H.N. 1978. Genetic restriction for the induction of suppressor T cells by hapten-modified lymphoid cells in tolerance to DNFB contact sensitivity. Role of the H-2D region of the major histocompatibility complex. J. Exp. Med. 147:788.
24. Wernet, D. and Lilly, F. 1975. Genetic regulation of the antibody response to $H\text{-}2D^b$ alloantigens in mice. I. Differences in activation of helper T cells in C57BL/10 and BALB/c congenic strains. J. Exp. Med. 141:573.
25. Wernet, D. and Lilly, F. 1975. Genetic regulation of the response to $H\text{-}2D^b$ alloantigens in mice. II. Tolerance to non-H-2 determinants abolishes the antibody response to $H\text{-}2D^b$ in B10.A(5R) mice. J. Exp. Med. 144:266.
26. Wernet, D., Shafran, H. and Lilly, F. 1976. Genetic response of the antibody response to H-2D alloantigens in mice. III. Inhibition of IgG response to noncongenic cells by preimmunization with congenic cells. J. Exp. Med. 144:654.
27. Hasek, M. and Chutna, J. 1979. Complexity of the state of immunological tolerance. Immunol. Rev. 46:3.
28. Tartakovsky, B., De Baetselier, P. and Segal, S. 1980. Serological detection of H-2K and H-2D gene products. I. Principal differences between T and B lymphocytes in expression of H-2D encoded alloantigens. Immunogenetics 11:585.
29. Lonai, P., Katz, E. and Haran-Ghera, N. 1981. Role of the major histocompatibility complex in resistance to viral leukemia. Its effect on the preleukemic stage of leukemogenesis. Cancer Biology Rev., in press.

GENETICS OF HUMAN CANCER WITH SPECIAL REFERENCE TO MODEL CANCERS

Alfred G. Knudson, Jr.

Fox Chase Cancer Center
Philadelphia, P.A. 19111

If the somatic mutation hypothesis of cancer is correct then an understanding of the fundamental nature of cancer depends upon the isolation and characterization of the genes in which the critical mutations occur. So far we have two classes of genes that may help in this respect. The first is associated with tumor-producing viruses, the so called oncogenes; the second, with host genes observed in man and animal, but particularly in man, which strongly predispose to the development of particular cancers. In passing we take note of the existence of some genes that are not directly tumor-producing genes but which can predispose to neoplasm. In mice for example there are genes that predispose to infection with known tumor viruses. In xeroderma pigmentosum in man the recessive gene that gives the disease is not per se a tumor gene but it is rather a gene that predisposes to ultra-violet induced mutation. The effect will depend upon which genes are mutated; if the mutated gene happens to be a tumor gene then tumor can result.

My intention here is to discuss a well known class of dominantly heritable genes that predispose to specific cancers in man. The common feature of these genes is that persons heterozygous for them have a high incidence of tumor and the spectrum of tumors produced is very narrow, in some cases essentially one type of tumor. Examples of such genes are those for polyposis of the colon, neurofibromatosis, retinoblastoma, and multiple endocrine neoplasia.

The genes that predispose to these heritable conditions are a relatively small part of the total human genome, being probably in the range of 100 to 1,000 in number. For each kind of tumor there is a small number of genes that produce tumor, perhaps even one in some cases. I hope to develop the thesis that, in the nonhereditary forms of the same tumor, initiation of the neoplasm starts with somatic mutation at the same gene as is affected in hereditary cases. I shall also discuss the implications of the fact that the presence of such a mutation does not by itself cause transformation of cells, that some further event is necessary, and that this further event may be the development of homozygosity. Finally, I shall discuss the possibility of isolating and characterizing human cancer genes.

For this discussion certain tumor genes are much more informative than others. We can state at the onset that the chromosomal location of human tumor genes is known with some assurance in only two cases, namely retinoblastoma and Wilms' tumor. There are other instances where we may have an identification. For example, there is a large pedigree in which persons with renal carcinoma carry a translocation between chromosomes 3 and 8, suggesting that in one of those two chromosomes lies a critical site for renal carcinoma.[1] From linkage studies it has been shown by King and coworkers[2] that a breast cancer gene is linked to the gene for glutamate pyruvate transaminase, which in turn is thought to be on the short arm of chromosome number 10. So far, there is no example of tumor gene mapping by direct analysis of DNA, but this technique will undoubtedly prove useful in the future.[3]

Retinoblastoma is still the best model for a dominantly heritable tumor in man.[4] Retinoblastoma occurs in both unilateral and bilateral forms. Genetic studies have clearly demonstrated that 50 per cent of the offspring of bilateral cases are at risk of tumor and we may therefore safely conclude that this form is invariably owing to a germinal mutation. Some of the offspring of unilateral cases are also at risk, from which it has been concluded that about 40 per cent of all cases are germinal in origin and that the germinal mutation usually produces bilateral disease, sometimes unilateral disease, and in rare instances no tumor. The sporadic unilateral case is regarded as nonhereditary. The distribution of number of tumors among cases closely approximates the Poisson distribution with a mean number of 3 to 4 tumors per gene carrier. It has been previously proposed that all tumors, whether heritable or not, are initiated by the same mutation and that some second event is necessary in order for tumor to occur. In the hereditary case the first event is present in all cells in the body; in the nonhereditary cases, only in some clone of mutant cells that arose somatically. Some second event, perhaps a mutation, is evidently necessary in order to complete tumor initiation. An implication of this model is that there is a "retinoblastoma gene" which must be altered in order to produce this tumor. Such a gene would be an example of a human cancer gene.

The chromosomal location of a retinoblastoma gene has been accomplished. There are more than 20 cases of retinoblastoma in which all somatic cells show deletion in the long arm of chromosome 13. These cases are associated with varying degrees of congenital abnormality, most commonly mental retardation. The somatic phenotype depends to a large extent upon the size of the deletion that is found. The cells are heterozygous for this deletion. All of the deletions have in common a deletion of band 13q14.

In one case in which the deletion is within band 13q14, as well as in cases with larger deletions, there is hemizygosity for an enzyme esterase D, long known to be located on chromosome 13.[6] The esterase D gene must be closely linked to that for retinoblastoma. This proximity may permit us to test the hypothesis that the hereditary cases of retinoblastoma involve a submicroscopic alteration in the same gene site. Since there is a polymorphism consisting of isozyme variants for the esterase D locus, it becomes possible in some instances in which retinoblastoma affects more than one member of a family to test for genetic linkage. At the present time the evidence favors the notion that heritable retinoblastoma cases are attributable to a gene at this same site, since linkage seems to be very close.[7] We can then imagine that all constitutional cases of retinoblastoma involve the same gene site with a spectrum of changes from the extensive deletion to changes not visible in the microscope.

There is one reported instance in which familial retinoblastoma has been associated with cytogenetic change.[8] In this family numerous cases of retinoblastoma have been noted but the transmission is not in typical dominant form. There are many skipped individuals who evidently transmit the susceptibility without being affected. These skipped individuals are carriers of a balanced translocation that involves part of chromosome 13 at the very site noted in the deletion cases. The individuals with retinoblastoma bear the unbalanced translocation; they are deletion cases. An individual who carries the balanced translocation may produce normal individuals with normal karyotype, normal individuals with balanced translocation, other persons with an unbalanced translocation that produces trisomy for a small segment of chromosome 13, and finally the deletion cases in which there is monosomy for this segment. One would anticipate under these conditions a 25 per cent incidence of retinoblastoma among the offspring of translocation cases. This indeed seems to be the case. This pedigree may explain a number of pedigrees in the literature that seem to constitute exceptions to the idea that all hereditary retinoblastoma is attributable to a simple dominant mutation. It may be that many other families, which are sometimes described as carrying a "premutation", can be explained in this way.

Our hypothesis further states that nonhereditary cases are also attributable to mutation at this locus. Some of these cases should be associated with deletion of the same 13q14 band, but the deletion would be present only in tumor cells, not in normal somatic cells. In other cases the alteration would be expected to be submicroscopic. We could expect a larger fraction of gross chromosomal abnormalities in nonhereditary cases, since selection would operate against some of these were they germinal. Indeed, even monosomy for chromosome 13 might be tolerated in the precursor retinal cells. One early report[9] suggested that chromosome 13 abnormality could be found in tumors of persons whose lymphocytes are cytogenetically normal. In a more recent study Balaban-Malenbaum and colleagues have shown a translocation in one case, deletion in three others and monosomy 13 in still another, while one case showed no abnormality.[10,11] It would seem then that this same site is affected in some proportion of retinoblastoma tumors of the nonhereditary type.

So far there are no data inconsistent with the model that all retinoblastoma is associated with an alteration at a specific site in the long arm of chromosome 13. It is a definite possibility that, whether the mutation occurs germinally or somatically, whether it is large enough to be seen microscopically or not, the path to tumor involves the same initial gene.

The other neoplasm for which similar data are available is Wilms' tumor. Again, this is a tumor that occurs in both hereditary and nonhereditary form, although there are not so many familial cases as with retinoblastoma.[12] The fraction that is heritable is probably in the range of 15 per cent or so. Wilms' tumor may be a more heterogeneous condition than is retinoblastoma. Some cases are associated with anomalies of the kidney. Some are associated with absence of the iris, or aniridia. Another condition that has been associated with Wilms' tumor is hemihypertrophy, and a related condition, the Wiedemann-Beckwith syndrome. The incidence of bilateral cases of Wilms' tumor is about 5 per cent, much lower than the 25 to 30 per cent observed for retinoblastoma. There is not as clear a presumption for Wilms' tumor that bilateral cases are always heritable since known hereditary cases do not have a very high incidence of bilaterality. Whether all Wilms' tumor is attributable to the same genetic change then is very much open to question.

The association of aniridia with Wilms' tumor invited speculation by Knudson and Strong[12] that the association might be attributable to chromosomal deletion, with a gene for Wilms' tumor and another gene for aniridia being located near each other on the same chromosome. This in fact has subsequently proved to be the case. All such cases so far have manifested a deletion of the short arm of chromosome 11: 11p13.[13] This deletion is found in lymphocytes and fibroblasts and is obviously a constitutional abnormality. So far there are no known deletion cases in which aniridia has not been reported.

The adjacent band on the short arm of chromosome 11, 11p12, is known to contain the site for the beta-globin complex of hemoglobin and this site may, therefore, prove to be a useful marker for linkage studies.[14] Thus, for example, in familial cases in which a hemoglobin abnormality is segregating with Wilms' tumor, in the absence of aniridia or deletion, a test of linkage could be accomplished. Very recently another neighboring genetic site has been identified, namely, that for catalase. It is possible to measure levels of catalase, thus permitting a test for hemizygosity. The result is that cases demonstrating a deletion with aniridia may be hemizygous for catalase.[15] If a suitable isozymic polymorphism could be found for catalase, this enzyme could be used in much the same way that esterase D has been used to test for linkage. Such a study would go far toward testing the hypothesis that Wilms' tumor is always associated with abnormality at the same genetic locus.

Just as with retinoblastoma there is also a reported family of more than one case of Wilms' tumor with aniridia.[16] There were three cases in one family in which these conditions were associated. One of them was available for study and revealed a typical deletion. The mother in this case was unaffected by either aniridia or Wilms' tumor and carried a balanced translocation. So again a failure of expression of tumor could be attributed to translocation. Transmission of a balanced translocation in this manner could possibly account for a report of Wilms' tumor in five cousins.[17] This extraordinary family involves many unaffected relatives as obligate carriers of a Wilms' tumor mutation. There is as yet no report of prophase banding to establish whether this family could be explained by balanced translocation in the carriers and unbalanced translocation in those affected. In that family aniridia was not present with the Wilms' tumor.

Unfortunately we do not have sound data yet for the presence or absence of 11p deletions in the tumors of persons whose somatic cells show no abnormality, as has been the case with retinoblastoma. However there is one report by Slater[18] of deletion, at the same site as in the constitutional cases, in two tumors out of a series of seven tumors studied. This supports the idea that there may be one or a small number of genes involved in the production of Wilms' tumor. Here again we may be able to provide a test of the hypothesis that Wilms' tumor begins with abnormality at the same genetic site, whether that abnormality is germinal or somatic in origin, and whether it constitutes a microscopic change or a submicroscopic alteration of the gene.

No other human cancer genes have been mapped. One possibility for future mapping is the neuroblastoma gene. Brodeur and colleagues[19] have investigated the tumors of individuals with neuroblastoma and found a high incidence of deletion in the short arm of chromosome number 1 in the tumors of persons whose somatic cells are cytogenetically normal. If the pattern follows as it does in retinoblastoma and Wilms' tumor, one could expect that a deletion might be found in some familial cases of neuroblastoma. In fact neuroblastoma is well known to have both a hereditary and a nonhereditary form.[20] So the same kinds of analysis might become available for that tumor as well. It so happens that the deletion in tumor cells involves band 1p34,[21] which is almost the same site to which the Rh locus is assigned, so it should be possible to do a test of linkage in familial cases of this tumor.

Let us suppose then that for each tumor there is at least one gene site which must be altered in order to initiate tumor. Since there are already 50 or so dominantly heritable conditions that predispose to tumors we may presume that there are at least 100 such genes and the total could well be several hundred. Nevertheless there is some specificity about tumor. We must also take into account the fact that some tumors may arise by other means than genetic change. Nevertheless a significant fraction would seem to be attributable to mutation, if we use that term broadly to include all kinds of genetic alteration. There remains the question, even if this is true, that, although such mutations may be necessary to create cancer, they are not sufficient. The existence of such a mutation obviously does not cause every cell that contains it to transform. It would seem that at least one other event is necessary. Such an event would seem to be rare. For example, even in the case of retinoblastoma there are only three or four tumors per gene carrier, although millions of target cells are present in the eye. Therefore the second event must be occurring at something less than once per million cell divisions.

The necessity for a second event also readily explains the well known latent period following exposure to known environmental carcinogens. A two event model interprets the exposure to carcinogen as causing a first event, with the subsequent emergence of a clone of cells, one of which later undergoes some secondary change. A model that can explain such phenomena has been constructed by Moolgavkar and Venzon[22] and has been applied to human cancers generally by Moolgavkar and Knudson.[23] In such a model the once-hit cell has a growth advantage over normal cells. Such a model has also been proposed by Potter[24] to account for observations in experimental animals. In fact hyperplastic lesions are seen in individuals who carry a germinal mutation for one of the cancers or cancer syndromes. A particularly well studied example is that of small hyperplastic lesions, C-cell hyperplasia, seen in individuals with hereditary medullary carcinoma of the thyroid.[25] Similarly undifferentiated nests of cells have been described in the kidneys of Wilms' tumor patients.[26] For neurofibromatosis Fialkow[27] has shown that neurofibromas have a multiple cellular origin, suggesting that cooperativity of a number of cells is necessary to produce intermediate lesions. On the other hand, when these tumors arise in normal individuals, they show a single cell origin, suggesting a somatic mutational event. Hyperplastic nodules are also found in the adrenals of patients with neuroblastoma and in occasional normal infants, and are referred to as neuroblastoma in situ. There is a peculiar form of neuroblastoma, manifest in the skin and liver particularly, called neuroblastoma IV-S, which is famous for its high frequency of spontaneous regression. It has been suggested[28] that these tumors are one-hit lesions rather than true neoplasms and that the germinal mutation that produced them has interfered with differentiation, but not caused true neoplasia. True neoplasia can be produced in any of these lesions, however, if some cell receives some "second hit". It might therefore be found that IV-S tumors are of multiple cell origin whereas true tumors arising in individuals with a germinal neuroblastoma mutation would show a single cell origin associated with their receipt of a "second hit". A single cell origin has been demonstrated for medullary carcinoma of the thyroid arising in individuals with multiple endocrine adenomatosis, one of the dominant tumor syndromes,[29] and for neurofibrosarcoma arising in a patient with neurofibromatosis.[30]

We have no evidence regarding the nature of a second event. Hence it is not known even whether it is genetic or epigenetic. If it is genetic it could theoretically occur in the same chromsome as did the first event, in the homologous chromosome at the same site affected by the first event, or at some other site. The first possibility seems to be ruled out by the fact that some deletions are very large and preclude a second event in the region of the first. The second possibility is a particularly interesting one because its implication is that tumor arises as a result of a single recessive gene defect. It also could be produced by other means than mutation. Thus, somatic recombination between homologous chromosomes, with subsequent segregation, could produce homozygous daughter cells.

It may be possible to test directly the hypothesis that the second event entails the development of homozygosity. For example, suppose a family with retinoblastoma is studied. There is a polymorphism for fluorescence of the short arm of chromosome 13. Consider a mating between an unaffected parent, both of whose chromosomes 13 are nonfluorescent, and an affected parent who is heterozygous for fluorescence. Now if a child develops retinoblastoma and has fluorescence of the chromosome 13 it is quite clear which chromosome carries the retinoblastoma gene. Balaban-Malenbaum and colleagues[10] have reported a deletion chromosome 13 in a tumor in which the karyotype in lymphocytes was normal. In addition the patient was bilaterally affected and we may therefore presume had the hereditary form of the disease. Therefore, this patient inherited a mutation that was not associated with visible change but in the tumor showed a chromosome abnormality. It is not immediately obvious whether the deleted chromosome is the one that was inherited with the abnormal gene or whether it is the other chromosome, which does not carry the gene. One possibility is that one chromosome is abnormal because of an inherited mutation and the other one became abnormal by a somatic deletion. If that were the case in our hypothetical family, the fluorescent 13 would be normal and the non-fluorescent one would be deleted. Such a finding would clearly indicate that both chromosomes 13 were abnormal and would support the hypothesis of homozygosity.

In the above example homozygosity would have occurred as a result of a new genetic event in the homologous chromosome. The same could be accomplished by recombination, of course. This may in fact be the reason that individuals with the recessively inherited condition, Bloom's syndrome, are so susceptible to different kinds of cancer.[31] It is well known that their cells have a great propensity for sister chromatid exchange and even homologous chromosome exchange. If a cell sustains an initiating somatic mutation then homozygosity is much more likely to occur than in a normal individual because of predisposition to somatic recombination and the conversion of a heterozygous cell to a homozygous cell.

Somatic recombination could also be produced by the mechanism of segregation of two mutant chromosomes from a tetraploid cell, the latter arising by either endoreduplication or cell fusion. Here the mechanism might be detected by finding two fluorescent chromosomes 13 in the tumor of a person heterozygous for this fluorescence.

No matter what the number of events on the path to cancer may be it is obvious that the initiating mutation, be it germinal or somatic, is very important. If this gene operates recessively to produce cancer, then cancer cells are the result of a single genetic defect. It would seem that the normal alleles of these cancer genes are important in tissue differentiation, a message imparted by their considerable specificity. Therefore the isolation and characterization of such a gene would have important implications for our knowledge of differentiation as well as for knowledge of the origin of cancer.

We may be in a position in the near future to isolate such a cancer gene. One direction lies with the ability to isolate chromosomes by flow cytometry. It is interesting to note that chromosome 13, carrier of the retinoblastoma gene, is said to be the chromosome best isolated by this method.[32] There are also hybrid cells between humans and rodents which contain only one human chromosome. One such hybrid contains only human chromosome 11, the carrier of the Wilms' tumor gene.[33] This simplifies, to some extent, the identification of segments of DNA peculiarly associated with these genes. It would be possible to provide the molecular geneticist with cells from normal individuals, from those whose cells contain a significant deletion in the area of interest, and with cells in which one of the chromosomes is altered submicroscopically. It might be that by restriction enzyme analysis one could identify the critical segments. Obviously the task would be dependent upon close association of the tumor gene with a gene, such as the beta-globin locus, for which there is a molecular probe. Thus if we had probes for the esterase D gene or the catalase gene the problem might be somewhat simpler in the case of retinoblastoma and Wilms' tumor. Even though this seems like a very large technical problem at the present time, we can at least imagine the isolation of such genes now.

SUMMARY

There is in man a set of dominant cancer genes that predispose strongly to a single tumor, or a narrow spectrum of tumors. The number of such genes is not known but measures at least 50 and may be of the order of several hundred. Two of these genetically determined cancers provide substantial evidence regarding the location of a responsible gene. One of these is retinoblastoma, the other Wilms' tumor. The existence of deletions has permitted the tentative conclusion that the same gene site is mutated in both hereditary and nonhereditary cases of the same tumor. The number of genes for a particular tumor may be as small as one but in any case may be a few. The genes that predispose in this way are not sufficient to produce cancer; a second event is apparently necessary. Intermediate lesions, often hyperplastic in nature, are found in some heritable tumors and may represent changes caused directly by the first inherited event. The nature of a second event is not known, but it is possible that it represents a change in the homologous chromosome, either by a new genetic event, or by recombination and segregation of a homozygous pair of mutant chromosomes. In such an instance a cancer would be caused by a recessive genetic abnormality. The new technology of molecular genetics offers the possibility that we may be able to locate, isolate, and characterize human cancer genes and identify their products, mechanisms of action, and roles in normal differentiation and in oncogenesis.

REFERENCES

1. A. J. Cohen, F. P. Li, S. Berg, D. J. Marchetto, S. Tsai, S. C. Jacobs, and R. S. Brown, Hereditary renal-cell carcinoma associated with a chromosomal translocation, N. Engl. J. Med. 301:592 (1979).

2. M. C. King, R. C. P. Go, R. C. Elston, H. T. Lynch, and N. L. Petrakis, Allele increasing susceptibility to human breast cancer may be linked to the glutamate-pyruvate transaminase locus, Science 208:406 (1980).
3. D. Botstein, R. L. White, M. Skolnick, and R. W. Davis, Construction of a genetic linkage map in man using restriction fragment length polymorphisms, Am. J. Hum. Genet. 32:314 (1980).
4. A. G. Knudson, Mutation and cancer: statistical study of retinoblastoma, Proc. Natl. Acad. Sci. USA 68:820 (1971).
5. J. J. Yunis and N. Ramsay, Retinoblastoma and subband deletion of chromosome 13, Am. J. Dis. Child 132:161 (1978).
6. R. S. Sparkes, M. C. Sparkes, M. G. Wilson, J. W. Towner, W. Benedict, A. L. Murphree, and J. J. Yunis, Regional assignment of genes for human esterase D and retinoblastoma to chromosome band 13q14, Science 208:1042 (1980).
7. R. S. Sparkes, Personal communication.
8. L. C. Strong, V. M. Riccardi, R. E. Ferrell, and R. S. Sparkes, Familial retinoblastoma and chromosome 13 deletion transmitted via an insertional translocation, Science 213:1501 (1981).
9. N. Hashem and S. H. Khalifa, Retinoblastoma: a model of hereditary fragile chromosomal regions, Hum. Hered. 25:35 (1975).
10. G. Balaban-Malenbaum, F. Gilbert, W. W. Nichols, R. Hill, J. Shields, and A. T. Meadows, A deleted chromosome no. 13 in human retinoblastoma cells: relevance to tumorigenesis, Cancer Genet. Cytogenet. 3:243 (1981).
11. G. Balaban, F. Gilbert, W. Nichols, A. T. Meadows, and J. Shields, Abnormalities of chromosome #13 in retinoblastomas from individuals with normal constitutional karyotypes, Cancer Genet. Cytogenet. 6:213 (1982).
12. A. G. Knudson and L. C. Strong, Mutation and cancer: a model for Wilms' tumor of the kidney, J. Natl. Cancer Inst. 48:313 (1972).
13. U. Francke, L. B. Holmes, L. Atkins, and V. M. Riccardi, Aniridia-Wilms' tumor association: evidence for specific deletion of 11p13, Cytogenet. Cell Genet. 24:185 (1979).
14. J. Gusella, A. Varsanyi-Breiner, F. T. Kao, C. Jones, T. T. Puck, C. Keys, S. Orkin, and D. Housman, Precise localization of human β-globin gene complex on chromosome 11, Proc. Natl. Acad. Sci. USA 76:5239 (1979).
15. C. Junien, C. Turleau, J. de Grouchy, R. Said, M. O. Rethoré, R. Tenconi, and J. L. Dufier, Regional assignment of catalase (CAT) gene to band 11p13. Association with the aniridia-Wilms' tumor-gonadoblastoma (WAGR) complex, Annales De Génétique 23:165 (1980).
16. J. J. Yunis and N. K. C. Ramsay, Familial occurrence of the aniridia-Wilms' tumor syndrome with deletion 11p13-14.1, J. Pediatr. 96:1027 (1980).

17. J. F. Cordero, F. P. Li, L. B. Holmes, and P. S. Gerald, Wilms' tumor in five cousins, Pediatrics 66:716 (1980).
18. R. M. Slater and E. M. Bleeker-Wagemakers, Aniridia, Wilms' tumour and chromosome number 11, in: "Proceedings, 12th Annual Meeting of the International Society of Pediatric Oncology, Budapest" (1980).
19. G. M. Brodeur, A. A. Green, and F. A. Hayes, Cytogenetic studies of primary human neuroblastomas, in: "Advances in Neuroblastoma Research," A. E. Evans, ed., Raven Press, New York (1980).
20. A. G. Knudson and L. C. Strong, Mutation and cancer: neuroblastoma and pheochromocytoma, Am. J. Hum. Genet. 24:514 (1972).
21. F. Gilbert, Personal communication.
22. S. H. Moolgavkar and D. J. Venzon, Two-event model for carcinogenesis: incidence curves for childhood and adult tumors, Math. Biosci. 47:55 (1979).
23. S. H. Moolgavkar and A. G. Knudson, Mutation and cancer: a model for human carcinogenesis, J. Natl. Cancer Inst. 66:1037 (1981).
24. V. R. Potter, Initiation and promotion in cancer formation: the importance of studies on intercellular communication, Yale J. Biol. Med. 53:367 (1980).
25. C. E. Jackson, M. A. Block, K. A. Greenawald, and A. H. Tashjian, The two-mutational-event theory in medullary thyroid cancer, Am. J. Hum. Genet. 31:704 (1979).
26. G. A. Machin, Persistent renal blastema (nephroblastomatosis) as a frequent precursor of Wilms' tumor; a pathological and clinical review, part 2, Am. J. Pediatr. Hematol. Oncol. 2:253 (1980).
27. P. J. Fialkow, Clonal origin and stem cell evolution of human tumors, in: "Genetics of Human Cancer," J. J. Mulvihill, R. W. Miller, and J. F. Fraumeni, Jr. eds., Raven Press, New York (1977).
28. S. B. Baylin, S. H. Hsu, D. S. Gann, R. C. Smallridge, and S. A. Wells, Inherited medullary thyroid carcinoma: a final monoclonal mutation in one of multiple clones of susceptible cells, Science 199:429 (1978).
29. A. G. Knudson and A. T. Meadows, Regression of neuroblastoma IV-S: a genetic hypothesis, N. Engl. J. Med. 302:1254 (1980).
30. P. J. Fialkow, Personal communication
31. R. S. Festa, A. T. Meadows, and R. A. Boshes, Leukemia in a black child with Bloom's syndrome: somatic recombination as a possible mechanism for neoplasia, Cancer 44:1507 (1979).
32. A. V. Carrano, J. W. Gray, R. G. Langlois, K. J. Burkhart-Schultz, and M. A. Van Dilla, Measurement and purification of human chromosomes by flow cytometry and sorting, Proc. Natl. Acad. Sci. USA 76:1382 (1979).

33. C. Waldren, C. Jones, and T. T. Puck, Measurement of mutagenesis in mammalian cells, *Proc. Natl. Acad. Sci. USA* 76:1358 (1979).

CHROMOSOME BREAKAGE AND SENSITIVITY TO DNA BREAKING AGENTS IN ATAXIA-TELANGIECTASIA AND THEIR POSSIBLE ASSOCIATION WITH PREDISPOSITION TO CANCER

Yechiel Becker, Meira Shaham, Yosef Shiloh and Ruth Voss

Departments of Molecular Virology and Human Genetics
The Hebrew University-Hadassah Medical Center
Jerusalem, 91 010 Israel

ABSTRACT

Ataxia-telangiectasia (A-T) is an autosomal recessive disease characterized by increased spontaneous chromosomal breakage and a predisposition to cancer in both A-T patients and their heterozygous relatives. We have demonstrated the presence of a clastogenic factor in plasma of A-T patients, in medium conditioned by A-T fibroblasts, and in amniotic fluid of an affected A-T fetus. Our A-T fibroblast strains were found to be as sensitive as normal strains to the cytotoxic effect of methylating carcinogens. We confirmed their hypersensitivity to ionizing radiation and found that they were also markedly sensitive to treatment with the antitumor antibiotic neocarzinostatin. Two A-T heterozygote strains had intermediate sensitivity to neocarzinostatin which was distinct from that of the homozygote and the normal cells. The intrinsic hypersensitivity to DNA-breaking agents, coupled with the presence of a clastogenic factor, may account for the high rate of chromosomal breakage in vivo and the predisposition to cancer.

Ataxia-telangiectasia (A-T) is a rare human autosomal recessive disease. This genetic disorder is expressed in the homozygotes as a progressive cerebellar ataxia, oculocutaneous telangiectases, severe immunodeficiencies which are associated with both humoral and cell-mediated immune response, and defective thymus (McFarlin et al., 1972; Paterson and Smith, 1979). A-T patients constitute a cancer-prone population, since they have a markedly increased tendency to develop lymphoreticular malignancies (Swift et al., 1976). There is a fivefold higher incidence of lymphomas, leukemias and other malignancies in heterozygous relatives of A-T patients than in the general population.

Cultured cells from A-T patients show increased chromosomal breakage with a high frequency of rearrangements, preferentially involving chromosome No. 14 (Cohen et al., 1975). We recently demonstrated that a clastogenic factor (chromosome-breaking factor) is present in plasma of A-T patients. Incubation of the patient's plasma with normal phytohaemagglutinin (PHA)-stimulated lymphocytes caused an increase in chromosome breakage in the normal cells (Shaham et al., 1980). The A-T clastogenic factor was also found in the medium used to culture A-T skin fibroblasts. Incubation of normal stimulated lymphocytes with cultured (conditioned) medium from A-T fibroblasts caused an increase in the chromosome breakage rate of the lymphocytes.

Further studies revealed that the A-T clastogenic factor is a small peptide with a molecular weight of 500-1000 (Shaham and Becker, 1981), and this factor has now also been shown to be present in the amniotic fluid of an A-T fetus (Shaham et al., 1981). The presence of the A-T clastogenic factor that causes increased spontaneous chromosome breakage and a specific chromosome translocation in amniotic cells, as well as in affected children, may be a clue to the development of malignancies. The A-T gene, which is probably responsible for the presence of the clastogenic factor in A-T homozygotes, may also be one of the human genes behind the predisposition to cancer.

A-T patients, suffering from a malignancy and treated with X-irradiation, were found to develop deep tissue necrosis, and the skin surface became pigmented and desquamated (Gotoff et al., 1967; Morgan et al., 1968; Cunliffe et al., 1975). This hypersensitivity to ionizing radiation was later confirmed in cultured cells (Taylor et al., 1975). Reduced DNA repair synthesis found in some A-T cell strains after gamma irradiation,and defective ability to remove unidentified gamma-ray products from the DNA, led to the assumption that a defect in DNA repair is responsible for this hypersensitivity (Paterson et al., 1976).

A-T cells were also found to be hypersensitive to the clastogenic and cytotoxic effects of bleomycin (Taylor et al., 1979; Lehmann and Stevens, 1979), and there are several reports on the unusual sensitivity of A-T fibroblasts to the methylating carcinogens methyl methane sulfonate (MMS) (Hoar and Sargent, 1976) and N-methyl-N'-nitro-N-nitrosoguanidine (MNNG) (Scudiero, 1980), and to mitomycin C (Hoar and Sargent, 1976). Studies in our laboratory (Shiloh et al., 1981a) which were made on a group of fibroblast strains from Middle-Eastern A-T patients revealed that they were all proficient in DNA repair synthesis induced by X-irradiation, MMS, MNNG and mitomycin C. We were able to demonstrate that our

A-T strains were hypersensitive to the lethal effect of X-rays similarly to A-T strains from Europe and North America. Our fibroblast strains were also found to be hypersensitive to the DNA breaking agent neocarzinostatin (NCS)(Shiloh et al., 1981b), but their sensitivity to treatment with MMS and MNNG was similar to that of normal fibroblasts. No defect was found in the ability of our A-T strains to repair purine alkylation adducts induced by MNNG. Our findings, therefore, indicate that A-T fibroblasts respond normally to DNA-damaging agents that induce the base alkylation type of modification, but are markedly hypersensitive to DNA breaking agents.

This paper describes our studies on cultured A-T cells,and their relevance to the study of human cancer is discussed.

The presence of a clastogenic factor in A-T fibroblast cultures

Conditioned media were obtained from monolayer cultures of fibroblasts derived from normal individuals and A-T patients. Normal human lymphocytes were incubated in the conditioned medium in the presence of PHA, and the chromosome spreads were analyzed. Table 1 shows that conditioned medium from A-T fibroblasts caused enhanced chromosome breakage (∿ 0.10 breaks/cell). A background level of <0.03 chromosome breaks/cell was noted with the conditioned medium from normal fibroblasts (Shaham et al., 1980).

The conditioned medium containing the clastogenic factor was filtered through Amicon-Diaflo filters, and the filtrates were tested for clastogenic activity (Table 2). The clastogenic factor was found to be a peptide with a molecular weight of 500-1000 (Shaham and Becker, 1981).

Table 1. Effect of fibroblast conditioned medium on chromosome breakage in normal lymphocytes

Source of medium	Chromosome breaks/cell in normal lymphocytes[a]
A-T (male; MI)	0.12
A-T (male; AA)	0.10
A-T (female; IB)	0.09
A-T (female; RA)	0.09
Normal (male; N=3)	<0.03

[a] At least 100 metaphases analyzed

Table 2. Properties of the clastogenic factor present in A-T plasma and in conditioned medium of A-T fibroblasts

Source of clastogenic factor	No. of cases	Treatment	Amicon-Diaflo filtration	Number of cells analyzed	Chromosome breaks/cell
Plasma	2	-	PM10+UM2	247	0.095-0.110
Conditioned medium	2	-	PM10	200	0.110-0.130
	3	-	PM10+UM2	334	0.110-0.134
	2	-	PM10+UM2+UM05	598	0.020-0.024
	1	Pronase	PM10+UM2	186	0.026
	1	RNase B	PM10+UM2	200	0.105

The presence of a clastogenic factor in the plasma of A-T patients

The plasma of A-T patients was found to contain a clastogenic factor with the same molecular weight as the clastogenic factor in conditioned medium of A-T fibroblasts (Table 2) (Shaham et al., 1980).

The presence of a clastogenic factor in the amniotic fluid of an A-T fetus

Prenatal diagnosis was performed on a fetus at risk for A-T. Clastogenic activity of the cell-free amniotic fluid (Table 3) produced between 0.09 to 0.17 breaks/cell in normal human peripheral blood lymphocytes cultured in the fluid. Lymphocytes grown in

Table 3. Chromosome breakage in normal lymphocytes cultured in cell-free amniotic fluids

Donor of amniotic fluid	Breaks/cell
Suspected A-T (I)[a]	0.09
Suspected A-T (II)[b]	0.17
A-T (III)[c]	0.11
Controls (N = 6)	<0.04

[a] Amniotic fluid obtained at first amniocentesis
[b] Amniotic fluid obtained at second amniocentesis
[c] Amniotic fluid obtained at termination of pregnancy

control amniotic fluids did not have an elevated breakage rate (<0.04 breaks/cell).

The rate of spontaneous chromosome breakage was found to be high (between 0.33 and 0.45 breaks/cell) in cultured amniotic fluid cells of the suspected A-T fetus, whereas in simultaneously cultured control amniotic cells a rate of <0.03 breaks per cell was found (Table 4). In addition, a clone of 14/5 translocation was observed in 15 cells (out of 100 analyzed). Thus the karyotype of the fetus is 46,XY/46,XY,t(14;5),confirming the diagnosis of A-T. Using this direct method, diagnosis can be performed within a week from amniocentesis (Shaham et al., 1981).

Chromosomal breakage is dominant in A-T human-mouse somatic cell hybrids

Somatic cell hybrids were constructed between skin fibroblasts from an A-T patient and a mouse cell line deficient in thymidine kinase [LM(TK^-)]. Selection of hybrids was done in medium containing aminopterin (HAT medium). Preliminary results indicated a higher breakage rate in A-T human-mouse hybrids compared to control hybrids constructed between normal human skin fibroblasts and the same mouse cell line. The breakage rate in the control hybrids was between 0-0.03 breaks/cell, while in A-T hybrids four out of five clones showed over 0.14 breaks/cell. It seems that the phenomenon of chromosome breakage is a dominant character in the hybrids. The clone with only 0.06 breaks/cell might indicate correlation with chromosome segregation, although a greater number of clones will be needed to substantiate this observation. Experiments are now under way in which the karyotype of the hybrid clones will be analyzed in order to identify the human chromosomes present in the hybrid cells.

Table 4. Spontaneous chromosome breaks in cultured amniotic fluid cells of fetus at risk

	Karyotype	No. of cells analyzed	Breaks/cell
AT (II)[a]	46,XY	85	0.45
	46,XY,t(14;5)	15	0.33
AT (III)[b]	46,XY	93	0.35
Controls (N=6)	46,XX & 46,XY	150	<0.03

[a] Amniotic fluid obtained at second amniocentesis
[b] Amniotic fluid obtained at termination of pregnancy

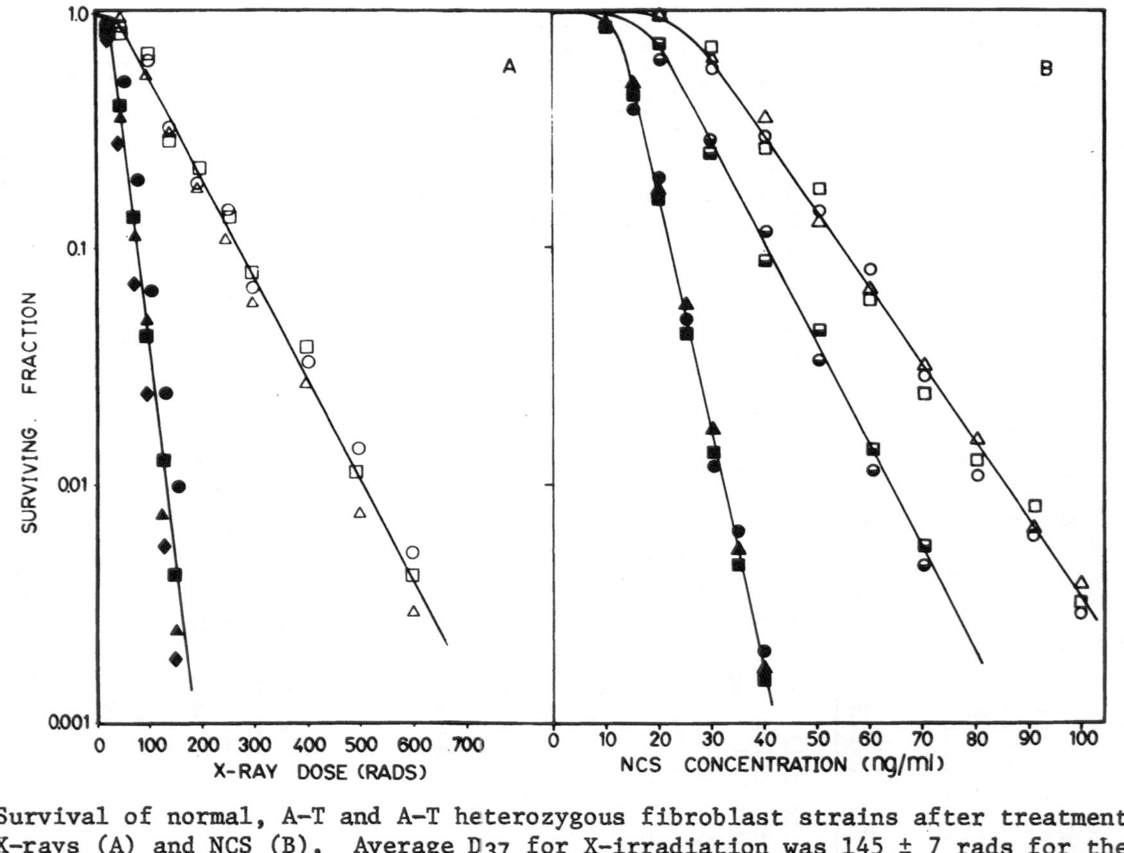

Fig. 1 Survival of normal, A-T and A-T heterozygous fibroblast strains after treatment with X-rays (A) and NCS (B). Average D_{37} for X-irradiation was 145 ± 7 rads for the three normal strains (○,□,△) and 52 ± 6 rads for the four A-T strains (●,■,▲,◆). Average D_{37} for NCS was 37.9 ± 0.8 ng/ml for the normal strains, 14.9 ± 0.25 ng/ml for the A-T strains and 26.9 ± 0.25 ng/ml for the A-T heterozygous strains (◐,◧). (Fig. 1B is from Shiloh et al., 1981b; with permission).

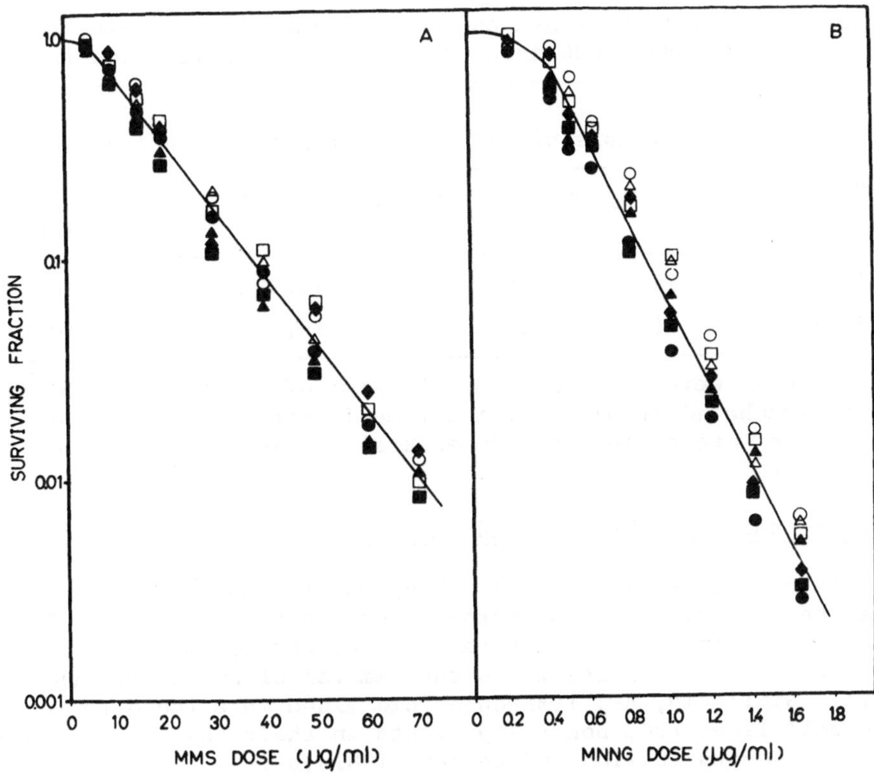

Fig. 2. Survival of normal (O,□, Δ) and AT (●,■, ▲, ♦) fibroblast strains after treatment with MMS (A) and MNNG (B). Average D_{37} for MMS and MNNG for all the cell strains was 18.0 ± 1.3 and 0.55 ± 0.031 µg/ml, respectively.

Cellular sensitivity to DNA damaging agents

Colony-forming ability served as the criterion for determining cellular survival following treatment with DNA damaging agents (Weichselbaum et al., 1980). Representative survival curves of A-T and normal fibroblast strains after treatments with X-rays, NCS, MNNG, and MMS are shown in Figs. 1 and 2. A marked hypersensitivity of the A-T cells to X-rays (Fig. 1A) and NCS (Shiloh et al., 1981b; Fig. 1B) is evident, while there is no difference between A-T and normal strains with regard to survival following treatment with either MMS (Fig. 2A) or MNNG (Fig. 2B). The A-T strains tend to cluster at the lower region of the variability zone of the MNNG survival curve, but we regard this possible slight difference as insignificant. The two A-T heterozygous strains demonstrate an intermediate degree of sensitivity to NCS, which is clearly distinct from that of the normal and the A-T homozygous strains (Shiloh et al., 1981b; Fig. 2B).

Table 5. Ability of A-T and normal fibroblast strains to repair MNNG induced purine methylation adducts

Cell strains	Repair capacity (fmole of DNA adduct repaired/10 hr/μg DNA		
	7-methylguanine	3-methyladenine	O^6-methyl-guanine
Normal (N=3)	11.77 ±0.048	3.41 ±0.021	3.61 ±0.49
A-T (N=4)	10.91 ±0.059	3.29 ±0.033	3.53 ±0.31

The cells were treated with 0.3 μg/ml of [^{3}H]MNNG for 3 min and incubated at 37°C for the time indicated. Zero time samples were cooled down immediately following treatment and lysed.

Removal of purine methylation adducts

Data on the ability of A-T and normal fibroblast strains to remove the three main purine methylation adducts induced by MNNG are summarized in Table 5. There is no difference between the two groups of strains with regard to the removal of any of the three adducts. Similarly, A-T lymphoblastoid lines were found to resemble cell lines from non-A-T patients in their ability to repair these DNA adducts (Shiloh and Becker, 1981a,b).

Is the A-T gene involved in the patient's predisposition to cancer?

Individuals homozygous for the A-T gene represent a subgroup in the population with a very high predisposition to cancer (Paterson and Smith, 1979). The most common cancers are lymphocytic leukemia and lymphomas, but other types of tumors were also reported (Hecht and McCaw, 1977). Chromosome No. 14 is preferentially involved in the chromosomal rearrangements occurring in A-T cells (Cohen et al., 1975) and a translocation involving this chromosome characterized a leukemic clone which developed in an A-T patient (McCaw et al., 1975). However, the general phenomenon of chromosomal breakage is not confined to a particular chromosome.

The finding that the A-T clastogenic peptide is capable of inducing chromosome breaks in normal lymphocytes may explain the phenomenon of increased chromosome breakage in A-T. It is of interest that a similar clastogenic factor is present in the plasma of A-T patients and in the amniotic fluid of an A-T affected fetus. A clastogenic factor was also observed in another chromosomal breakage syndrome predisposing to cancer, namely Bloom's syndrome (Emerit and Cerutti, 1981).

Cellular hypersensitivity to DNA damaging agents is another facet of A-T cells. The results presented here show that A-T cells repair alkylation damage normally and do not support previous data (Hoar and Sargent, 1975; Scudiero, 1980) on hypersensitivity of A-T cells to methylating carcinogens, but rather give further support to the hypothesis that DNA breaks of a certain type are the critical DNA lesion which accounts for the differential response of A-T cells to ionizing radiation. With our A-T strains we confirmed the hypersensitivity to X-rays, and preliminary results (data not shown) indicate their cellular hypersensitivity to bleomycin. We also found that these strains are markedly sensitive to the cytotoxic effect of another DNA breaking agent, NCS. This antitumor antibiotic induces primarily DNA strand scissions in vitro and in vivo (Beerman and Goldberg, 1974; Sarma et al., 1976). These data, together with the previously reported normal sensitivity of A-T cells to UV irradiation and "bulky" carcinogens (Paterson and Smith,1979) point to a specific defect in the repair of DNA breaks in A-T cells, although previous studies failed to show any deficiency in the repair of DNA single-stranded and double-stranded breaks induced by ionizing radiation (Vincent et al., 1975; Lehmann and Stevens, 1977, Sheridan and Huang, 1979). In these studies, however, common laboratory techniques for the measurement of DNA breaks with limited resolution were used, and a minor type of DNA breaks unrepaired in A-T cells may have escaped their detection. The proficiency of A-T strains in DNA repair synthesis may indicate that the possible unrepaired lesion constitutes only a minor fraction of the total damage induced by X-rays (Shiloh et al., 1981a).

Recently, a new explanation for the radiosensitivity of A-T cells was put forward based on the finding that, unlike normal cells, A-T cells fail to halt their DNA synthesis following X-irradiation (Edwards and Taylor, 1980; Painter and Young, 1980), or treatment with bleomycin (Cramer and Painter, 1981). We have similar findings with our A-T strains (Y. Shiloh, unpublished data) and assume that this phenomenon may either be linked to a defect in a DNA repair mechanism or may reflect a change in the DNA polymerizing system that allows a nascent strand to be synthesized on a damaged template, probably with a reduced fidelity.

A-T cells are therefore not only abnormally sensitive to DNA breaking agents, but also produce a clastogenic factor which may be a DNA breaking agent by itself. This situation may account for the continuously high breakage rate and rearrangements of chromosomes in vivo. It is possible that this high chromosomal breakage rate in vivo is linked to the increased tendency of A-T patients to develop tumors. These tumors may arise only when a specific cellular oncogene (Hayward et al., 1981) is activated as a result of chromosomal breakage or transfer of chromosome fragments from one chromosome to another.

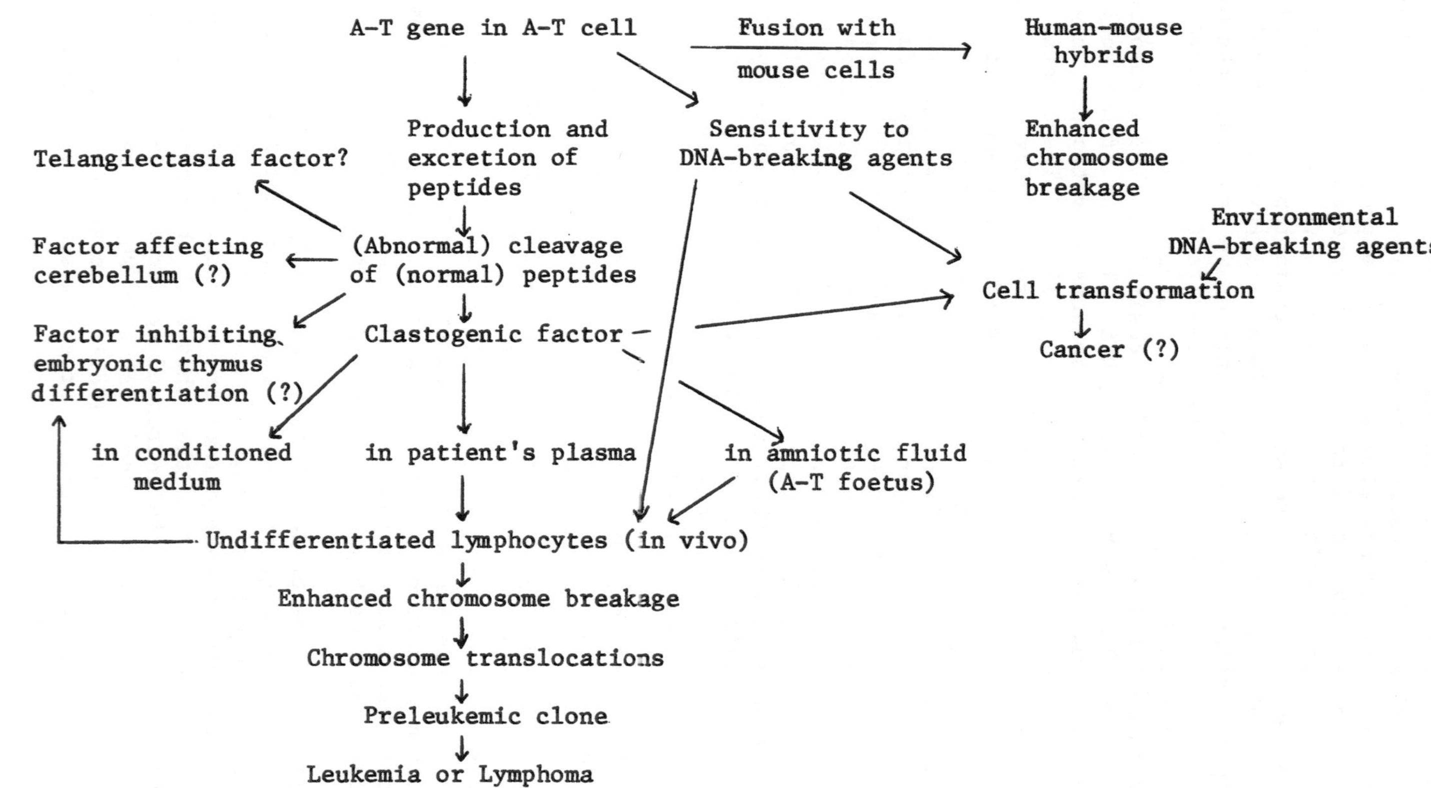

Fig. 3. Predisposition to cancer in ataxia-telangiectasia

Fig. 3 describes the possible involvement of the A-T clastogenic factor in the development of cancer in A-T patients. It is of interest that, since thymus development and differentiation in A-T patients is inhibited, the peripheral lymphocytes are not only immature but also have induced chromosomal breaks. The combination of the two, and possibly other, unknown factors could lead to the development of lymphatic and other tumors in A-T patients.

The heterozygote carriers of the A-T gene constitute another distinct cancerprone subpopulation (Swift et al., 1976). To date, no satisfactory laboratory test exists for the identification of the heterozygous state (Paterson et al., 1979). The data presented here point to a possibility of detecting the A-T heterozygotes on the basis of their differential sensitivity to DNA breaking agents. A-T heterozygotes may be more sensitive to DNA-breaking agents than people not carrying the A-T gene, and may also be producers of low levels of the A-T clastogenic factor. This may account for their cancer proneness. The results obtained with the human-A-T-mouse hybrid cells indicate that the breakage effect can be demonstrated in a "heterosomatic" state. The supposedly normal genes contributed by the mouse genome do not extinguish the gene responsible for the clastogenic effect in the A-T genome.

Understanding the function which the A-T gene controls may, therefore, contribute to our understanding of the mechanisms for genetic predisposition to cancer in humans.

ACKNOWLEDGMENTS

Supported by grants from The Leukemia Research Foundation, Inc., Chicago, USA, The Israel Cancer Association, the United States-Israel Binational Science Foundation, and The Leonard Wolfson Foundation for Scientific Research

REFERENCES

Beerman, T.A. and Goldberg, I.H., 1974, DNA strand scission by the antitumor protein neocarzinostatin, Biochem. Biophys. Res. Commun., 59:1254.

Cohen, M.M., Shaham, M., Dagan, J., Shmuelli, E. and Kohn, G., 1975, Cytogenetic investigations in families with ataxia-telangiectasia, Cytogenet. Cell Genet. 15:338.

Cramer, P. and Painter, R.B., 1981, Bleomycin-resistant DNA synthesis in ataxia-telangiectasia cells, Nature, 291:671.

Cunliffe, P.N., Mann, J.R., Cameron, A.H. and Roberts, K.D., 1975, Radiosensitivity in ataxia-telangiectasia, Br. J. Radiol., 48:374.

Edwards, M.J. and Taylor, A.M.R., 1980, Unusual levels of (ADP-ribose)$_n$ and DNA synthesis in ataxia-telangiectasia cells following X-ray irradiation, Nature, 287:745.

Emerit, E. and Cerutti, P., 1981, Clastogenic activity from Bloom's syndrome fibroblast cultures, Proc. Natl. Acad. Sci. USA, 78:1868.

Gotoff, S.P., Amirmokro, E. and Liebner, E.J., 1967, Ataxia-telangiectasia. Neoplasia, untoward response to X-irradiation and tuberous sclerosis, Amer. J. Dis. Child, 114:617.

Hayward, W.S., Neel, B.G. and Astrin, S.M., 1981, Activation of a cellular onc gene by promoter insertion in ALV-induced lymphoid leukosis, Nature 290:475.

Hecht, F. and McCaw, B.K., 1977, Chromosome instability syndrome, Progr. Cancer Res. Ther., 3:105.

Hoar, D.I. and Sargent, P., 1976, Chemical mutagen hypersensitivity in ataxia telangiectasia, Nature, 261:590.

Lehmann, A.R. and Stevens, S., 1977, The production and repair of double-strand breaks in cells from normal humans and from patients with ataxia-telangiectasia. Biochim. Biophys Acta 474:49.

Lehmann, A.R. and Stevens, S., 1979, The response of ataxia-telangiectasia cells to bleomycin, Nucleic Acids Res. 6:1953.

McCaw, B.K., Hecht, F., Harnden, D.G. and Teplitz, R.L., 1975, Somatic rearrangement of chromosome 14 in human lymphocytes, Proc. Natl. Acad. Sci. USA, 72:2071.

McFarlin, D.E., Strober, W., and Waldmann, T.A., 1972, Ataxia-telangiectasia. Medicine 51:281.

Morgan, J.L., Holcomb, T.M. and Morrissey, R.W., 1968, Radiation reaction in ataxia telangiectasia, Am. J. Dis. Child 116:557.

Painter, R.B., and Young, B.R., 1980, Radiosensitivity in ataxia-telangiectasia: a new explanation, Proc. Natl. Acad. Sci. USA 77:7315.

Paterson, M.C., Anderson, A.K., Smith, B.P. and Smith, P.J., 1979, Enhanced radiosensitivity of cultured fibroblasts from ataxia-telangiectasia heterozygotes manifested by defective colony-forming ability and reduced DNA repair replication after hypoxic X-irradiation, Cancer Res. 39:3725.

Paterson, M.C., Smith, B.P., Lohman, P.H.M., Anderson, A.K. and Fishman, L., 1976, Defective excision repair of X-ray damaged DNA in human (ataxia-telangiectasia) fibroblasts. Nature 260:444.

Paterson, M.C. and Smith, P.J., 1979, Ataxia-telangiectasia. An inherited human disorder involving hypersensitivity to ionizing radiation and related DNA damaging chemicals. Ann. Rev. Genet. 13:291.

Sarma, D.S.R., Pazalakshmi, S. and Samy, T.S.A., 1976, Studies on the interaction of neocarzinostatin with rat liver DNA in vivo and in vitro, Biochem. Pharmacol., 25:789.

Scudiero, D.A., 1980, Decreased DNA repair synthesis and defective colony-forming ability of ataxia-telangiectasia fibroblast cell strains treated with N-methyl-N'-nitro-N-nitrosoguanidine, Cancer Res.,40:984.

Shaham, M. and Becker, Y., 1981, The ataxia-telangiectasia factor is a low molecular weight peptide, Human Genet. 58:422.

Shaham, M., Becker, Y. and Cohen, M.M., 1980, A diffusable clastogenic factor in ataxia telangiectasia, Cytogenet. Cell Genet. 27:155.

Shaham, Kohn, G., Yarkoni, S., Becker, Y. and Voss, R., 1981, Prenatal diagnosis of ataxia-telangiectasia in cell-free amniotic fluid, J. Pediat. (in press).

Sheridan, R.B., and Huang, P.C., 1979, Ataxia-telangiectasia:further considerations of the evidence for single-strand break repair, Mutation Res., 61:415.

Shiloh, Y. and Becker, Y., 1981a, Ataxia-telangiectasia lymphoblastoid cell lines resemble non-A-T cell lines in their ability to remove MNNG-induced purine methylation adducts, in: "Ataxia-telangiectasia: A Cellular and Molecular Link Between Cancer, Neuropathology and Immune Deficiency", B.A. Bridges and D.G. Harnden, eds., John Wiley and Sons, London (in press).

Shiloh, Y. and Becker, Y., 1981b, Kinetics of O^6-methylguanine repair in human normal and ataxia-telangiectasia cell lines, and correlation of repair capacity with cellular sensitivity to methylating agents, Cancer Res. (in press).

Shiloh, Y., Cohen, M.M. and Becker, Y., 1981a, Ataxia-telangiectasia: studies on DNA repair synthesis in fibroblast strains, in: "Chromosome Damage and Repair", K. Kleppe and E. Seeberg, eds., NATO/EMBO Advanced Studies Institute, Bergen, Norway, 1980, Plenum Press, New York (in press).

Shiloh, Y., Tabor, E. and Becker, Y., 1981b, Cellular hypersensitivity to neocarzinostatin in ataxia-telangiectasia skin fibroblasts

Swift, L., Sholman, L., Perry, M. and Chase, C., 1976, Malignant neoplasms in the families of patients with ataxia-telangiectasia, Cancer Res. 36:209.

Taylor, A.M.R., Harnden, D.G., Arlett, C.F., Harcourt, S.A., Lehmann, A.R., Stevens, S. and Bridges, B.A., 1975, Ataxia-telangiectasia: a human mutation with abnormal radiation sensitivity, Nature 258:427.

Taylor, A.M.R., Rosney, C.M. and Campbell, J.B., 1979, Unusual sensitivity of ataxia-telangiectasia cells to bleomycin, Cancer Res. 39:1046.

Vincent, R.A., Sheridan, R.B. and Huang, P.C., 1975, DNA strand breakage and repair in ataxia-telangiectasia fibroblast-like cells, Mutation Res. 33:357.

Weichselbaum, R.R., Nove, J. and Little, J.B., 1980, X-ray sensitivity of fifty-three human diploid fibroblast cell strains from patients with characterized genetic disorders, Cancer Res. 40:920.

DEREPRESSION AND MODULATION OF PLASMINOGEN ACTIVATOR IN HUMAN CELL HYBRIDS AND THEIR SEGREGANTS

L. Larizza[1], M.L. Tenchini[1], E. Rampoldi[1]
L. De Carli[2], M. Colombi[3], S. Barlati[3]

[1]Istituto di Biologia Generale
Faculty of Medicine, Univ.of Milan,Italy
[2]Istituto di Genetica,Univ.of Pavia, Italy
[3]Istituto di Genetica Bioch.ed Evol.,CNR
Pavia, Italy

INTRODUCTION

Cell fusion has proven to be an useful tool for studying the genetic control of neoplastic transformation and its modulation *in vitro* (for review see,[1-4]). Several transformation parameters have been investigated in different systems of somatic cell hybrids produced from normal and tumorigenic parental cells. Increased fibrinolytic activity has been extensively studied as a transformation marker[5-8]. However, to our knowledge, no data are available on the dominance relationship of plasminogen activator (PA) in human intraspecific cell hybrids. We therefore investigated the expression of this marker in human cell hybrids between PA^+ HeLa cells and PA^- diploid fibroblasts. Furthermore, in order to check whether different degrees of chromosome loss could give rise to clones showing reversion or modulation of this transformation marker, segregation experiments were carried out on one of isolated hybrid lines. Since as with other transformed phenotypes, increased fibrinolytic activity has not been demonstrated to be sufficient or even necessary for transformation[8,9], we also attempted to define the relationship between the level of PA secretion and anchorage-independent growth.

MATERIALS AND METHODS

Parental cells

D98/AH-2 cells are a thymidine kinase-deficient variant of HeLa cells. GM 1362 are a strain of fibroblasts from a male with the Lesch-Nyhan syndrome, provided by the Human Genetic Mutant Repository (Camden; New Jersey). Cells were maintained in Eagle's minimum essential medium (MEM, Gibco) supplemented with 10% fetal or calf serum (Gibco).

Cell fusion

Hybridization was carried out with polyethyleneglycol (PEG 1000, Merck) at 50%, according to Davidson and Gerald[10]. Hybrids were selected in MEM supplemented with 10% fetal calf serum and HAT (hypoxanthine, aminopterin and thymidine). Colonies isolated by means of cloning cylinders were propagated and maintained in HAT medium until performing the PA assay and anchorage independence test.

Segregation experiments

For the segregation experiments, one of the hybrid lines, B5, was propagated in MEM (instead of HAT) supplemented with 10% calf serum (Gibco) for 2-3 passages. Cells, either treated with Griseofulvin (GF, Glaxo) as reported[11] or untreated, were then plated in medium containing 20 µg/ml of 8-Azaguanine (8AG, Sigma) at an inoculum of 5 x 10^4 or 10^5 cells/100 mm Petri dish. 8AG-resistant colonies were isolated by means of cloning cylinders 20-30 days after plating. The frequencies of spontaneous and GF-induced $8AG^R$ clones were 35 x 10^{-5} and 15 x 10^{-5}, respectively.

PA assay

Extracellular PA secretion was measured in 18 hr-serum free media, collected from semiconfluent cell cultures, by radial caseinolysis assay[12]. PA levels were quantified using human Urokinase (UK, Sigma) as a standard and activity was expressed as UK Sigma equivalents.

Growth in soft agar

Cells were tested for growth in MEM containing 0.3% agar (Noble

Agar, Difco), by the method of Mac Pherson and Montagnier[13]. Three replicate wells of a 24-well plate (Nunc), inoculated with 10^3 cells, were incubated in a humidified CO_2 incubator at 37°C with weekly feeding. The efficiency of colony formation was scored after 21 days' incubation.

RESULTS

The fusion between D98/AH-2 cells, homogenous at the clonal level with respect to PA activity, and diploid fibroblasts (GM 1362),

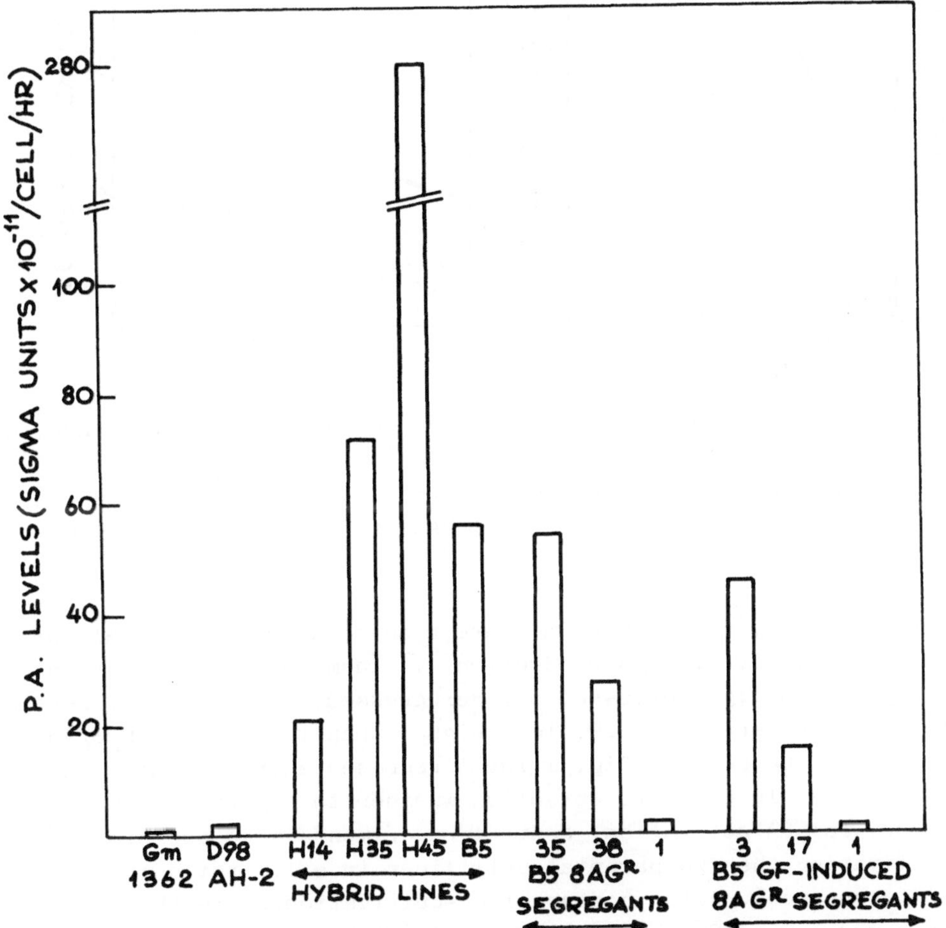

Fig. 1. Plasminogen activator levels in parental, hybrid and segregant lines. The secreted PA activity is given as Sigma Units/cell/h.

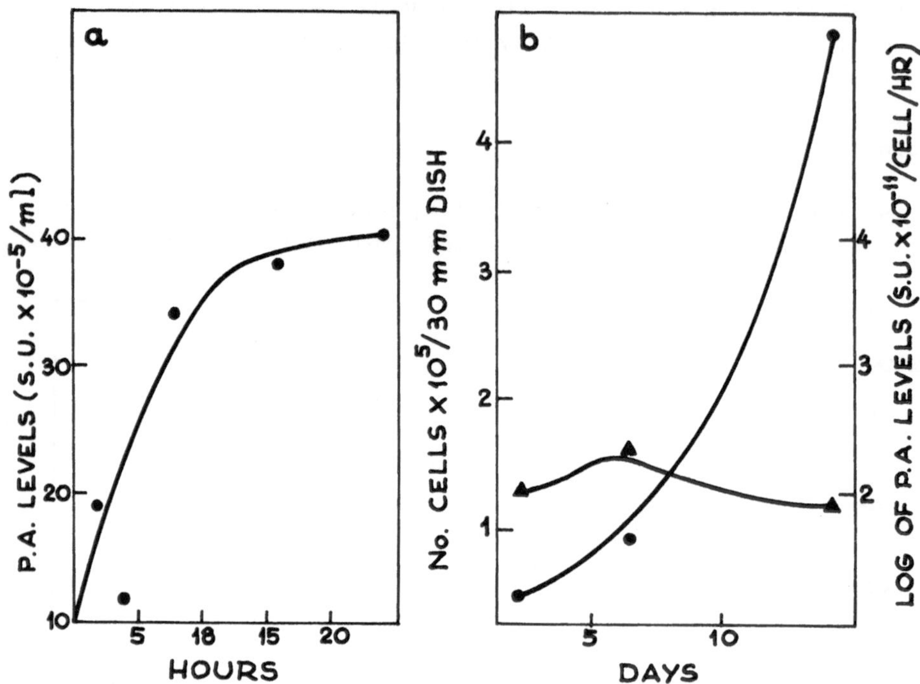

Fig. 2. a) Kinetics of PA release into serum-free medium by hybrid clone B5.
b) Production of PA as a function of cell growth: (■) cell counts; (▲) PA activity released/cell/h.

not producing any detectable PA, gave rise to hybrid clones all of which were positive for PA production[14]. Plasminogen activator was not only dominantly expressed, but derepressed, as the enzyme levels in the hybrids were up to 100-fold higher than that of the D98/AH-2 cells, as can be seen in Fig. 1 for 4 representative isolates (H14, H35, H45, B5). In order to establish whether the hybrid formation is necessary for PA activation, control experiments were performed by co-cultivating the two parental cell lines in a 1:1 ratio for periods of up to one week. After 5 days of co-cultivation, PA in the cell mixture was 1.6 x 10^{-5} S.U./ml, a level intermediate between that of the PA^+ parent (2.7 x 10^{-5} S.U./ml) and that of the PA^- parent (< 1.5 x 10^{-5} S.U./ml) grown independently.

In order to rule out the possibility that the higher PA levels

Table 1. Plasminogen Activator Levels and Growth in Soft Agar of Parental, Hybrid and Segregant Lines

Cells	Secreted PA Sigma Units/cells/h $x10^{-11}$	Plating efficiency in agar %
Parental cells:		
GM 1362	<0.06	$< 5 \times 10^{-5}$
D98/AH-2	2.7	38.72
Hybrid line:		
B5	113	12.16
B5 $8AG^R$ segregants:		
1	5	32.54
35	107	20.68
38	53	0.68
B5 GF-induced $8AG^R$ segregants:		
1	3	6.25
3	92	0.17
17	30	7.73

observed in the hybrid clones could be due to different culture conditions, the extracellular fibrinolytic activity of B5 cells was studied as a function of incubation time and cell concentration. As shown in Fig. 2a, PA reaches a plateau after 15 hrs incubation and its production is not dependent on cell density (Fig. 2b).

The PA^+ hybrid clones exhibited the transformed phenotype, since they were able to form colonies in soft agar, albeit with lower plating efficiencies than the transformed parental cells[14].

In order to derive secondary hybrid clones with a reduced number of chromosomes which were possibly segregating for the transformation traits observed, the B5 hybrid line was subcloned in medium containing 8AG. This was also accomplished using the drug GF which has been found to increase segregation to 8AG resistance in intraspecific human hybrids[11]. Back-selection in medium containing 8AG, either with or without GF pretreatment proved to be effective in recovering clones with reduced chromosome sets (data not shown). Both spontaneous and GF-induced $8AG^R$ isolates displayed greatly varying PA levels (Fig. 1), and cloning efficiencies in soft agar either increased or decreased as compared to B5 (Table 1).

DISCUSSION

One of the features of somatic cell hybrids derived from normal and tumorigenic cells appears to be the expression of the traits exhibited by the transformed parent[15,16]. In this study we have shown that this also applies to PA. It should be however emphasized that the level of PA production by D98/AH-2/fibroblast hybrids is up to 100-fold higher than that of their PA^+-transformed parent. Since D98/AH-2 cells have been demonstrated to be homogenous at the clonal level in PA activity, derepression of this function has probably occurred by cell fusion.

Activation of PA gene has been observed in murine intraspecific hybrids produced from PA^- tumor cells and PA^+ normal fibroblasts[17].

Activation of the human PA gene has been reported in interspecific hybrids derived from fusion of PA^- human diploid fibroblasts and a PA^+ mouse cell line[18]. We do not know whether this is the case with our intraspecific hybrids, because it is not possible to distinguish between derepression of the already active gene(s) of the PA^+ parent (D98/AH-2) and *de novo* activation of the fibroblast PA gene(s). Results obtained from segregation experiments provide clear evidence of modulation in the PA expression, which may be due either to a variation in gene dosage or to regulation of gene activity. However we did not observe any correlation between the level of fibrinolytic activity and cloning efficiency in agar in agreement with the results obtained by Jones et al.[7] and Barrett et al.[9]. Indeed, segregant clones could be isolated which expressed comparable levels of PA (B5-35 and B5-GF3), but which displayed a 100-fold difference in cloning efficiency in semisolid medium. On the other hand, clones with similar cloning efficiency in agar (B5-GF1 and B5-GF17) showed a 10-fold difference in PA production. These differences could not be ascribed to experimental variations, since there is only a slight difference in PA levels as a function of culture conditions as compared to the differences observed among parental, hybrid and segregant lines.

ACKNOWLEDGEMENTS

This work was supported by Progetto Finalizzato: Controllo della crescita neoplastica RFP Biol 1 (Contratto No. 81.01389.96) C.N.R., Rome, Italy.

REFERENCES

1. H. L. Ozer, and K. K. Jha, Malignancy and transformation: expression in somatic cell hybrids and variants, Adv. Cancer Res. 25:53 (1977).
2. G. Barski, and J. Belehradek Jr., Inheritance of malignancy in somatic cell hybrids, Somat. Cell Genet. 5:897 (1979).
3. H. Harris, Some thoughts about genetics, differentiation and malignancy, Somat. Cell Genet. 5:923 (1979).
4. E. Sidebottom, The analysis of malignancy by cell fusion, In vitro 16:77 (1980).
5. L. Ossowski, J. C. Unkeless, A. Tobia, J. P. Quigley, D. B. Rifkin, and E. Reich, An enzymatic function associated with transformation of fibroblast by oncogenic viruses. II. Mammalian fibroblasts cultures transformed by DNA and RNA tumor viruses, J. exp. Med. 137 : 112 (1973).
6. R. Pollack, R. Risser, S. Conlon, and D. Rifkin, Plasminogen activator production accompanies loss of anchorage regulation in transformation of primary rat embryo cells by simian virus 40, Proc. nat. Acad. Sci. (Wash.) 71:4792 (1974).
7. P. A. Jones, J. S. Rhim, and H. Isaacs, The relationship between tumorigenicity, growth in agar and fibrinolytic activity in a line of human osteosarcoma cells. Int. J. Cancer 16:616 (1975).
8. W. E. Laug, P. A. Jones, and W. Benedict, Relationship between fibrinolysis of cultured cells and malignancy, J. nat. Cancer Inst. 54:173 (1975).
9. J. C. Barrett, S. Sheela, O. Kazunori, and T. Kakunaga, Reexamination of the role of plasminogen activator production for growth in semisolid agar of neoplastic hamster cells, Cancer Res. 40:1438 (1980).
10. R. Davidson, and P. Gerald, Improved techniques for the induction of mammalian cell hybridization by polyethylene glycol. Somat. Cell Genet. 2:165 (1976).
11. L. Larizza, G. Simoni, M. Stefanini, and L. De Carli, Spontaneous and Griseofulvin induced segregation for 8-Azaguanine resistance in hybrids from a human heteroploid line. Exp. Cell Res. 120:405 (1979).
12. H. Saksela, Radial caseinolysis in agarose: a simple method for detection of plasminogen activator in the presence of inhibitory substances and serum. Analytical Biochemistry 111:276 (1981).
13. I. MacPherson, and L. Montagnier, Agar suspension culture for

the selective assay of cells transformed by polyoma virus. Virology 23:291 (1964).

14. L. Larizza, M. L. Tenchini, A. Mottura, E. Rampoldi, M. Colombi, and S. Barlati, Expression of fibronectin and plasminogen activator in human cell hybrids, in "Protides of the Biological Fluids", Pergamon Press, in press.
15. K. K. Jha, J. Cacciapuoti, and H. L. Ozer, Expression of transformation in cell hybrids. II. Non suppression of the transformed phenotype in hybrids between a chemically transformed and nontransformed derivatives of BALB/3T3, J. Cell. Physiol. 97:147 (1978).
16. E. J. Stanbridge, and J. Wilkinson, Analysis of malignancy in human cells: malignant and transformed phenotypes are under separate genetic control, Proc. nat. Acad. Sci. (Wash.) 75: 1466 (1978).
17. D. S. Straus, J. Jonasson, and H. Harris, Growth in vitro of tumor cell x fibroblast hybrids in which malignancy is suppressed, J. Cell Sci. 25:73 (1977).
18. R. Kuckerlapati, R. Tepper, A. Granelli-Piperno, and E. Reich, Modulation and mapping of a human plasminogen activator by cell fusion, Cell 15:1331 (1978).

BIOCHEMICAL MARKERS FOR HUMAN LEUKEMIA AND CELL DIFFERENTIATION

P. S. Sarin, M. Virmani, P. Pantazis and R. C. Gallo

Laboratory of Tumor Cell Biology
National Cancer Institute
Bethesda, Maryland 20205

SUMMARY

Biological markers that can be utilized in following patients during treatment and remission and are differentiation specific have been examined. Two such markers are: (1) terminal deoxynucleotidyl transferase (TdT) and (2) a histone polypeptide (HP). TdT is a unique DNA polymerase that can carry out DNA synthesis on an initiator molecule in the absence of a template. The potential usefulness of this enzyme in predicting the onset of relapse before any morphological indications has been demonstrated in chronic myelogenous leukemia patients in blast phase of the disease. In order to be able to detect low levels of TdT activity especially during remission phase, we have used cell separation techniques which can enrich cell populations containing TdT activity. We have used the techniques of unit gravity sedimentation and free flow electrophoresis to achieve enrichment of TdT positive cell populations. Our results show that up to 30 fold enrichment of TdT activity in normal human bone marrow can be accomplished by using cell separation techniques. With the use of free flow electrophoresis, we have achieved enrichment of TdT positive cell populations from normal human bone marrow, cells from patients with acute lymphoblastic leukemia and chronic myelogenous leukemia in blast phase of the disease. No TdT positive cells were detected in patients with acute myelogenous leukemia. These cell separation techniques should prove to be useful in early detection of relapse in patients in remission. A differentiation specific histone polypeptide (HP) with an apparent molecular weight of 12,500 has been identified among the acid extractable chromosomal proteins of a human myeloid cell line, HL60, treated with dimethyl sulfoxide (DMSO). The relative increase in the amount of HP in DMSO treated HL60 cells correlated with a concomitant decrease in histone H2A. Fingerprint analysis of tryptic digests shows HP and histone H2A are related, HP is generated from endogenous and exogenous H2A and is probably produced by a proteolytic enzyme associated with the chromatin of the differentiated HL60 cells. HP is not detected in a number of human or mouse hematopoietic tissues and cell lines. However, it is present in leukocytes from patients with acute leukemia. These results indicate that the observed level of HP in HL60 cells is related to the stage of differentiation and that HP is a potential biological markers for human acute leukemia.

INTRODUCTION

Attempts to obtain biological markers that could be useful in the identification of individuals at high risk of developing cancer, early detection of cancer and for following patients during treatment and remission have been made by a number of workers. We have been involved in developing biological and biochemical markers that could be useful in following progression of the disease in human leukemic patients, and markers that are specific for stage of differentiation in myeloid cells. Two such markers that we have studied are: (1) terminal deoxynucleotidyl transferase (TdT) and (2) a histone polypeptide (HP). TdT is an enzyme that can catalyze the polymerization of deoxyribonucleotides on the 3'-OH ends of oligo- and polydeoxyribonucleotide initiators in the absence of a template. This enzyme was discovered in thymus tissue and was considered to be specific for this tissue[1]. High levels of this enzyme have since been detected in various forms of human leukemia[2], including acute lymphoblastic leukemia (ALL)[3,5,7-10], acute myelomonocytic leukemia (AMML)[4], chronic myelogenous leukemia in blast phase of the disease[6-8], and acute undifferentiated leukemia[11]. High levels of terminal transferase have also been detected in cell lines with T cell characteristics, such as Molt-4 and 8402, derived from leukocytes of patients with acute lymphoblastic leukemia[12,13]. Detection of TdT in various forms of leukemia raises a question as to whether TdT is present in all cells or only in a select population of cells. Cell separation techniques can be useful in enriching cell population that contain the marker enzyme. One of the techniques of cell separation called free flow electrophoresis has been used in the past for separation of human[14] and mouse[15] bone marrow cells, and for the separation of cell membranes and organelles[16]. We have utilized this technique for separation of human leukemic cells into TdT positive cell populations. With this technique it is possible to obtain up to 30 fold enrichment in TdT positive cells thus making the detection of low levels of enzyme simpler, and providing a tool for selection and identification of cell types containing this enzyme for biochemical and biological studies.

A histone polypeptide (HP) was detected in acute promyelocytic cells (HL60) induced to differentiate with DMSO. Histones have been implicated in the structural integrity of chromatin by a number of workers[17-19]. The relative amounts of histone variants appear to be tissue specific[20] and have been shown to be associated with various stages of differentiation in mouse erythroid cells[21]. A human myeloid cell line (HL60) which can undergo differentiation in the presence of dimethylsulfoxide (DMSO)[22] provides a suitable model system for investigating changes in chromatin induced during differentiation. We have detected a histone related polypeptide (HP) in HL60 cells which is increased after induction of differentiation. HP is absent in histone complements of a number of mouse and human tissues including chronic leukemias. HP is, however, present in leukocytes from patients with acute leukemia.

MATERIALS AND METHODS

Materials

Tritium labeled deoxyribonucleoside triphosphates were obtained from New England Nuclear, Boston, Mass. Oligo- and polydeoxyribonucleotides, deoxyribonucleoside triphosphates and other chemicals used in gel electrophoresis were obtained from P. L. Biochemicals, Sigma Chemicals, and Biorad. Calf thymus histones were obtained from Boehringer-Mannheim and were labeld with [^{14}C]-formaldehyde according to the procedures of Rice and Means[23].

Source of Cells

Cultured cells from T (8402, CCRF-CEM, Molt 4) and B (8392) cell lines established from patients with acute lymphoblastic leukemia were grown at Biotech Laboratories. Fresh human cells were obtained by leukaphoresis from patients with leukemia, and were obtained from M. D. Anderson Hospital, Houston, Texas and NIH Clinical Center. HL60 cells used in the present study were from early passages 19-25 and were grown as described[22].

K562 is an erythroid cell line derived from a patient with chronic myelogenous leukemia[24]. A Friend virus-transformed mouse cell line (line T, clone 2) was treated with DMSO as described by Friend *et al.*[25]. BALB/c mice were obtained from Charles River Breeding Laboratories (Wilmington, Mass.). Normal human liver was obtained from a fresh autopsy of an individual killed in an accident. Logarithmically grown HL60 or DMSO-treated HL60 cells were treated with [H^3]-leucine (4 μCi/ml) for 24 hrs to yield radiolabelled proteins. Friend erythroleukemia cells were also labeled with [^{14}C]-leucine under similar conditions.

Processing of Cells

Human leukemic cells (2 x 10^9 cells) were washed with RPMI-1640 at 4°C. The process was repeated three times. The cells were layered on top of isolymph (ficol/hypaque) and centrifuged at 400g at room temperature. The separated leukocytes at the interface were removed, washed with phosphate buffered saline (PBS, pH 7.4) twice and used for the cell separation studies.

Cell Separation by Free Flow Electrophoresis

Separation of human normal and leukemic cells was performed with a free flow electrophoresis apparatus model FF5 (Bender and Hobein, Munich, Germany) using conditions similar to the ones already described[14-16]. Briefly, 2-5 x 10^8 cells were suspended in 10 ml of chamber buffer (15 mM triethanolamine, pH 7.4, 10 mM glucose and 4 mM potassium acetate) (TGK buffer), and applied through the entry porthole of the electrophoresis chamber at a dosage pump speed of 350. The Suction pump speed for the buffer in the chamber was set at 140. The separation chamber was maintained at 5°C. Voltage was adjusted to 550 volts, and the current was 150 amp. The electrode buffer consisted of 75 mM triethanolamine (pH 7.4) and 4 mM potassium acetate (TK buffer). At these settings approximately 2 x10^8 cells can be separated per hour. Cell fractions were collected in the fraction collector maintained at 4°. Cells were pelleted and washed with RPMI 1640. A portion of the cell pellet was used for making cytospin slides and the other was extracted in 0.5 ml buffer A (50 mM Tris-HCl (pH 7.5), 5 mM dithiothreitol and 20% glycerol) containing 0.3 M KCl and 0.3% triton X100.

Cell Extraction

Cell pellets suspended in buffer A containing KCl and triton X100 were sonicated at maximum output of a Branson sonifer (4 x 30 seconds) using a microtip. The extract was stirred in the cold for two hrs. and centrifuged at 100,000 xg for one hr. The supernatant was analyzed for the presence of terminal transferase and cellular DNA polymerase activity.

Enzyme Assays

Terminal transferase activity was assayed at 37°C for one hr. as described[6,7,13] in a standard reaction mixture (0.05 ml). Cellular DNA polymerase activity was measured under conditions similar to those described for terminal transferase except $(dT)_{15}(dA)_n$ was used as the primer-template and [^3H] dTTP was used as the radiolabeled triphosphate. DNA polymerase activity was assayed both in the presence of 0.6 mM Mn^{2+} or 10 mM Mg^{2+}. Cell fractions containing enzyme activity were pooled and TdT purified by various column chromatographic techniques as described[7,13].

Histone Extraction and Gel Electrophoresis

Chromatin was prepared from tissue culture cells or tissues and histones were extracted with acid as described[26]. Total cell protein from [^{3}H]-labelled cells was analyzed on 7.5 to 15% acrylamide gradient slab gels[26]. Each gel slot was loaded with approximately 10^5 cpm of sample. Changes in histone masses were studied on cylindrical gels by using the SDS-urea-phosphate acrylamide gel system which gives an excellent resolution of histones. Enzymatic cleavage products of histone H2A were analyzed by a two-dimensional electrophoresis. Gels were analyzed either after staining or after fluorography[26].

Peptide Analysis

Cylindrical polyacrylamide-SDS-urea-phosphate gels were used to obtain individual histones for peptide analysis. Stained gel-cuts with the histone-bands were extensively washed with 30% ethanol to remove SDS before lyophilization to dryness. The samples were labeled with ^{125}I, treated with trypsin (50 μg/ml) and the peptides analyzed by two-dimensional electrophoresis as described by Elder[27].

Digestion Experiments

For these experiments nuclei were prepared fresh and used immediately. The digestion conditions for autodigestion and digestion of exogenous labeled H2A were similar to those published[26].

RESULTS

Distribution of TdT in Human Leukemia

Examination of leukocytes from patients with different forms of leukemia has shown that high levels of this enzyme are present in the leukocytes of patients with acute lymphoblastic leukemia[3], and some patients with chronic myelogenous leukemia in blast phase of the disease[6-8]. As shown in Table 1, high levels of terminal transferase have been observed in leukocytes from 374/407 patients with ALL and 91/236 CML patients in the blast phase of the disease. Leukocytes from patients with other diseases, including CLL, CML stable phase, Sezary syndrome, xeroderma pigmentosa, systemic lupus erythrematosus, and infectious mononucleosis were found to be negative for this enzyme. Low levels of terminal transferase have also been observed in approximately 8% of the AML patients. The levels of terminal transferase activity observed in thymus tissue and T and B cell lines derived from patients with acute lymphoblastic leukemia is shown in Table 2. As is evident from this table high levels of terminal transferase are present in human and calf thymus tissue and T cell lines whereas B cell lines and PHA stimulated human peripheral blood lymphocytes do not contain any detectable terminal transferase activity.

Table 1: Distribution of Terminal Transferase in Various Cells

#	Diagnosis	TdT	# Positive/ # Tested	Mean TdT Level (nm/10^9 cells)	% TdT Positive
1.	ALL	+++	374/407	90	92
2.	CML (BC)	++++	91/236	300	39
3.	CML (stable)	-	0/80	<0.5	0
4.	AML	±	24/310	5	8
5.	CLL	-	0/25	<0.5	0
6.	Sezary syndrome	-	0/28	<0.5	0
7.	Xeroderma pigmentosa	-	0/6	<0.5	0
8.	Systemic lupus erythematosus	-	0/12	<0.5	0
9.	Infectious mononcuelosis	-	0/14	<0.5	0
10.	Hodgkin's disease	-	0/6	<0.5	0
11.	Hairy cell leukemia	-	0/4	<0.5	0

Table 2: Distribution of Terminal Deoxynucleotidyl Transferse in Various Cells

#	Tissues	Source	TdT	Mean TdT Level (nmoles/10^9cells)
1.	Thymus	Human/Calf	++++	400-500
2.	8402	ALL	++++	400
3.	Molt 4	ALL	++++	350-400
4.	SB	ALL	-	<0.5
5.	8392	ALL	-	<0.5
6.	PHA stimulated human blood lymphocytes	Human (Normal)	-	<0.5

Morphology of Terminal Transferase Positive Cells

From the studies carried out with cells from patients with ALL at the time of diagnosis and continued serially throughout induction chemotherapy, hematological remission, and relapse, it appears that there is no direct correlation between the presence of T cell like surface markers and terminal transferase in such cells[10]. This is probably due to the presence of terminal transferase in immature or primitive cells which are precursors of mature T cells[7,10,28]. Studies with acute undifferentiated leukemia also show a lack of lymphoid cell surface markers on these leukemic cells[11].

Cell Separation

The unique biochemical properties of TdT make it a novel enzyme for easy detection and characterization. To be effective as a biological marker, the enzyme should be detectable in low amounts so as to have a predictive value for early detection of relapse. Our studies on TdT positive patients with chronic myelogenous leukemia (CML) in blast crisis indicate that TdT positive CML patients respond to vincristine and prednisone therapy[29,30] and that TdT levels begin to increase 5-6 months before any morphological indications of relapse[29].

In our earlier attempts to enrich TdT positive cell populations from human bone marrow, we have used the technique of unit gravity sedimentation in sucrose[28]. In order to be able to obtain a large number of cells for biochemical and biological studies, it is important to utilize a separation technique that can separate large quantitites of cells in a relatively short period. Free flow electrophoresis is one such technique by which it is possible to separate up to 2 x 10^8 cells per hour. This method separates cells based on their net surface charge differences. This method could be used both for the detection of low levels of TdT as well as for obtaining large numbers of TdT positive cells for biochemical and biological studies. The protocol used for the free flow electrophoresis (FFE) of fresh human normal and leukemic cells and cell lines is outlined in Table 3.

Separation of T and B Cells

The cells from a patient with acute lymphoblastic leukemia have been used to establish a T (8402) and a B cell line (8392). This T cell line contains terminal transferase and is a useful cell line to determine the distribution of cells positive for terminal transferase after free flow electrophoresis. We have used these cell lines to check the distribution of T and B cells on free flow electrophoresis. Figure 1 shows the separation of T cells (8402) on free flow electrophoresis. As can be seen from this figure, majority of the terminal transferase and cellular DNA polymerase activity is contained in cells in fractions 20-30. These cells migrate toward the anode and carry a net negative surface charge. Figure 2 shows the distribution of B cells (8392) on free flow electrophoresis. A major portion of the cellular DNA polymerase activity is distributed between fractions 25 and 30. Separation of a mixture of T and B cells (a mixture of $2x10^8$ cells each of 8402 and 8392) by free flow electrophoresis is shown in Figure 3. In this figure the leading and the trailing edge of the various fractions contain terminal transferase and cellular DNA polymerase activities. This is in contrast to the distribution of terminal transferase and cellular DNA polymerase activities in T and B cells shown in Figures 1 and 2. The distribution observed in Figure 3 suggests the possibility that net surface charge carried by the cells present in a mixed population is somewhat different than the net surface charge carried by a homogeneous population of cells.

Table 3: Separation of Human Fresh Cells and Cell Lines by Free Flow Electrophoresis

Fresh Cells (Suspend in RPMI 1640)

|

Separate on Isolymph (Ficoll-Hypaque)

|

Interface Cells/or T or B Cells

|

$1-5x10^8$ Cells Resuspended in 10 ml. of Chamber Buffer

|

8 ml. Applied for Free Flow Electrophoresis

|

Fractions Count for Cell Number

|

Centrifuge at Low Speed

|

Make Slides from each Fraction for Morphological Study.
Assay for Terminal Transferase and Cellular DNA Polymerase Activity

Separation of Human Normal Bone Marrow

We have analyzed human normal bone marrow to determine if it is possible to enrich cell populations that may be positive for TdT. We have shown in the past that low level of TdT activity present in normal bone marrow can be enriched in a subpopulation of cells by simple cell separation techniques[28]. Figure 4 shows

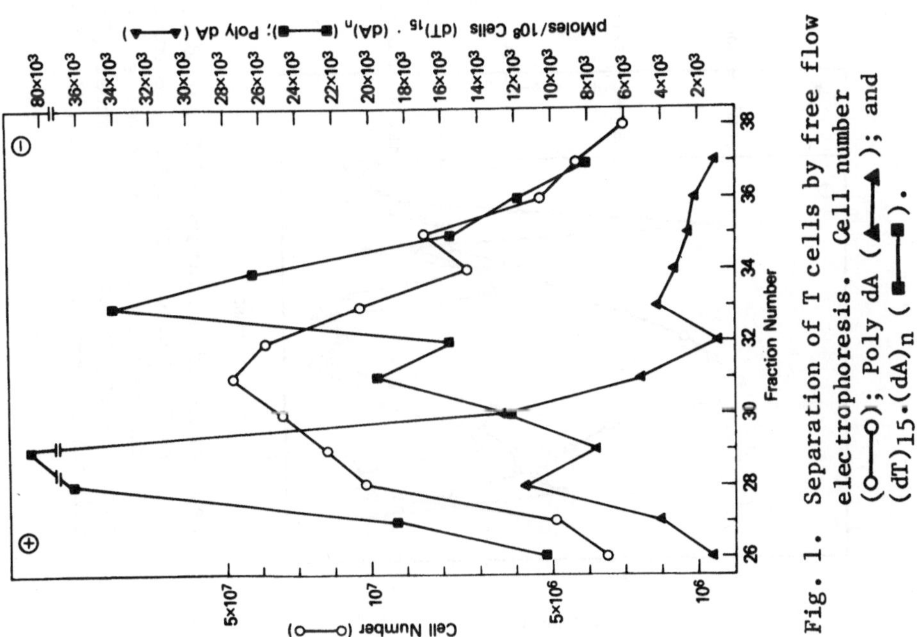

Fig. 1. Separation of T cells by free flow electrophoresis. Cell number (○—○); Poly dA (▲—▲); and $(dT)_{15}\cdot(dA)_n$ (■—■).

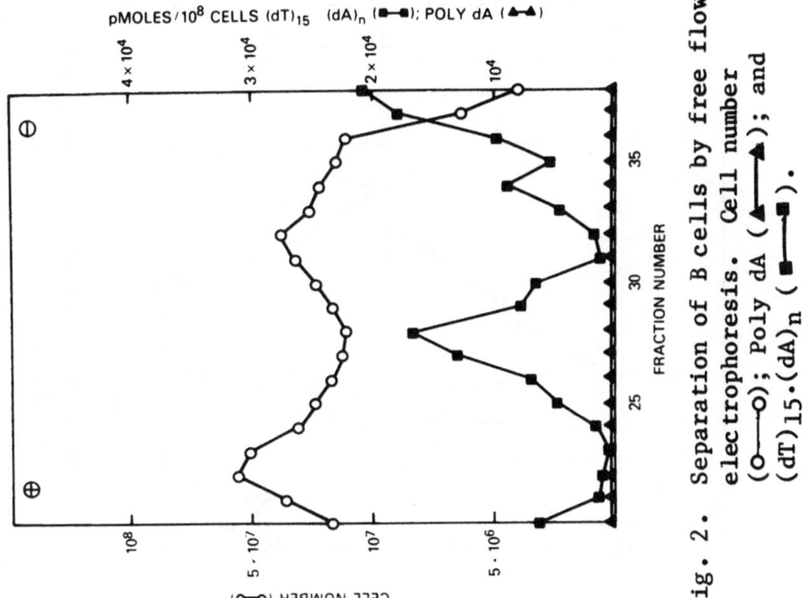

Fig. 2. Separation of B cells by free flow electrophoresis. Cell number (○—○); Poly dA (▲—▲); and $(dT)_{15}\cdot(dA)_n$ (■—■).

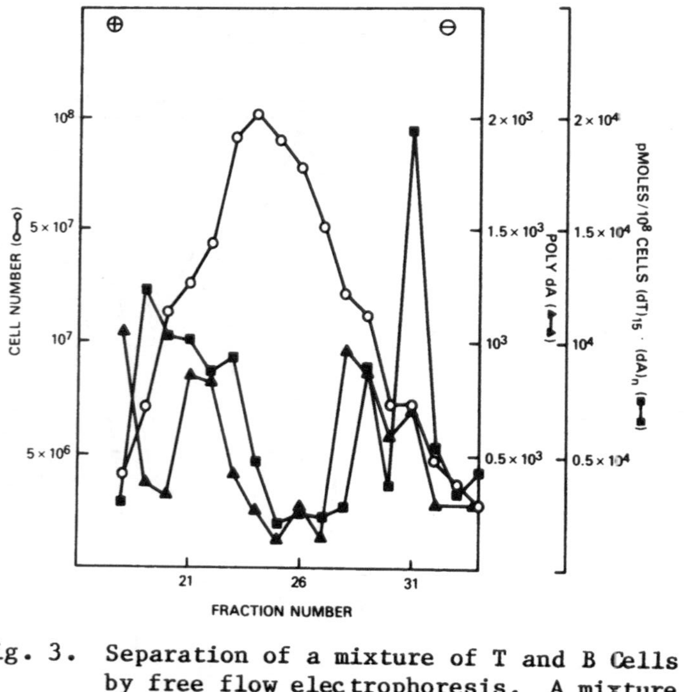

Fig. 3. Separation of a mixture of T and B Cells by free flow electrophoresis. A mixture of equal numbers of T and B cells was used for this experiment. Cell number (O——O); Poly dA (▲——▲); and $(dT)_{15} \cdot (dA)_n$ (■——■).

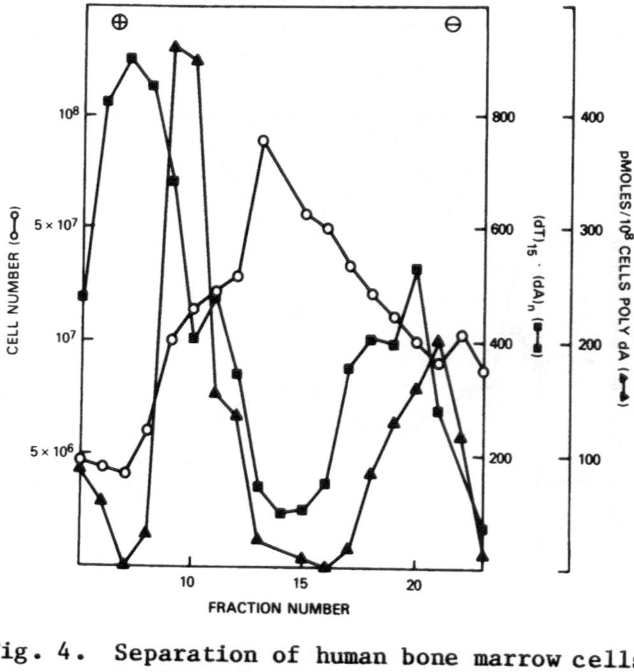

Fig. 4. Separation of human bone marrow cells by free flow electrophoresis. Cell number (O——O); Poly dA (▲——▲); and $(dT)_{15} \cdot (dA)_n$ (■——■).

the distribution of human normal bone marrow into two distinct cell populations which contain TdT and cellular DNA polymerase activity. This is somewhat similar to that observed with a mixture of T and B cells shown in Figure 3. A large increase in TdT activity seen in a subpopulation of bone marrow cells indicates that it is possible to enrich for subpopulations of cells which carry the same biological marker.

Separation of Human Leukemic Cells

Representative examples of separation of human leukemic cells by free flow electrophoresis are shown in Figures 5-8. Separation of cells from patients with ALL is shown in Figure 5. As can be seen from this figure there is an enrichment of subpopulation of cells that contain TdT as a biochemical marker. Analysis of cells from patients with AML is shown in Figure 6. No terminal transferase positive cell populations were seen in the AML patient. Figure 7 shows the enrichment of TdT positive cells from a CML patient in blast crisis, whereas Figure 8 shows the absence of TdT positive cell populations in CML patients in chronic phase of the disease. Tables 4 and 5 show that up to 30 fold enrichment in the terminal transferase or DNA polymerase activity can be achieved by cell separation using free flow electrophoresis technique. Thus it appears that cell separation techniques may prove to be useful for the detection of biochemical markers that may be expressed in these cells. This technique could be very useful in following patients during remission phase for early detection of biological markers that may signal the onset of relapse. This early detection may prove useful in giving treatment ealier than otherwise may be possible by detection of relapse by cell morphology alone.

Response to Vincristine and Prednisone

The detection of high levels of TdT in patients with ALL and in some patients with CML in blast phase suggests that these patients have cells blocked early in their differentiation. Moreover, the detection of TdT in some CML patients in blast phase has suggested that these cells may be early lymphoblasts rather than myeloblasts [7,9,31]. Since, steroid therapy is commonly used in ALL, attempts to use vincristine and prednisone for treatment of CML patients in blast phase has been successful in producing remission in the past[31,32]. Recent studies on TdT positive CML patients in blast phase have shown excellent correlations with the presence of TdT and response to treatment with vincristine and prednisone[29,30,33]. High levels of TdT present in CML blast phase revert to background levels during remission. Table 6 summarizes the chemotherapeutic response of CML blast crisis patients positive and negative for TdT. Approximately 80% of the patients positive for TdT responded to the chemotherapeutic treatment that included vincristine and

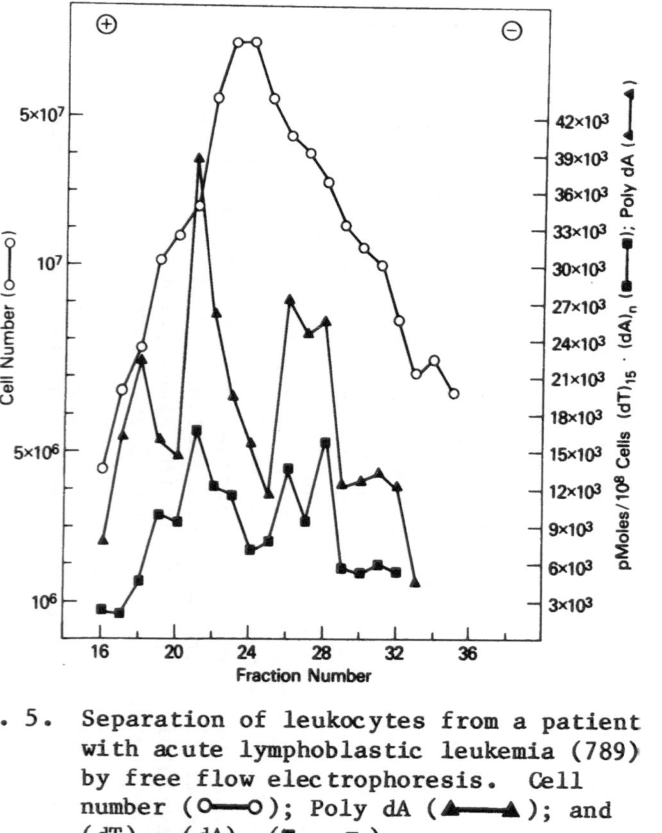

Fig. 5. Separation of leukocytes from a patient with acute lymphoblastic leukemia (789) by free flow electrophoresis. Cell number (○—○); Poly dA (▲—▲); and $(dT)_{15} \cdot (dA)_n$ (■—■).

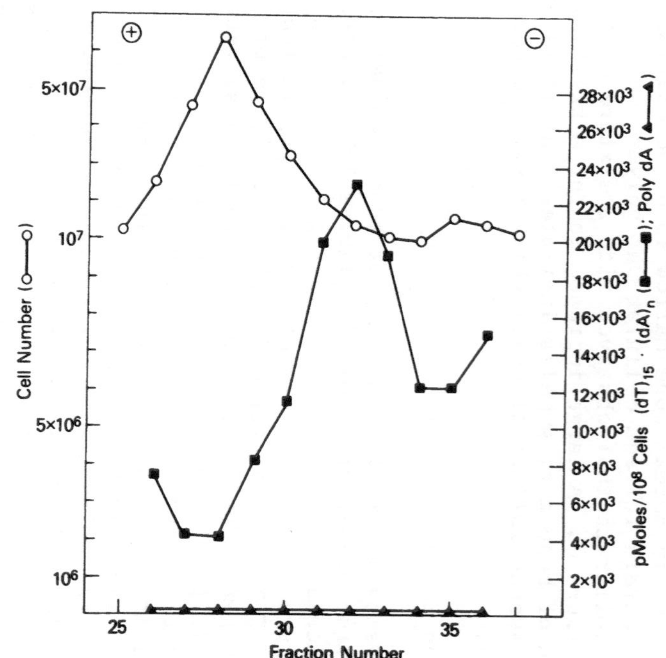

Fig. 6. Separation of leukocytes from a patient with acute myelogenous leukemia (725) by free flow electrophoresis. Cell number (○—○); Poly dA (▲—▲); and $(dT)_{15} \cdot (dA)_n$ (■—■).

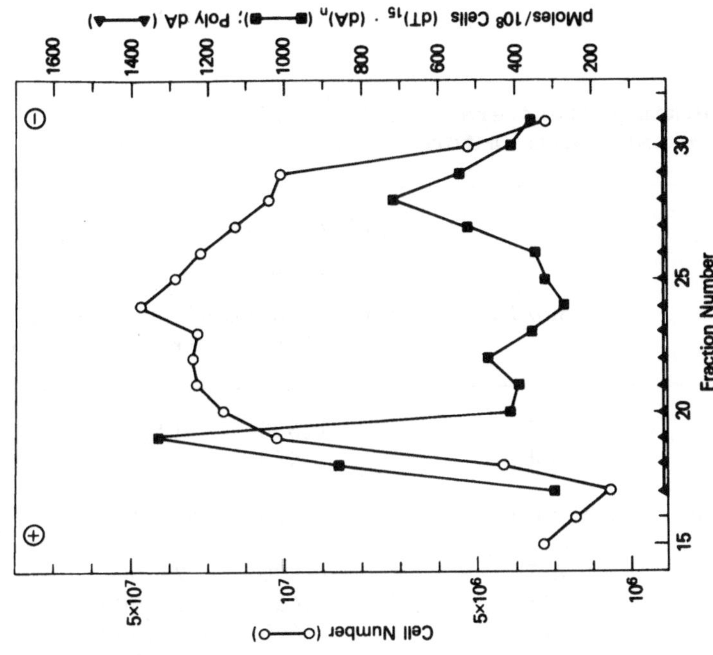

Fig. 8. Separation of leukocytes from a patient with chronic myelogenous leukemia (739) by free flow electrophoresis. Cell number (○—○); Poly dA (▲—▲); and $(dT)_{15} \cdot (dA)_n$ (■—■).

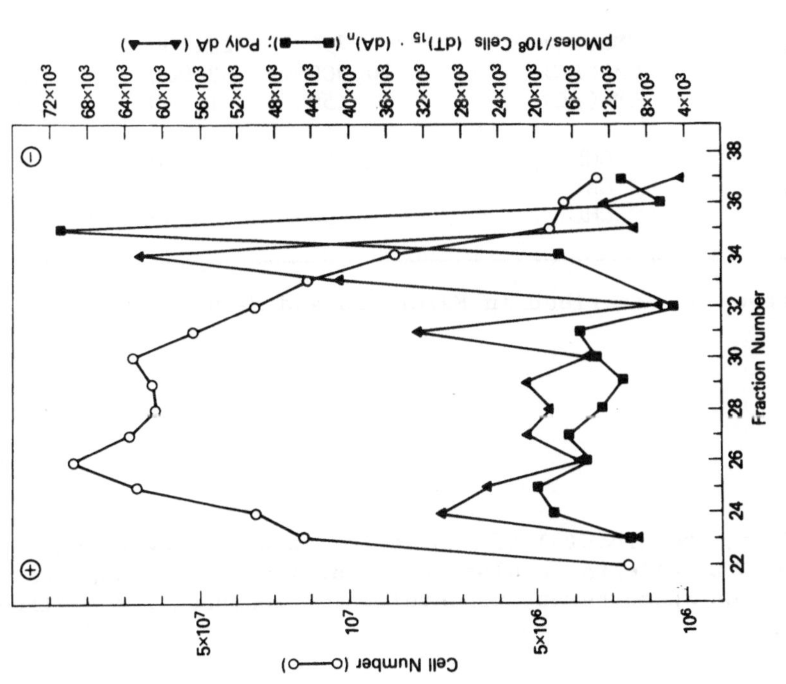

Fig. 7. Separation of leukocytes from a patient with chronic myelogenous leukemia in blast phase of the disease (FM) by free flow electrophoresis. Cell number (○—○); Poly dA (▲—▲); a $(dT)_{15} \cdot (dA)_n$ (■—■).

Table 4: Terminal Transferase Activity in Human Normal and Leukemic Cells Before and After Cell Separation

#	Source of Cells	Diagnosis	TdT* Activity (pmoles/10^8 cells)		
			Unfractionated Cells	Peak Fraction	Enrichment (Fold)
1.	T (8402)	ALL	1,390	12,900	9
2.	B (8392)	ALL	<50	<50	-
3.	Bone Marrow	Normal	70	500-1400	7-20
4.	719	ALL	1,830	22,500	12
5.	789	ALL	5,430	38,900	7
6.	792	ALL	460	8,530	30
7.	FG	ALL	250	7,540	18
8.	711	AML	<50	<50	-
9.	725	AML	<50	<50	-
10.	7101	AML	<50	<50	-
11.	VH	CML(BC)	710	23,400	33
12.	FM	CML(BC)	10,900	62,900	6
13.	7100	CML(BC)	<50	1,080	20
14.	737	CML	<50	<50	-
15.	739	CML	<50	<50	-
16.	799	CML	<50	<50	-

*TdT was assayed as described in Materials and Methods.

prednisone. Thus, the procotol of choice for inducing remission in TdT positive CML patients in blast phase of the disease appears to be a combination chemotherapy procotol which includes vincristine and prednisone.

Table 5: DNA Polymerase Activity in Human Normal and Leukemic Cells Before and After Cell Separation

#	Source of Cells	Diagnosis	DNA Polymerase Activity* ($pm/10^8$ cells) Unfractionated Cells	Peak Fraction	Enrichment (Fold)
1.	T (8402)	ALL	6,380	82,600	13
2.	B (8392)	ALL	3,300	75,400	23
3.	Bone Marrow	Normal	130	900-3,000	6-25
4.	719	ALL	4,570	15,600	3
5.	789	ALL	5,430	13,400	3
6.	792	ALL	7,300	113,000	15
7.	FG	ALL	6,200	47,300	8
8.	711	AML	590	6,670	11
9.	725	AML	1,130	23,000	20
10.	7101	AML	2,200	30,500	14
11.	VH	CML(BC)	760	21,200	28
12.	FM	CML(BC)	18,670	79,700	4
13.	7100	CML(BC)	90	600	7
14.	737	CML	150	1,450	10
15.	739	CML	110	4,670	42
16.	799	CML	150	2,370	16

*DNA polymerase activity was assayed with $(dT)_{15} \cdot (dA)_n$ as described in Materials and Methods.

Table 6: Response of Chronic Myelogenous Leukemia (Blast Crisis) Patients to Vincristine/Prednisone Therapy

Reference	TdT (+)	TdT (-)	Ph_1
Oken *et al.* (1978)	1/1	NT	+
Marks *et al.* (1978)	8/13	0/9	+
Ross *et al.* (1979)	1/1	NT	+
Janossy *et al.* (1979)	6/7	0/14	+
Total (# +/# Tested)	16/22	0/23	

Detection of Histone Polypeptide (HP)

Proteins extracted with acid form chromatin of HL60 cells, on analysis by SDS-PAGE showed gel profiles typical of histones except for the presence of HP. HP has a molecular weight of 12,500. After DMSO treatment of HL60 cells, the amount of HP was increased with a parallel decrease in histone H2A. Increase of HP after DMSO treatment is not directly due to DMSO per se, since in control experiments DMSO treatment of other cells does not result in the appearance of HP. Analysis of [^{3}H]-leucine labeled HL60 cells before and after treatment with DMSO showed the presence of HP in DMSO treated HL60 cells (Fig. 9) thus suggesting that HP is an integral part of the differentiated cells and is not an artifact of isolation.

Relatedness of HP to H2A

In order to determine the relatedness of HP to other histones including H2A, tryptic digests of radioiodinated HP and H2A were analyzed by two dimensional gel electrophoresis[27]. Fingerprint analysis of the tryptic digests showed HP to be similar to H2A (Fig. 10). Since the electrophoretic patterns and the amino acid sequence of H2A is significantly different from other human histones, the fingerprint analysis strongly suggests that HP is derived from H2A.

Generation of HP from H2A

Two dimensional electrophoretic analysis of autodigests of nuclei from radiolabeled HL60 cells treated with DMSO showed almost complete digestion of H2A to HP after incubation at 37°C for 4 hr (Fig. 11). Analysis of autodigests of nuclei from HL60 cells showed very little digestion of endogenous H2A. Incubation of radiolabeled mouse H2A with sonicated nuclei of DMSO treated HL60 cells produced HP (Fig. 12), whereas no HP was generated when radiolabeled mouse H2A was mixed with nuclear lysates from mouse or human cells. These results suggest that HP is a degradation product of H2A and not a product of an independent gene, and that the appearance of HP in autodigests of nuclei of DMSO-treated cells was not due to a direct effect of DMSO on H2A. The involvement of H2A or its variants in cell differentiation has been reported for mouse erythroid cells[21] and aging mammalian cells[20]. The present results indicate that H2A and HP are associated with changes in differentiation state of HL60 cells.

Distribution of HP in Cells and Tissues

HP was not detected in the histone complement of (a) human leukemic T cell lines (CCRF-CEM and Molt-4); (b) human erythroid cell line (k562); (c) mouse Friend erythroleukemia cells before

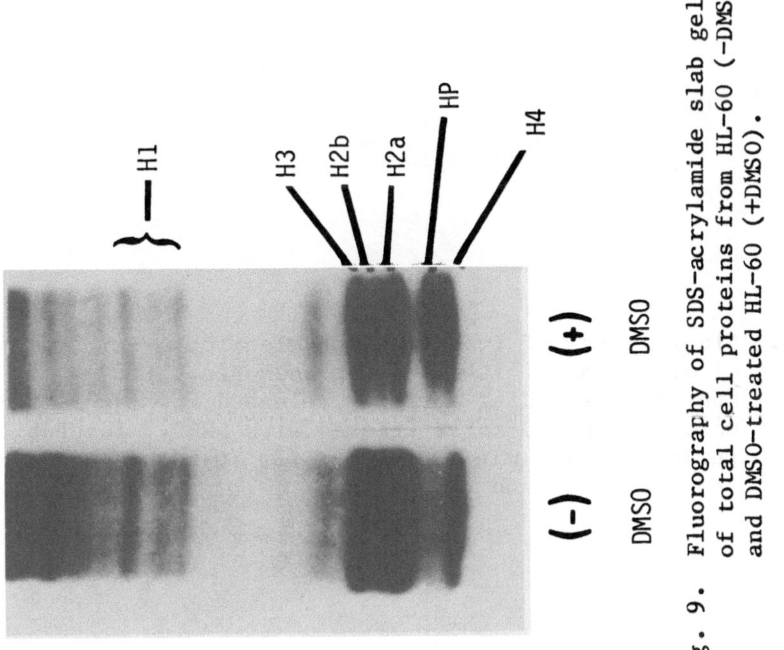

Fig. 9. Fluorography of SDS-acrylamide slab gels of total cell proteins from HL-60 (-DMSO) and DMSO-treated HL-60 (+DMSO).

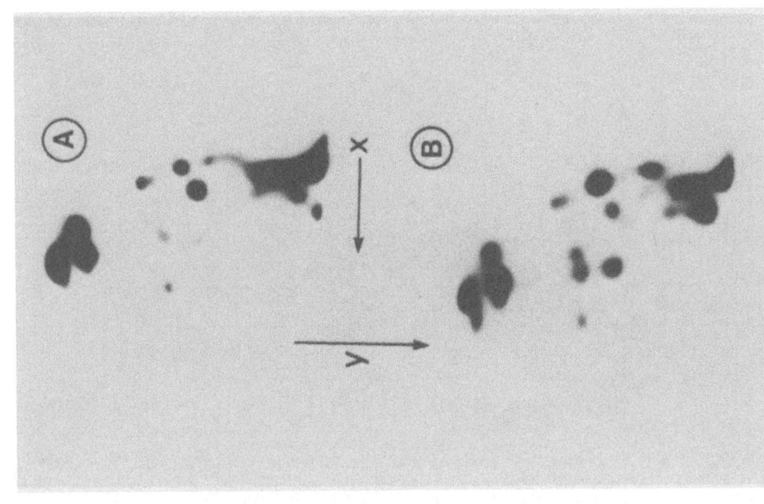

Fig. 10. Two-dimensional electrophoresis of (I^{125})-labeled tryptic digests from H2A (panel A) and HP (panel B). First dimension electrophoresis was done in X-direction followed by second dimension electrophoresis in Y-direction.

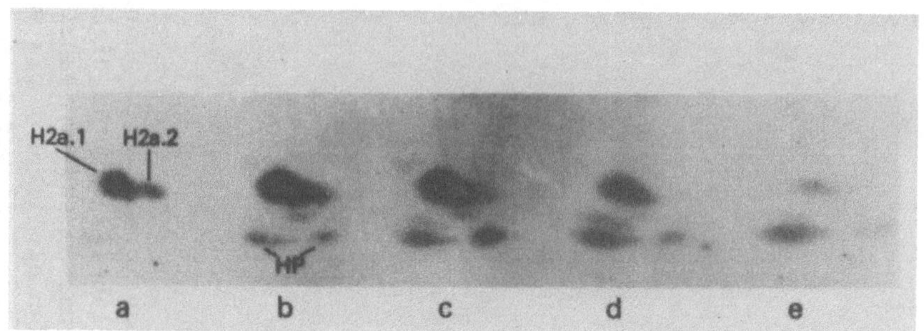

Fig. 11. Fluorography of two dimension electrophoretic patterns of autodigestion of H2a from HL-60 cells treated with DMSO. Incubation was at 37°C for: (a) 0 min., control; (b) 15 min.; (c) 60 min.; (d) 120 min.; and (e) 360 min.

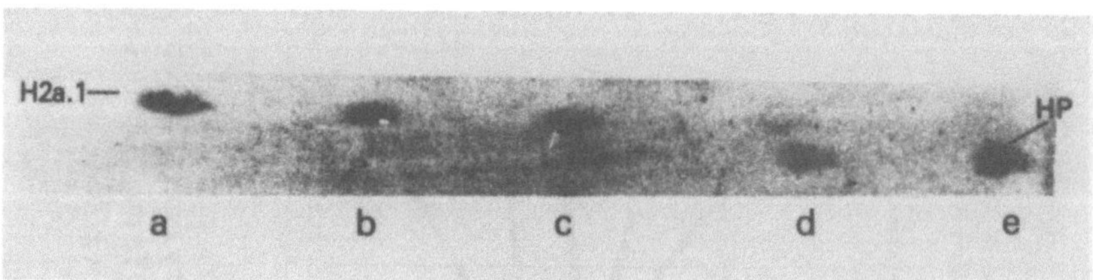

Fig. 12. Fluorography of two dimension electrophoretic patterns of digestion of labeled exogenous H2a.1 (mouse) by H2A-specific degrading activity released from nuclei of unlabeled HL-60 cells treated with DMSO. Incubation periods: (a) 0 min., control; (b) 10 min.; (c) 30 min.; (d) 120 min.; and (e) 360 min.

and after DMSO treatment; and (d) mouse spleen and liver. Low levels of HP were detected in fresh human granulocytes and lymphocytes before and after stimulation with phtohemagglutinin (Table 7). Since HL60 cell line is derived from a patient with acute promyelocytic leuekmia, we have examined leukocytes from patients with different forms of leukemia for the presence of HP. HP was present in leukocytes from patients with acute lymphoblastic leukemia (4/4) and acute myelogenous leukemia (6/6) whereas leukocytes from patients with chronic myelogenous leukemia (0/5) and chronic lymphocytic leukemia (0/3) did not contain any HP[26] (Table 8). This selective association of HP with acute leukemias suggests the possible usefulness of HP as a biological marker for detection and classification of acute leukemias.

DISCUSSION

Previous studies in evaluating human leukemic cells for leukemia specific molecules, including retrovirus related proteins and nucleic acids, have only been successful in a limited number of patients. This may be due to the fact that these leukemia specific molecules are present only in a small population of leukocytes, thereby making detection difficult and variable. In the past, we have reported on the enrichment of terminal transferase activity in human bone marrow cells[28] by using the techniques of ficoll-hypaque followed by unit gravity sedimentation in sucrose. We have utilized the techniques of free flow electrophoresis for separation of human leukemic cells into different cell populations for detection of terminal transferase. This technique of free flow electrophoresis has been used in the past for separation of human[14] and mouse[15] bone marrow cells, and for the separation of cell membranes and organelles[16].

With the use of cell separation techniques it is possible to enrich TdT positive cell populations for early detection of relapse and for biochemical and biological studies. In our studies on the followup of TdT positive CML patients during treatment and remission, we have observed that the TdT positive CML patients respond to vincristine and prednisone therapy[29,30], and their TdT level start going up 5-6 months before any morphological indications of relapse[29]. These studies point to the usefulness of techniques to enrich cell populations containing biological markers for early detection of relapse or for early indications of the onset of a disease state. As is evident from the results obtained on cell separation, up to 30 fold enrichment in TdT activity can be achieved by using the technique of free flow electrophoresis.

The detection of a histone polypeptide (HP) in DMSO treated HL60 cells and its specificity to acute leukemias point to the usefulness of HP as a potential marker for the stage of differentiation of human

Table 7: HP Distribution in Various Tissues and Cells

Cells or Tissue	HP
A. Human	
HL-60	+
HL-60/DMSO	+++
NC37	-
NC37/SSV*	-
NC37/BaEV	-
NC37/GALV	-
A204	-
A204/SSV	-
A204/BaEV	-
A204/GALV	-
8402 (T Cells)	-
8392 (B Cells)	-
Molt-4 (T Cells)	-
CCRF-CEM (T Cells)	-
CCRF-CEM/SSV (T Cells)	-
Lymphocytes	±
PHA-stimulated Lymphocytes	±
Granulocytes	±
Liver (Normal)	-
K562	-
B. Mouse	
T3 #2	-
T3 #2 + DMSO	-
BALB/c Spleen	-
BALB/c Liver	-

*SSV, Simian sarcoma virus; BaEV, baboon endogenous virus; GALV, gibbon ape leukemia virus; A204, human rhabdomyosarcoma cells; A204/SSV represents A204 cells with simian sarcoma virus. See "Materials and Methods" for more details.

Table 8: HP Distribution in Human Leukemia Patients

Patient	Diagnosis	Sex	Age	HP
A. Acute Leukemias				
JT 834	ALL*	M	28	++
AB 858	ALL	M	30	++
JL 865	ALL	M	28	++
KG 874	ALL	F	29	++
FF 711	AML	F	47	++
FM 7105	AML	M	45	++
LL 875	AML	F	48	++
GH 842	AML	M	46	++
EF 849	AML	M	49	++
PB 856	AML	M	48	++
B. Chronic Leukemias				
EB 7113	CML	F	48	-
WB 7117	CML(BC)	M	53	-
CM 897	CML(BC)	M	52	-
JP 898	CML(BC)	M	50	-
IN 8103	CML	M	67	-
LM 835	CLL	F	69	-
RW 867	CLL	F	69	-
JK 8101	CLL	M	69	-

*ALL, acute lymphocytic leukemia; AML, acute myelogenous leukemia; CML, chronic myelogenous leukemia; CML(BC), chronic myelogenous leukemia with blast crisis; and CLL, chronic lymphocytic leukemia.

myeloid cells and for classification of acute leukemias. Several reports have suggested the involvement of H2A or its variants in cell differentiation in sea urchin[34], newt embryos[35], mouse erythroid cells[21] and aging mammalian cells[20]. The preparation of a monoclonal antibody against this histone polypeptide, and the availability of a fluorescence activated cell sorter (FACS) should be useful in the analysis of different cell types for the presence of differentiated cells and their selective removal from a mixed population. The availability of the antibody should also be useful in the subtyping of acute leukemias. It will be of interest to isolate and analyze specific protease responsible for the degradation of histone H2A. Antibody against this protease may be useful in the detection of differentiated cells by direct immunofluorescence or with the help of a cell sorter (FACS). The availability of FACS and suitable monoclonal antibodies against various antigens including cell surface antigens could prove useful in the identification, separation and classification of T and B cells, and other cell types.

REFERENCES

1. F. J. Bollum, Terminal deoxynucleotidyl transferase, in: "The Enzymes" P. D. Boyer, ed., Academic Press, N.Y., p. 145 (1974).
2. P. S. Sarin, Terminal transferase as a biological marker for human leukemia, in: "Recent Studies in Cancer Research" R. Gallo, ed., CRC Press, Cleveland, p. 131 (1977).
3. R. McCaffrey, D. F. Smoler and D. Baltimore, Terminal deoxynucleotidyl transferase in a case of childhood acute lymphoblastic leukemia, Proc. Natl. Acad. Sci. USA 70:521 (1973).
4. M. S. Coleman, J. J. Hutton, P. D. Simone, and F. J. Bollum, Terminal deoxynucletidyl transferase in human leukemia, Proc. Natl. Acad. Sci. USA 71:4404 (1974).
5. S. Khan, J. Minowada, E. Henderson and I. Tabowski, Terminal deoxynucleotidyl transferase activity and blast cell characteristics in adult acute leukemias, Leukemia Res. 4:209 (1980).
6. P. S. Sarin and R. C. Gallo, Terminal deoxynucleotidyl transferase in chronic myelogenous leukemia, J. Biol. Chem. 249:8051 (1974).
7. P. S. Sarin, P. N. Anderson and R. C. Gallo, Terminal deoxynucleotidyl transferase activities in human blood leukocytes and lymphoblast cell lines. High levels in lymphoblast cell lines and in blast cells of some patients with chronic myelogenous leukemia in acute phase, Blood 47:11 (1976).
8. P. S. Sarin and R. C. Gallo, Terminal deoxynucleotidyl transferase as a biological marker in human leukemia, in: "Modern Trends in Human Leukemia" R. Neth, ed., J. F. Lehmann's Verlag, Munich, Vol. 2, p. 491 (1976).
9. R. McCaffrey, T. Harrison, R. Parkman and D. Baltimore, Terminal deoxynucleotidyl transferase activity in human leukemic cells and in normal human thymocytes, N. Engl. J. Med. 292:775 (1975).
10. M. S. Coleman, M. F. Greenwood, J. J. Hutton, F. J., Bollum, B. Lampkin and P. Holland, Serial observations on terminal deoxynucleotidyl transerase activity and lymphoblast surface marker in acute lymphoblastic leukemia, Cancer Res. 36:120 (1976).
11. S. L. Marcus, S. W. Smith, C. L. Jarowski and M. J. Modak, Terminal deoxyribonucleotidyl transferase activity in acute undifferentiated leukemia, Biochem. Biophys. Res. Commun. 70:37 (1976).

12. B. Srivastava and J. Minowada, Terminal deoxynucleotidyl transferase in a cell line (Molt-4) derived from the peripheral blood of a patient with acute lymphoblastic leukemia, Biochem. Biophys. Res. Commun. 51:529 (1973).
13. P. S. Sarin and R. C. Gallo, Characterization of terminal deoxynucleotidyl transferase in a cell line (8402) derived from a patient with acute lymphoblastic leukemia, Biochem. Biophys. Res. Commun. 76:673 (1975).
14. J. C. F. Schubert, F. Walther, E. Holzberg, G. Pascher and K. Zeiller, Preparative electrophoretic separation of normal and neoplastic bone marrow cells, Klin. Wschr. 51:327 (1973).
15. K. Zeiller, J. C. F. Schubert, F. Walther and K. Hanning, Electrophoretic distribution analysis of in vivo colony forming cells in mouse bone marrow, Hoppe-Seyler's Z. Physiol. Chem. 353:95 (1972).
16. K. Hanning and H. G. Heidrich, The use of continuous preparative free flow electrophoresis for dissociating cell fractions and isolation of membranous components, in: "Methods in Enzymology" S. Fleischer and L. Packer, L. eds., Academic Press, New York, Vol. 31, p. 746 (1974).
17. R. Axel, W. Melchoir, B. Sollner-Webb and G. Felsenfeld, Specific sites of interaction between histones and DNA in chromatin, Proc. Natl. Acad. Sci. USA 71:4101 (1974).
18. G. Felsenfeld, Chromatin, Nature 271:115 (1978).
19. R. E. Kornberg and J. D. Thomas, Chromatin structure: oligomers of the histones, Science 184:865 (1974).
20. A. Zweidler, M. Urban and P. Goldman, The origin and significance of tissue-specific histone variant patterns in mammals. in: "Miami Winter Symposia", F. Ahmad, T. R. Russel, J. Schultz and R. Werner, eds., Academic Press, New York, Vol. 15, p. 531 (1978).
21. L. A. Blankenstein and S. B. Levy, Changes in histone F2A2 associated with proliferation of Friend leukemic cells, Nature 260:638 (1976).
22. S. J. Collins, F. W. Ruscetti, R. E. Gallagher and R. C. Gallo, Terminal differentiation of human promyelocytic leukemia cells induced by dimethyl sulfoxide and other polar compounds, Proc. Natl. Acad. Sci. USA 75:2458 (1978).
23. R. H. Rice and G. E. Means, Radioactive labelling of proteins in vitro, J. Biol. Chem. 246:831 (1971).
24. C. B. Lozzio and B. B. Lozzio, Human chronic myelogenous leukemia cell line with positive Philadelphia chromosome, Blood 43:321 (1975).
25. C. Friend, W. Scher, J. G. Holland and T. Sato, Hemoglobin synthesis in murine virus-induced leukemic cells in vitro: stimulation of erythroid differentiation by dimethyl sulfoxide, Proc. Natl. Acad. Sci. USA 68:378 (1971).

26. P. Pantazis, P. S. Sarin and R. C. Gallo, Detection of a histone-2A related polypeptide in differentiated human myeloid cells (HL-60) and its distribution in human acute leukemia, Int. J. Cancer 27:585 (1981).
27. J. H. Elder, F. C. Jensen, H. L. Bryant and R. A. Lerner, Polymorphism of the major envelope glycoproteins (gp70) of murine C-type viruses: virions associated with differentiation antigens encoded by multi-gene family, Nature 267: 23 (1977).
28. R. D. Barr, P. S. Sarin and S. Perry, Terminal transferase in human bone marrow lymphocytes, Lancet 1:508 (1976).
29. M. M. Oken, P. S. Sarin, R. C. Gallo, et al. Terminal transferase levels in chronic myelogenous leukemia in blast crisis and in remission, Leukemia Res. 2:173 (1978).
30. D. D. Ross, P. H. Wiernik, P. S. Sarin and J. Whang-Peng, Loss of terminal deoxynucleotidyl transferase activity as a predictor of emergence of resistance to chemotherapy in a case of chronic myelogenous leukemia, Cancer 44:1566 (1979).
31. D. R. Boggs, Hematopoietic stem cell theory in relation to possible lymphoblastic conversion of chronic myeloid leukemia, Blood 44:449 (1974).
32. G. P. Canellos, V. T. DeVita, J. Whang-Peng and P. P. Carbone, Hematologic and cytogenetic remission of blastic transformation in chronic granulocytic leukemia, Blood 38:671 (1971).
33. S. M. Marks, D. Baltimore and R. McCaffrey, Terminal transferase as a predictor of initial responsiveness to vincristine and prednisone in blast crisis myelogenous leukemia. New Eng. J. Med. 298:812 (1978).
34. L. Cohen, K. Newrock and A. Zweidler, Stage-specific switches in histone synthesis during embryogenesis of the sea urchin, Science 190:994 (1975).
35. H. Imoh, Re-examination of histone changes during development of newt embryos, Exp. Cell Res. 113:23 (1978).

ADENOSINE DEAMINASE AND TERMINAL TRANSFERASE IN LEUKEMIC CELLS

U. Bertazzoni[1], A.I. Scovassi[1], S. Torsello[1]
E. Brusamolino[2], P. Isernia[2], C. Bernasconi[2]
E. Ginelli[3], N. Sacchi[3]

[1]Istituto CNR Genetica Biochimica Evoluzionistica, Pavia
[2]Divisione di Ematologia, Policlinico S. Matteo, Pavia
[3]Istituto di Biologia Generale, Università di Milano

INTRODUCTION

The characterization of leukemic blasts has been recently greatly improved by the use of biochemical markers, such as Terminal deoxynucleotidyl Transferase (TdT) and Adenosine Deaminase (ADA).

TdT catalyzes the random addition of deoxynucleotides to 3' termini of DNA. Its function is still unknown though its restricted distribution to immature cell populations of the lymphocyte series (1) and its *in vitro* mechanism of action have suggested a role as a somatic generator of immunological diversity (2, 3). The enzyme proved to be a very useful marker for identifying lymphocyte progenitors and hence for studying T- and B-cell differentiation (4, 5, 6). High levels of TdT were reported in acute leukemic lymphoblasts (ALL), in about 30% of patients with Chronic Myelogenous Leukemia (CML) in blastic transformation, and on rare cases of acute myeloid leukemia (AML) (see ref. 1 and 7 for a review). These types of leukemic cells are believed to be the neoplastic counterparts of the normal hematopoietic cells frozen at a certain stage of their differentiation (8). The availability of a monospecific antiserum against TdT has recently made possible the analysis of

This publication is contribution no. 1819 of the Radiation Protection Programme of the Commission of the European Communities. U.B. is a scientific official of CEC.

individual cells in bone marrow of leukemic patients, thus permitting a more accurate diagnosis especially when used in combination with other markers (9).

ADA catalyzes the conversion of adenosine and deoxyadenosine to inosine and deoxyinosine, respectively. The enzyme appears to play a very important role in purine nucleotide catabolism and in the development of the immune systems in humans. The inherited absence of ADA is associated with a severe combined immune deficiency characterized by thymic involution and T- and B-lymphocyte dysfunction (10). A considerable accumulation of dATP has been found in erythrocytes and lymphocytes from ADA-deficient children, which could be responsible for the observed inhibition of DNA synthesis in lymphoid cells and possibly result in the preferential utilization of this nucleotide by TdT (11). The analysis of ADA level in animal tissues has shown that the enzyme, at variance with TdT, is present in all different organs though lymphoid cells present higher ADA activity in comparison with other tissues; in particular, the level of ADA is elevated in thymus and the enzyme is preferentially associated with cortical thymocytes (12, 13). In effects, it has been recently shown that the human thymus-leukemia-associated antigen corresponds to the low molecular weight form of ADA (14). It has also been shown that coformycin, a potent inhibitor of ADA, is able to block the maturation of T cells (15), thus suggesting that the enzyme is an important factor in the differentiation of precursor lymphocytes into T cells. Larger amounts of ADA are contained in lymphoblasts derived from patients with acute leukemia (16, 17) whereas subnormal levels of the enzyme have been reported in chronic lymphocytic leukemia (18, 19).

We have analyzed the levels of TdT and ADA in several hundred cases of acute lymphoid and myeloid leukemias with the aim of finding a possible correlation between the two markers that would permit a more precise attribution of the leukemic blasts to their lymphoid or myeloid origin, and allow a better characterization of the blastic transformations of CML. ADA and TdT activities were also measured in human hematopoietic cell lines derived from lymphoid and myeloid leukemias. We have followed the variations of these two enzymes during the differentiation induced in the T leukemic cell line RMPI-8402 by the tumor promoter TPA.

MATERIALS AND METHODS

Patients

All adult patients involved in this study were admitted to the Division of Hematology of Policlinico San Matteo in Pavia between January 1979 and March 1981. All children were admitted to the De Marchi Pediatric Clinic of the University of Milan between January

1979 and March 1981. Analysis of results was blind. Clinical features, cytochemistry, surface markers and enzyme assay were analyzed separately by different investigators. The classification of acute leukemia was based on conventional morphology and cytochemical findings, as stated in 1976 by an international panel (FAB).

Enzyme Assays

Mononuclear cells, separated on Phicoll-Hypaque gradient, were washed with PBS, counted and frozen at -20°C. The extraction was made by resuspending the cells at a density of 1-2 x 10^8 in 0.25 M K phosphate, 1 mM mercaptoethanol and sonicating 2-4 times for 15 sec. The cellular debris was removed by centrifugation at 40,000 RMP for 60 min at 2°C and the supernatant was used as enzymatic extract. The TdT reaction mixture contained in a vol. of 0.25 ml: 0.2 M K cacodylate pH 7.5, 0.5 mM ^{3}H dGTP (100 cmp/pmole), 4 mM $MgCl_2$, 0.01 mM poly d$(pA)_{50}$, 1 mM mercaptoethanol and enzyme extract. Acid insoluble radioactivity was measured as already described (7). One unit of TdT activity catalyzes the incorporation of 1 nmole of nucleotide in 1 hour. The ADA assay measures the conversion of adenosine into inosine and is essentially as described by Coleman and Hutton (20). The reaction mixture contained in a vol. of 0.1 ml: 0.05 M K phosphate pH 7.5, 0.6 mM ^{14}C adenosine (6000 cpm/nmole), 1 mM 2-mercaptoethanol and enzyme extract. One unit of enzyme activity corresponds to 1 nmole of inosine produced in 1 min.

TdT Immunofluorescence (IF)

The assay was performed essentially as described by Stass *et al.* (21). Testing was made on smears of bone marrow aspirates, bone marrow touch preparations, cytocentrifugal spread of cerebro-spinal fluid, which were air-dried and kept at room temperature for no longer than 10 days. The primary(monospecific rabbit antibodies to homogenous calf TdT) and the secondary (purified goat anti-rabbit IgG coupled to FITC) reagents were a gift from Prof. F.J. Bollum. The specimens were examined with a Leitz Orthoplan microscope equipped with a 40 Phaco objective and IR filter for FITC. The field was observed under phase microscopy for counting the total number of white cells and epifluorescent light for scoring the percentage of cells with fluorescent nuclei.

Cell Cultures

The following human hematopoietic cell lines were used: the T leukemic line RPMI-8402 (22), the normal B line RPMI-8392, the non-T, non-B leukemic line NALM-1 (23) and the leukemic myeloid line HL-60 (24). The cells were cultured in RPMI 1640 medium supplemented with 15% heat inactivated fetal calf serum, L-glutamine 2 mM, penicillin-streptomycin 100 U/ml at 37°C in 5% CO_2 atmosphere. In the *in vitro* induction experiments the RPMI-8402 cells were grown in the presence of 10^{-7} M 12-O-tetradecanoylphorbol 13-acetate (TPA) for 96 hours.

RESULTS

TdT and ADA in leukemia

The cases of adults with acute leukemia at presentation were classified by conventional methods as: ALL, 38 cases; AML, 55 cases and Acute Unclassifiable Leukemia (AUL), 15 cases. The 58 cases of childhood acute leukemia presented as ALL (51 cases), AML (6 cases) and a single case as AUL. The values of TdT and ADA enzymatic activities obtained in acute leukemias are reported in Fig. 1 and the results of TdT immunofluorescence (IF) are shown in Fig. 2. By using cell surface marker determinations, the ALL cases were subdivided into T-ALL, non-T, non-B ALL and B-ALL. The TdT and ADA values were reported according to this subclassification. TdT was tested in peripheral blood and bone marrow with both the biochemical and/or the IF method in adults, and mainly by the IF test in children. ADA was determined only in peripheral blood cells. A close correlation between enzymatic and IF assays was found for TdT. In the positive cases the percent of fluorescent cells ranged between 40 and 100% of total nucleated cells. The ADA enzymatic assay was performed in 27 adults and 9 children with ALL (see Fig. 1). The range of ADA varied between 20 and 7580 U/10^8 cells, with a mean of 1458±393 U/10^8 cells in adults and 745±334 U/10^8 cells in children. These values were significantly higher ($P<0.001$) than those found in non-neoplastic controls (113±16). The mean value of ADA in adult T-ALL (3996 ±931 U/10^8 cells) is significantly higher ($P<0.001$) than the mean value of 607±78 U/10^8 cells in adult non-T, non-B ALL. In adult B-ALL the mean is 74±27 U/10^8 cells and differs significantly from both T-ALL and non-T, non-B ALL ($P<0.001$) but not from normal values. TdT and ADA values found in peripheral lymphoblasts of adult ALL at presentation were plotted against one another on logarithmic scales. Three distinct biochemical groups of ALL in adults are recognized (see Fig. 3). T-ALL group is characterized by positivity to TdT and by significantly higher levels of ADA; the group of B-ALL presents the distinctive features of TdT absence and ADA activities within the range of normal controls. The cases of non-T, non-B ALL are more heterogeneous since the TdT values extend on a larger scale and ADA activities range between T-ALL and B-ALL cases. The TdT biochemical activity in peripheral blood of patients in remission was non-detectable and in bone marrow fell within normal control values; in 2 adults showing only partial remission the TdT activity presented values of 2 and 3.5 U/10^8 cells. As shown in Fig. 2, in all cases of adults with apparent complete remission the percent of TdT fluorescent cells was below 2% with a few exceptions. In children the fluctuation was much wider, ranging from 0 to 13%; however, no direct correlation was found between cases showing TdT positivity higher than 2% and the observed relapses. The 10 cases

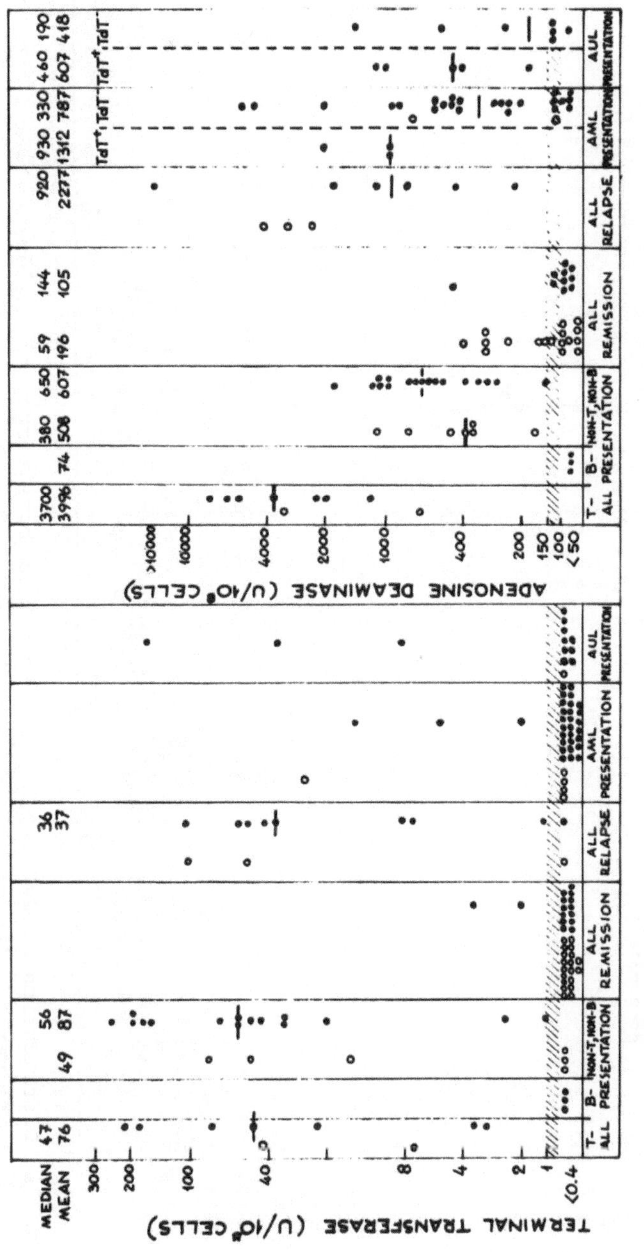

Fig. 1. Distribution of TdT and ADA enzymatic activities in childhood (O) and adult (●) acute leukemias. The values are plotted on a log scale. Bars represent medians. Shaded area separates normal from pathological values.

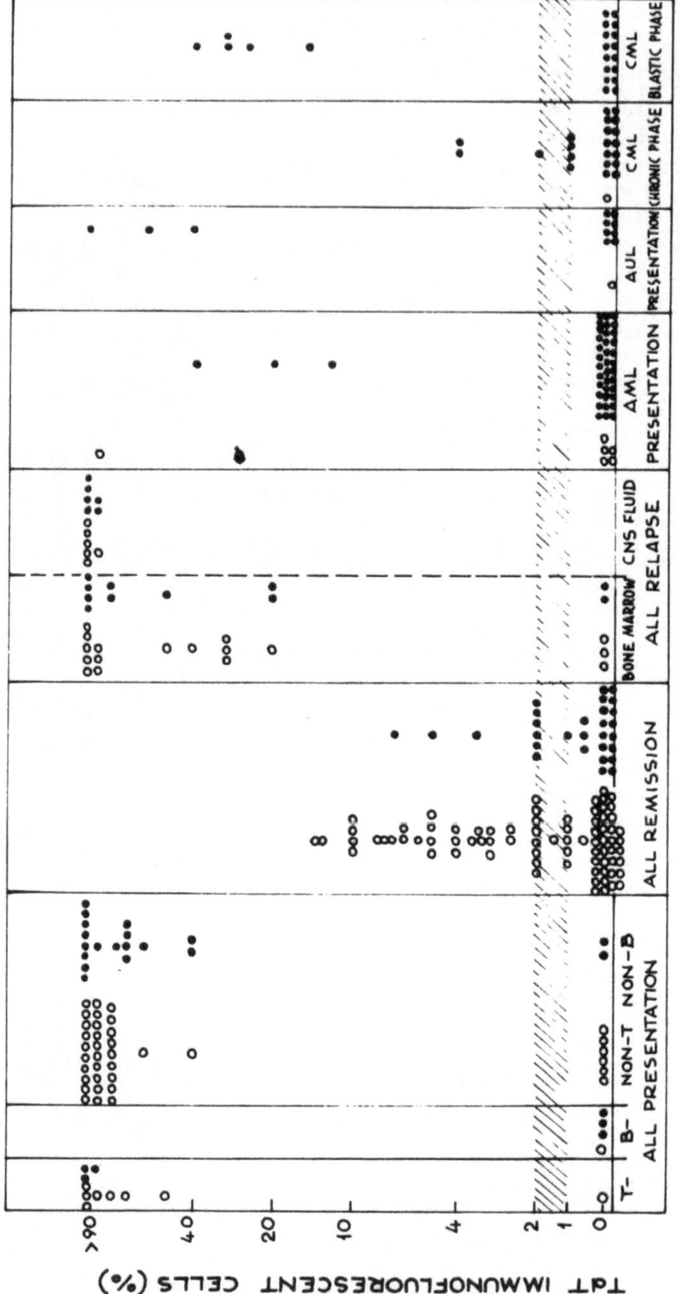

Fig. 2. Distribution of TdT immunofluorescence in bone marrow of children (O) and adults (●) with various leukemias. See also legend of Fig. 1.

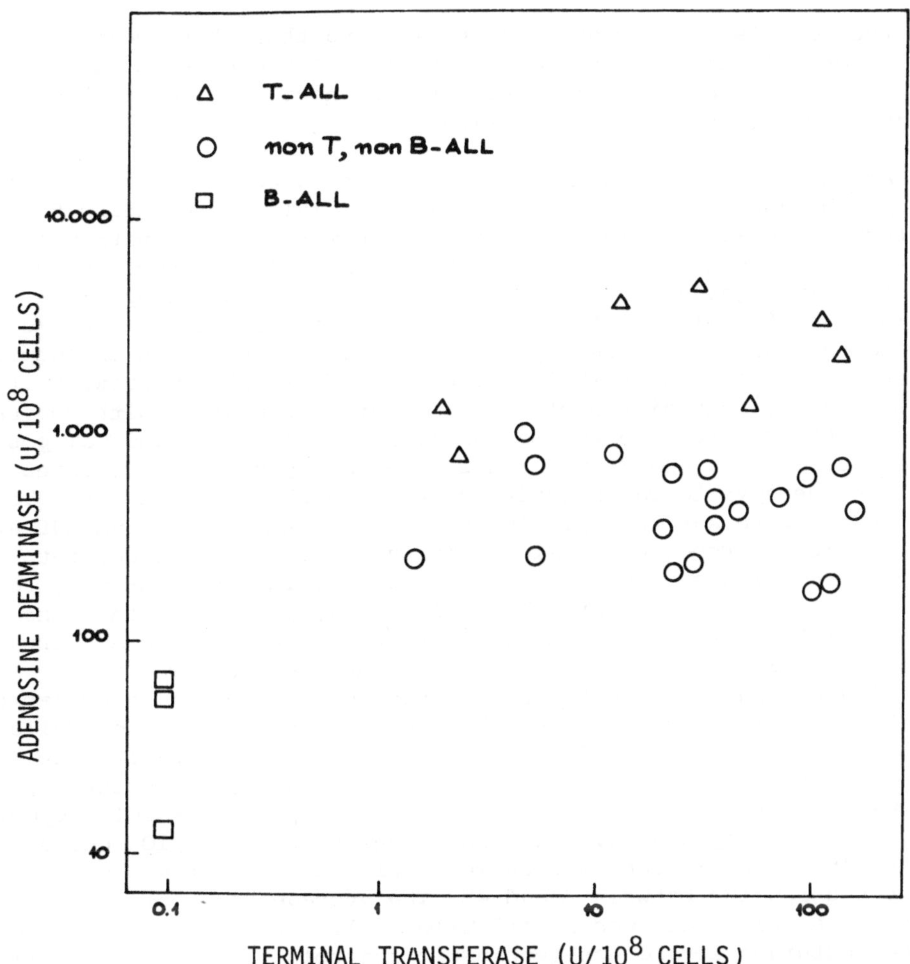

Fig. 3. Correlation between TdT and ADA in adult ALL at presentation. The values of enzymatic activities measured in the peripheral blasts of each patient have been plotted against one another on logarithmic scales.

of children off therapy showed TdT IF values of 1-2%. A similar pattern was obtained for the ADA determinations in remission (see Fig. 1): the mean of values was significantly lower than that found for ALL at presentation both for children (196±36 U/10^8 cells) and for adults (105±41 U/10^8 cells), and the children showed a higher degree of variation than the adult cases. The observations in 40 cases with ALL at the time of clinical relapse are reported in Figs. 1 and 2. The biochemical values found for TdT and ADA were similar to those found for ALL at presentation, showing means significantly higher than the normals and a wide range reflecting their heterogeneity. TdT IF determinations at relapse (see Fig. 2) presented a pattern close to that of presentation and were diagnostic in cases showing CNS involvement as the only localization.

Among the cases with AML (55 adults and 6 children) four adults and one child were found to be positive for TdT, representing 8% of the tested population. The percent of bone marrow cells with nuclear fluorescence ranged between 15 and 90%. Immunofluorescence analysis was strictly negative in all other patients studied, except in a single case of childhood AML in relapse showing 5% fluorescent cells in bone marrow. The IF test, performed in 10 cases of AML in complete remission gave less than 1% fluorescent cells in bone marrow. Fifteen cases of Acute Leukemias unclassifiable at presentation with conventional methods were also tested for TdT. Five cases, found TdT positive (see Figs. 1 and 2) were classified as lymphoid and treated with ALL therapy; 10 TdT negative cases were classified as myeloid and treated with AML therapy. The analysis of ADA in AML and AUL at presentation is reported in Fig. 1 by subdividing ADA values into two different lanes according to their positivity or negativity to TdT. Although the number of TdT cases is low, it appears that the medians of ADA activity of the TdT^+ cases in both AML and AUL are about three times higher than those obtained in the TdT^- cases. Morphological and cytogenetical analyses were performed in 40 cases of CML in stable phase, showing the presence in all cases of Philadelphia chromosome. TdT and ADA were analyzed biochemically (see Fig. 4) in the peripheral blood of 18 and 25 cases, respectively; the TdT IF test was performed in 22 cases (Fig. 2). The TdT enzymatic assay was strictly negative, giving values below 0.4 U/10 cells and the IF values in bone marrow were below 1%, with the exception of 3 cases ranging between 2 and 4%. The average value of ADA activity in stable phase was 103±15 $U/10^8$ cells, which is very close to that obtained for 10 non-neoplastic controls (113±16 $U/10^8$ cells). The values of ADA were increased over the control line only in 4 cases showing the early signs of transformation (see Fig. 4). The observations in 31 adult patients in acute phase of Ph'+ leukemia are given in Figs. 2 and 4. The morphology was lymphoid in 8 cases, myeloid in 14 cases; in 9 cases it presented a mixed type since in peripheral blood and bone marrow granulated blasts with clear myeloid morphology and agranulated blasts with lymphoid appearance were present. Twenty-three patients were analyzed for TdT by the enzymatic assay, 21 patients by the IF test and 10 by both methods. Ten of 31 patients (32%) were found positive for TdT. These included 7 patients developing blastic transformation following CML and 3 patients with no history of chronic phase and presenting as Ph'+ ALL. The mean value of TdT enzymatic activity obtained in bone marrow of TdT^+ cases was 82±34 $U/10^8$ cells. This value is close to the mean value found in adult ALL at presentation. The percent of bone marrow cells with nuclear fluorescence ranged between 15 and 45% of total nucleated cells. The ADA determinations are of particular in-

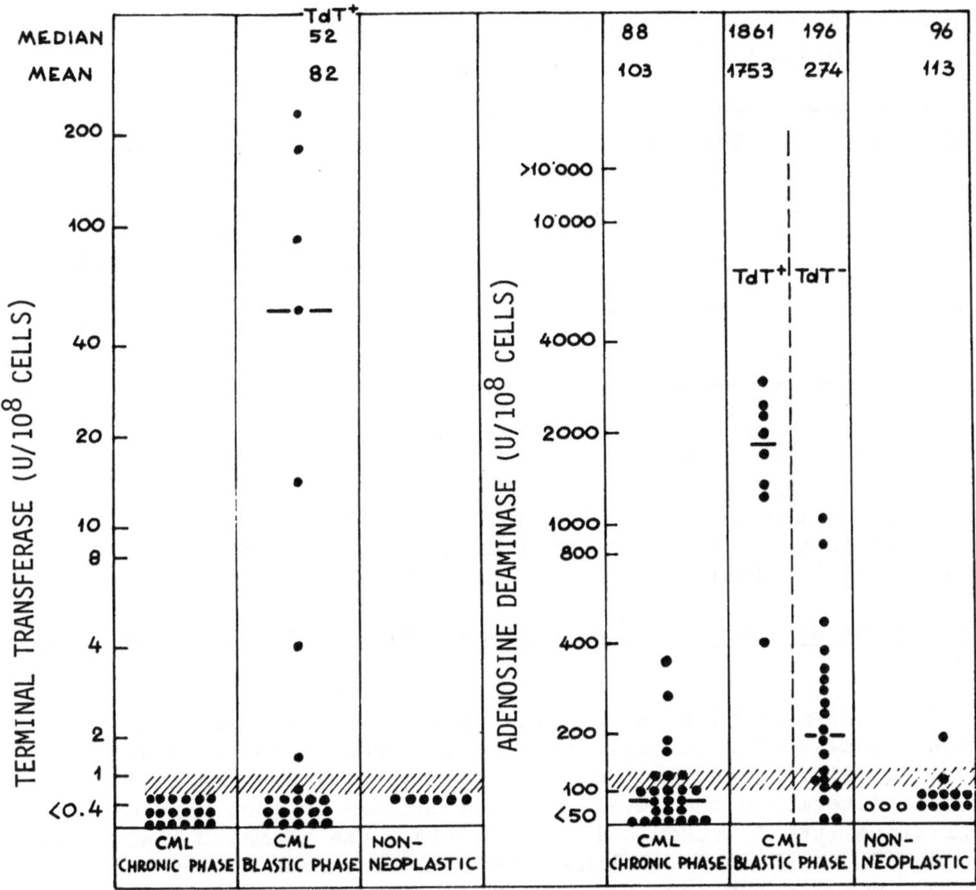

Fig. 4. Distribution of TdT and ADA enzymatic activities in Ph'+ leukemias and in non-neoplastic controls. See also legend of Fig. 1.

Table 1. TdT and ADA activities in some human hematopoietic cell lines

Cell line	Type of cells	TdT ($U/10^8$ cells)	ADA ($U/10^8$ cells)
RMPI 8402	T-ALL	105	2889
RMPI 8392	normal B	<0.4	248
NAML-1	non-T,non-B ALL	112	1282
HL 60	myeloid	<0.4	353

Table 2. Range of TdT and ADA activities in 8402 cell after 96 hrs of growth with or without TPA

Cell line	Treatment	TdT ($U/10^8$ cells)	ADA ($U/10^8$ cells)
RPMI 8402	Control	93-115	2256-3288
RPMI 8402	10^{-7} M TPA	12-15	2229-2558

terest since the mean value of the TdT^+ group (1753±275 $U/10^8$ cells) is significantly higher ($P<0.001$) than that found in the TdT^- group (274±58 $U/10^8$ cells).

TdT and ADA activities in leukemic cell lines. Effect of TPA induction.

In agreement with the results found in leukemias, human cell lines established from patients with different forms of leukemias also show differences in the levels of both TdT and ADA (25, 26). We have determined the levels of both enzymes in four representative cell lines (the T leukemic line RPMI-8402, the normal B line 8392, the non-T, non-B leukemic line Nalm-1 and the leukemic myeloid line HL-60). Results are reported in Table 1. These values just exemplify how the leukemia lines often reflect the target cell from which they are derived: the T line, as opposed to the B line, is positive for TdT and presents a very high level of ADA. The non-T, non-B line, also positive for TdT, shows an intermediate ADA activity. This is consistent with the picture observed in lymphoblasts from patients with non-T, non-B ALL. We have recently reported (27) that treatment of the T cell line RPMI-8402, which is phenotypically related to an early cortical thymocyte, with the potent tumor promoter TPA induces a certain T-differentiation. We have now measured both TdT and ADA activities in the extracts of cells grown for 96 hours in the presence of 10^{-7} M TPA. In the treated cells we observed that TdT activity is decreased to about 12% of the control value, whereas ADA levels do not decrease significantly (Table 2). The TdT immunofluorescence assay confirmed the biochemical data: in TPA treated cells, the disappearance of nuclear fluorescence was evident. These results are consistent with the differentiation observed on the basis of other parameters such as the appearance of E rosettes in a fraction of the treated population, the disappearance of EAC rosettes and a drop in DNA synthesis rate (27).

DISCUSSION

TdT is certainly important in assessing the lymphoid nature of leukemic blasts; 85% of our cases of acute leukemia classified as lymphoid by conventional methods (morphology and cytochemistry) were TdT positive, with the exception of B-cell derived leukemias and of a few cases of non-T, non-B ALL. ADA activity at presentation was raised several times with respect to controls, with the exception of the B-ALL subgroup, which ranged within control values; significantly higher values were observed in T-ALL than in non-T, non-B ALL classes. The diagnostic relevance of TdT and ADA correlation in adult ALL was evaluated by plotting on logarithmic scales the values obtained for each patient. The sub-types defined by membrane markers fell into three distinct biochemical groups: the group of T-ALL with positivity to TdT and very high levels of ADA; the group of B-ALL which is TdT negative and presents low levels of ADA; the cases of non-T, non-B ALL with positivity to TdT and intermediate values of ADA. A similar pattern was obtained by Coleman _et al._ (28) in childhood ALL. Very high levels of ADA could be of diagnostic interest in the identification of T-ALL devoid of E^+ blasts, with Acid Phosphatase reactivity and TdT positivity. Most cases of acute leukemias classified as myeloid by conventional methods were TdT negative. However, in 5 (8%) AML cases the presence of TdT was ascertained by biochemical and IF assays. All positive cases had a high percentage of blast cells in bone marrow, presented cytochemical features of myeloid origin and were Philadelphia chromosome negative. The finding of TdT in AML would indicate the presence of a mixed lymphoid and myeloid cell population, as recently reported in human Ph' negative acute leukemias other than CML in blast crisis (29, 30). This could result from the evolution of the original leukemic cells into clones presenting different phenotypes. The comparison of the ADA values found in ALL and in AML indicates that this marker is not useful in distinguishing lymphoid and myeloid leukemias. However, when the ADA values found in AML and AUL are divided into the two TdT^+ and TdT^- subgroups, it is evident that in both cases the medians of the TdT^+ groups are several times higher than those of the TdT^- groups. The possibility of detecting, with the TdT IF assay, residual or relapsing leukemic cells after therapy has been exploited for the diagnosis of ALL diseases in the CNS fluid, whereas the observed wide fluctuation of TdT IF values prevented a reliable monitoring of remissions as well as the detection of residual disease in bone marrow. In this respect, the determination of ADA in remission is recommended since the level of this enzyme is well correlated to the rate of cellular proliferation. In fact, the mean ADA value was raised by at least 10 times in our cases of ALL at relapse. In Ph'+ leukemias TdT and ADA analyses provided new elements for monitoring the chronic phase of CML, by recognizing the appearance of early transformation and yielding prognostic information. The stable phase of CML was characterized by TdT absence (7) and ADA ranging within normal controls in the majority of the cases studied. The ADA values were increased over the control line only in cases showing the early signs of transformation.

In the acute phase of Ph'+ leukemias, the TdT assay was positive in 32% of the cases, a value similar to that reported by other authors (8, 31). An interesting observation concerns the significant increase of ADA values in the TdT^+ patients in acute phase with respect to the TdT^- transformations. This represents an additional useful tool in the definition of the two, TdT^+ and TdT^-, groups which are known to be treated with different protocols (31).

The characterization of Ph'+ leukemia presenting as ALL is of particular interest to possibly enlighten the relationship between Ph'+ ALL and Ph'+ CML. Our three cases of Ph'+ ALL were all TdT positive, had high values of ADA, responded to ALL therapy reverting to a stable phase of CML; the Ph' chromosome was present in all the metaphases at onset and did not disappear in chronic phase after therapy. This altogether suggests that these two forms of leukemia are strictly related, arising from a common "target" cell.

The results we have obtained by analyzing for TdT and ADA several hematopoietic cell lines established in culture are also of interest in order to make a possible relationship between the phenotypes of the leukemic cells and their normal counterparts. Thus the T-ALL line 8402 is characterized by very high levels of both TdT and ADA, as reported for early cortical thymocytes (12, 13); the normal B-line 8392, which is phenotypically related to a rather mature stage of the B-lineage is TdT^- and presents low levels of ADA, as it occurs for the B-ALL leukemias; NALM-1, a non-T, non-B line derived from a patient with CML in blast crisis of the lymphoid type, contains a high amount of TdT and an "intermediate" level of ADA, similarly to what is observed for the patients with non-T, non-B ALL and with CML in lymphoid blastic transformation; finally, low levels of ADA were detected in the myeloid line HL-60.

Exposure of 8402 cells to the tumor promoter TPA resulted in a certain extent of T-differentiation. The enzymatic analysis of these cells showed that the TdT activity tended to decrease dramatically, its nuclear fluorescence disappearing after 96 hours of TPA treatment. On the contrary, no significant variations were observed for ADA. This could be due to the known fluctuation of ADA levels during the growth cycle of 8402 (25) and to the reduced cell proliferation following TPA treatment.

We have tried to summarize the observations obtained for these two markers in normal and in leukemic cells, in the scheme outlined in Fig. 5. It is evident that these enzymes appear to be particularly useful in the characterization of the T-line. The existence of a double thymocyte population, one TdT^+ and another TdT^-, as pos-

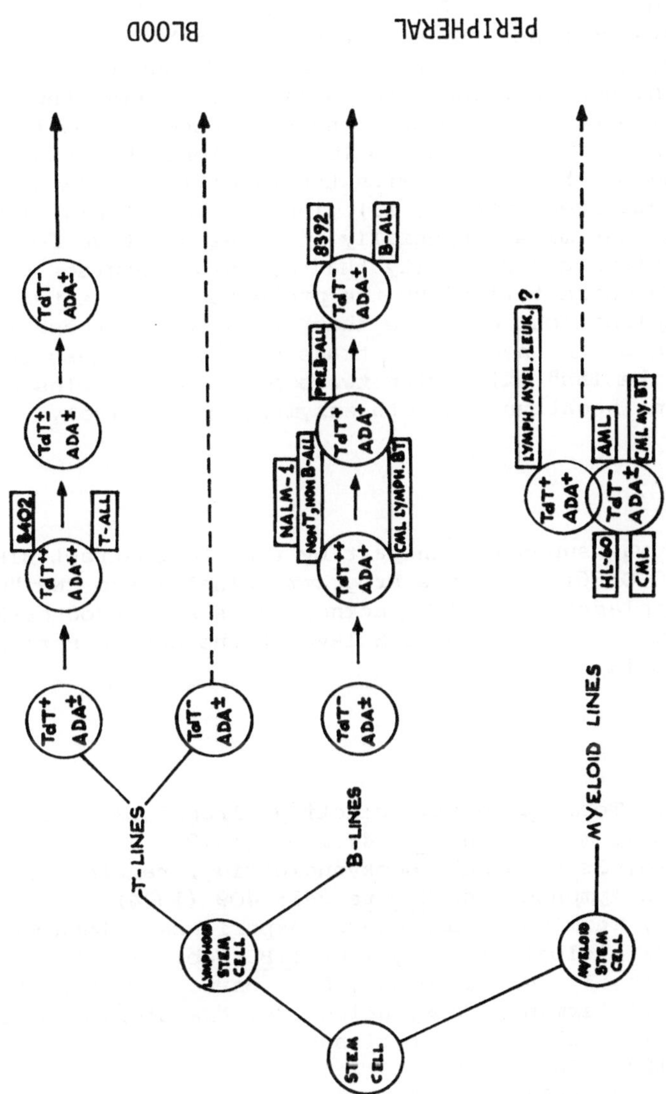

Fig. 5. TdT and ADA in possible hemopoietic precursors and leukemic cells.

sible progenitors of cortical and medullary thymocytes has been recently suggested by Goldschneider *et al.* (6). This possibility is also supported by the different ADA contents of cortical and medullary thymocytes (12, 13). The variations in both TdT and ADA might reflect metabolic differences among the different stages of T-cell maturation. The presence of TdT is not uniquely restricted to the normal T precursors since the enzyme is detectable also in pre-B lymphocytes (3, 4), in pre B-ALL and in non-T, non-B ALL (1). High levels of ADA have been found also in leukemic cells other than T-lymphoblasts. The possibility that the rare bone marrow TdT^+ cells could be also positive for ADA should be explored by using the double labeling of these cells with the recently available anti-TdT and anti-ADA immunoreagents (1, 13). The finding of high levels of ADA in myeloid leukemias, especially in cases positive for TdT, raises the intriguing possibility that myeloid precursor cells exist which are characterized by the presence of these two markers. Alternative explanations could be that the noted increase in ADA is the consequence of the neoplastic transformation process itself or that "mixed leukemias" exist with two populations of blasts, one positive and one negative for both enzymes, respectively.

ACKNOWLEDGMENTS

This work was supported in part by the CEC contract BIE428I and by grant 80-01561-96 of the Programma Finalizzato CNR "Controllo crescita neoplastica". A.I.S. acknowledges a postdoctoral fellowship from Istituto Sieroterapico Sclavo at the Scuola Perfezionamento Genetica, Pavia

REFERENCES

1. F.J. Bollum, Terminal deoxynucleotidyl Transferase as a hematopoietic cell marker, *Blood* 54: 1203 (1979)
2. D. Baltimore, Is terminaly deoxynucleotidyltransferase a somatic mutagen in lymphocytes? *Nature* 248: 409 (1974)
3. F.J. Bollum, Terminal transferase: experienced biochemical reagent seeks biological assignment, *TIBS* Feb: 41 (1981)
4. K.E. Gregoire, I. Goldschneider, R.W. Barton and F.J. Bollum, Ontogeny of terminal deoxynucleotidyl transferase-positive cells in lymphohemopoietic tissues of rat and mouse, *J. Immunol* 123: 1347 (1979)
5. R. Sasaki, F.J. Bollum and I. Goldschneider, Transient populations of terminal transferase positive (TdT^+) cells in juvenile rats and mice, *J. Immunol.* 125: 2501 (1980)
6. I. Goldschneider, A. Ahmed, F.J. Bollum and A.L. Goldstein, Induction of terminal deoxynucleotidyl transferase and LyT

antigens with thymosin: Identification of multiple subsets of prothymocytes in mouse bone marrow and spleen, Proc. Natl. Acad. Sci. 78: 2469 (1981)

7. E. Brusamolino, U. Bertazzoni, P. Isernia, E. Ginelli, A.I. Scovassi, M.G. Zurlo, P. Plevani, N. Sacchi and C. Bernasconi, Clinical relevance of terminal transferase and adenosine deaminase in leukemia. In "Terminal Transferase in Immunobiology and leukemia" (U. Bertazzoni and F.J. Bollum eds.), Plenum Press, in press.

8. A.V. Hoffbrand, H. Ganeshaguru, P. Llewelin and G. Janossy, Biochemical markers in Leukaemia and Lymphoma. Recent Results in Cancer Research (ed by R. Gross and K.-P. Hellriegel) 69: 25, Springer-Verlag, Berlin, Heidelberg, New York (1979)

9. G. Janossy, F.J. Bollum, K.F. Bradstock and J. Ashley, Cellular phenotypes of normal and leukemic hemopoietic cells determined by analysis with selected antibody combinations, Blood 56: 430 (1980)

10. E.R. Giblett, J.E. Anderson, F. Cohen, B. Pollara and H.J. Menwissen, Adenosine deaminase deficiency in two patients with severely impaired cellular immunity, Lancet 2: 1067 (1972)

11. M.S. Coleman, J. Donofrio, J.J. Hutton and L. Hahn, Identification and quantitation of adenine deoxynucleotides in erythrocytes of a patient with adenosine deaminase deficiency and severe combined immunodeficiency, J. Biol. Chem. 253: 1619 (1978)

12. R. Barton, F. Martinink, R. Hirschhorn and I. Goldschneider, The distribution of adenosine deaminase among lymphocyte populations in the rat, J. Immunol. 122: 216 (1979)

13. B.E. Chechik, W.P. Schrader and J. Minowada, An immunomorphologic study of adenosine deaminase distribution in human thymus tissue, normal lymphocytes and hematopoietic cell line, J. Immunol.126: 1003 (1981)

14. B.E. Chechik, W.P. Schrader and P.E. Daddona, Identification of human thymus-leukemia-associated antigen as a low-molecular-weight form of adenosine deaminase, J. Natl. Cancer Inst. 64: 1077 (1980)

15. J.J. Ballet, R. Insel and E. Merier, Inhibition of maturation of human precursor lymphocytes by coformycin, an inhibitor of the enzyme adenosine deaminase, J. Exp. Med. 143: 1271 (1976)

16. J.F. Smyth, K.R. Harrap, Adenosine Deaminase activity in leukaemia, Brit. J. Cancer 31: 544 (1975)

17. J. Meier, M.S. Coleman and J.J. Hutton, Adenosine deaminase activity in peripheral blood cells of patients with haematological malignancies, Brit. J. Cancer 33: 312 (1976)

18. R. Tung, R. Silver, F. Quagliata, M. Conklyn, J. Gottesman and R. Hirschhorn, Adenosine Deaminase activity in chronic lymphocytic leukemia. Relationship to B- and T-cell subpopulations, J. Clin. Invest. 57: 756 (1976)

19. B. Ramot, F. Brok-Simoni, N. Barnea, I. Bank and F. Holtzmann, Adenosine Deaminase (ADA) activity in lymphocytes of normal individuals and patients with chronic lymphatic leukemia, Brit. J. Haematology 36: 67 (1977)
20. M.S. Coleman and J.J. Hutton, Micromethod for quantitation of adenosine deaminase activity in cells from human peripheral blood, Biochem. Med. 13: 46 (1975)
21. S.A. Stass, A.R. Schumacher, T.P. Keneklis and F.J. Bollum, Terminal Deoxynucleotidyl Transferase immunofluorescence on bone marrow smears: experience in 156 cases, Am. J. Clin. Pathol. 72: 898 (1979)
22. C.C. Huang, Y. Howy, L.K. Woods, G.E. Moore and T. Minowada, Cytogenetic study of human lymphoid T cell lines derived from lymphocytic leukemia, J. Natl. Cancer Inst. 53: 665 (1974)
23. J. Minowada, T. Tsubota, M.F. Greaves and T.R. Walters, A non-T, non-B human leukemia cell line (NALM-1): establishment of the cell line and presence of leukemia associated antigens, J. Natl. Cancer Inst. 59: 83 (1977)
24. S.I. Collins, R.C. Gallo and R.E. Gallagher, Continuous growth and differentiation of human myeloid leukaemic cells in suspension culture, Nature 270: 347 (1977)
25. G.L. Tritsch and J. Minowada, Differences in purine metabolizing enzyme activities in human leukemia T-cell, B-cell, and null cell lines, J. Natl. Cancer Inst. 60: 130L (1977)
26. E.B. Chechik and J. Minowada, Quantitation of human thymus-leukemia associated antigen in established hematopoietic cell line by radioimmune assay, J. Natl. Cancer Inst. 63: 609 (1979)
27. N. Sacchi, U. Bertazzoni, D. Breviario, P. Plevani, G. Badaracco and E. Ginelli, Disappearance of nuclear TdT in RPMI-8402 following TPA treatment. In "Terminal transferase in Immunobiology and Leukemia" (U. Bertazzoni and F.J. Bollum, eds.) Plenum Press, New York, in press
28. M.S. Coleman, M.F. Greenwood, J.J. Hutton, Ph. Holland, B. Lampkin, C. Krill and J.E. Kastelic, Adenosine Deaminase, Terminal Deoxynucleotidyl Transferase (TdT) and Cell Surface Markers in Childhood Acute Leukemias, Blood 52: 1125 (1978)
29. R. Mertelsmann, B. Koziner, P. Ralph, D. Filippa, S. McKenzie, Z.A. Arlin, T.S. Gee, M.A.S. Moore and B.D. Clarkson, Evidence for distinct lymphocytic and monocytic populations in a patient with Terminal Transferase positive acute leukemia, Blood 51: 1051 (1978)
30. A.G. Prentice, A.G. Smith and K.F. Bradstock, Mixed Lymphoblastic and Myelomonoblastic Leukemia in treated Hodgkin's disease, Blood 56: 129 (1980)
31. S.M. Marks, D. Baltimore and R. McCaffrey, Terminal Transferase as a predictor of initial responsiveness to vincristine and prednisone in blastic chronic myelogenous leukemia, N. Engl. J. Med. 298: 812 (1978)

POLY(A)-POLYMERASE LEVELS IN LYMPHOID CELLS FROM ACUTE AND CHRONIC LYMPHOCYTIC LEUKEMIAS

C.M. Tsiapalis, C. Ioannides, T. Trangas, N. Courtis, G.A. Pangalis+, H.V. Cosmides* and M. Papamichael

Depts of Biochemistry and Immunology, Papanikolaou Research Center, +First Dept. of Internal Medicine and * Second Dept. of Pediatrics, University of Athens, Greece.

Poly(A)-polymerase is an enzyme found in a variety of eukaryotic cells and there is evidence that polyadenylation of HnRNA and mRNA is an early post-transcriptional process presumably mediated by poly(A)-polymerase. It has also been shown that the population of polyadenylated RNA molecules is constantly higher in the poorly differentiated and slowly growing human acute leukemic blast cells than normal and PHA-stimulated lymphocytes (Torelli,U and Torelli, G. Nature New Biol. 1973, 244,134). In the present investigation we have shown that the poorly differentiated human acute leukemic blast cells also have higher levels of poly(A)polymerase activity than the better differentiated chronic leukemic lymphocytes. Measurements of poly(A)-polymerase with poly$(A)_n$ and oligo$(A)_{10}$ as initiators revealed differences in the enzyme activity with a level of significance 99%. The mean poly(A)-polymerase unit per mg of soluble protein in 10 CLL cases was 3.85, while in 10 ALL cases was 19.42.

INTRODUCTION

Polyadenylate polymerizing and degrading activities have been described in a number of eukaryotic cells[2,3]. The polymerizing enzyme activity preferentially catalyzes the addition of AMP residues from ATP to the 3'-hydroxyl group of preformed oligo-or polynucleotide chains with no requirement for a template chain and has been detected, purified and characterized in the nucleus[4,5,6] and cytoplasm[6,7,8,9], while the degrading enzyme activities both nuclear and cytoplasmic preferentially catalyze the hydrolysis of polyadenylic acid exo-and endonucleolytically [6,10,11,12,13] Recent interest in the regulation of polyadenylate polymerizing and degrading activities stems from their biological role in the processing of hnRNA and mRNA, where polyadenylation is apparently an early step mRNA maturation[3,14]. Published reports on polyadenylate polymerizing and degrading enzymic activity levels in response to various physiological and pathological conditions are often contradictory[3] and only recently efforts to fit the data have met with some success[15,16]. In the interest of further clari fying this matter we have made quantitative biochemical measurements of these enzyme activities in crude soluble extracts of human lymphoid cells from acute and chronic lymphocytic leukemias. We have found highly significant differences in the levels of oligo (A) and poly(A) initiated polymerase between chronic lymphocytic leukemia (CLL) and acute lymphocytic leukemia (ALL). Furthermore in the analysis of our findings we recognized that quantitative biochemical measurements of these enzymes may provide useful clini cal information for the lymphoid leukemias.

MATERIALS

[^{3}H] ATP (specific activity 30 Ci/mmol), [^{3}H] ADP (specific activity 714 GBq/mmol) [^{3}H] poly (U) (specific activity 777 GBq/mmol were purchased from the Radiochemical Centre (Amersham, England); venom phosphodiesterase, spleen phosphodiesterase, microccal nuclease, unlabeled 3'-AMP, 5'-AMP, ADP, ATP, oligoriboadenylates, poly(A), poly(U), poly(C) and yeast tRNA from Boehringer Mannheim (Tutzing, Germany); GF/C and DE-81 sheets were a Whatman/Reeve Agnel product. Dialysis tubing, bovine serum albumin, Folin-Ciocalteus Phenolreagent and enzyme grade sucrose, urea, ammonium sulfate from Serva (Heidelberg, Germany); G-50, G-75 from Pharmacia (Upsalla, Sweden); polynucleotide phosphorylase and Trizma base from Sigma (St. Louis, USA). Some [^{14}C] poly(A) and [^{3}H] poly(A) were synthe sized with polynucleotide phosphorylase from M. lysodecticus[20]. All other reagents were commercial reagent grade.

METHODS

Isolation and characterization of human lymphocytes

Venous blood from patients with lymphoproliferative disorders was drawn into heparin and mononuclear cells were isolated on lymphoprep (Nyegaard and Co., Oslo, Norway) according to the manufacturer's instructions.

Preparation of soluble cell extracts and protein determination

The isolated total lymphocytes from each patient were disrupted and protein solubilized by centrifugation over a layer of 9% sucrose containing 1% NP-40 according to the method of Berger and Cooper[17] Protein determinations were made by the Lowry method[18].

Enzyme Assays: Poly(A) Polymerase

Poly(A) polymerase (Mn^{2+} dependent) activity was measured by incorporation of [^{3}H] ATP into acid insoluble product using poly(A) as initiator, unless otherwise specified[7,8,9]. In all enzyme assays [^{3}H] ATP was used at a specific activity of about 30cpm/pmole. Acid insoluble material from the reactions was spotted on Whatman GF/C glass fiber disks and processed as has been previously described[19]. The radioactivity was determined in a liquid-scintillation spectro meter. The reaction mixture contained in a final volume of 100µl: 200mM Tris-HCl (pH 8.3), o.5mM $Mncl_2$, 0.5mM [^{3}H] ATP, 4mM 2-mercapto ethanol, ImM (Pi) or about 1µM (3'-OH) of poly(A) and 5 to 10µl of the enzyme to be assayed. Reactions initiated by oligo$(A)_{10}$, poly (C) or yeast tRNA contained final concentrations 40µM (3'-OH), 1mM (Pi) and 500 g/ml, respectively. One unit of enzyme activity is defined as 1nmole of radioactive ribonucleotide incorporated per hour. Specific activity is expressed as untis of activity per mg of protein.

Poly(A) Exoribonuclease

The assay measures the conversion of [^{3}H] poly(A) into acid soluble material as has been previously described[7]. The reaction was carried out at 35^0 in 200mM Tris-HCl (pH 8.6), unless otherwise indicated containing 0.5mM $MnCl_2$, 4mM 2-mercaptoethanol, 0.05 to 0.25mM (Pi) [^{3}H] poly(A) (specific activity 500 to 1500 cpm/nmol) and 5 to 10µl enzyme. All of the aliquots from the reaction mixture were placed on filter disks (GF/C) and batch washed and counted[8]. One unit of enzyme is defined as the amount which forms 1nmol AMP per hour under the reaction conditions. Specific activity is expressed as units of activity per mg protein.

RESULTS

Poly(A)-polymerase activity in soluble extracts of lymphoid cells from acute and chronic lymphocytic leukemia.

Table 1 shows that poly-(A) polymerase activity from 10 ALL patients was higher than the activity from 10 CLL patients. The mean polymerase units per mg of soluble protein in 10 CLL cases was 3.85 while in 10 ALL cases was 19.42. The polyadenylation reaction showed an absolute requirement for divalent cations (Mn^{2+} being better than Mg^{2+}) and exogenous initiator as shown in Fig. 1 One cannot observe differences in the level of poly(A)-polymerase initiated with oligo$(A)_{10}$ or poly$(A)_n$ with soluble extracts from ALL and CLL patients. The oligo$(A)_{10}$- initiated polymerase activity from ALL soluble cell extracts shows linear incorporation of AMP for more than one hour, while from CLL soluble cell extracts there is linear incorporation of AMP only for the first 15 min. In

Table 1. Poly(A)-polymerase activity in soluble extracts of lymphoid cells from acute and chronic lymphocytic leukemia

Donors	Diagnosis	Units Enzyme per mg soluble protein
1	CLL	1.28
2	"	1.41
3	"	1.98
4	"	1.13
5	"	1.19
6	"	3.10
7	"	1.98
8	"	6.00
9	"	4.72
10	"	6.54
11	ALL	15.45
12	"	15.62
13	"	19.73
14	"	18.21
15	"	33.97
16	"	32.01
17	"	25.51
18	"	15.57
19	"	15.45
20	"	12.56

contrast, the poly(A)-initiated polymerase activity from both ALL and CLL soluble cell extracts shows linear incorporation of AMP for more than one hour, with an apparent initial lag phase only in the ALL extracts. Similar kinetics were obtained in all ALL and CLL cases examined. The following experiments were designed to examine the possibility of inhibitors and/or activators of poly(A)-polymerase present in the soluble cell extracts from human lymphoid cells.

Mixing Experiment for the detection of activators and/or inhibitors of poly(A)-polymerase activity in soluble extracts of lymphoid cells from acute and chronic lymphocytic leukemia.

As it may be seen in Fig. 2, there is a linear incorporation of AMP with 5µl ALL ① and 5µl CLL ② and with one order of magnitude difference in the enzyme levels as it may be seen also in Fig.1 Upon mixing 5µl ALL with 20µl CLL ③ Fig. 2, it is clear that the activity measured at 35^0 up to 90min is not the sum total of the mixing experiment. In fact, it is shown that after 90 min incubation only half of the polymerase activity of 5µl ALL ① can be detected

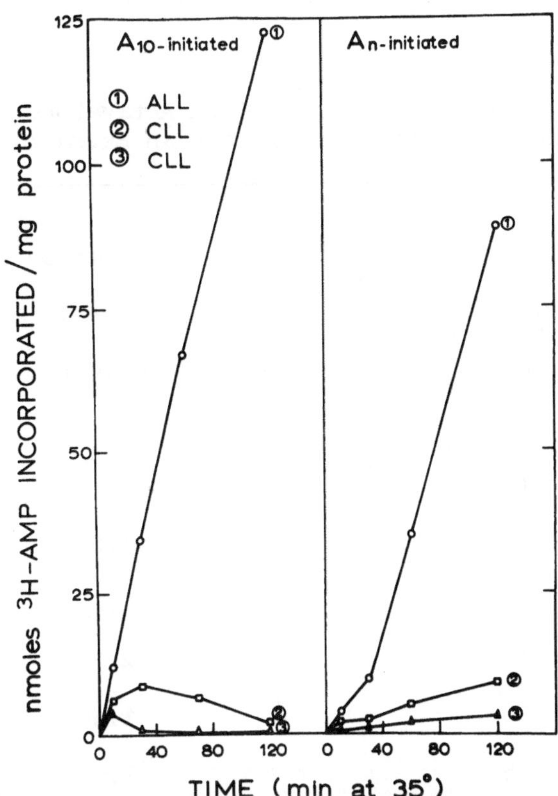

Fig. 1. Oligo$(A)_{10}$- and poly$(A)_n$-initiated polymerase activity in the soluble cell extracts from acute and chronic lymphocytic leukemia at 300µg of final protein concentration of each cell extract.

It is suggested that soluble cell extracts from CLL may contain inhibitors of AMP polymerization. In the same Fig. 2, reaction ④ is identical to reaction ③, except there was a 30min preincubation of the reaction mixture in the absence of the substrate of poly(A) polymerase at 35^0. This preincubation (see reaction ④), resulted in essentially complete inhibition of poly(A)-polymerase of 5µl ALL used in this mixing experiment. The results of the same mixing experiment are also presented in Table 2. Examination of the calculated and measured enzyme units in all cases indicated that the soluble cell extracts from human chronic lymphoid cells contain inhibitors of poly(A)-polymerase activity. Experiments in progress are designed to establish the structure and function of this inhibitor(s) detected in soluble extracts from human chronic lymphoid cells.

Table 2. Mixing Experiments to Detect Activators and/or inhibitors of Poly(A)-polymerase in Extracts of Cells.

Type and Amount of Extract Added (μl)		Polymerase Measured (units)	Activity Calculated (units)
ALL cell extract	CLL cell extract		
5	0	4.01	——
5	5	2.27	5.94
5	10	2.20	8.72
5	20	1.60	11.73
0	5	1.93	——
0	20	1.18	——
5	5	4.55	5.94
5	10	3.32	8.72
5	20	1.36	11.73

Conditions for the assay of poly(A)-initiated enzyme were as described in Materials and Methods. Total reaction volume was 100μl including cell extract as indicated in the table. Incubation was at 35^{0}C, and 10-μl aliquots of the reaction mixture were removed at 5, 15, 45 and 90min for determination of incorporation of nucleotide into poly(A) except for the last three reaction mixtures that were first preincubated at 35^{0}C for 30min in the absence of substrate. Calculated enzyme units in the reaction are the sum of activities in the two extracts assayed separately and represent the activity that would be expected in the mixture if no inhibitors or activators were present. Results based on initial rates of each reactions.

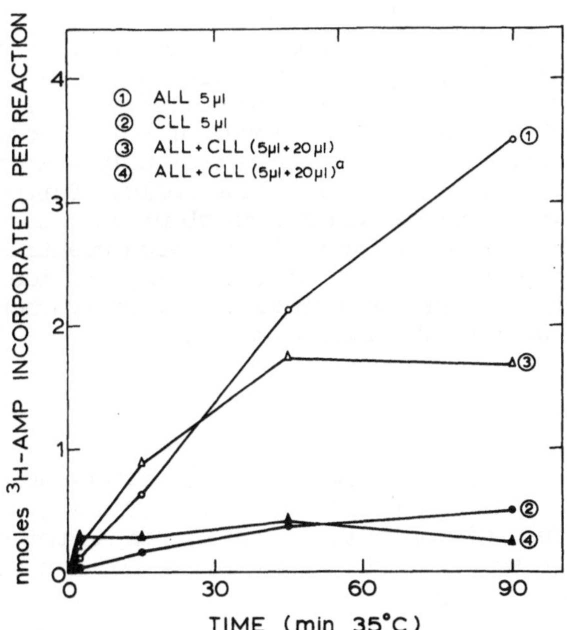

Fig. 2. Mixing experiments to detect activators and/or inhibitors of poly(A)-initiated polymerase in cell extracts from human acute and chronic leukemic patients. Conditions for the assay of soluble enzyme were as described in Materials and Methods. The total reaction was 100μl including tissue extracts as indicated in this Fig. a. In this case, a preincubation of each reaction mixture in the absence of the substrate of the poly(A)polymerase at 35^0 for 30 min, preceded, the standard conditions of the assay of soluble enzyme.

DISCUSSION

The studies reported here clearly demonstrated that there is a statistically significant difference of 99% in the levels of poly (A)-polymerase activity between the soluble cell extracts of ALL and CLL (Table 1 and Fig. 1). The majority of lymphocytes from CLL patients are B type cells, while the majority of lymphocytes from ALL patients are T type cells. Furthermore, it is known that B type cells are more differentiated than T type cells. In this respect, it is interesting to note that the population of poly(A)-containing hnRNA molecules was found higher in such poorly differentiated human acute leukemic blast cells than both in normal and PHA-stimulated lymphocytes[1]. There is, also, evidence that poly(A) chains are added to hnRNA by a post-transcriptional process[21-25] presumably mediated by poly(A)-polymerase[8] an enzyme found in a variety of prokaryotic and eukaryotic cells[2,3]. The results presented here show that a higher level of poly(A)-polymerase activity accompanies the higher content of poly(A)-containing RNA molecules in human acute leukemic blast cells. Furthermore, in a detailed analysis of our findings with about 65 cases of human lymphocytic leukemias we recognised that such quantitative measurements of poly(A)-polymerase, in addition, may provide a useful clinical adjunct for a better understanding of lymphoproliferative disorders as it will be reported elsewhere.

ACKNOWLEDGEMENTS

This work was supported by the Hellenic Anticancer Institute and in part by a NATO Grant 1848 to C.M.T. The authors thank Dr. Helen Cosmides, Mis Antonia Gounaris, Mis Andromachi Laskaridou for assistance in some aspects of this work and Mrs Despo Nicolaou for typing the manuscript.

REFERENCES

1. Torelli, U., and Torelli, G.,(1973) Nature 244, 134-136
2. Edmonds, M. and Winders, M.A. (1976) In Progress in Nucleic Acids Research and Molecular Biology, Davidson, I.N. and Cohn, W.E.Eds.Vol. XVII, pp. 149-179, Academic Press, New York.
3. Edmonds, M.(1981) in the Enzymes in press, Academic Press, New York.
4. Winders , M.A. and Edmonds, M. (1973) J.Biol. Chem. 248, 4763-4768.
5. Winders, M.A. and Edmonds, M. (1973). J.Biol. Chem. 248,4756-4762.
6. Nevins J.R.and Joklik, W. (1977). J.Biol. Chem. 252.6939-6947.
7. Tsiapalis, C.M., Dorson, J.W., De Sante, D.M., and Bollum, F. J. (1973). Biochem.Biophys. Res.Commun, 50, 737-743.
8. Tsiapalis, C.M., Dorson, J.W., and Bollum, F.J. (1975). J.Biol. Chem. 250, 4486-4496.
9. Bollum, F.J., Chang, L.M.S., Tsiapalis, C.M and Dorson, J.W. (1974), in Methods in Enzymology vol. 29, pp. 70-81
10. Schröder, D.C., Zahn, R.K., Dose, K., and Müller, W.E.G. (1980). J.Biol. Chem. 255, 4535-4538.

11. Müller, W.E.G., (1976). Eur. J.Biochem. 70, 241-248.
12. Kimagai, H., Igarashi, K., Takayama, T. Watanabe , K., Sugimoto K., and Hirose, S. (1980). Biochim. Biophys. Acta 608, 324-331.
13. Kumagai, H., Igarashi, K., Tanaka, K., Nakao, H., and Hirose, S. (1979). Biochim. Biophys. Acta 566, 192-199.
14. Jacob, S.T., and Rose, K.M. (1978) in Methods Cancer Research vol. XIV, pp. 191-241.
15. Matts, R.L. and Siegel, F.L. (1979).J.Biol.Chem.254, 11228-11233.
16. Tsiapalis,C.M., Ioannides,C., Trangas,T., Courtis,N.Pangalis, G.A., Cosmides,H.V., and Papamichail,M. (1981) in Comperative Research on Leukemia and Related Diseases, Yohn,D.Ed. in press Elsevier North Holland Inc. New York.
17. Berger, S.L., and Cooper, H.L. (1975). Proc.Natl.Acad.Sci.(USA) 72, 3877-3878).
18. Lowry , O.H., Rosebrough, N.J., Farr, A.L., and Randall, R.J. (1951). J.Biol. Chem. 193, 265-275.
19. Bollum, F.J.(1959). J.Biol.Chem. 234, 2733-2734.
20. Klee, C.B. (1967). J.Biol.Chem. 242, 3579-3580.
21. Walters, R.A., Yandel, P.M., and Enger, M.D. (1979). Biochemistry 18, 4254-4261.
22. Shepherd, G.W., and Flickinger, R. (1979). Biochim.Biophys. Acta 563, 413-421.
23. Shepherd, G.W., and Nemer, M. (1980). Proc. Natl. Acad. Sci. USA 77, 4653-4656.
24. Guyette, W.A., Matusik, R.J., and Rosen, J.M., (1979). Cell 17, 1013-1023.
25. Harpold, M.M., Evans, R.M., Salditt-Georgieff, M., and Darnell, J.E., (1979) Cell, 17, 1025-1035.

ARE POLYAMINES USEFUL MARKERS FOR MONITORING CANCER THERAPY?

Victor R. Villanueva

Institut de Chimie des Substances Naturelles, C.N.R.S.
91190 Gif-sur-Yvette
France

INTRODUCTION

Polyamines-Putrescine, Spermidine and Spermine-are a group of organic polycations which are ubiquitous in all living systems. Prokaryotic organisms normally contain putrescine and spermidine; eukaryotic cells also contain spermine.

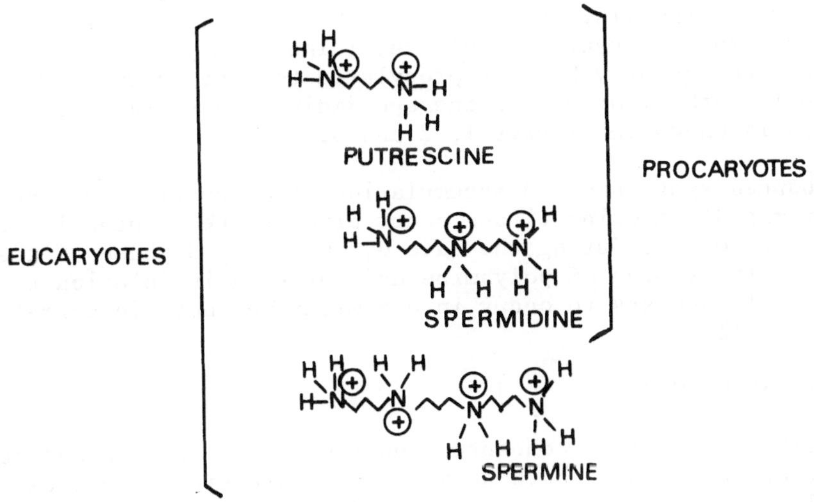

EUCARYOTES

PUTRESCINE

SPERMIDINE

PROCARYOTES

SPERMINE

Owing to their polycationic nature polyamines are able to interact with polyanionic cellular components such as DNA, RNA phospholipids, etc. These interactions form the basis for the multiple roles of polyamines in different cellular processes related to cell division, cell differentiation and cell growth. Polyamines have been reported to be involved in DNA synthesis, chromosomal condensation, protein synthesis, gene activation in early embryonic development, cytokinesis and in a considerable number of other cell metabolism process[1-4].

Polyamine content vary with cell type and also with the physiological state of the cell. Rise of polyamine synthesis precedes in general, organ regeneration and elevation of hormonal levels in animals following growth stimuli[5]. Polyamines are also increased in rapidly growing normal and neoplastic tissue. For this reason polyamines present a great interest as they can be used as biological markers in a number of physiopathological cases like cancer, psoriasis, cystic fibrosis, etc..., where an increase of the polyamine levels in the human serum and urine has been shown[6].

POLYAMINE BIOSYNTHESIS

The primary precursors of polyamines in both, microorganisms and animal tissue are L-ornithine and L-methionine. Ornithine is converted to putrescine by decarboxylation while methionine is first activated, in the presence of ATP, to give S-adenosyl-methionine, which after decarboxylation is used as the donor of the propyl amine moiety for the synthesis of spermidine from putrescine. Spermidine can subsequently be aminopropylated to give spermine[1]. The biosynthetic pathway of putrescine, spermidine and spermine in animal tissue is shown in figures 1, 2 and 3.

Enhanced synthesis and accumulation of putrescine and spermidine in rapidly growing tissue occur prior to the synthesis of DNA, that is to say, during the late G_1 and early S phase of the cell cycle. The timing of polyamine accumulation in relation to the cell cycle appears to occur in a similar way both in normal and neoplastic cells [6].

POLYAMINE AND CANCER

The high polyamine concentration found in embryonic and other rapidly growing tissues, both normal and neoplastic, also caracterises malignant tumors. As early as 1853, Charcot and Robin[7] reported that leukemic spleen was rich in spermine. This observation was confirmed almost a hundered years later by Hämäläinen[8] who found increased concentrations of this polyamine in the liver and bone marrow of patients who had died of leukemia. Rosenthal and Tabor[9] have demonstrated the presence of polyamines in lymphome, sarcoma,

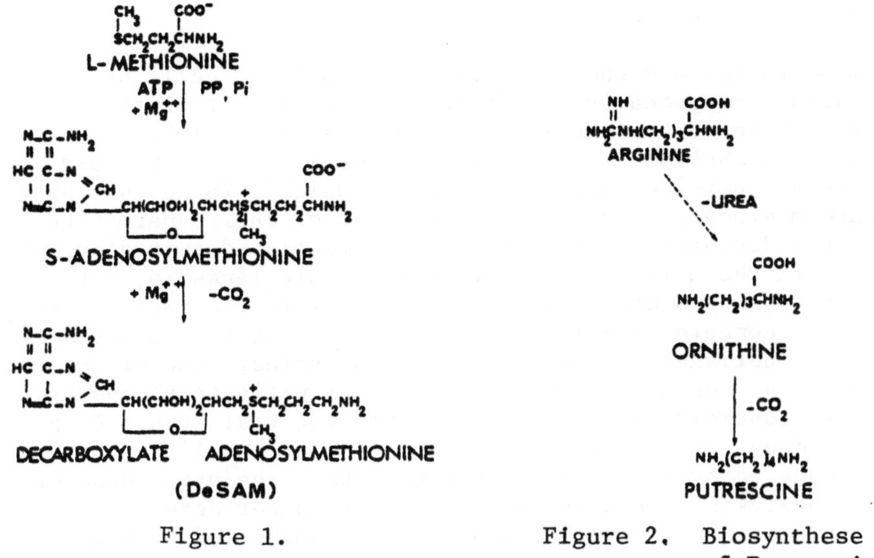

Figure 1.

Figure 2. Biosynthese of Putrescine

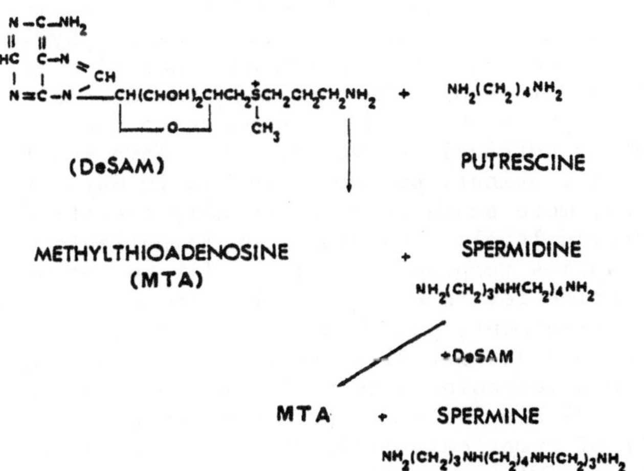

Figure 3. Biosynthesis of Spermidine and Spermine

hepatome and carcinome tissues of mice while Kremzner et al.[10] found polyamines in human brain tumors : neurofibroma, meningioma, glioblastoma and astrocytoma.

Although the presence of polyamines in tumor cells has been demonstrated unequivocally, it is not yet clear whether they occur as the free bases or as complexes. Thus Kosaki[11] in 1958 reported that cancerous, but not normal tissues, contain a phospholipid derivative of spermine, malignolipin, and Walle[12] found that patients with acute myelocytic leukemia excrete most of their spermidine as a monoacetyl derivative. Tsuji et al.[13] have also detected acetyl spermidine in the urine of patients with cronic leukemia, reticullum cell-sarcoma and other tumors. But certainly the most spectacular report concerning polyamines and cancer was that of Russell[14] in 1971 who published data showing that the urinary excretion of polyamines was considerably higher in cancer patients as compared with normal controls. This first published material, consisting mainly of blood cancers, lymphomas and metastatic solid cancers, probably gave an over-optimistic picture about the usefulness of polyamine analysis as a diagnostic tool for cancer detection and aroused interest in this area. Since then a number of laboratories have been engaged in the development of assay methods and analysis of clinical cancer material[15]. Studies were expanded to include, a larger number of patients and a variety of different types of cancers. Polyamines were also found to be augmented in the serum of a majority of patients with cancer[16]. Based on data from animal and patients studies it was proposed that spontaneous cellular death and lysis of tumor resulted in high levels of polyamine excretion in untreated patients. Also cell killing due to chemoterapy and radiotherapy increased the release of polyamines from tumor cell. Russell[4] has suggested that polyamines might serve as markers of tumor kinetics, the concentration of putrescine in physiological fluids reflecting tumor growth fraction and that of spermidine cell turnover. Two types of changes in the polyamine levels of physiological fluids have been observed in response to therapy : successful chemotherapy or surgical removal of tumor have resulted in a decrease of elevated urinary polyamine levels in days or weeks following therapy, more acute changes, notably elevations in serum or urinary polyamine levels have been seen in patients with hematological malignancies responding successfully to chemotherapy, whereas the polyamine levels changed much less in unresponsive patients. Before treatment, polyamine levels in these patients seemed to reflect disease activity and tumor burden. It has been proposed that a considerable increase in the serum and/or urinary spermidine levels within 24-48h after commencement of therapy is an excellent marker of tumor cell kill. Therefore, polyamine determinations might be useful in rapidly assessing tumor response to chemotherapy or to multimodality therapy.

DISCUSSION AND GENERAL REMARKS

At the present time the published data on polyamine levels in human malignancies is unsufficient to make a final evaluation about the worth of polyamines as biochemical markers of cancer. Elevations in urinary polyamines have been observed in pernicious and hemolytic anemias, polymyositis, pulmonary tuberculosis, psoriasis and other non malignant diseases[17]. These results showed that increased excretion of polyamines is not specific for cancer. However as the excretion pattern of polyamines was different from that found in cancer patients the specificity might be increased by taking into account combinations of polyamines levels. For example spermine seems to be a fairly specific indicator of colorectal carcinoma[18]. However this obviously needs more systematic work.

On the other hand, polyamine concentration in human physiological fluids may also be influenced by factors such as age, diet, sex, hormonal balance, medication and other parameters that could influence polyamine metabolism[19]. This must be kept in mind when chosing normal controls and for the interpretation of results as the factors mentioned can introduce considerable variability.

Also the choice of specimen, urine, serum, plasma, that is to say the fraction isolated for study and the mode of sample preparation, before analysis, may result in profound differences in the kinds of polyamine information. Actually work from our laboratory and from others seems to indicate that polyamine analysis from red cells may give more reliable results.

Another important remark is that most of the studies reported at present concern only putrescine, spermidine and spermine. In our laboratory, we are currently pursuing multiphase studies from cancer single samples. This method consists of (i) simultaneous quantification of polyamine precursors, putrescine, spermidine and spermine ; (ii) evaluation of the conversion of arginine to ornithine by measuring arginase activity ; and (iii) exploration of bound-compartmentalized polyamines using appropriated antibodies. The quantification of these different aspects of polyamine analysis from a single sample could contribute to broaden our understanding of polyamine homeostasis and should permit the development of a more selective and highly amplified detection of differences between normal and neoplastic samples.

Finally, full clinical application of polyamine determinations must await the development of a simple, reliable, inexpensive assay system for polyamines in biological fluids. In this respect the introduction of automated analysers is certainly an improvement. Nevertheless the equipement is expensive and the number of samples which can be analysed is limited. Radioimmunological assay appears to be the best analytical approach for routine analyses and must be considered as the most promising for mass screening. Thus clinical application of polyamines await the development of a simple, reliable, sensitive and inexpensive assay system.

In conclusion it seems appropriate that polyamines with their ubiquitous relationship to growth in all living tissues would be general markers of cell growth and cell death processes and their possible use as markers of desease activity for clinical applications looks promising. Further basic work on polyamines and their associated enzyme system is necessary to elucidate the exact intracellular functions of these organic polycations.

REFERENCES

1. H. Tabor, and C.W. Tabor, Spermidine, spermine and related amines, Pharmacol.Rev., 16:245-300 (1964).
2. S.S. Cohen, "Introduction to the polyamines," Prentince-Hall, Inc., Englewood Cliffs, New Jersey, (1971).
3. U. Bachrach, "Function of naturally occuring polyamines." Academic Press, New-York, (1973).
4. D.H. Russell, and B.G.M. Durie, "Polyamines as biochemical markers of normal and malignant Growth", Raven Press, New-York, (1978).
5. A. Raina, T. Eloranta, R.-L. Pajula, R. Mäntyjärvi, and K. Tuomi, Polyamines in rapidly growing animal tissues , in "Polyamines in Biomedical Research," J.M. Gaugas, ed., John Wiley and Sons, Sussex, England, (1980).
6. O. Heby, G.P. Sarna, L.J. Marton, M. Omine, S. Perry and D.H. Russell, Cancer Res., 33:2959-2964 (1973).
7. J.M. Charcot, and C.P. Robin, Observation de leucocythémie, C.R.Soc.Biol. (Paris), 5:44-50 (1853).
8. R. Hämäläinen, Über di quantitative Bestimmung des Spermins im Organismus und sein Vorkomen in neusehlichen Geweben und Körper-lüssigkeiten, Acta Soc. Med., "Duodecim," Ser.A 23:97-165 (1941).
9. S.M. Rosenthal and C.W. Tabor, The pharmacology of spermine and spermidine. Distribution and excretion, J.Pharmacol. Exp.Ther., 116:131-138 (1956).
10. L.T. Kremzner, R.E. Barret and M.J. Terrano, Polyamine metabolism in the central and peripheral nervous system, Ann.N.Y.Acad.Sci., 171:735-748 (1970).
11. T. Kosaki, T. Ikoda, Y. Kotani, S. Nakagawa and T. Saka, A new phospholipid, malignolipin, in human malignant tumors, Science, 127:1176-1177 (1958).
12. T. Walle, Gas chromatography-mass spectrometry of di- and polyamines in human urine : identification of monoacetylspermidine as a major metabolic product of spermidine in a patient with acute myelocytic leukemia in "Polyamines in normal and neoplastic growth," D.H. Russell, Raven Press, New-York, pp 355-365 (1973).

13. M. Tsuji, T. Nakajiama and I. Sano, Putrescine spermidine, N-acetyl spermidine and spermine in the urine of patients with leukemias and tumors, Clin.Chim.Acta, 59:161-167 (1975).
14. D.H. Russell, Increased polyamine concentrations in the urine of human cancer patients, Nature New Biology, 233:144-145 (1971).
15. V.R. Villanueva, Polyamine and related compounds : assay Methods in: "Methods in Biogenic Amine Research," S. Parvez, T. Nagatsu, I. Nagatsu and H. Parvez, eds. Elsevier/North Holland Medical Press, Amsterdam and New-York (in press).
16. L.J. Marton, J.G. Vaughn, I.A. Hawk, C.C. Levy and D.H. Russell, Elevated polyamine levels in serum and urine of cancer patients : detection by a rapid automated technique utilizing an aminoacid analyzer, in :"Polyamines in normal and neoplastic growth," D.H. Russell, ed. Raven Press, New-York (1973).
17. B.G.M. Durie, S.E. Salomon and D.H. Russell, Polyamines as markers of response and desease activity in cancer chemotherapy, Cancer Res., 37:214-221 (1977).
18. K. Nishioka and M.M. Romsdahl, Preliminary longitudinal studies of serum polyamines in patients with colorectal carcinoma , Cancer Lett., 3:197-202 (1977).
19. T.P. Waalkes, G.W. Gehrke, D.C. Tormey, R.W. Zumwalt, J.N. Hueser, K.C. Kuo, D.B. Lakings, D.L. Ahmann and C.G. Moertel, Urinary excretion of polyamines by patiens with advanced malignancy , Cancer Chemother.Rep., 59:1103-1116 (1975).

ADP-RIBOSYLATION OF PROTEINS IN NORMAL AND NEOPLASTIC TISSUES

Helmuth Hilz, Peter Adamietz,
Reinhard Bredehorst, and Klaus Wielckens

Institut für Physiologische Chemie
Universität Hamburg, Germany

INTRODUCTION

In 1966, Mandel and coworkers in Strasbourg described an enzyme located in the nucleus of chicken liver, which is able to transfer active ADPR groups from NAD to form a homopolymer (1). In this polymer ADPR residues are linked O-glycosidically. Shortly later, Hayaishi and coworkers (2) as well as Sugimura's group (3) reported on similar findings. Recent evidence indicates that the structure can also be branched (4). The nuclear ADPR transferase is apparently present in all eukaryotes (5-7). It differs basically from ADPR transferase subsequently found to be associated with certain bacterial toxins and with phages: The prokaryotic enzymes transfer single ADPR residues either to one acceptor protein as in the case of Diphteria toxin, where the ribosomal elongation factor 2 is the target (8), or to multiple proteins which serve as acceptors of the T4 and the N4 phage-induced ADPR transferase (cf. 5-7). By contrast, the nuclear enzyme forms preferentially oligo- and poly(ADPR) conjugates with various protein acceptors.

In broken cell preparations, the nuclear poly(ADPR) synthetase becomes artificially activated. Under these conditions, multiple acceptors will be ADP-ribosylated leading, however, to conjugates that differ from conjugates obtained from intact tissues (cf. 9). This led us to develop procedures for the quantitation of conjugate levels in intact tissues. The procedures worked out in our laboratory are based on specific chemical and enzymic reactions of $(ADPR)_n$ groups released from the acceptor proteins by alkaline treatment, followed by quantitation of the products using specific radioimmunoassays (10, 11, 12).

DIFFERENT TYPES OF $(ADPR)_n$ PROTEIN CONJUGATES

Application of the tests to intact tissues allowed to discriminate three types of eukaryotic $(ADPR)_n$ protein conjugates (fig. 1):

- poly(ADPR) proteins carrying oligomers or polymers, at least some being linked via carboxyl groups of amino acids as recently reported for histone H1 and H2B conjugates formed in vitro (13, 14).
- NH_2OH sensitive mono(ADPR) protein conjugates, the linkage possibly being of the same ester type as above (15).
- NH_2OH resistant mono(ADPR) protein conjugates. No solid information as to the nature of the acceptor amino acids is presently available, but some conjugates may contain arginine-linked ADPR residues (cf. 15, 16).

When the procedures were used to quantitate endogenous levels of these conjugates, the following basic information was obtained (17):

- In all eukaryotic tissues studied, low levels of protein-bound ADPR residues were found. They range from 1 ADPR per 8000 DNA bases (Yoshida hepatoma AH 130) to 1 ADPR per 250 DNA bases (adult rat liver).

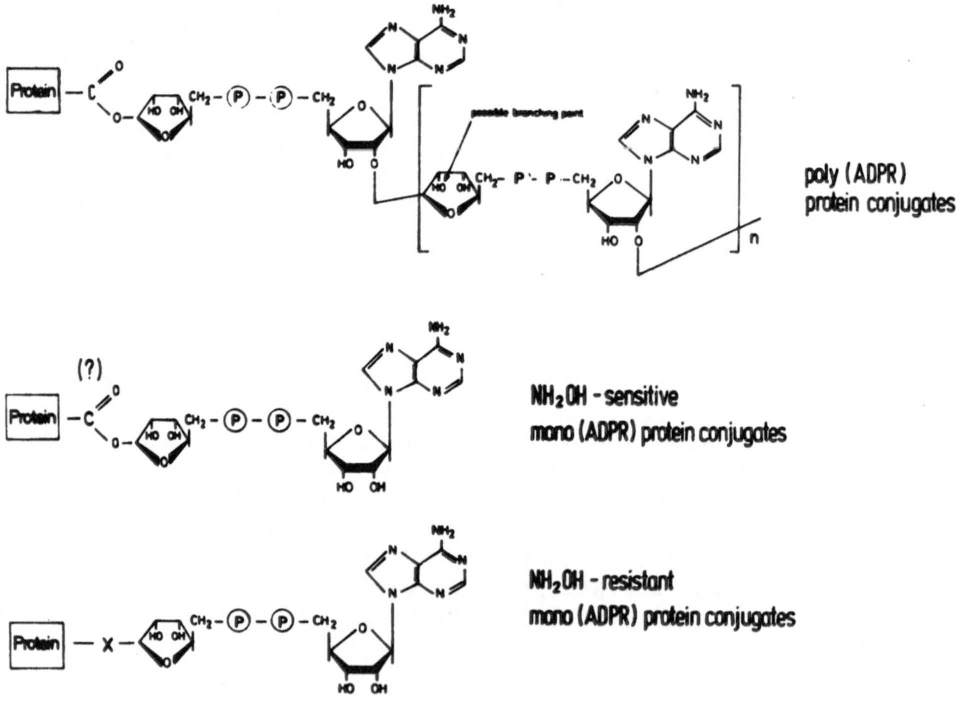

Fig. 1. Different types of ADP-ribosylated proteins

- Substrate levels (NAD + NADH) exceed the amounts of the protein-bound ADPR residues 50 - 200 fold.
- In all tissues analyzed much more ADPR residues are bound to proteins as single ADPR groups ("mono(ADPR) residues") than as oligomeric or polymeric $(ADPR)_n$ chains (table 1): In neonatal rat liver, the ratio of monomeric : polymeric ADPR residues is about 70 :1, in adult liver it is 320 :1, and in human lymphocytes 50 :1. Assuming a mean chain length of 10 for polymeric ADPR, 700 - 3200 times more acceptor sites are occupied by single mono(ADPR) residues than by poly(ADPR) chains.
- Changes of mono(ADPR) and poly(ADPR) residues were independent: In developing rat liver (table 1) mono(ADPR) conjugates increase steadily, while identical amounts of poly(ADPR) protein conjugates were found in adult and neonatal liver. If protein-bound mono(ADPR) residues were to function solely as the starting points for chain elongation, an inverse relationship of mono and poly(ADPR) residues should be found under changing conditions, polymeric ADPR groups increasing at the expense of the starting monomeric ADPR group, and vice versa.
- Independence is also found with respect to mono(ADPR) conjugate subclasses. In Physarum polycephalum NH_2OH-sensitive conjugates were synthesized at a time of the cell cycle distinctly different from that of the NH_2OH-resistant subfraction (18).

INTRACELLULAR DISTRIBUTION OF MONO(ADPR) PROTEIN CONJUGATES

Although ADP-ribosylation was originally detected as an exclusively nuclear function, studies of the subcellular distribution of rat liver mono(ADPR) conjugates indicated that over 95% of these conjugates are not located in the nucleus but found in other compartments (19). Interestingly a different distribution was found in a tumor cell line: In these cells nearly 40% of the total mono(ADPR) proteins was associated with the cell nucleus (19). However, this divergence comes about by the strongly reduced amounts of mono(ADPR) conjugates in the cytoplasm of these undifferentiated tumor cells, while nuclear values were in the same range as in liver. As the cell line is an undifferentiated highly malignant tumor, the loss of cytoplasmic conjugates might be indicative of the low degree of differentiation.

Subfractionation of the liver homogenate demonstrated an uneven distribution of mono(ADPR) conjugate subclasses. Two thirds of the hydroxylamine-resistant subclass were associated with the 12,000 g pellet fraction, mitochondria containing most of the NH_2OH-resistant conjugates. Both mitochondria and plasma membranes were characterized by low proportions of the NH_2OH-sensitive subfraction. Mono(ADPR) proteins found in the post-mitochondrial supernatant were associated mainly with the endoplasmic reticulum, while polysomes and the cytosol had low amounts of ADPR proteins. The distribution of mono(ADPR)

Table 1 Hepatic protein-bound monomeric and polymeric ADPR residues in relation to each other and to NAD(H) levels (modified from (23)).

Hepatic Tissues	ADPR Residues (pmol / mg DNA)		Ratio	Ratio
	monomeric	polymeric	monomeric / polymeric	NAD(H) / polymeric
neonatal day 1	2 300	32	72	3 275
neonatal day 17	3 180	10	318	24 400
adult (> day 150)	12 570	39	322	13 794
hepatoma 130 proliferating	1 060	61	18	803
hepatoma 130 stationary	1 100	25	44	1 800
hepatoma H35 proliferating	2 890	60	48	1 950

protein conjugates in rat liver is reflected to a certain extent by the distribution of mono(ADPR) transferase activities (unpublished results). The data so far obtained do indicate that mono(ADPR) protein conjugates are constituents of all major compartments of differentiated eukaryotic cells, the two subclasses being distributed, however, unevenly.

NUCLEAR POLY(ADPR) PROTEINS

Other than the mono(ADPR) proteins, the conjugates carrying poly(ADPR) appear to be confined to the nucleus (cf. 19). There, they constitute a heterogeneous group of modified acceptor proteins, including histones and non-histone proteins as well as the auto ADP-ribosylated form of poly(ADPR) synthetase (cf.5-7 , 20, 21). We have isolated, in pure form and with a yield of over 85%, the nuclear $(ADPR)_n$ protein conjugates formed from [^{3}H]NAD in EAT cell nuclei (22). Separation was based on covalent chromatography of the perchloric acid-insoluble fraction on immobilized aminophenyl boronic acid. Due to the presence of cis diol groups in the unsubstituted ribose moieties mono(ADPR) and poly(ADPR) proteins bind to the boronate columns and can be freed from unmodified proteins, DNA and RNA. When subjected to SDS gel electrophoresis, these $(ADPR)_n$ protein conjugates separated into multiple bands exhibiting a distinct pattern. The pattern was similar when stained with Coomassie Blue or made visible by fluorography of the [^{3}H]ADPR residues. Differences in the intensities of staining indicated preferential ADP-ribosylation of certain acceptor proteins. Selective detachment of the modifying groups by alkaline treatment, or by digestion with phosphodiesterase I abolished the affinity of the boronate matrices, and the bands in SDS gel electrophoresis were shifted to a large extent to lower molecular sizes.

Isolated conjugates were separated into a histone and a non-histone fraction by conventional methods. The histone fraction resembled, but was not identical with EAT cell histones, as judged from the electrophoretic mobility. Upon release of the $(ADPR)_n$ residues, however, the protein bands moved to the position of the four core histones which were present in unequal amounts (H1 was not included in this perchloric acid-insoluble fraction). Surprisingly, release of the modifying groups from the "non-histone" fraction of the conjugates yielded an even higher amount of histones. This immediately showed that modification of proteins by oligo and poly(ADPR) can change the properties of the acceptor fundamentally, shifting histones into the non-histone fraction. On the basis of protein content 60% of the total acceptor proteins in these tumor cell nuclei were non-histone proteins.

Subsequent to the isolation of the total nuclear $(ADPR)_n$ conjugates, we were able to isolate the first single acceptor protein in the form of its $(ADPR)_n$ conjugates. By combination of cation exchange and boronate chromatography, pure histone H1 conjugates free of nonmodified proteins could be obtained, which after SDS gel electrophoresis separated into multiple protein bands. The bands differed from each other by one ADPR residue. Specific removal of the ADPR residues with phosphodiesterase yielded a single band of histone H1. Close inspection of the conjugate bands revealed an identical microheterogeneity of each of these bands corresponding to the state of phosphorylation. The data are consistent with the interpretation that histone H1 can serve as an acceptor for ADPR transfer independent of the degree of phosphorylation, and that the acceptor amino acids for phosphoryl groups are not likely to serve also as acceptors for ADPR transfer. The results from EAT cells differ to some extent from reports on histone H1 formed in rat liver nuclei (13, 14), indicating different distribution of ADPR residues in H1 in different tissues.

POSSIBLE FUNCTIONS OF ADP-RIBOSYLATION

Mono(ADPR)Protein Conjugates

Preliminary experiments had indicated that several tumors have low levels of mono(ADPR) protein conjugates compared to their parent tissue. When the ADP ribosylation status in developing rat liver was analyzed it was found that low levels of mono(ADPR) protein conjugates were not restricted to malignant tissues. Very low amounts were also seen in fetal liver, while higher values were found after birth,and highest levels in adult liver. These data suggested a role of protein modification by mono(ADPR) in processes associated with proliferation and / or differentiation.

When both subclasses of mono(ADPR) protein conjugates were quantified separately interesting differences between the subfractions became apparent (23). In the fetal stage, hydroxylamine-resistant mono(ADPR) proteins were extremely low. They showed a continuous increase during liver development reaching the highest values when the organ had acquired its full metabolic capacity. By contrast, hydroxylamine-sensitive mono(ADPR) protein conjugates were present in fetal liver (8th prenatal day) at concentrations 25 times higher than the resistant conjugates. Their level changed with the unusual growth pattern of postnatal liver, indicating that high rates of tissue proliferation were associated with low levels of hydroxylamine-sensitive mono(ADPR) proteins. This relationship was also found in the hepatoma series. When the amount of hydroxylamine-sensitive mono(ADPR) protein conjugates in normal and neoplastic hepatic tissues was plotted against the average doubling time, a linear correlation was obtained. Since most of the hydroxylamine-sensitive conjugates were found in the endoplasmic reticulum (19), it appears that it is the ADP-ribosylation of microsomal proteins that change with the hepatic growth rate.

In contrast, the hydroxylamine-resistant conjugate subfraction did not relate to the proliferation rate. Instead, a correlation of this subclass to the degree of terminal differentiation was found, which applied to the hepatoma series as well. In hepatomas the hydroxylamine-resistant fraction, too, related to the degree of differentiation, being low in the poorly differentiated hepatomas AH 7974 and AH 130 but elevated in the well-differentiated Reuber H35 or Morris 9618A tumors (23).

Nevertheless, the undifferentiated hepatic tumors differ from the "undifferentiated" fetal tissue in one aspect of ADP-ribosylation. The absolute amounts of hydroxylamine-resistant conjugates in hepatomas were never decreased to such an extent as for instance in fetal liver. This phenomenon becomes especially evident when the ratios of the hydroxylamine-resistant over the hydroxylamine-sensitive conjugates (R/S ratios) are plotted against the degree of terminal differentiation: Fetal tissue had extremely low R/S ratios, while undifferentiated hepatomas exhibited considerably higher ratios than adult liver, and the well-differentiated tumors approached the values found in the adult organ. This situation is characteristic of hepatic tissues, and seen in stationary as well as in 'resting' hepatic tissues (23).

Indications that the hydroxylamine-resistant mono(ADPR) conjugate subclass related to terminal differentiation came also from a comparative analysis of normal blood lymphocytes and lymphocytes from patients with chronic lymphocytic leukemia (24). Neither the normal nor the leukemic lymphocytes do proliferate to a significant extent. However, they differ in their degree of differentiation, the tumor cells exhibiting an extended life span and a retarded or missing response to mitogens. Total mono ADP-ribosylated conjugates in leukemic lymphocytes were markedly lower than in normal blood lymphocytes. The decrease was mainly at the expense of the hydroxylamine-resistant subfraction and seen in all patients so far studied (fig. 2).

Mitochondria contain most of the hydroxylamine-resistant mono (ADPR)protein conjugates (19). The low levels of this subclass in poorly differentiated tissues may therefore at least in part result from a lower content of mitochondria as well as a lower degree of protein modification in this compartment. Several individual acceptor proteins in mitochondria and the plasma membrane of the NH_2OH-resistant conjugate subclass have been found - at least in vitro (cf.25,26). Their relation to these changes is presently under investigation.

Poly ADP-Ribosylation - a Function of the Nucleus

Nuclei of eukaryotic cells contain a poly(ADPR) synthetase (5-7). When analyzed in nuclear preparations (or better so in permeabilized cells) the enzyme is strongly stimulated by the addition of endonucleases, leading to DNA fragmentation (27-29). Fragmentation of DNA by alkalyting agents or X-rays is also accompanied by an increase of ADPR transferase activity in permeabilized cells - as shown first by Berger's and by Shall's groups (28,29). Alkylating agents also lower the cellular NAD content and thereby interfere with glycolysis, a phenomenon known for many years (cf. 30-32). It was suggested that the decrease in NAD under these conditions was caused by an increased demand for poly(ADPR) synthesis (33). It should be taken into consideration, however, that the amounts of NAD(H) in most cells are three to four orders of magnitude higher than the levels of ADPR residues present in polymeric form. Even a hundred-fold increase in poly(ADPR) could therefore "consume" only a small fraction of the cellular NAD. Only when an extremely high turnover of poly(ADPR) residues during repair, exceeding by far resynthesis of NAD, is assumed, the above mentioned explanation may hold.

ADPR transferase activity in permeabilized cells, however, is a rather indirect measure of protein-bound poly(ADPR) and not identical with the actual level of its product in the intact cell (34). We therefore determined the amounts of poly(ADPR) conjugates in response to the alkylating aziridine derivative Trenimon. A rapid in-

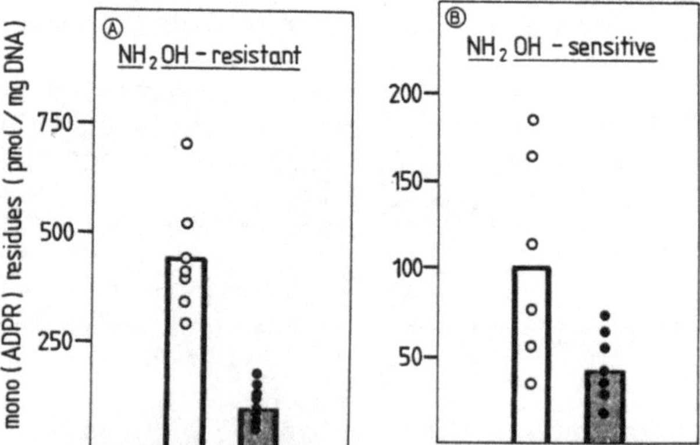

Fig. 2 Mono(ADPR) protein conjugates in normal and leukemic (chronic lymphocytic leukemia) lymphocytes. - open bars: normal; hatched bars: leukemic. From (24).

crease of poly(ADPR) levels concomitant with the decline of NAD was seen (fig. 3). Under the same conditions mono(ADPR) conjugates showed only a slight increase.

In agreement with a possible role of poly ADP-ribosylation in DNA repair is the observation made in different laboratories, that NAD depleted cells are able to proliferate but unable to repair DNA damage (33, 35). It was also shown, that permeabilized cells from patients with Xeroderma pigmentosum (deficient in UV-induced DNA incision) did not react to UV-irradiation with an increase in poly(ADPR) synthetase activity but did so when treated with nucleases or with alkylating agents (36).

Otto Warburg maintained that the primary cause of cancer is a defect in respiration, and he proposed treatment of cancer patients with active groups of respiratory enzymes, including nicotinamide

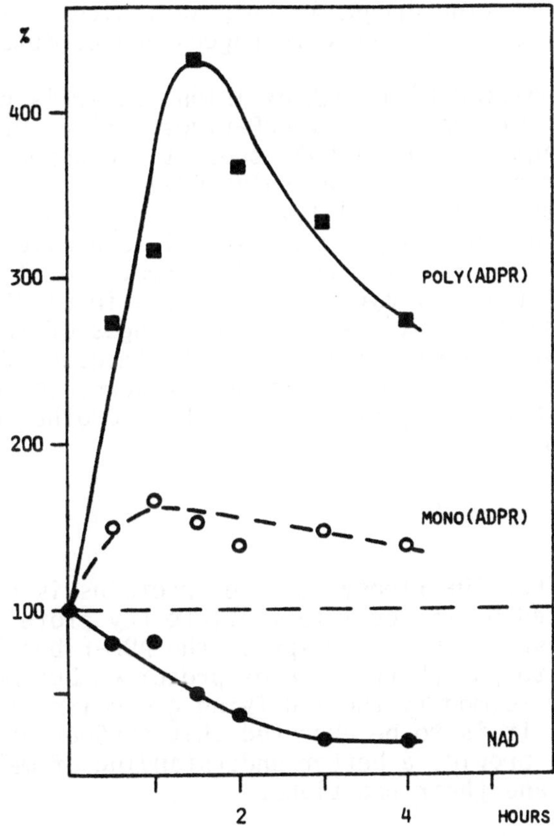

Fig. 3. Kinetics of Trenimon-induced $(ADPR)_n$ formation and NAD depletion (from (34)).

(37). Nicotinamide could indeed become a value in the destruction of tumor cells, though not as a respiration-improving co-factor. Nicotinamide and some analogs like benzamide are potent inhibitors of the nuclear poly(ADPR) synthetase (38, 39). If poly ADP-ribosylation is indeed involved in DNA repair, combination of nicotinamide or is analogs with repair-inducing agents should increase their cytostatic efficiency. When HeLa cells were treated with a limiting amount of the alkylating agent dimethyl sulfate, cell proliferation was reduced by 50% due to residual damage that could not be repaired. Addition of the nicotinamide analog benzamide did not significantly alter cell multiplication. However, when combined with DMS, benzamide showed a highly synergistic effect virtually eliminating proliferation. Similar findings were reported by Shall's group using mouse leukemia cells and plating efficiency determination (33). Thus the vitamin nicotinamide and its analogs when present at higher concentrations proved to act as co-cytostatics. Since alkylating agents can also transform normal cells presumably by inducing damage to DNA, suppression of repair by nicotinamide analogs may again increase the efficiency of the alkylating drugs. In this way, nicotinamide (analogs) might turn out to be also co-carcinogens and co-mutagens.

Involvement of poly(ADPR) in DNA excision repair is probably not the only function of nuclear poly ADP-ribosylation. Other processes leading to fragmentation of DNA appear to be associated with a modification of nuclear proteins by poly(ADPR) as well. This can be inferred from altered poly(ADPR) levels during changes of transcriptional activity in renal tubular cells (40), and altered ADPR transferase activity in cells induced to differentiate, like in Friend leukemia cells (41), in myoblasts (42) and in L1-3T3-preadipocytes (43). In all these situations profound changes of the physicochemical properties of acceptor proteins may be imposed by transfer of poly(ADPR) residues. Such modifications may serve to change the interaction of these acceptor proteins with DNA and other constituents of the chromatin.

CONCLUSION

It now appears that ADP-ribosylation of proteins is a multifunctional process used by the cell to modulate key proteins for many divergent processes. In this respect, the ADP-ribosylation system is comparable to phosphorylation of proteins, but more versatile by the ability to modify the modifying group itself, to form oligo or poly(ADPR). It is to be expected that further progress in this field will also provide a better understanding of malignant cells, their markers and their reactions.

REFERENCES

1. P. Chambon, J.D. Weill, J. Doly, M.T. Strosser, and P. Mandel, Biochem. Biophys. Res. Commun. 25:638 (1966).
2. Y. Nishizuka, K. Ueda, K. Nakazawa, and O. Hayaishi, J. Biol. Chem. 242:3164 (1976).
3. T. Sugimura, S. Fujimura, S. Hasegawa, and Y. Kawamura, Biochem. Biophys. Acta 138:438 (1967).
4. M. Miwa, N. Saikawa, Z. Yamaizumi, S. Nishimura, and T. Sugimura, Proc. Natl. Acad. Sci. USA 76:595 (1979).
5. H. Hilz and P.R. Stone, Rev. Physiol. Biochem. Pharmacol. 76:1 (1976).
6. O. Hayaishi and K. Ueda, Ann. Rev. Biochem. 46:95 (1977)
7. M.R. Purnell, P.R. Stone, and W.J.D. Whish, Biochem. Soc. Trans. 8:215 (1980).
8. T. Honjo, Y. Nishizuka, O. Hayaishi, and I. Kato, J. Biol. Chem. 243:3553 (1968)
9. P. Adamietz, R. Bredehorst, and H. Hilz, Eur. J. Biochem. 91:317 (1978).
10. R. Bredehorst, A.M. Ferro, and H. Hilz, Eur. J. Biochem. 82:115 (1978)
11. R. Bredehorst, M. Schlüter, and H. Hilz, Biochim. Biophys. Acta 652:16 (1981).
12. K. Wielckens, R. Bredehorst, P. Adamietz, and H. Hilz, Eur. J. Biochem. 117:69 (1981).
13. R.T. Riquelme, L.O. Burzio, and S.S. Koide, J. Biol. Chem. 254: 3018 (1979).
14. N. Ogata, K. Ueda, H. Kagamiyama, and O. Hayaishi, J. Biol. Chem. 255:7616 (1980).
15. R. Bredehorst, K. Wielckens, A.L. Gartemann, H. Lengyel, K. Klapproth, and H. Hilz, Eur. J. Biochem. 92:129(1978).
16. J. Moss and S.J. Stanley, J. Biol. Chem. 256:7830 (1981).
17. H. Hilz, K. Wielckens, and R. Bredehorst. in "ADP-ribosylation Reactions", O. Hayaishi and K. Ueda, eds., Academic Press, New York (in press).
18. K. Wielckens, W. Sachsenmaier, and H. Hilz, Hoppe-Seyl. Z. Physiol. Chem. 360:39 (1979).
19. P. Adamietz, K. Wielckens, R. Bredehorst, H. Lengyel and H. Hilz, Biochem. Biophys. Res. Commun. 101:96 (1981).
20. M. Smulson, T. Butt, N. Nolan, D. Jump, and B. Decoste, in Novel ADP-ribosylations of Regulatory Enzymes and Proteins", M.E. Smulson and T. Sugimura, eds., Elsevier / North Holland, Amsterdam / New York (1980).
21. K. Ueda, M. Kawaichi, J. Oka, and O. Hayaishi, in "Novel ADP-ribosylation of Regulatory Enzymes and Proteins", M.E. Smulson and T. Sugimura, eds., Elsevier / North Holland, Amsterdam / New York (1980).
22. P. Adamietz, K. Klapproth, and H. Hilz, Biochem. Biophys. Res. Commun. 91:1232 (1979).

23. R. Bredehorst, K. Wielckens, P. Adamietz, E. Steinhagen-Thiessen, and H. Hilz, Eur. J. Biochem. (in press).
24. K. Wielckens, M. Garbrecht, M. Kittler and H. Hilz, Eur. J. Biochem. 104:279 (1980).
25. E. Kun, P.H. Zimber, A.C.Y. Chang, B. Puschendorf, and H. Grunicke, Proc. Nat. Acad. Sci. USA 72:1436 (1975).
26. T.M. Reilly, S. Beckner, E.M. McHugh, and M. Blecher, Biochem. Biophys. Res. Commun. 98:1115 (1981).
27. E.G. Miller, Biochim. Biophys. Acta 395:191 (1975).
28. H. Halldorsson, D.A. Gray, and S. Shall, FEBS Lett. 85:349 (1978)
29. N.K. Berger, G. Weber, and A.S. Kaichi, Biochim. Biophys. Acta 519:87 (1978).
30. I.M. Roitt, Biochem. J. 63:300 (1956).
31. H. Holzer, P. Glogner, and O. Sedlmayr, Biochem. Z. 330:59 (1958).
32. H. Hilz, P. Hlavica, and B. Bertram, Biochem. Z. 338:283 (1963).
33. B.W. Durkuacz, N. Nduka, O. Okidiji, S. Shall, and A. Zia'ee, in "Novel ADP-ribosylations of Regulatory Enzymes and Proteins", M.E. Smulson and T. Sugimura, eds., Elsevier / North Holland, Amsterdam / New York (1980).
34. K. Wielckens, A. Schmidt, R. Bredehorst, and H. Hilz, in preparation.
35. E.L. Jacobson and G. Narashimhan, Fed. Proc. 38:619 (1979).
36. N.A. Berger, G.W. Sikorski, S.J. Petzold, and K.K. Kurohara, Biochem. 19:289 (1980).
37. O. Warburg, Biochem. Z. 142:317 (1923).
38. J. Preiss, R. Schlaeger and H. Hilz, FEBS Letters 19:244 (1971).
39. M.P. Purnell and J.D. Whish, Biochem. J. 185:775 (1980)
40. A. Gartemann, R. Bredehorst, K. Wielckens, W.H. Strätling and H. Hilz, Biochem. J. 198:37 (1981).
41. E. Rastl and P. Swetly, J. Biol. Chem. 253:4333 (1978).
42. F. Farzaneh, S. Shall and R. Zalin, in "Novel ADP-ribosylations of Regulatory Enzymes and Proteins", M.E. Smulson and T. Sugimura, eds., Elsevier / North Holland, Amsterdam / New York (1980).
43. P.H. Pekala, M.D. Lane, P.A. Watkins and J. Moss, J. Biol. Chem. 256:4871 (1980).

ALTERATIONS IN tRNA METABOLISM AS MARKERS OF NEOPLASTIC TRANSFORMATION

Ronald W. Trewyn, Mark S. Elliott,
Ronald Glaser, and Michael R. Grever

Department of Physiological Chemistry (R.W.T., M.S.E.)
Department of Medical Microbiology & Immunology (R.G.)
Department of Medicine (M.R.G.)
The Ohio State University
Columbus, Ohio 43210

INTRODUCTION

Numerous changes in tRNA modification and catabolism are observed when cells undergo neoplastic transformation. The nature of certain of these changes is well established, while others require further characterization. Enzymes involved in the macromolecular modification of tRNA, the tRNA methyltransferases, exhibit idiosyncratic alterations during neoplastic transformation. These alterations include increases in enzyme specific activity as well as the appearance of different tRNA methyltransferases in the malignant tissue[1,2]. The increased tRNA methyltransferase activity and capacity observed *in vitro* for the enzymes isolated from transformed cells can also be correlated to increased methylation of specific tRNA isoaccepting species, although not total tRNA, *in vivo*[3].

Elevated turnover of tRNA is another characteristic of neoplastic transformation[4,5], and the increased rate of tRNA catabolism may explain the lack of extensive hypermethylation of total tRNA in malignant cells. Most modified nucleosides in tRNA cannot be salvaged when the macromolecules are degraded, and therefore, they are excreted[4]. Monitoring the elevated excretion of these tRNA catabolites by cancer patients is being investigated to determine the usefulness of these components as biochemical markers for cancer[6-8].

The enhanced generation of methylated tRNA catabolites by cancer patients may also have a fundamental role in the neoplastic process. Chronic exposure of normal mammalian cells in culture to specific methylated purine RNA catabolites can lead to neoplastic transformation[9,10]. However, the mode of action by which these natural products elicit such a response has not been established.

The appearance of many unique tRNA isoaccepting species is another common feature established for malignant cells[4,11,12]. In some cases, these species appear to differ from their normal counterparts with respect to their macromolecular modifications. Hypomodification for Y-base adjacent to the anticodon in phenylalanine tRNA and for Q-base in the first position of the anticodons for histidine, tyrosine, asparagine, or aspartic acid tRNA's is responsible for the appearance of some of the different isoacceptors in transformed cells[12-14]. Again, these tRNA aberrations offer biochemical markers for neoplastic transformation.

In this report, we examine certain of the alterations in tRNA metabolism associated with neoplasia. Potential interrelationships between the changes in tRNA modification and catabolism are explored, and a role for these aberrations in the expression of carcinogenesis is postulated.

MATERIALS AND METHODS

Nucleosides in urine were resolved and quantitated using reversed-phase high performance liquid chromatography[8] following clarification on a boronate column[15]. Quantitation was relative to the creatinine content in random urine specimens[7].

Establishment and propagation of primary cultures of Chinese hamster embryo cells were as previously described[9,16] except that the culture medium was supplemented with only 5% fetal bovine serum. These cells typically exhibit a finite lifetime in culture of 10 to 12 passages under the conditions employed. The methods for transforming these cells by chronic exposure to selected methylated purines have also been published[9,10]. The concentration of methylated purine utilized was always 10 μM.

The assay for Q-deficient tRNA makes use of the enzyme tRNA transglycosylase from *Escherichia coli*[14]. This enzyme can utilize mammalian tRNA's for histidine, tyrosine, asparagine, and aspartic acid as substrates only if they are hypomodified, i.e., the tRNA's have guanine in the first position of the anticodon instead of Q-base[14]. Transfer RNA from proliferating Chinese hamster cells treated with 10 μM 7-methylguanine was isolated utilizing published protocols[17], and it was evaluated as a substrate for the *E. coli* enzyme. The assay for Q-hypomodification involves an exchange reaction with [8-^{3}H]guanine. Previously published methods were employed[14,18]. Yeast tRNA is Q-deficient, and therefore, was utilized as a positive control. E. coli tRNA is Q-sufficient, so it was used as a negative control. A tRNA transglycosylase from rabbit erythrocytes[19] was used to assess enzyme inhibition by 7-methylguanine.

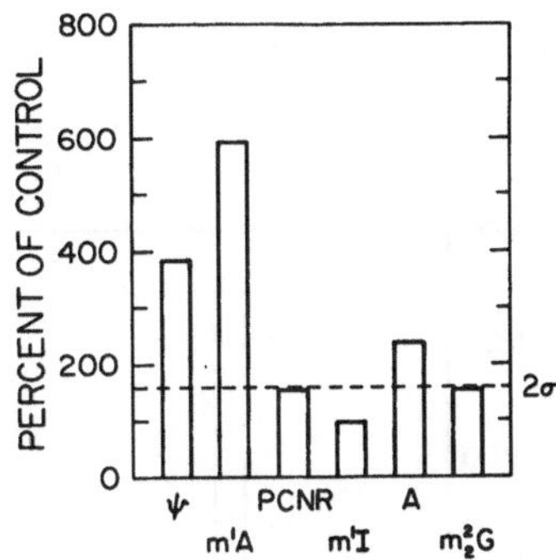

Fig. 1. Excretion of nucleoside markers by a patient at the time of of diagnosis of NPC. The results were calculated as nmoles nucleoside/μmole creatinine, and are expressed relative to normal values as percent of control. The dashed line denotes the position of two standard deviations above normal for adenosine, the marker exhibiting the largest relative standard deviation. Excretion levels above the dashed line represent significant ($P<0.02$) increases for adenosine and highly significant ($P<0.01$) for the other nucleosides. The abbreviations are: ψ, pseudouridine; m^1A, 1-methyladenosine; PCNR, 2-pyridone-5-carboxamide-N'-ribofuranoside; m^1I, 1-methylinosine; A, adenosine; and m^2_2G, N^2,N^2-dimethylguanosine.

RESULTS

Nucleoside Excretion by Cancer Patients

Urinary nucleoside excretion has been quantitated for patients with nasopharyngeal carcinoma (NPC) and leukemia, and the results have been compared to normal excretion levels. The normal excretion values (nmoles nucleoside/μ mole creatinine) used for comparison were as follows[8]: pseudouridine, 24.8; 1-methyladenosine, 2.02; 2-pyridone-5-carboxamide-N'-ribofuranoside, 1.14; 1-methylinosine, 0.96; 1-methylguanosine, 0.70; adenosine, 0.23; and N^2,N^2-dimethylguanosine, 1.05. The greatest relative standard deviation for the controls was 30.4% for adenosine.

The relative nucleoside excretion pattern for a Caucasian NPC patient at the time of diagnosis is presented in Fig. 1. At that time, the excretion levels of pseudouridine and 1-methyladenosine

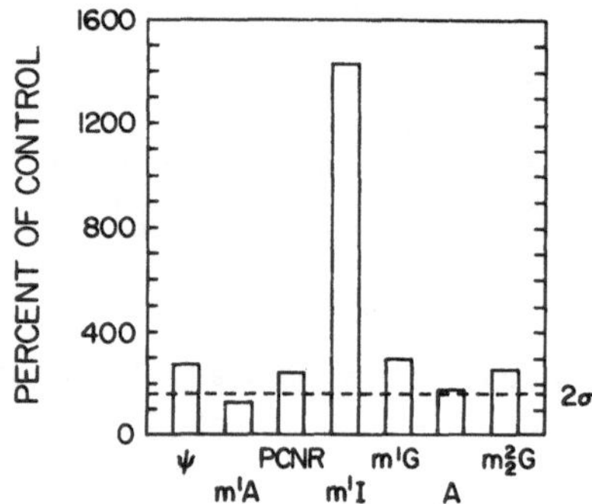

Fig. 2. Excretion of nucleoside markers by a patient at the time of diagnosis of AML. See legend to Fig. 1 for details. The additional abbreviation is: m^1G, 1-methylguanosine.

were elevated 4-fold and 6-fold respectively. Cells from tumor tissue biopsies contained the Epstein-Barr virus (EBV) genome (12.5 equivalents/cell)[20], and the patient's serum contained very high levels of antibodies to an EBV-specific DNase (10.2 units neutralized/ml serum).

The nucleoside excretion pattern for an individual at the time of diagnosis of acute myelogenous leukemia (AML) is shown in Fig. 2. In this case, 1-methylinosine was the primary marker with an increase of greater than 14-fold. Four of the other nucleosides (pseudouridine, 2-pyridone-5-carboxamide-N'-ribofuranoside, 1-methylguanosine, and N^2,N^2-dimethylguanosine) were elevated approximately 3-fold. This patient also had significantly elevated adenosine deaminase levels in his peripheral blood cells (21.0 units/10^6 cells) compared to normal (8.4 units/10^6 cells).

Cell Transformation by Methylated Purines

The significant increase in the excretion of modified RNA catabolites by cancer patients led to the examination of the response of normal mammalian cells to these components[9,10]. Certain methylated purines were found to transform Chinese hamster embryo cells *in vitro*, with neoplastic transformation being demonstrated in some cases[10]. A summary of methylated purines evaluated and those transforming the cells for proliferative capacity (finite to continuous lifetime in culture) can be seen in Table 1. Certain of the RNA catabolites (1-methylguanine and 7-methylguanine) greatly enhance the generation of continuous cell lines, while others do not. Other naturally occurring methylated purines (7-methylxanthine and 1,3,7-trimethylxanthine) are also quite effective in transforming the cells.

The expression of various transformed phenotypes appearing during continuous exposure to the methylated purines can be reversed by removal of the methylated purine. An example showing increased saturation density of a 7-methylxanthine-transformed cell line is presented in Fig. 3. Removal of 7-methylxanthine at passage level 15, a passage level exceeding the normal number of passages obtainable before senescence and cell death, resulted in a significant decrease in the cell density of subsequent passage levels. In 3 of 4 independent transformation experiments with 7-methylxanthine, the cultures went through such a "crisis"

Table 1. Methylated Purines Generating Continuous Chinese Hamster Cell Lines

Methylated Purine	Continuous/Treated
None (Control)	2/30
Guanine (Control)	0/2
1-Methyladenine	0/2
1-Methylguanine	15/16
3-Methylguanine	0/2
7-Methylguanine	11/12
1-Methylhypoxanthine	0/2
1-Methylxanthine	1/4
3-Methylxanthine	0/2
7-Methylxanthine	4/4
1,3,7-Trimethylxanthine	4/6

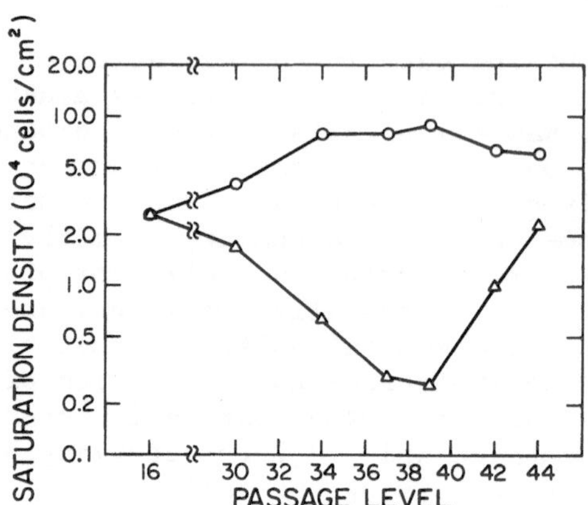

Fig. 3. Saturation densities for a 7-methylxanthine-transformed Chinese hamster cell line. The cells had been treated continuously with 10 μM 7-methylxanthine since the first passage of the primary culture. When the transformed, "continuous" cell line was subcultured at passage 15, duplicate cultures were maintained thereafter in the presence (o) or absence (Δ) of 7-methylxanthine. Saturation densities were determined after confluent cultures were split 1:4 and allowed to grow for 7 days. Duplicate cultures were trypsinized, and cells were counted with a hemacytometer.

period after removal of the methylated purine. With a 1,3,7-trimethylxanthine-transformed cell line, the cloning efficiency in soft agar decreased 6-fold after removal of the methylxanthine (unpublished observation), and tumorigenicity in *nude* mice was reversed by removal of 1-methylguanine from a corresponding cell line[21]. In the latter case, there was no change in the cloning efficiency in soft agar or any other *in vitro* characteristic related to transformation.

Q-Hypomodification of Cellular tRNA

The enzyme tRNA transglycosylase from mammalian sources catalyzes the reaction depicted in Fig. 4. Transfer RNA isolated from normal cells is mainly in the Q-modified form, while tRNA from transformed cells is Q-deficient[14]. The possibility that 7-methylguanine, a structural analog of Q-base, might inhibit Q-modification of tRNA was examined by treating normal Chinese hamster cells with 10 μM 7-methylguanine; the same concentration and conditions used for transformation. Transfer RNA was isolated from treated and untreated normal cells after 6 population doublings and assayed for Q-deficiency. Transfer RNA was also isolated from cells treated for 4 population doublings followed by no treatment for 2 more doublings to assess reversibility of any 7-methylguanine-induced Q-hypomodification. As can be seen in Table 2, 7-methylguanine induced Q-hypomodification of tRNA in the cells, and the Q-deficiency was reversible.

A tRNA transglycosylase isolated from rabbit erythrocytes was also shown to be inhibited by 10 μM 7-methylguanine _in vitro_. In 4 separate experiments with Q-deficient yeast tRNA, the percent inhibition obtained was 60.1 ± 7.6 (mean ± standard deviation) when the guanine substrate concentration was 1 μM.

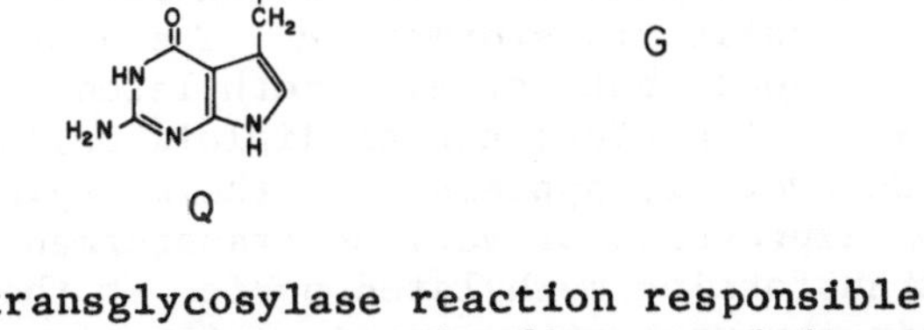

Fig. 4. tRNA transglycosylase reaction responsible for exchanging Q-base for guanine in the first position of the anticodon of tRNA's for tyrosine, histidine, asparagine, and aspartic acid. The abbreviations are: Q, 7-(3,4-_trans_-4,5-_cis_-dihydroxy-1-cyclopenten-3-ylaminomethyl)-7-deazaguanine; and G, guanine.

Table 2. Q-Hypomodification of tRNA in Chinese Hamster Cells Induced 7-Methylguanine

tRNA Source	Guanine Incorporation (pmoles/hr/A_{260} unit)
Chinese hamster cells	1.86
Plus 7-methylguanine	3.46
Minus 7-methylguanine	2.09
Escherichia coli	<0.2
Yeast	6.58

DISCUSSION

The potential value of modified nucleosides as biochemical markers for cancer can be seen in Fig. 1 and Fig. 2. Even at the time of cancer diagnosis, nucleoside excretion was elevated significantly for the NPC patient (Fig. 1), and this correlated with high serum antibodies to EBV antigens including antibody to the EBV DNase, a marker for NPC[22]. The AML patient exhibited even higher nucleoside excretion levels at the time of diagnosis (Fig. 2), although the pattern of elevated excretion was different. Unique excretion patterns may offer additional means to characterize specific cancers. The patient with AML also exhibited elevated peripheral blood cell adenosine deaminase activity, another potential biological marker for leukemia. The clinical value of monitoring various markers for leukemia and NPC is being assessed for both diagnostic and prognostic purposes.

The early increases in tRNA catabolism associated with neoplasia led to an examination of the potential role of the catabolites in neoplastic transformation. The discovery that chronic exposure to some, but not all, methylated purines derived from cellular RNA can transform normal diploid cells was quite perplexing[9,10]. However, it appears that the methylated purines may influence the expression of various transformed phenotypes. Removal of the transforming methylated purine at the appropriate time can result in reversal of expression of various transformed phenotypes, e.g., increased proliferative capacity (Fig. 3), anchorage independent growth, and tumorigenicity[21]. All transformed phenotypes are not reversed by removing the methylated purine from a particular culture. However, it was demonstrated previously that the methylated purine-transformed cells exhibit elevated tRNA methyltransferase activity[9], and therefore, the endogenous methylated purine level may negate the need for an exogenous source.

The results obtained with the methylated purines suggested similarities to dedifferentiation associated with carcinogenesis. Since dedifferentiation reportedly involves changes in gene regulation at the post-transcriptional level[23], we have attempted to identify cellular targets for the methylated purines that might alter phenotypic expression by similar means. A proposed target was tRNA transglycosylase, since the enzyme from *E. coli* is inhibited by the methylated purine 7-methylguanine[18]. It was presumed that a major structural change (guanine *vs* Q-base) in transformed tRNA's for histidine, tyrosine, asparagine, and/or aspartic acid generated by inhibiting the transglycosylase might allow the altered tRNA isoaccepting species to translate disparate mRNA's more efficiently. As we have now found, 7-methylguanine does inhibit tRNA transglycosylase from a mammalian source, and it induces Q-hypomodification of cellular tRNA (Table 2) under conditions leading to the expression of transformation.

The question of whether Q-deficient tRNA's actually have some role in the expression of transformed phenotypes remains to be answered. However, it has been reported that reversing tRNA Q-deficiency in tumor cells by administration of purified Q-base was associated with diminution of tumor cell growth *in vivo*[24].

The numerous alterations in tRNA metabolism associated with neoplasia have led us to devise a scheme by which they may interrelate in the expression of carcinogenesis, and the proposed sequence of events is presented in Fig. 5. The induction (initiation) of carcinogenesis could be by any means. The subsequent events are then predicted to have cause and effect relationships, i.e., each change depicted would occur in order and be caused by the previous change. Therefore, soon after the induction event there would be an increase in tRNA methyltransferase activity which would result in an increase in methylated RNA catabolites. The higher endogenous levels of methylated purines would then modulate tRNA modification by inhibiting tRNA transglycosylase. The methylated purines might also modulate tRNA modification by acting as an alternate substrate for the transglycosylase or by other, as yet unidentified, means. Both the modulation and methylation steps would be involved in generating altered tRNA isoaccepting species, and these species might allow the translation of different mRNA's that are not translated efficiently by the normal tRNA population. It is then assumed that some of these translation products would be onco-developmental proteins. If

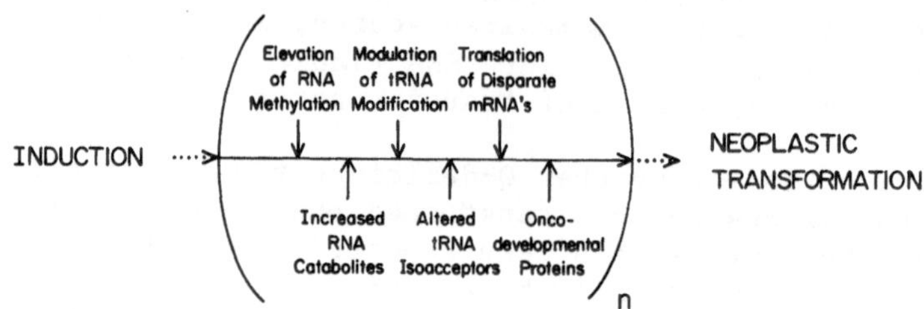

Fig. 5. Proposed model for the role of altered tRNA metabolism in the expression of carcinogenesis. Each event in the sequence is predicted to influence subsequent events, and the last may reinitiate the first.

any of these proteins were tRNA methyltransferases, and it is known that enzymes with different specificities appear[1], the cycle would repeat at a more aberrant level. By this means, the cycle could continue to generate accruing phenotypic alterations until neoplastic transformation is attained.

The proposed model offers an explanation for the general staging process of carcinogenesis. It also allows interpretation of the phenotypic reversibility phenomenon demonstrated for the exogenous methylated purines. A step back to the previous cycle might be possible by such a withdrawal, but the increased generation of endogenous methylated purines would block any further phenotypic reversion.

Certain of the individual points outlined in Fig. 5 have been proposed by other investigators to have a role in neoplastic transformation. However, linking the increased RNA catabolites to the expression of transformed phenotypes as well as the induction of tRNA hypomodification allowed us to formulate the comprehensive scheme presented. The hypothesis is being tested using a variety of model systems, and the similarities to promotion of neoplastic transformation are being studied. The numerous biochemical markers involved should greatly facilitate these investigations.

ACKNOWLEDGEMENTS

We wish to thank Holly Gatz and Jane Holiday for expert technical support, and Aline Davis for manuscript preparation and editing. This work was supported in part by grants and contracts from the following agencies: the Air Force Office of Scientific Research (AFOSR-80-0283), the American Cancer Society-Ohio Division, and the National Cancer Institute (N01-CP81021 and CA-16058-08).

REFERENCES

1. S. J. Kerr and E. Borek, The tRNA methyltransferases, Adv. Enzymol. 36:1 (1972).
2. S. J. Kerr, tRNA methyltransferases in normal and neoplastic tissues, in: "Isozymes: Developmental Biology," Vol. III, pp. 855, Academic Press. New York (1975).
3. Y. Kuchino and E. Borek, Tumour-specific phenylalanine tRNA contains two supernumerary methylated bases, Nature 271:126 (1978).
4. E. Borek and S. J. Kerr, Atypical transfer RNA's and their origin in neoplastic cells, Adv. Cancer Res. 15:163 (1972).
5. E. Borek, B. S. Baliga, C. W. Gehrke, K. C. Kuo, S. Belman, W. Troll and T. P. Waalkes, High turnover rate of transfer RNA in tumor tissue, Cancer Res. 37:3362 (1977).

6. J. Speer, C. W. Gehrke, K. C. Kuo, T. P. Waalkes and E. Borek, tRNA breakdown products as markers for cancer, Cancer 44:2120 (1979).
7. C. W. Gehrke, K. C. Kuo, T. P. Waalkes and E. Borek, Patterns of urinary excretion of modified nucleosides, Cancer Res. 39:1150 (1979).
8. R. W. Trewyn, R. Glaser, D. R. Kelly, D. G. Jackson, W. P. Graham and C. E. Speicher, Elevated nucleoside excretion by patients with nasopharyngeal carcinoma: Preliminary diagnostic/prognostic evaluations, Cancer (in press).
9. R. W. Trewyn and S. J. Kerr, Altered growth properties of Chinese hamster cells exposed to 1-methylguanine and 7-methylguanine, Cancer Res. 38:2285 (1978).
10. R. W. Trewyn, J. M. Lehman and S. J. Kerr, Cell transformation by exogenous methylated purines, Adv. Enz. Reg. 16:335 (1978).
11. Y. Kuchino and E. Borek, Changes in transfer RNA's in human malignant trophoblastic cells (BeWo line), Cancer Res. 36:2932 (1976).
12. J. T. Muchinski and M. Marini, Tumor-associated phenylalanyl transfer RNA found in a wide spectrum of rat and mouse tumors but absent in normal adult, fetal, and regenerating tissues, Cancer Res. 39:1253 (1979).
13. J. R. Katze, Alterations in SVT2 cell transfer RNA's in response to cell density and serum type, Biochim. Biophys. Acta 383:131 (1975).
14. N. Okada, N. Shindo-Okada, S. Sato, Y. H. Itoh, K. Oda and S. Nishimura, Detection of unique tRNA species in tumor tissues by Escherichia coli guanine insertion enzyme, Proc. Natl. Acad. Sci. USA 75:4247 (1978).
15. C. W. Gehrke, K. C. Kuo, G. E. Davis, R. D. Suits, T. P. Waalkes and E. Borek. Quantitative high performance liquid chromatography of nucleosides in biological materials, J. Chromatog. 150:455 (1978).
16. J. M. Lehman and V. Defendi, Changes in deoxyribonucleic acid synthesis regulation in Chinese hamster cells infected with Simian virus 40, J. Virol. 6:738 (1970).
17. R. Wilkinson and S. J. Kerr, Alteration in tRNA methyltransferase activity in mengovirus infection: Host range specificity. J. Virol. 12:1013 (1973).
18. N. Okada and S. Nishimura, Isolation and characterization of a guanine insertion enzyme, a specific tRNA transglycosylase, from Escherichia coli, J. Biol. Chem. 254:3061 (1979).
19. N. K. Howes and W. R. Farkas, Studies with a homogeneous enzyme from rabbit erythrocytes catalyzing the insertion of guanine into tRNA, J. Biol. Chem. 253:9082 (1978).
20. R. Glaser, M. Nonoyama, R. T. Szymanowski and W. Graham, Human nasopharyngeal carcinomas positive for Epstein-Barr virus DNA in North America, J. Natl. Cancer Inst. 64:1317 (1980).

21. R. W. Trewyn, S. J. Kerr and J. M. Lehman, Karyotype and tumorigenicity of 1-methylguanine-transformed Chinese hamster cells, J. Natl. Cancer Inst. 62:633 (1979).
22. Y. C. Cheng, J. Y. Chen, R. Glaser and W. Henle, Frequency and levels of antibodies to Epstein-Barr virus-specific DNase are elevated in patients with nasopharyngeal carcinoma, Proc. Natl. Acad. Sci. USA 77:6162 (1980).
23. K. H. Ibsen and W. H. Fishman, Developmental gene expression in cancer, Biochim. Biophys. Acta 560:243 (1979).
24. J. R. Katze and W. T. Beck, Administration of queuine to mice relieves modified nucleoside queuosine deficiency in Ehrlich ascites tumor tRNA, Biochem. Biophys. Res. Commun. 96:313 (1980).

21. R. W. Trewyn, [illegible] Kerr and J. M. Lehman, [illegible] and [illegible] specificity of [illegible] Simian virus 40-transformed [illegible] hamster cells, J. Natl. Cancer Inst. 62:633 (1979).
22. [illegible] C. Cheng, [illegible] Chen, [illegible] and [illegible] [illegible] levels of antibodies to [illegible] elevated in patients with [illegible] Proc. Natl. Acad. Sci. USA [illegible] (1980).
23. [illegible] and [illegible] cancer, Biochim. Biophys. Acta 560:[illegible] (1979).
24. [illegible] patients [illegible] ability [illegible] CEA, Biochem. Biophys. Res. Commun. 96:[illegible] (1980).

ANTIBODIES TO CHROMOSOMAL NON-HISTONE PROTEINS IN MALIGNANT CELLS

Lubomir S. Hnilica, Warren N. Schmidt, Marta Stryjecka-Zimmer, David M. Duhl, Zainy Banjar and Robert C. Briggs

Department of Biochemistry and the
A.B. Hancock, Jr. Memorial Laboratory of the
Vanderbilt University Cancer Center
Vanderbilt University School of Medicine
Nashville, Tennessee 37232

INTRODUCTION

The exceptional selectivity of the immune response makes the use of antibodies to chromosomal proteins attractive for studies on the structure and function of chromatin. It is believed that alterations of cellular phenotype during differentiation and carcinogenesis are accompanied by changes in the composition of nuclear proteins, some of which may serve gene regulatory functions. Development of highly specific antisera would not only facilitate the detection, isolation and characterization of such regulatory proteins, but may also provide additional tools for the immunodetection of cancer. Isolated nuclear nonhistone proteins,[1-3] nucleolar fractions,[4-6] intact or dehistonized chromatins[7-10] were all used as immunogens to elicit antisera which exhibited various degrees of specificity. Some antisera recognized antigens which changed during differentiation and development[7,11] or carcinogenesis.[9,12] A considerable cell specificity was demonstrated for antisera to dehistonized chromatin preparations.[2,7-14]

NOVIKOFF HEPATOMA

We have found, several years ago, that antibodies to dehistonized preparations of chromatin from Novikoff hepatoma can distinguish between normal rat liver and hepatoma chromatins.[8,9] Such antibodies were relatively weak and non-precipitating,

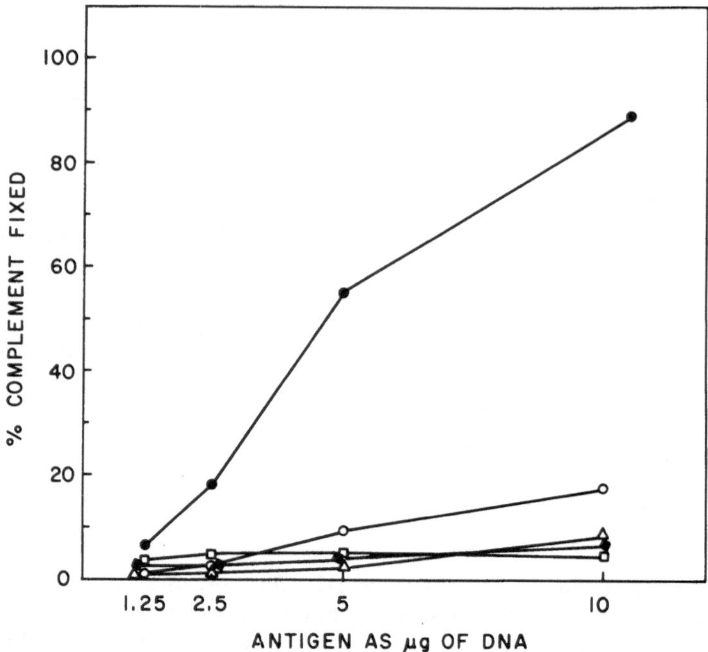

Fig. 1. The specificity of antiserum to dehistonized Novikoff hepatoma chromatin. The complement fixation assays were performed in the presence of increasing amounts of various chromatins. The antiserum was optimally reactive at a dilution of 1:200. The symbols are Novikoff hepatoma (●), normal rat liver (△), rat Sertoli cells (◆), rat thymus (O) and HeLa cells (□).

detectable by the microcomplement fixation assay of Wasserman and Levine.[15] Prolonged storage at -20° or repeated freezing and thawing destroyed the specific complement fixing activity of the antisera. Complement fixation of antiserum to dehistonized Novikoff hepatoma chromatin assayed in the presence of chromatins isolated from several rat tissues is shown in Figure 1. As can be seen, the antiserum reacted with its immunogen and was not reactive with the other chromatins. This is in good agreement with our previous results where the antiserum to Novikoff hepatoma reacted by complement fixation with chromatins of various malignant tumors transplantable in rats but only marginally or not at all with normal rat liver chromatin.[16]

Since complement fixation assays recognize preferentially large antigenic complexes, it is conceivable that the specificity may reflect the interactions of antibody with complexes of chromosomal nonhistone proteins with DNA. Indeed, dissociation and reconstitution experiments where dehistonized or total Novikoff hepatoma chromatins were separated by ultracentrifugation into protein and DNA components showed that neither one fixed complement in presence of the appropriate antiserum. However, when reconstituted, the

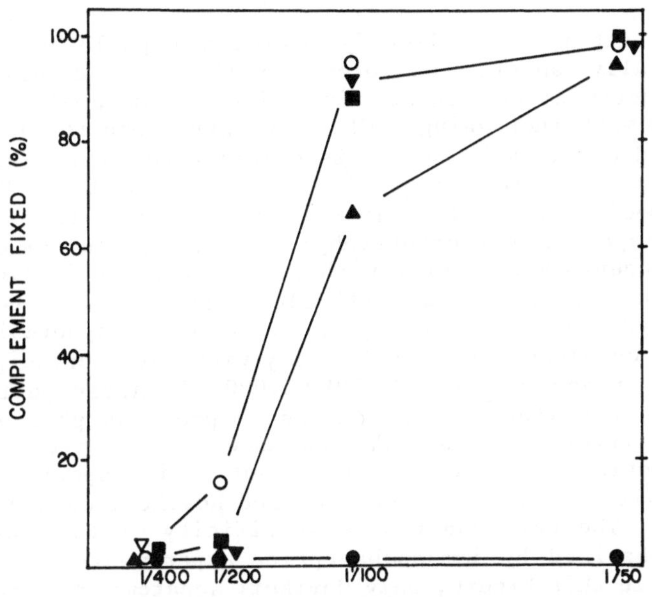

Fig. 2. Complement fixation of antiserum to the m.w. 40-60,000 region of Novikoff hepatoma fraction obtained by hydroxylapatite chromatography. The assays were performed with increasing dilutions of this antiserum in the presence of 10 μg (as DNA) of various chromatin preparations. Novikoff hepatoma (●), fetal rat brain and fetal rat kidney (▲), fetal and mature rat liver (■), regenerating (24 hr) rat liver (▼). The same symbols are used for two different tissues where the experimental results were essentially identical.

resulting complexes were again reactive.[8,9,16] Furthermore, hybrid experiments in which hepatoma or liver DNAs were reconstituted with the homologous or heterologous proteins,[17] showed the protein fraction to be the immunological specificity donor to the complexes (i.e. only Novikoff hepatoma protein reconstituted with either liver or Novikoff hepatoma DNA reacted with the Novikoff antiserum).

Administration of a hepatocarcinogen, N,N-dimethyl-p-(tolylazo) aniline resulted in immunological changes in liver which could be detected as early as 3 weeks after the initiation of hepatocarcinogenic diet.[9] Similar results were obtained with another system using 1,2-dimethylhydrazine induced carcinogenesis of the large bowel in rats.[12] These experiments suggest the existence of specific interactions between a selected group of proteins and DNA. These interactions can be detected immunologically and seem to change qualitatively during differentiation and carcinogenesis.

In an attempt to identify proteins capable of establishing immunologically specific complexes with DNA, we have initiated several fractionation schemes in which chromosomal proteins were dissociated with increasing NaCl concentrations in the presence of concentrated urea. A fraction dissociating between 15 mM and 100 mM phosphate in 5 M urea at pH 7.6 contained most of the Novikoff hepatoma specific activity. This fraction was further fractionated by hydroxylapatite chromatography;[18] most of the immunologically active components were eluted with 50 mM phosphate in 5.0 M urea - 2 M NaCl. This fraction was still electrophoretically heterogeneous (in SDS polyacrylamide gels); preparative electrophoresis localized most of the complement fixing activity with three principal proteins in the m.w. range between 40,000-60,000.[18] After purification on BioGel P200, a mixture of the antigenic proteins p39, p49 and p56, was reconstituted with rat DNA and used to immunize rabbits. The resulting antiserum exhibited complement fixing specificity similar to that of the antiserum to dehistonized Novikoff hepatoma chromatin (Figure 2). The cell and tissue specificity of this antiserum was further emphasized by immunoabsorption experiments (Figure 3). As can be seen in this Figure, only Novikoff hepatoma chromatin was able to absorb the immunoactivity of this antiserum; absorption with chromatins from normal or regenerating rat liver as well as rat kidney did not change its reactivity.

Using large amounts of starting material, the three principal proteins comprizing the active m.w. 40,000-60,000 area of preparative electrophoresis were purified by chromatography on hydroxylapatite followed by gel filtration on BioGel P200. The gel filtration was performed in buffers containing 1% SDS to prevent aggregation and precipitation of the principal protein fractions. The proteins were recovered in virtually homogeneous (electrophoretically) form: m.w. 39,000, 49,000 and 56,000. All the three

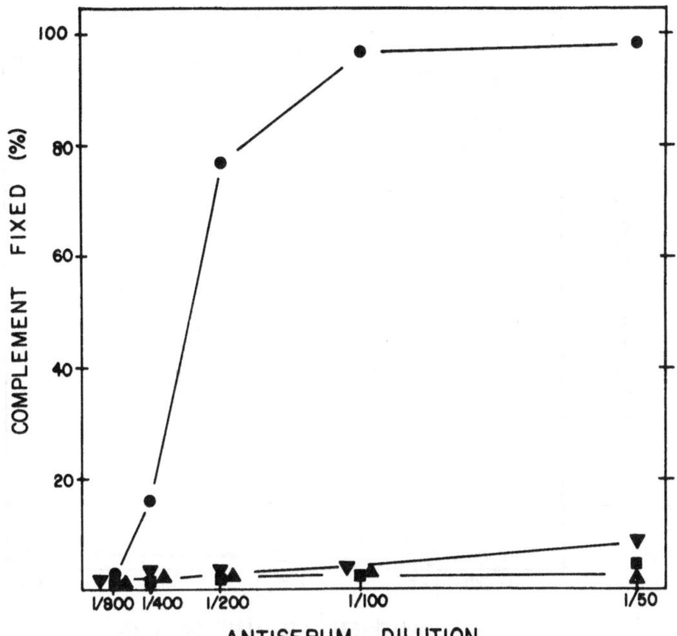

Fig. 3. Complement fixation of antiserum to the m.w. 40-60,000 Novikoff hepatoma proteins which was absorbed twice with various chromatin preparations. The assays were performed in the presence of 10 μg (as DNA) of Novikoff hepatoma chromatin. The symbols indicate antiserum absorbed with chromatin from Novikoff hepatoma (●), 24 hr regenerating rat liver (□), normal rat liver (▲), rat kidney (▼), and unabsorbed antiserum control (O).

proteins reacted with antiserum to dehistonized Novikoff hepatoma chromatin (i.e. the original immunogen). A 50 mM phosphate buffer fraction prepared identically from normal rat liver was inactive in the complement fixation assays with the antiserum to Novikoff hepatoma although it contained three principal proteins of electrophoretic mobilities very similar to those of the Novikoff hepatoma proteins p39, p49 and p56 (Figure 4). One interesting difference was noted, however. The Novikoff hepatoma proteins p39, p49 and p56 stained positively with periodic acid-Schiff reagent[19] while the corresponding liver proteins did not. The possibility that both the liver and Novikoff hepatoma chromatins contain identical proteins

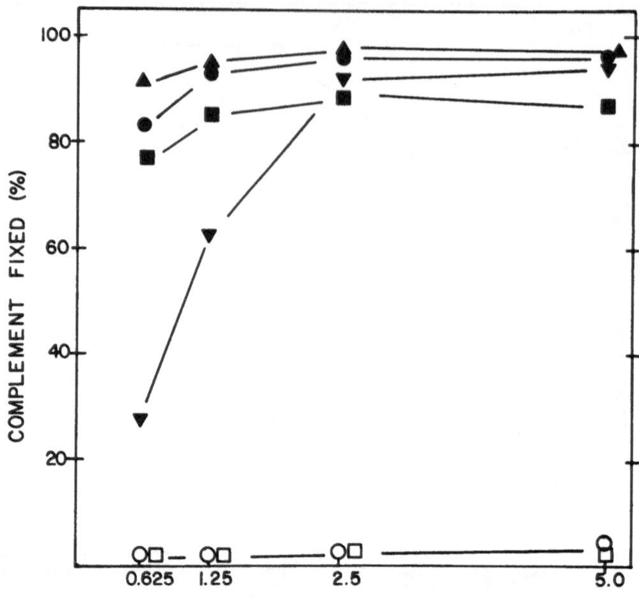

Fig. 4. Complement fixation of antiserum to the m.w. 40-60,000 Novikoff hepatoma hydroxylapatite fraction proteins in the presence of Novikoff hepatoma or rat liver chromatins and Novikoff hepatoma proteins p39, p49 and p56. The proteins were reconstituted to purified rat spleen DNA in the ratio of protein:DNA = 0.4:1.0 and used in the assays. The dilution of antiserum was 1:200. Novikoff hepatoma protein p39 (■), p49 (▲), p56 (●), Novikoff hepatoma chromatin (▼), rat liver chromatin (□) and rat liver 50 mM hydroxylapatite fraction (O).

where immunological specificity is altered in the latter by glycosylation is being investigated in our laboratory.

To facilitate detection of the p39, p49 and p56 proteins in electrophoretograms of chromatins from various cell types as well as to aid the positive identification of these proteins during their further fractionation and characterization we took the advantage of our observations that although the antisera to dehistonized Novikoff hepatoma chromatins did not form immunoprecipitates, they did immunolocalize the antigens aided with the peroxidase-antiperoxidase

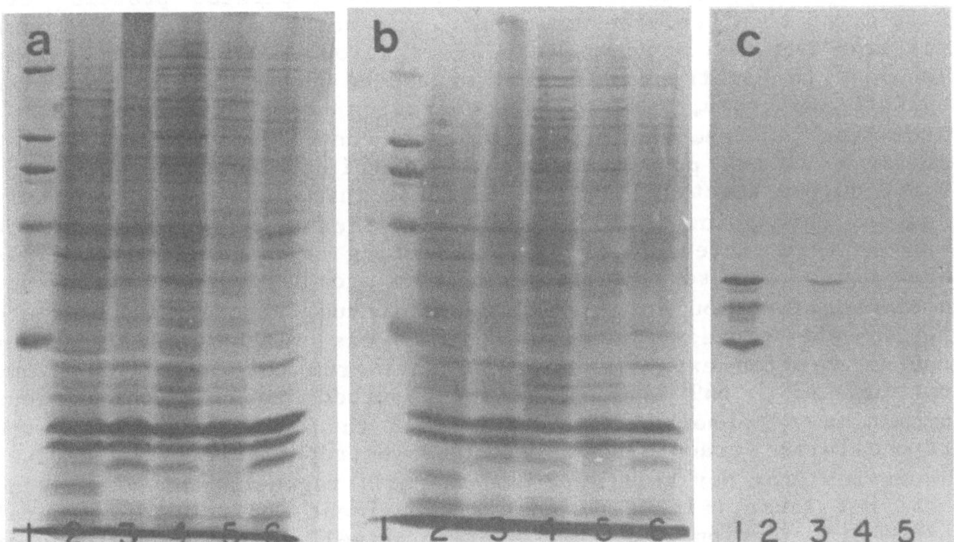

Fig. 5. Identification of Novikoff hepatoma antigens in chromosomal proteins transferred to nitrocellulose sheets. A: SDS-PAGE of various chromatins. Chromatins were prepared in SDS gel sample buffer, sonicated, and electrophoresed. Each lane contained 25 μg (as DNA) of the respective chromatin. Lane 1, high molecular weight standards (BioRad) (myosin (200,000), β-galactosidase (116,500), phosphorylase β (94,000), albumin (68,000) and ovalbumin (43,000)); lane 2, Novikoff hepatoma; lane 3, mature rat liver; lane 4, 24 hr regenerating rat liver; lane 5, fetal rat liver; and lane 6, mature rat kidney chromatins. The origin is at the top of the gels. B: SDS-PAGE of various chromatins electrophoretically transferred to nitrocellulose and stained with Amido black. Chromatins were electrophoresed as in A and then transferred to nitrocellulose and stained with Amido black. Lanes are as described in A. C: localization of immunoreactive antigens on nitrocellulose containing SDS-PAGE-separated chromatins. Chromatins were electrophoresed as in A, transferred to nitrocellulose, and then the immunoreactive antigens were localized by incubation with antiserum to Novikoff hepatoma proteins p39, p49 and p56 (1:500 dilution) followed by the PAP procedure. Lane 1, Novikoff hepatoma; lane 2, mature rat liver; lane 3, 24 hr regenerating rat liver; lane 4, fetal rat liver; and lane 5, mature rat kidney chromatins.

(PAP) reaction of Sternberger.[20] A method was developed in our laboratory[21] by which electrophoretically separated proteins are transferred to nitrocellulose sheets,[22] reacted with the appropriate antiserum and developed by the PAP reaction. Application of this technique to electrophoretically separated chromatin proteins from Novikoff hepatoma, mature, regenerating and fetal rat liver chromatins is presented in Figure 5. As can be seen in this Figure, antiserum to the p39, p49 and p56 protein (electrophoretic m.w. 40,000-60,000 fraction) recognized these three proteins in Novikoff hepatoma chromatin. Smaller, but significant amounts of the p56 protein were detectable in regenerating and normal rat liver chromatins. These results are in a good agreement with the complement fixation data (absorbed antiserum, Figure 3). Presence of the p56 protein in normal liver was further confirmed by immunoabsorption experiments. The antiserum to Novikoff hepatoma proteins p39, p49 and p56 was absorbed twice with various chromatins. Absorbed antisera were then incubated with the nitrocellulose transfers containing electrophoretically separated chromatins from Novikoff hepatoma, normal, regenerating (24 hr) and fetal rat liver (Figure 6). This experiment confirms the general tissue distribution of the p56 protein as well as the specificity of the proteins p39 and p49 for Novikoff hepatoma.

Because of the anticomplementarity of most cytoplasmic fractions, we were unable to assay with confidence the cellular distribution of the complement fixing Novikoff hepatoma antigens detected in our chromatin preparations. However, the nitrocellulose transfer technique made such experiments possible. We have prepared the following cellular fractions from Novikoff hepatoma: 10,000 x g cytoplasmic pellet from the 800 x g supernatant, 100,000 x g cytoplasmic supernatant and pellet, purified nuclei and chromatin. Proteins of these fractions were separated electrophoretically, transferred to nitrocellulose sheets and stained with antiserum to the three proteins. The results, illustrated in Figure 7, show the presence of all the three Novikoff hepatoma proteins (p39, p49 and p56) in every cellular fraction assayed. Since the electrophoretic separation was based on equal protein concentrations, the intensity of PAP staining, although not quantitative, indicates that while the cytoplasm, nuclei and chromatin contain similar quantities of the p39 protein, the p49 band appears enriched in the cytoplasm. Since the antiserum was absorbed with normal rat liver chromatin prior to the immunolocalization, the p56 protein is not detected by this antiserum. The 10,000 x g pellet, containing mitochondria stained only weakly, presumably because of the presence of nuclear fragments contaminating this fraction.

It is noteworthy that the fractionation and purification of the Novikoff hepatoma proteins was accompanied with progressive loss of dependence on DNA for the specific complement fixation reaction. We

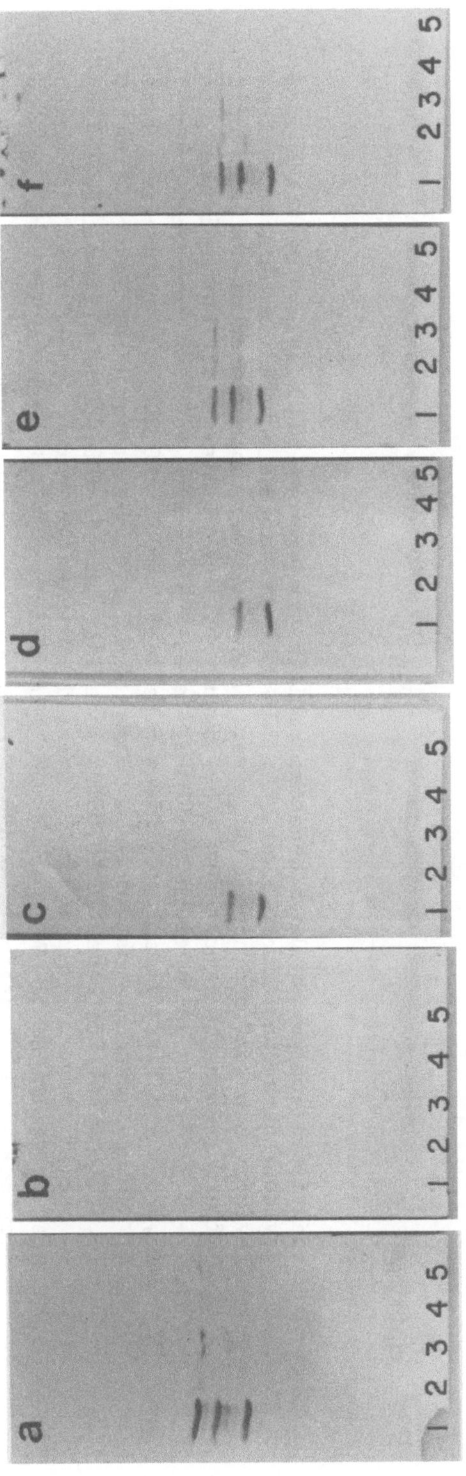

Fig. 6. Immunoabsorption of antiserum to Novikoff hepatoma proteins p39, p49 and p56 with various chromatins. Each antiserum was absorbed twice. Absorbed antisera were then incubated with nitrocellulose sheets containing SDS-PAGE-separated chromatins (25 μg as DNA of each chromatin), and immunoreactive species were then visualized with the PAP reaction. Each nitrocellulose sheet contains (lanes left to right), Novikoff hepatoma, rat liver, 24 hr regenerating rat liver, fetal rat liver, and rat kidney chromosomal proteins with the origin at the top of the sheet. The absorbed antisera used to stain each sheet were: A) unabsorbed; B) absorbed with Novikoff hepatoma chromatin; C) absorbed with rat liver chromatin; D) absorbed with 24 hr regenerating rat liver chromatin; E) absorbed with rat kidney chromatin; F) absorbed with fetal rat liver chromatin.

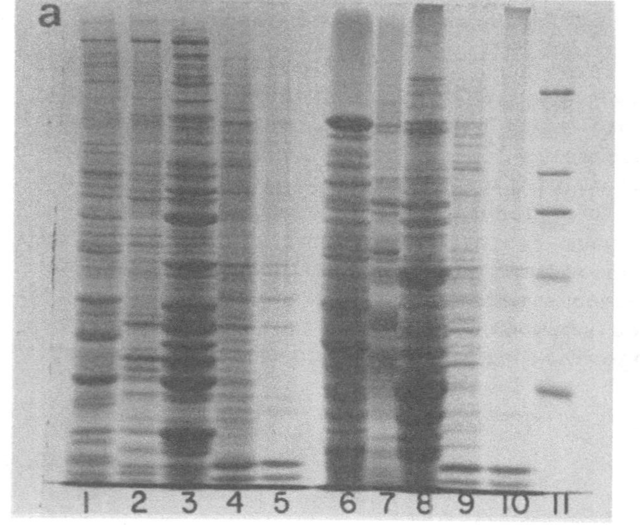

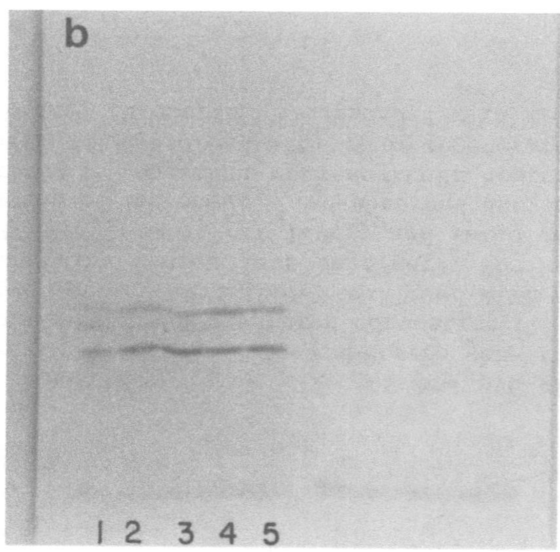

Fig. 7. Identification of Novikoff hepatoma antigens in various subcellular fractions from Novikoff hepatoma and rat liver. A: SDS-PAGE of various subcellular fractions from Novikoff hepatoma and normal rat liver. Fractions were solubilized in SDS gel sample buffer. Each lane contained the amount of subcellular fraction equivalent to 5.4×10^5 cells in a volume of 30 µl. Lanes 1-5 are from Novikoff hepatoma (lane 1, 10,000 x g pellet; lane 2, 100,000 x g pellet; lane 3, 100,000 x g supernatant; lane 4, nuclei; lane 5, chromatin). Lanes 6-10 are from normal rat liver (lane 6, 10,000 x g pellet; lane 7, 100,000 x g pellet; lane 8, 100,000 x g supernatant; lane 9, nuclei; lane 10, chromatin). Lane 11 contains molecular weight standards which are as described in Figure 5. B: localization of immunoreactive antigens on nitrocellulose containing SDS-PAGE separated proteins from various subcellular fractions of Novikoff hepatoma or normal rat liver. Fractions were prepared and electrophoresed as in A, transferred to nitrocellulose, and then the immunoreactive

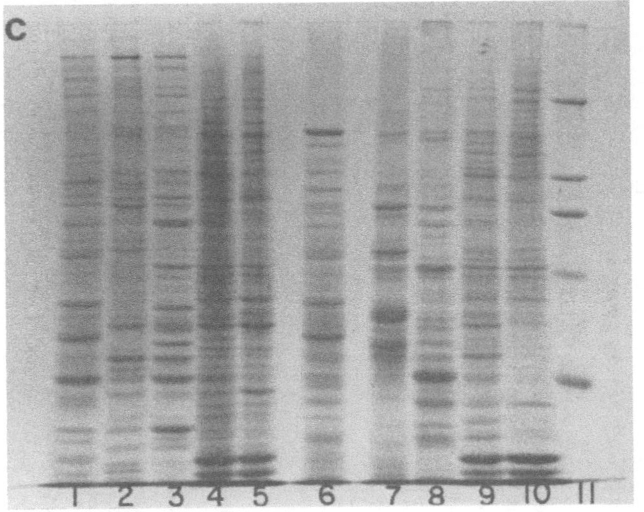

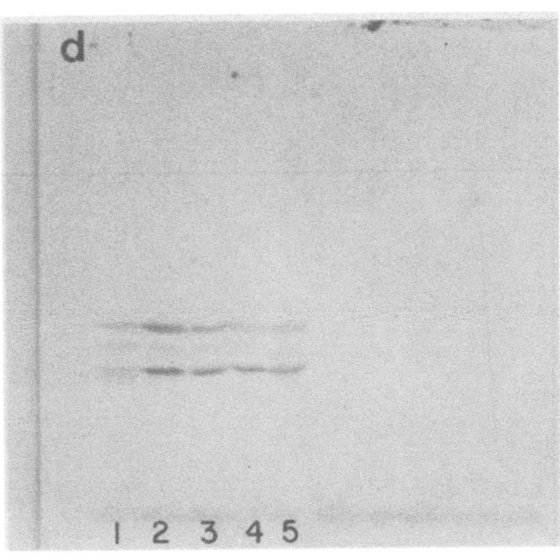

(Fig. 7 continued)
antigens localized with antiserum to Novikoff hepatoma antigens p39, p49 and p56 (1:500 dilution) and the PAP procedure. Antiserum was absorbed twice with liver chromatin before use. Lanes are as shown in A. C: SDS-PAGE of various subcellular fractions from Novikoff hepatoma and normal rat liver. Fractions were prepared and electrophoresed as described in A; however, each lane contained an equivalent amount of protein from the respective fraction (20 μg). The identity of each lane is as described in A. D: localization of immunoreactive antigens on nitrocellulose containing SDS-PAGE-separated proteins from various subcellular fractions of Novikoff hepatoma or normal rat liver. Fractions were prepared and electrophoresed as in C, transferred to nitrocellulose, then the immunoreactive antigens were localized with antiserum to Novikoff hepatoma antigens, p39, p49 and p56 (1:500 dilution) and the PAP procedure. Antiserum was absorbed twice with liver chromatin before use. Lanes are as indicated in A.

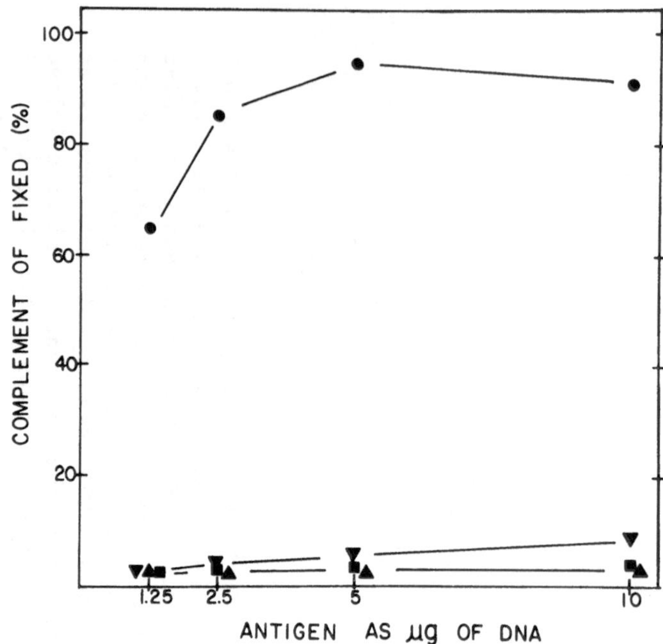

Fig. 8. The specificity of antiserum to dehistonized HeLa chromatin. The complement fixation assays were performed in the presence of increasing amounts of chromatins from various human tissues. The antiserum dilution was 1:400. The symbols are: HeLa (●), W138 cells (■), human term placenta (▼), normal human lung or lung carcinoma (▲).

attribute this phenomenon to the heterogeneity of our antisera to dehistonized chromatins. Such antisera apparently contain antibodies to the complexes of chromosomal proteins with DNA in addition to antibodies to the individual protein antigens. The second generation antiserum which was elicited mainly to the three partially purified proteins p39, p49 and p56 contains very little, if any, of the activity which recognizes the nonhistone protein:DNA complexes.

HUMAN TUMORS

Since our earlier results with human tumors and HeLa cells (Figure 8) showed specificities similar to those observed for Novikoff hepatoma,[14,23] we have initiated experiments aimed at a

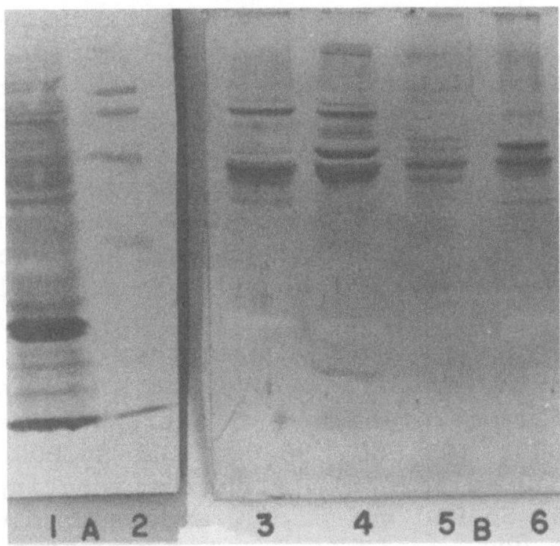

Fig. 9. Localization of nuclear protein antigens transferred to nitrocellulose sheets. A: Amido black stain of total proteins transferred from polyacrylamide gels. Lane 1, HT-29 chromatin; lane 2, molecular weight standards (myosin (200,000), β-galactosidase (116,500), phosphorylase β (94,000), albumin (68,000) and ovalbumin (43,000)). All standards were part of a kit purchased from BioRad, Richmond, CA. B: Immunological staining of: lane 3, LoVo; lane 4, HT-29; lane 5, non-neoplastic colon; lane 6, human placenta. All chromatin samples were applied to the gel as 40 μg of DNA and the antiserum to dehistonized HT-29 chromatin was diluted 1:400.

better characterization of cell specific antigens associated with human tumor chromatins. Although still preliminary, our experiments with antisera to dehistonized chromatin from tissue cultured human colon adenocarcinoma showed the presence in chromatin of at least two specific antigenic proteins.

Using experimental strategy described for the Novikoff hepatoma, antisera were obtained to dehistonized chromatins isolated from human colon adenocarcinoma cell lines HT-29 (provided by Dr. Jorgen Fogh, Sloan Kettering Institute for Cancer Research) and LoVo (obtained from Dr. Benjamin Drewinko, The University of Texas System Cancer Center, M.D. Anderson Hospital). Both the HT-29 and LoVo antisera reacted with chromatins of its own as well as the other cell line. By complement fixation, the antiserum to HT-29 did not show specificity when reacted with normal human colon chromatin, thereby indicating the presence of major complement fixing antigens in both normal and malignant cells.

The electrophoretic distribution of chromosomal proteins from the HT-29, LoVo and normal colon chromatins exhibited a considerable heterogeneity reflected by the presence of numerous protein bands in range extending from 25,000 to over 200,000 m.w. Most major bands fell into a m.w. range of 105,000-150,000 and 62,000-81,000, with molecular weights of 148,000, 135,000, 120,000 and 106,000 prominent in the former. Additional major bands had molecular weights corresponding to 53,000, 42,000 and 37,000. When chromatin proteins were transferred to nitrocellulose sheets and antigens localized with the PAP method, only antibodies to a small population of proteins could be seen, indicating that many chromosomal proteins were either not immunogenic or that they were removed during dehistonization. Both cell lines LoVo and HT-29 as well as normal colon showed the presence of high m.w. antigens, the most prominent ones at m.w. 113,000, 100,000, 96,000 and 86,000. In all human chromatins (including placenta), strongly reacting antigens with m.w. 73,000, 70,000, 66,000 and 62,000 were common while additional antigens of m.w. 92,000 and 67,000 were detected in the tumor cell lines only (Figure 9). When antiserum was absorbed with normal human colon chromatin, the antigens of m.w. 67,000 and 92,000 remained as the major immunoreactive bands demonstrating that they are indeed absent from normal colonic tissue (Figure 10). The isolation and characterization of the m.w. 67,000 and 92,000 proteins is in progress.

CONCLUSIONS

As was already mentioned in the Introduction, the antigenicity of chromosomal nonhistone proteins is well documented in the literature. A complex immunogen, such as dehistonized chromatin, elicits reaction which results in the production of antibodies to many of the numerous proteins present in the preparations. This type of immunological response can be assayed with ease and accuracy by separating the chromosomal proteins by polyacrylamide gel electrophoresis and detecting the reactive antigens, after their transfer to nitrocellulose sheets, by the peroxidase-antiperoxidase reaction of Sternberger.[20] This extremely useful method can be modified in various ways, e.g. to include two dimensional electrophoresis or to detect electrophoretically separated glycoproteins.[24] Using this analytical approach, we have found that individual rabbits can differ in their immunological responses to the same immunogen. In other words, immunization with dehistonized Novikoff hepatoma chromatin produced antisera which reacted strongly with the p39, p49 and p56 proteins described here. Antisera from other animals recognized preferentially antigens in the high molecular weight protein spectrum.[21] Irradiation of chromatin stabilized some of the antigenic proteins[25] and resulted in a more uniform immunological response.

Proteins of molecular weights similar to the three antigens described here have been reported in the literature. Of particular interest are the B1 antigen (m.w. 37,000) of Yeoman et al.,[26] the NoAg1 antigen of Novikoff hepatoma nucleoli (m.w. 60,000) reported by Marashi et al.[5] and the m.w. 45,000 protein purified from Novikoff hepatoma cytosol.[27] However, although the molecular weights of these antigenic proteins are similar or identical to those of the Novikoff hepatoma proteins p39, p49 and p56, their behavior during extraction, isolation and immunization as well as their physicochemical properties show that the p39, p49 and p56 proteins are different.

As indicated by the immunological analysis of antisera to dehistonized human colon adenocarcinoma chromatins, the specific antigens found in the nuclei of these cells differ significantly from the Novikoff hepatoma proteins p39, p49 and p56. Only two

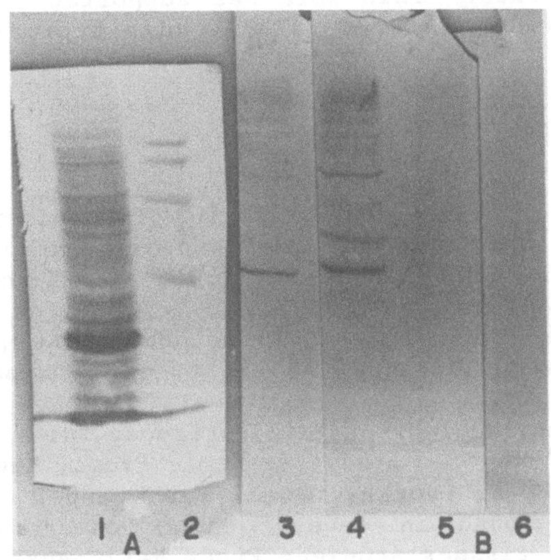

Fig. 10. Localization of nuclear protein antigens with antiserum absorbed with normal colon chromatin. A: Amido black staining of electrophoretically separated chromatin proteins transferred to nitrocellulose sheets. Lane 1, HT-29; lane 2, molecular weight standards. B: Immunochemical staining of tumor-associated antigens. Lane 3, LoVo; lane 4, HT-29; lane 5, non-neoplastic colon; lane 6, human placenta. All chromatin samples represent 40 µg (as DNA) applied to the polyacrylamide gel. The antiserum to dehistonized HT-29 chromatin was absorbed 4x with normal colon chromatin. The antiserum dilution was 1:400.

antigens of approximate molecular weights 67,000 and 92,000 remained detectable after immunoabsorption with normal colon. Both these protein antigens differ in their molecular weights from the two colon carcinoma specific proteins described by Boffa et al.[28] (m.w. 44,000 and 62,000). We do not know whether these latter proteins are not antigenic or whether they were removed by the dehistonization of chromatin.

From the preliminary work in our laboratory, it appears that the nitrocellulose transfer and immunological identification of electrophoretically separate protein antigens may not be suitable for the identification of immunogens and complement-fixing antibodies which recognize complexes between DNA and chromosomal nonhistone proteins. These complexes become separated during electrophoresis and seem to loose their immunological specificity. Additional experimentation, perhaps using nitrocellulose transfers overlaid with DNA is necessary to resolve this complex problem.

Acknowledgments: This work was supported by National Cancer Institute Grants CA 26412, CA 27338 and a Training Grant CA 90313 (DMD).

REFERENCES

1. L. M. Silver and S. C. R. Elgin, Immunological analysis of protein distributions in Drosophila polytene chromosomes, in: "The Cell Nucleus, Vol. V," H. Busch, ed., Academic Press, New York (1978).
2. L. S. Hnilica and R. C. Briggs, Nonhistone protein antigens, in: "Cancer Markers," S. Sell, ed., Humana Press, Clifton (1980).
3. L. C. Yeoman, Nuclear protein antigens, in: "The Cell Nucleus, Vol. V," H. Busch, ed., Academic Press, New York (1978).
4. F. M. Davis, F. Gyorkey, R. K. Busch and H. Busch, Nucleolar antigen found in several human tumors but not in the nontumor tissues studied, Proc. Natl. Acad. Sci. USA 76:892 (1979).
5. F. Marashi, F. M. Davis, F. M. Busch, H. E. Savage and H. Busch, Purification and partial characterization of nucleolar antigen-1 of the Novikoff hepatoma, Cancer Res. 39:59 (1979).
6. P. K. Chan, A. Feyerabend, R. K. Busch and H. Busch, Identification and partial prufication of human tumor nucleolar antigen 54/6.3, Cancer Res. 40:3194 (1981).
7. F. Chytil and T. C. Spelsberg, Tissue differences in antigenic properties of nonhistone protein-DNA complexes, Nature New Biol. 233:215 (1971).

8. K. Wakabayashi and L. S. Hnilica, The immunospecificity of nonhistone protein complexes with DNA, Nature New Biol. 242:153 (1973).
9. J. F. Chiu, M. Hunt and L. S. Hnilica, Tissue-specific DNA-protein complexes during azo dye hepatocarcinogenesis, Cancer Res. 34:913 (1975).
10. J. F. Chiu, S. Wang, H. Fujitani and L. S. Hnilica, DNA-binding nonhistone proteins isolation, characterization, and tissue specificity, Biochemistry 14:4552 (1975).
11. F. Chytil, S. R. Glasser and T. C. Spelsberg, Alterations in liver chromatin during perinatal development of the rat, Develop. Biol. 27:295 (1974).
12. J. F. Chiu, D. Pumo and D. Gootnick, Antigenic changes in nuclear chromatin in 1,2-dimethylhydrazine-induced colon carcinogenesis, Cancer 45:1193 (1980).
13. L. S. Hnilica, J. F. Chiu, K. Hardy, H. Fujitani and R. C. Briggs, Antibodies to nuclear chromatin fractions, in: "The Cell Nucleus, Vol. V," H. Busch, ed., Academic Press, New York (1978).
14. R. C. Briggs, A. Campbell, J. F. Chiu, L. S. Hnilica, G. Lincoln, J. Stein and G. Stein, Specificity of DNA-associated nuclear antigens in HeLa cells and distribution during the cell cycle, Cancer Res. 39:3683 (1979).
15. E. Wasserman and L. Levine, Quantitative microcomplement fixation and its use in the study of antigenic structure by specific antigen-antibody inhibition, J. Immunol. 87:290 (1960).
16. K. Wakabayashi, S. Wang and L. S. Hnilica, Immunospecificity of nonhistone proteins in chromatin, Biochemistry 13:1027 (1974).
17. J. F. Chiu, C. Craddock, H. P. Morris and L. S. Hnilica, Immunospecificity of chromatin nonhistone protein-DNA complexes in normal and neoplastic growth, FEBS Letters 42:94 (1974).
18. H. Fujitani, J. F. Chiu and L. S. Hnilica, Purification of nuclear antigens in Novikoff hepatoma, Proc. Natl. Acad. Sci. USA 75:1943 (1978).
19. R. M. Zacharius, T. E. Zell, J. H. Morrison and J. J. Woodlock, Glycoprotein staining following electrophoresis on acrylamide gels, Anal. Biochem. 30:148 (1969).
20. L. A. Sternberger, "Immunocytochemistry," Prentice-Hall, Engelwood Cliffs (1974).
21. W. F. Glass, R. C. Briggs and L. S. Hnilica, Identification of tissue-specific nuclear antigens transferred to nitrocellulose from polyacrylamide gels, Science 211:70 (1981).
22. H. Towbin, T. Staehelin and J. Gordon, Electrophoretic transfer of proteins from polyacrylamide gels to nitrocellulose sheets: procedure and some applications, Proc. Natl. Acad. Sci. USA 76:4350 (1979).

23. J. F. Chiu, L. S. Hnilica, F. Chytil, J. T. Orrahood and L. W. Rogers, Tissue-specific antibodies against human lung and breast carcinoma dehistonized chromatins, J. Natl. Cancer Inst. 59:151 (1977).
24. W. F. Glass, R. C. Briggs and L. S. Hnilica, Use of lectins for detection of electrophoretically separated glycoproteins transferred into cellulose sheets, Anal. Biochem., in press.
25. R. Olinsky, R. C. Briggs and L. S. Hnilica, Gamma radiation-induced crosslinking of the cell-specific chromosomal non-histone protein-DNA complexes in HeLa chromatin, Radiation Res. 86:102 (1981).
26. L. C. Yeoman, L. M. Woolf, C. W. Taylor and H. Busch, Nuclear antigens of tumor cell chromatin, in: "Biological Markers of Neoplasia: Basic and Applied Aspects," R. W. Ruddon, eds., Elsevier-North Holland, New York (1978).
27. K. S. Raju, H. P. Morris and H. Busch, Purification and characterization of cytosol protein 45/7.8 present in rapidly growing hepatomas, Cancer Res. 40:1623 (1980).
28. L. C. Boffa, G. Vidali and V. G. Allfrey, Changes in nuclear nonhistone protein composition during normal differentiation and carcinogenesis of intestinal epithelial cells, Exptl. Cell Res. 98:396 (1976).

TRANSFORMATION-ENHANCING FACTORS AND FIBRONECTIN DEGRADATION PRODUCTS AS POSSIBLE TUMOUR MARKERS

Sergio Barlati[1], Giuseppina De Petro[1,2], Tapio Vartio[2,3], and Antti Vaheri[2]

1) Istituto di Genetica Biochimica ed Evoluzionistica, C.N.R., 27100 Pavia, Italy and
2) Department of Virology and [3)] Pathology, University of Helsinki, Helsinki 00290, Finland

INTRODUCTION

The medium of chick embryo fibroblasts (CEF) transformed by Schmidt-Ruppin strain of Rous sarcoma virus (SR-RSV) contains factors which complement the expression of some transformation parameters depending on the *src* gene. Notably, they reverse the block by puromycin and/or cycloheximide of morphological transformation of cells infected with three ts-T mutants after shift-down from restrictive (41.5°) to permissive (37°) temperature. This reversal is not due to the release of inhibition of protein synthesis and is accompanied by the expression of two other *src*-dependent transformation parameters: disorganization of the cytoskeleton and loss of cell surface-associated fibronectin. These factor(s) were operationally called transformation-enhancing factors (TEF). TEF is lacking in media of untransformed cells, uninfected or infected with a nontransforming virus (RAV-1), and its production by RSV-infected cells seems to depend on the acquisition of the transformed phenotype, therefore on the expression of the *src* gene. Similar activities are found in the medium of mammalian (rodent) cells transformed by SR-RSV and by other RNA and DNA oncogenic viruses, but not in the medium of untransformed controls[1,2,3,4].

A "TEF-like" activity has been also evidenced in the plasma cryoprecipitate from patients affected with different types of neoplastic diseases[5]. The screening of 200 cases of different neoplastic diseases and controls, healthy subjects or patients affected with other non-neoplastic diseases indicates that TEF activity is generally related to the presence of neoplasia. Furthermore, a follow-up of patients from the onset of the disease through its evolution during therapy suggests that variations of TEF activity in the plasma cryoprecipitate correlate well with the clinical and pathological conditions, thus indicating TEF as a potential marker for monitoring cancer patients[6,7].

We recently discovered that a similar TEF activity is also effected by defined fragments of human plasma fibronectin obtained by limited digestion with major humoral or tissue proteinases. TEF activity can be obtained from plasminolytic fragments and from cathepsin G-treated fibronectin. No activity can be recorded from intact dimeric fibronectin or its reduced and alkylated subunits, from fibrinogen or its plasminolytic fragments, or from plasmin or cathepsin G-treated or untreated with proteinase inhibitors. All of the TEF activity of the proteolytic fragments of fibronectin is located on the gelatin-binding peptides, active in nanamolar concentrations. TEF activity of proteinase-treated fibronectin is inhibited by gelatin and by intact dimeric fibronectin[8,9].

In view of the latter findings it was therefore of importance to verify whether the various TEF activities found in _in vitro_ cell cultures and in the plasma cryoprecipitates could be related, as previously proposed,[5] to the generation of proteolytic fibronectin fragments which might be produced either _in vitro_ or _in vivo_, _via_ the activation of the fibrinolytic system at the cellular or plasma level. This possibility was also suggested by the findings that fibronectin was known to be present in plasma cryoprecipitate and to be quite susceptible to various proteinases[10,11].

In the present paper we report on the similarities

between the properties of TEF activity found in the plasma cryoprecipitate of cancer patients and TEF active fibronectin degradation products (FNdP).

MATERIALS AND METHODS

Assay for TEF activity

TEF assay was as described[5]. Secondary cultures of chicken embryo fibroblasts were infected in suspension with the temperature-sensitive mutant PA 1[12] of the Schmidt-Ruppin strain of Rous sarcoma virus, at a low multiplicity of infection to give about 100 focus-forming units per $3X10^5$ cells seeded per 8-cm^2 plate in medium 199 supplemented with 5% (vol/vol) newborn calf serum, 10% (vol/vol) tryptose phosphate broth, and antibiotics. The liquid medium was replaced 24 hrs after seeding by semisolid medium containing 0.6% agar. After 5 days of incubation at 37°C, when foci of morphologically transformed cells became detectable, the cultures were shifted to 41°C (non-permissive temperature for transformation) for 48 hrs. At this time, when no foci were detectable, the gel medium was removed and replaced with liquid medium 199 containing 5% decomplemented (incubated for 30 min at 56°C) newborn calf serum, 10 μg of cycloheximide per ml, and the appropriate concentration of the sample to be tested or corresponding control preparation. (Use of the inhibitor of protein synthesis, cycloheximide, is not mandatory but gives more reproducible results in the TEF assay). Immediately afterwards the cultures were shifted to 37°C for 4 hours, observed microscopically, and stained supravitally with methylene blue, which selectively stains foci of virus-transformed cells[13]. The number of foci were calculated in each plate, and TEF activity was recorded as the factor of enhancement (FE) the ratio of the number of foci scored in triplicate test cultures to that in triplicate control cultures. Evaluation of the standard error and the statistical significance of the FE were calculated as described[14].

Preparation of plasma cryoprecipitate and gelatin-binding fractions

Plasma cryoprecipitate was prepared as previously described[5,7]. To separate gelatin-binding proteins from other peptides, 200 µl (≃1mg) of plasma protein cryoprecipitate were mixed with 1 vol of 50% (wt/vol) gelatin-Sepharose,[15] prepared as described,[11] in 50 mM Tris HCl, pH 7.5 buffer containing 0.02% sodium azide. The mixture was rotated in an end-over mixer for 6 hrs at room temperature and then applied to a column and the flow-through was collected (gelatin-nonbinding fraction). The elution of the gelatin-binding fraction was carried out with 1 M arginine[16] in the above buffer. Both fractions were dialyzed at room temperature against phosphate buffered saline before analysis and tested at the same concentration as that of the total cryoprecipitate (0.2% vol/vol) in the TEF assay.

Gel electrophoresis and protein transfer analysis

Plasma cryoprecipitate proteins were separated in 8% $NaDodSO_4$ polyacrylamide gels in the presence of β-mercaptoethanol[17] and were then transferred electrophoretically to nitrocellulose sheets[18]. The nitrocellulose filters were saturated with 3% (wt/vol) bovine serum albumin and the transferred proteins were then allowed to react with polyclonal rabbit anti-fibronectin antibodies (1:200) and, after extensive washing, treated with peroxidase-conjugated antirabbit IgG to visualise the polypeptide bands which reacted with the antibodies. The conjugated (DAKO, Copenhagen) was used at a 1:200 dilution and 3,3[1] -dimethoxy benzidine diHCl (Sigma, St. Louis, MO) served as the substrate.

RESULTS

TEF activity from plasma cryoprecipitates binds to gelatin

Nine TEF positive cryoprecipitates from patients affected by different types of tumours (3 lymphomas, 3 carcinomas, 2 myelomas and one acute leukemia) were subjected to gelatin-Sepharose fractionation and the TEF activity of gelatin binding and non binding fraction was measured. The same was done with plasma cryoprecipitates from four patients affected by nonmalignant diseases

Table 1. TEF activity in plasma cryoprecipitates and its gelatin binding and non binding fractions

DISEASE[a]	TEF ACTIVITY (F.E.)		
	Total cryo	Gelatin Binding	Gelatin non Binding
1 Lymphoma	2.2 (+)[b]	2.0 (+)	1.0 (-)
2 "	2.4 (+)	2.4 (+)	1.0 (-)
3 "	1.5 (+)	1.1 (-)	1.0 (-)
4 Carcinoma	1.8 (+)	1.7 (+)	0.8 (-)
5 "	2.1 (+)	1.8 (+)	1.1 (-)
6 "	2.3 (+)	2.7 (+)	1.2 (-)
7 Myeloma	2.1 (+)	2.2 (+)	1.0 (-)
8 "	1.4 (+)	1.8 (+)	0.8 (-)
9 Leukemia	1.8 (+)	2.5 (+)	2.2 (+)
10 Cirrhosis	1.7 (+)	1.5 (+)	1.0 (-)
11 Hepatitis	1.2 (-)	1.0 (-)	1.3 (-)
12 Icterus	1.0 (-)	0.9 (-)	1.0 (-)
13 Diabetes	1.0 (-)	1.2 (-)	0.9 (-)
14 Healthy	0.9 (-)	0.8 (-)	0.7 (-)
15 "	1.0 (-)	1.2 (-)	0.9 (-)
16 "	1.1 (-)	0.9 (-)	1.0 (-)
17 "	1.0 (-)	0.9 (-)	0.9 (-)

a) 1 to 9 Malignant diseases; 10 to 13 nonmalignant diseases 14 to 17 Healthy controls.

b) (+) indicates TEF-positive sample with $p \leq 0.05$ (or lower) while (-) indicates a TEF-negative sample.

(one of which was TEF-positive) and from that of four healthy subjects (all TEF-negative).

The results obtained are reported in Table 1. In

nine out of ten TEF-positive cryoprecipitates, the activity was recovered in the gelatin binding fraction; in one case (leukemia), the activity was also present in the gelatin non-binding fraction. The eight TEF-negative plasma cryoprecipitates, from patients affected by non-malignant diseases or healthy controls, gave rise to TEF-negative fractions.

Detection of fibronectin fragments in plasma cryoprecipitates

In order to verify whether high levels of FNdP could be present in the TEF-positive cryoprecipitates, two TEF-positive samples from patients affected by mammary carcinoma and four TEF-negative samples from healthy controls were subjected to $NaDodSO_4$-polyacrylamide gel

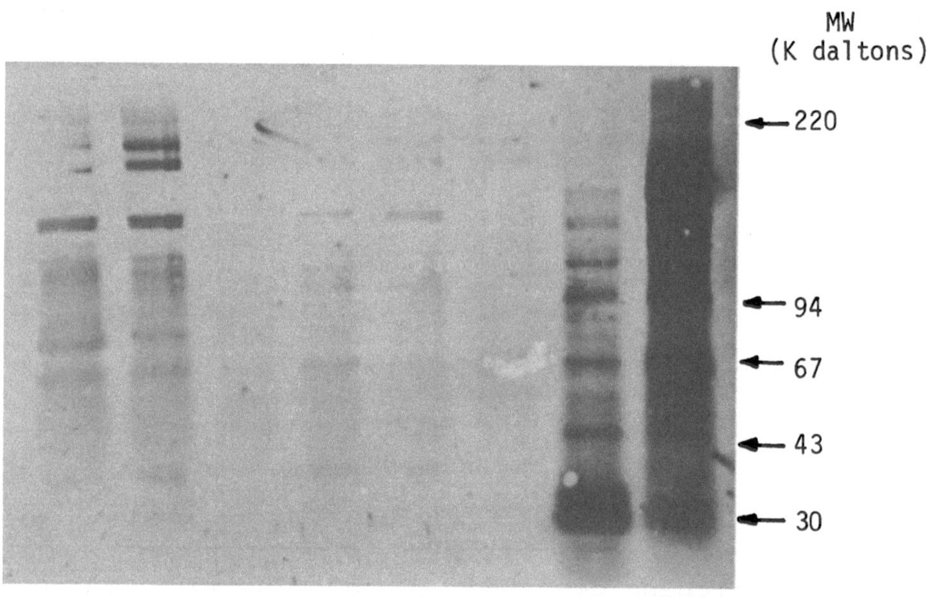

Fig. 1. Anti IgG peroxidase reaction of electrophoretically transferred plasma cryoprecipitates (1 to 6) and gelatin nonbinding (7) and binding (8) plasmin-cleaved fragments of fibronectin. The same volume (1 µl) of cryoprecipitate was applied to each lane. Lanes 7 and 8 contained both about 15 µg protein.

electrophoresis and, after protein transfer,[18] analyzed for the presence of proteins recognised by polyclonal rabbit anti-fibronectin antibodies[16].

The results obtained after peroxidase reaction are reported in Fig. 1.

It can be seen that the anti-fibronectin antibodies are able to recognize plasmin generated fibronectin fragments binding and non binding to gelatin (lane 8 and 7 respectively) as well as intact fibronectin (not shown). Furthermore, it is evident that there is a higher level of fibronectin and fibronectin fragments in the plasma cryoprecipitate from the two TEF-positive cancer patients (lane 1 and 2) as compared to those found in TEF-negative samples obtained from healthy controls (lanes 3 to 6).

DISCUSSION

The presence of TEF activity in biological fluids, either <u>in vitro</u> and <u>in vivo</u>, seems to be associated with the presence of transformed or malignant cells[1-7]. Both activities, after gel filtration, are found associated with protein fractions mainly of large apparent MW (150-300.000 daltons)[3,4,5,6], which are also detected after fractionation of plasmin degraded fibronectin analyzed under similar conditions[9].

These results, together with the findings that many tumour cells <u>in vivo</u> are surrounded by a FN-containing pericellular matrix or stroma,[19] that they produce high levels of plasminogen activators[20] and that increased levels of "fibronectin like" proteins are found in the serum of patients affected by certain malignant diseases,[21] suggest a possible relationship between FNdP and TEF activity as previously proposed[5].

This is further strengthened by the present results showing that all TEF-positive cryoprecipitates from patients affected by different types of tumours (except one for which the activity was lost: lymphoma N° 3), have a gelatin binding TEF activity in agreement with what was found for TEF-positive FN fragments[8].

It should be however pointed out that the TEF activity released *in vitro* by Rous sarcoma virus transformed cells does not seem to bind to gelatin[4] and that other proteins, including plasminogen activators[22] and a 70 000 dalton glycoprotein produced by various normal and malignant adherent cells[23] are able to bind to gelatin-Sepharose.

Further studies are therefore required before concluding that FNdP are the only proteins responsible for TEF activity.

In conclusion we have presented data favouring the hypothesis that, at least in the case of the *in vivo* activity, TEF shares several properties in common with FNdP; these fragments might therefore play an important role not only as potentially useful tumour markers but also, because of the associated TEF activity, might act as natural *in vivo* tumour promoters.

ACKNOWLEDGEMENTS

We wish to express our gratitude to Prof. E. Ascari of the University of Pavia and Dr. E. Orefice of the Istituto Tumori, Milano for providing most of the plasma samples used in this study. We also thank Mrs. M. Bensi, L. Kostamovaara, A. Virtanen and Mr. F. Tredici for the excellent technical assistance. This work was supported by Consiglio Nazionale delle Ricerche, Progetto Finalizzato: Controllo della Crescita Neoplastica RFP metastasi and by grants awarded by the Finnish Cancer Foundation the Medical Research Council of the Academy of Finland and the National Cancer Institute (CA24605). S.B. was recepient of a short term EMBO fellowship and G.D.P. of a fellowship of the Sigfrid Juselius Foundation.

REFERENCES

1. S. Barlati, C. Kryceve and P. Vigier, Stimulation of transformation by factor(s) present in the medium of Rous Sarcoma Virus transformed cells, Abstracts XIth Int. Cancer Congress, 86 (1974).

2. C. Kryceve, P. Vigier and S. Barlati, Transformation enhancing factor(s) produced by virus-transformed and established cells, Int. J. Cancer. 17:370 (1976).
3. C. Kryceve, J. M. Biquard, S. Barlati, D. Lawrence and P. Vigier, Characteristics of the transformation enhancing factor(s) produced by RSV-transformed and other transformed cells, in: "Avian RNA Tumour Viruses", S. Barlati and C. De Giuli, ed., Piccin Medical Books, Padua (1978).
4. C. Kryceve-Martinerie, J. M. Biquard, D. Lawrence, P. Vigier, S. Barlati and P. Mignatti, Transformation-enhancing factor(s) released from chicken Rous sarcoma cells: effect on some transformation parameters, Virology. 112:436 (1981).
5. S. Barlati, P. Mignatti, A. Brega, G. De Petro and E. Ascari, Utilization of Rous sarcoma virus for the detection of transformation enhancing and inhibiting factors in human plasma, in: "Avian RNA Tumour Viruses", S. Barlati and C. De Giuli, ed., Piccin Medical Books, Padua (1978).
6. P. Mignatti, E. Ascari and S. Barlati, Potential diagnostic and prognostic significance of the transformation-enhancing factor(s) in the plasma cryoprecipitate of tumour patients, Int. J. Cancer. 25:727 (1980).
7. S. Barlati and P. Mignatti, Transformation enhancing factors in the plasma of cancer patients, in: "Control Mechanisms in Animal Cells",L.Jimenez, de Asua et al. ed., Raven Press, New York (1980).
8. G. De Petro, S. Barlati, T. Vartio and A. Vaheri, Transformation-enhancing activity of gelatin-binding fragments of fibronectin, Proc. Natl. Acad. Sci. U.S.A. 78:4965 (1981).
9. S. Barlati, G. De Petro, T. Vartio and A. Vaheri, Role of fibronectin degradation products on cellular transformation, in: "Protides of the Biological Fluids", E. Peeters, ed., Pergamon Press (in press) (1981).
10. A. Vaheri, E. Ruoslahti and D. F. Mosher, (eds.) "Fibroblasts Surface Protein", Ann. Acad. Sci., New York (1978).
11. T. Vartio, H. Seppa and A. Vaheri, Susceptibility of

soluble and matrix fibronectins to degradation by tissue proteinases, mast cell chymase and cathepsin G, J. Biol. Chem.256:471 (1981).

12. J. M. Biquard and P. Vigier, Characteristics of a conditional mutant of Rous sarcoma virus defective in the ability to transform cells at high temperature, Virology. 4:444 (1972).

13. S. Barlati, C. Kryceve and P. Vigier, Different stages of transformation of Rous virus-infected cells evidenced by the methylene blue staining, Intervirology. 4:23 (1974).

14. N. M. Blackett, Statistical accuracy to be expected from cell colony assays; with special reference to the spleen colony assay, Cell Tissue Kinet. 7:407 (1974).

15. E. Engvall and E. Ruoslahti, Binding of soluble form of fibroblast surface protein, fibronectin to collagen, Int. J. Cancer. 20:1 (1977).

16. M. Vuento and A. Vaheri, Purification of fibronectin from human plasma by affinity chromatography under nondenaturing conditions, Biochem.J. 183:331 (1979).

17. U. K. Laemmli, Cleavage of structural proteins during the assembly of the head of bacteriophage T4, Nature. 227:680 (1970).

18. H. Towbin, T. Staehelin and J. Gordon, Electrophoretic transfer of proteins from polyacrylamide gels to nitrocellulose sheets: Procedure and some applications, Proc. Natl. Acad. Sci. U.S.A. 76:4350 (1979).

19. S. Stenman and A. Vaheri, Fibronectin in human solid tumours, Int. J. Cancer. 27:427 (1981).

20. J. P. Quigley, Proteolytic enzymes of normal and malignant cells, in: "Surface of Normal and Malignant cells", R. Hynes, ed., John Wiley & Sons New York (1979).

21. H. D. Todd, M. S. Coffee, T. P. Vaalkes, M. D. Abeloff and R. G. Parsons, Serum levels of fibronectin and a fibronectin-like DNA-binding protein in patients with various diseases, J. Natl. Cancer Inst. 65:901 (1980).

22. K. Huber, J. Kirchheimer and B. R. Binder, Rapid isolation of native urokinase from normal human

urine, Thrombos. Haemostas. 46:11 (1981).

23. T. Vartio and A. Vaheri, A gelatin binding 70.000 dalton glycoprotein synthesized distinctly from fibronectin by normal and malignant adherent cells, J. Biol. Chem. (in press).

AFFINITY MODULATION OF EPIDERMAL GROWTH FACTOR MEMBRANE RECEPTORS BY BIOLOGICALLY ACTIVE PHORBOL AND INGENOL ESTERS

Mohammed Shoyab

Laboratory of Viral Carcinogenesis
National Cancer Institute, NIH
Frederick, Maryland 21701

INTRODUCTION

Tumour promoters are compounds which are not themselves carcinogenic but which can induce tumours in mice previously treated with a suboptimal dose of certain chemical carcinogens. One of the most potent tumour promoting agents is TPA (12-0-tetradecanoylphorbol-13-acetate), initially isolated from croton oil. Tumour promoters modulate an array of biochemical functions in mouse skin, including the stimulation of DNA, RNA and protein synthesis. TPA also evokes various biological and biochemical changes when added to cultured cells, including the stimulation of DNA synthesis and cell proliferation, induction of plasminogen activator and ornithine decarboxylase, loss of surface-associated fibronectin, either inhibition or stimulation of differentiation, and alteration in cell permeability (1-6).

The amphiphatic nature of the TPA molecule, and several other lines of evidence, suggest that the initial site of action of TPA and other related promoting agents is the membrane of the target cell. The nature of the biological and biochemical responses initiated by TPA *in vivo* as well as *in vitro*, has many similarities with those of growth stimulating polypeptide hormones such as epidermal growth factor (EGF) and sarcoma growth factor (SGF) (2-4).

The above considerations led to the investigation of the effects of TPA and related tumour promoters on the interactions between EGF and its membrane reaceptors. The studies described here suggest that TPA reduces the binding of EGF to its receptors by reducing the affinity of the EGF receptor-ligand interaction, while not affecting the total number of available EGF receptors.

The affinity modulation of EGF receptor by biologically active phorbol and ingenol esters seems to be mediated by the perturbation of membrane phospholipid by these agents. There is also a good correlation between the degree of modulation of the EGF-receptor interactions by the various derivatives of phorbol and ingenol and their ability to act as promoters of carcinogenesis both in cell culture and in mouse skin (7-10).

TPA TREATMENT OF CELLS AND EGF RECEPTORS

TPA elicited one round of cell division in confluent BALB/3T3 cells. This effect was observed at TPA concentrations of 0.1-0.5 $\mu g\ ml^{-1}$. When sparse cultures of 3T3 were treated with the optimal doses of TPA, they reached a saturation density two times higher than untreated cells although the doubling time, before confluence, of both the treated and untreated cells was the same. The untreated BALB/3T3 cells had a regular growth pattern as a flat monolayer, whereas treated cells were more refractile, exhibited a less discernible overall growth pattern, crisscrossed randomly, and showed some tendency to grow over one another. The TPA-induced changes in morphology and growth behaviour were reversible; when the treated cells were subcultured in the absence of TPA, their morphology and final cell density were not recognisably different from the untreated controls. Some of these effects on cell cultures, including BALB/3T3, have been previously described (2-4).

The effect of various concentrations of TPA on the binding of ^{125}I-EGF to BALB/3T3 cells in culture was studied. Cells in the log phase of growth were tested for binding 48 hours after subconfluent seeding. Cells were treated with 0-5 $\mu g\ ml^{-1}$ of TPA for 2 hours before the binding of ^{125}I-EGF. Incubations were for 40 minutes at 37°C, at an EGF concentration of 2 ng ml^{-1}. Exposure of cells to TPA markedly reduced the binding of ^{125}I-EGF. A maximal reduction, approximately 90%, was observed at TPA concentrations of 0.2-2 $\mu g\ ml^{-1}$ (0.32-3.2 x 10^{-6} M), while 50% inhibition occurred at approximately 2-4 ng ml^{-1} of TPA (3.2-6.4 x 10^{-9} M). TPA exhibited slight toxicity at the highest dose used in this experiment. The TPA elicited inhibition of EGF binding to receptors is a rapid phenomenon. The maximum effect was observed 0.5-2 hours after exposure, but substantial inhibition (>50%) is seen within minutes. A puzzling feature of TPA inhibition of EGF binding (TIEB) is that if cells are continuously exposed to TPA for a longer time, the degree of TIEB starts to gradually decrease after a few hours and cells become refractory after a few days. This loss of TIEB is not due to the conversion of TPA into an inactive form because if TPA containing media from these cells is transferred to new cells, a 85-90% inhibition of EGF binding is observed. However, EGF receptors on cells continuously exposed to TPA for a few days are susceptible to down regulation by EGF.

The comparative binding of labelled EGF to TPA treated cells (0.1 μg/ml, 2 hours) following removal of TPA by washing as a function of time after the addition of TPA-free media was studied. EGF bound progressively more to cells with the increasing times after the removal of TPA. Two days after replacement of TPA-containing media with TPA-free media, previously treated cells bound almost the same amount of EGF as control cells not exposed to TPA. It is impossible to completely remove TPA from the cell surface by washing with media because TPA is a highly lipid-soluble compound and rapidly interacts with plasma membranes. The slow reversal of the TPA-induced modulation of EGF binding may be related to the lipid soluble nature of TPA.

TIEB is not significantly affected by steroid hormones, mouse interferon, cytoskeleton disrupting agents, modulators of polyamine metabolisms, protease inhibitors, inhibitors of DNA, RNA, and protein synthesis, by various mucopolysaccharides, water soluble vitamins, vitamins D, E and K1. However, it is partially reversed by retinoids and oubain (Na-K ATPase inhibitor).

GENERALITY OF TPA MODULATION OF EGF BINDING

We investigated the effects of TPA exposure on the binding of ^{125}I-labelled EGF (2 ng ml^{-1}) to various normal cell lines derived from mouse, rat, mink, cat, hamster, rabbit, human and chicken and to two murine cell lines transformed by the DNA tumour viruses, simian virus 40 (SV40) and polyoma (Table 1). Mouse sarcoma virus-transformed cells were not tested as they already produce an EGF-related polypeptide and do not have available EGF receptors. Approximately 1.2 x 10^5 cells of each cell type were seeded in 35-mm dishes. Specific binding of ^{125}I-EGF to the cells was determined 48 hours after seeding. Cells were exposed to 0.1 μg ml^{-1} of TPA for 2 h at 37°C before binding assays.

TPA inhibition of EGF binding was demonstrated in all types of cells investigated. Murine BALB/3T3-A31 clone 7 cells bound 3,616 d.p.m. of ^{125}I-EGF per 10^6 cells and TPA exposure of these cells reduced the EGF binding to 560 d.p.m. per 10^6 cells. Thus, the degree of TPA induced inhibition of EGF binding was approximately the same in both the normal and SV40-transformed murine cells. Of the two normal human ell strains tested, the TPA inhibition of EGF binding was more pronounced in CRL 1224, a skin fibroblast (85% inhibition), than in CCL 137, a lung fibroblast strain (26% inhibition). The rat epithelial cell clone 52E, derived from normal rat kidney cells, binds about five times more EGF than the rat fibroblastic clone 49F, derived from the same stock of normal rat kidney cells; TPA exposure, however, affected both the fibroblastic and epitheloid cells. Thus, the action of TPA on EGF receptors seems to be a general property of vertebrate cells in culture and is not restricted to untransformed cells or to rodent cells.

TABLE 1

Effects of TPA on the Binding of ^{125}I-EGF to Normal and Transformed Cells

Species	Cells	Designation	^{125}I-EGF (dpm bound per 10^6 cells) Control	TPA-Treated
Mouse	Balb/3T3	A31-Clone 7	3616	560
	SV40-Transformed	SV-A31-clone 6	4343	489
	Swiss 3T3	Swiss 3T3 clone 42	5597	963
	Polyoma-Transformed	Py-3T3 clone 4A	3243	869
	Epidermal cells	MEC	9011	3347
Rat	NRK (Fibroblasts)	clone 49F	785	372
	NRK (Epithelial)	clone 54E	3951	1066
Mink	Lung	MV1Lu (CCL 64)	14656	2973
Cat	Embryo	FFc60	4791	550
Hamster	Kidney	BHK-21	2939	986
Rabbit	Cornea	SIRC (CCL 60)	2740	464
Human	Skin Fibroblasts	CRL 1224	3846	590
	Embryo Lung Fibroblasts	CCL 137	5419	4006
Chicken	Embryo Fibroblasts	CEF (Second passage)	209	96

SPECIFICITY OF TPA INHIBITION OF EGF BINDING

To test the specificity of the effect of TPA treatment on membrane receptor binding, we studied various ligands which are known to bind to cell membranes. Concanavalin A (Con A), multiplication stimulating activity (MSA), insulin, murine type C ecotropic viral glycoprotein (gp70), nerve growth factor (NGF) and low density lipoprotein (LDL) also specifically bind to cell membranes of normal and transformed cells. As shown in Table 2, the exposure of cells to TPA did not reduce the binding of Con A, MSA, insulin, gp70, NGF or LDL to cells, whereas in the same conditions the binding of EGF was reduced to 17% of the untreated cells.

TABLE 2

Specific Reduction of EGF Binding Produced by TPA

	^{125}I-ligand	Cell	^{125}I binding (dpm per 10^6 cells) Control	TPA-Treated
Expt 1	EGF	Murine 3T3	2,947 ± 8	508 ± 21
	Con A	Murine 3T3	15,682 ± 861	16,454 ± 767
	MSA	Murine 3T3	334 ± 29	369 ± 38
	gp70	Murine 3T3	6,128 ± 569	6,048 ± 610
	Insulin	Murine 3T3	795 ± 37	880 ± 71
Expt 2	MSA	Rat kidney fibroblasts (49F-MSV)	1,619 ± 11	1,933 ± 18
	NGF	Human melanoma (A875)	10,416 ± 536	10,054 ± 195
	LDL	Human foreskin	10,543 ± 890	10,426 ± 257

MEMBRANE FLUIDITY AND TIEB

Effects of temperature on TPA-induced reduction in the binding of EGF to its receptors are summarised in Table 3. The level of EGF binding to both treated and untreated murine or mink cells was similar when TPA treatment and the EGF binding assays were performed at 0°C. However, both TPA treatment at 37°C and EGF binding at 0°C, and TPA treatment at 0°C followed by EGF binding at 37°C, markedly reduced EGF binding in both cell systems. TPA-induced inhibition of EGF binding was maximal when both TPA treatment and EGF binding assays were performed at 20-22°C. Thus, the modulation of EGF binding to its receptor which is elicited by TPA is a temperature-dependent phenomenon. TIEB is also not exhibited with fixed cells or isolated plasma membranes. Although EGf binds to fixed cells and plasma membranes. Thus it appears that TIEB is related to membrane fluidity and treatments which affect the transmembrane mobility of receptors also alter the TIEB.

TABLE 3

Effects of Temperature on the TPA Inhibition of the Binding of ^{125}I-EGF to Cells

Cells	Temp. of TPA-Treatment (°C)	Temp. of EGF Binding assay (°C)	^{125}I-binding (dpm per 10^6 cells) Control	TPA-treated	% Inhibition
Mouse	0	0	587	542	7.7
BALB/3T3-	37	0	447	57	87.2
A31	37	37	3,130	657	81.0
clone 7	0	37	2,487	440	82.3
Mink	0	0	1,530	1,533	0
Lung	37	0	1,501	245	83.7
Mv1Lu-64	37	37	4,425	725	83.6
	0	37	3,517	486	86.2

TPA EXPOSURE DOES NOT REDUCE THE NUMBER OF AVAILABLE EGF RECEPTORS

The effect of EGF concentration on the extent of EGF binding to control cells and TPA treated cells was investigated. These lines were tested for EGF binding either after a 2 hour pretreatment period with TPA (mouse BALB/3T3 cells and mink CCL64 cells), or simultaneously with TPA treatment (HeLa, human cervical carcinoma cells). The inhibitory effects of TPA were much greater at lower concentrations of ligand. As the concentration of EGF was increased, decreasing the ratio of receptors to EGF molecules, the TPA-elicited inhibitory effects lessened or vanished at an EGF concentration of 500 ng ml^{-1} in the case of murine cells. At this concentration, the same quantity of EGF bound both to the TPA-treated cells and to the untreated cells. These data produced curvilinear Scatchard plots, indicating the presence of a receptor population in both cell types with varying affinities for EGF. TPA-treated cells showed a marked decrease in ^{125}I-EGF binding at the lower concentrations of added ligand. However, the differences between TPA-treated cells and control cells decreased as the concentration of EGF increased. In contrast, the treatment of mink lung cells with SGF reduced the binding of EGF to its receptors by decreasing the number of available receptors. The Scatchard plots for control and SGF-treated mink cells are almost parallel. At saturating concentrations of EGF approximately 5.5×10^5 molecules bound per mink lung cell compared with 4.1×10^4 bound per BALB/3T3 cell and 5.5×10^5 bound per HeLa cell. HeLa, a human carcinoma cell line and MvlLu, a mink lung cell line are of epithelial origin whereas BALB/3T3 cells are of fibroblastic origin. As in the rat kidney-cell system, epithelial cells have considerably more receptors than fibroblastic cells.

The major murine leukemia virus glycoprotein, gp70, has high affinity receptors on BALB/3T3 cells. Treatment of these cells with TPA did not alter either the apparent number of gp70 receptors or the affinity of gp70 for its receptors. Similarly, treatment of murine or rat cells, human melanoma cells, and human fibroblastic cells with TPA did not change either the apparent number of receptors or the affinity of receptors for MSA, NGF and LDL, respectively. When the amount of gp70, MSA or NGF bound to cells was plotted as a function of the ligand concentrations, identical results were obtained for the control and TPA-treated cells. Thus, reduction in receptor affinity is not a general property of TPA-treated cells.

RELATIONSHIP BETWEEN STRUCTURE OF TUMOUR PROMOTERS AND MODULATION OF EGF BINDING

There are now available several natural and synthetic analogues of TPA with various degrees of promoting activity in the "two-stage tumorigenesis" model. We studied their effects on EGF binding (Table 4). TPA was the most potent inhibitor of EGF binding among the phorbol, ingenol and mezerein derivatives tested. The dose required for 50% inhibition of EGF binding to cells by for TPA, phorbol-12,13-dibutyrate (PDBu), phorbol-12,13-dibenzoate (PDB), 12-deoxyphorbol-13-tetradecanoate (DPTD), ingenol-3-hexadecanoate (IHD), mezerein (MZ), phorbol-12,13-diacetate (PDA) and ingenol 3,5,20-triacetate (ITA) were 2.9×10^{-9} M, 6.5×10^{-8} M, 8.6×10^{-8} M, 1.8×10^{-7} M, 3.8×10^{-8} M, 1.7×10^{-7} M, 1.7×10^{-8} M, 6.4×10^{-6} M and 9.8×10^{-6} M, respectively. 4α-PDD, 4-O-methyl TPA (MeTPA), phorbol and ingenol had negligible effect on EGF binding. The relative potency of these agents in competing EGF binding was: TPA > MZ > DPTD > PDBu > PDD > IHD > PDD > PDA > ITA > MeTPA > 4 α-PDD > Phorbol > ingenol. The inhibition of EGF binding to its receptors by different phorbol and ingenol derivatives correlated very well with their tumor promoting activity.

TABLE 4

Correlation between the Potency of Phorbol and Ingenol Derivatives for Promoting Skin Tumors and their Ability to Inhibit the Binding of ^{3}H-PDBu or ^{125}I-EGF to Cells

Derivative	Dose Required for 50% Inhibition (ng/ml)		
	PDBu Binding to its receptor	EGF Binding to its receptor	Tumor Promoting Activity
12-O-tetradecanoyl-phorbol-13-acetate (TPA)	5.9	1.8	+++
Phorbol-12,13-di-butyrate (PDBu)	50.2	32.8	++
Phorbol-12,13-di-decanoate (PDD)	83.5	53.8	++
Phorbol-12,13-di-benzoate (PDB)	104.0	105.0	+
4 α-Phorbol-12,13-di-decanoate (4 α PDD)	>100,000	>10,000	non-promoter
Phorbol-12,13-di-acetate (PDA)	3,100	>10,000	non-promoter
4-O-Methyl-TPA (Me-TPA)	>100,000	>10,000	non-promoter
Phorbol	>100,000	>10,000	non-promoter
12-deoxyphorbol-3-tetradecanoate (DPTD)	13.8	21.1	++
Mezerein (MZ)	42.5	9.8	?
Ingenol-3-hexa-decanoate (IHD)	34.2	101	+
Ingenol-3,5,20-triacetate (ITA)	3,400	3,000	non-promoter
Ingenol	>100,000	>10,000	non-promoter

AFFINITY MODULATION OF EGF RECEPTORS BY PERTURBATION OF MEMBRANE PHOSPHOLIPIDS

We studied the effect of lipases (lipase, phospholipase A, C, and D), glycosidases (neuraminidase, α-fucosidase, α- and β-glucosidases, α- and β-galactosidases, and β-N-acetylglucosaminidase), nucleases (DNAase 1, RNAase A, and RNAase T1), alkaline protein phosphatase, cyclic AMP dependent protein kinase and trypsin treatment of mink lung cells on the binding of ^{125}I-EGF (2 ng/ml) to its receptors in the presence or absence of 4 ng/ml of TPA (Table 5). Control untreated cells bound 11,690 and 3120 dpm per 10 cells in the absence and presence of TPA, respectively. As expected, trypsin treatment (20 μg/ml) reduced the binding to 5106 and 972 dpm per 10^6 cells in the absence and presence of TPA, respectively, indicating the protein nature of EGF receptors. Other enzymes tested, except for phospholipase C, did not significantly affect the binding of EGF to receptors. Phospholipase C either from *B. cereus* (Sigma Chemical Co.) or from *C. perfringens* (Calbiochem) reduced the EGF binding to treated cells in a dose- and time-dependent manner. At a concentration of 10 μg/ml of phospholipase C (for 30 min), treated cells bound 5611 and 1630 dpm per 10^6 cells (*B. cereus* enzyme) and 5036 and 1414 dpm (*C. perfringer* enzymes) in the absence and presence of TPA, respectively (Table 5). We used only *C. perfringer* enzyme in other studies described herein. Phospholipase C used in this study was found to be free from detectable proteolytic activity. The phospholipase C inhibition of EGF binding remained effective when incubation was performed in the presence of up to 50 μg/ml of phenylmethylsulfonyl fluoride, a potent inhibitor of proteases. Phospholipase C inhibition of EGF binding was almost maximal by 30 min (10 μg/ml enzyme at 37°C). The reduction in the binding increased with increasing concentration of enzyme up to 10 μg/ml. Beyond this concentration, treated cells started to come off from dishes during washing with binding buffer, thus presenting problems in determining accurate binding. It should be noted that the nonspecific binding of ^{125}I-EGF (the binding in the presence of 10^4-fold excess of unlabeled EGF to treated cells) was about three times higher than to control cells. Thus, disruption of membrane phospholipids by phospholipase C apparently alters the organization and the environment of EGF-membrane receptors which leads to the reduced ligand binding. Phospholipase A selectively removes the acyl group from carbon-2 of phospholipids and phospholipase D cleaves the linkage between 1,-3-phosphatidic acid and the nitrogenous base of phospholipids. Phospholipase C cleaves the linkage between diglyceride and phosphate thus removing the nitrogeneous base phosphate from phospholipids. The modulation of EGF receptor interaction by phospholipase C but not by two other phospholipases would suggest that phosphate moiety of membrane phospholipids is very critical in the ligand receptor interaction. Interestingly, some compounds

TABLE 5

Effect of Various Lipases on the Binding of ^{125}I-EGF to Mink Lung Cells

Enzyme	Activity and source	Concentration used (μg/ml)	Binding of ^{125}I-EGF (dpm per 10^6 cells) Control	TPA treated (4 ng/ml)
None	--	--	11,690	3120
Phospholipase A (bee venom)	1:1000 (Calbiochem)	20	12,490	3286
		50	12,040	3190
Phospholipase C (*B. cereus*)	12 units/mg protein (Sigma)	10	5,611	1630
		20	5,377	1695
Phospholipase C (*C. perfringer*)	8.8 units/mg protein (Calbiochem)	10	5,209	1444
		20	5,036	1402
Phospholipase D (cabbage)	24 units/mg protein (Sigma)	20	12,858	3268
		50	12,391	3039
Lipase (*R. arrbizus*)	1818 units/mg protein (Sigma)	20	11,527	2645

(vitamin K3 and digitonin) that are capable of interacting with membrane lipids caused reduction in the binding of EGF to cells. The effects of phospholipase C treatment and vitamin K3 were additive whereas the addition of digitonin to phospholipase C-treated cells did not reduce ^{125}I-EGF binding further. Polyene antibiotics (nystatin, amphotericin B) which preferentially disrupt membrane systems containing cholesterol as well as fillipin and kanaidin which selectively interact with membrane phospholipids reduced the EGF binding to both the phospholipase C-treated and to control cells (Table 6). Addition of ganglioside, cholesterol, or various phospholipids to the binding mixture did not significantly affect EGF binding either to the control or enzyme-treated cells.

To test the selectivity of the effect of phospholipase C treatment on EGF-receptor interaction, we studied various other ligands that are known to bind to cell membranes. Multiplication stimulation activity (MSA), insulin, concanavalin A (Con A), α-2-macroglobulin (α-2M), and murine type C ecotropic viral glycoprotein (gp70) also specifically bind to cell membranes. The phospholipase C treatment of either mink lung cells or murine 3T3 cells did not alter the binding of insulin, Con A, α-2M, and gp70, while under the same conditions, the binding of EGF was reduced to 52% in the case of mink lung cells and to 66% for 3T3 cells of the untreated control. MSA binding was increased by approximately 50% above the control in both cell systems.

The effect of EGF concentration on the extent of EGF binding to control and phospholipase C treated cells was studied. The inhibitory effects of enzyme treatment were much greater at lower concentrations of ligand. As the concentration of EGF was increased, decreasing the ratio of receptor to ligand, the phospholipase C-elicited inhibitory effects lessened. The Scatchard plots of the effect of EGF concentration onthe EGF binding to control and enzyme-treated mink lung cells were drawn. These data produced a curvilinear plot for control cells and a slight curvilinear plot for enzyme-treated cells. However, available data do not demonstrate whether these results are due to heterogeneity or negative cooperativity. At saturating concentrations of EGF, approximately 2.8×10^5 molecules bound to either control or enzyme-treated cells. The apparent K_d values were calculated to be 5×10^{-10} M for the control cells and 2.3×10^{-9} for treated cells. Thus, phospholipase C treatment of mink lung cells reduces by about fivefold the affinity of receptor for its ligand.

The present studies indicate that an important property (binding function) of EGF receptors of mink lung cells and murine 3T3 cells is markedly affected by the removal of phosphorylated amine groups from membrane phospholipids by phospholipase C digestion of these cells. The removal of phosphorylated amines

TABLE 6

Effects of Various Lipid-Interacting Compounds on the Binding of ^{125}I-EGF to Mink Lung Cells

Compounds	Concentration (μg/ml)	Binding of ^{125}I-EGF (dpm per 10^6 cells)	
		Control	Phospholipase C treated
None	--	12,580	6038
Vitamin K_3	1	10,864	5003
	10	3,541	1712
Digitonin	1	11,082	5934
	10	5,772	5639
Nystatin	1	10,989	5921
Amphotericin B	1	11,295	5767
	10	6,658	3319
Fillipin	1	8,116	5053
	10	4,522	2429
Kanacidin	1	9,985	5138
	10	5,707	2878
Mellitin	1	6,764	3509
Gangliosides	10	12,783	6172
	100	12,453	6098
Cholesterol	1	12,612	6265
	10	13,089	6887

from membrane phospholipids probably alters the conformation of EGF binding sites leading to the reduction of affinity for their ligands. It is quite possible, therefore, that the components of phospholipids thus removed may normally be involved in EGF-receptor interaction. Alternatively, the removal of phospholipids may alter the structure of other membrane constituents which in turn are responsible in altering the affinity of EGF for its receptors. Interestingly, the solubilization of EGF membrane receptors by Triton X-100 treatment of plasma membranes results in a 10-fold decrease in the affinity of the solubilized receptors for EGF (11). This result is consistent with our conclusion of an important role for lipids in EGF-receptor interaction.

Both the TPA exposure of cells and phospholipase C digestion of cells preferentially reduce the binding of EGF to its receptors by decreasing the affinity of receptors for ligands. Therefore, it is tempting to speculate that the initial action of tumor-promoting phorbol and ingenol esters may involve the perturbation of the organization of membrane phospholipids.

CONCLUSION

The EGF-competing activity of the phorbol esters parallels the tumour-promoting activity *in vivo*. The phorbol derivatives lacking tumour-promoting activity also lack EGF-competing activity. TPA treatment seems to modulate EGF binding by decreasing the affinity of the receptors on the treated cells for EGF, rather than by decreasing the number of receptors per cell. This affinity modulation is reversible and dependent on time, temperature and TPA concentration. The effect appears to be specific, for the EGF receptor system, as four other receptor-ligand systems tested in the same TPA-treated cells and three receptor-ligand systems in the other cells did not show any alterations in receptor affinity. TPA modulation of EGF binding is observed with doses of promoter comparable to those required to elicit biological responses *in vivo* as well as *in vitro*. The above data suggest that TPA-mediated alterations in growth factor(s)-receptor interactions might be related to the underlying mechanism by which tumour-promoting agents initiate a chain of events cauusing alterations in cellular growth and function. Interestingly, EGF has been reported to enhance the tumorigenesis induced by a chemical carcinogen (12). SGF, produced by mouse sarcoma virus-transformed cells also interacts with EGF receptors, stimulates cell growth and anchorage-independent growth (13). The putative endogenous growth factor(s) produced in response to the exposure of cells to tumour-promoting agents, may, then, activate a programme of gene expression in those cells that have already been genetically altered by initiating agents.

There is need for rapid cell culture assays for tumour-promoting agents both of exogenous and endogenous origin. The data presented here, showing the consequences of promoter treatment of EGF-receptor interations and the specificity, sensitivity and rapidity of this response, might provide a means for the qualitative and quantitative detection of other classes of tumour-promoting agents.

REFERENCES

1. B. L. Van Duuren, Prog. Exp. Tumour Res. 11:31 (1969).
2. E. Hecker, Meth. Cancer Res. 6:439 (1971).
3. R. K. Boutwell, Crit. Rev. Tox. 2:419 (1974)
4. T. J. Slaga, A. Sivak, and R. K. Boutwell, (eds) "Mechanisms of Tumor Promotion and Cocarcinogenesis", Raven, New York (1978).
5. A. Sivak, Biochim. Biophys. Acta 560:67 (1979).
6. L. Diamond, T. G. O'Brien, and W. M. Baird, Adv. Cancer Res. 32:1 (1980).
7. M. Shoyab, J. E. DeLarco, and G. J. Todaro, Nature 279:387 (1979).
8. M. Shoyab and G. J. Todaro, J. Biol. Chem. 255:8735 (1981).
9. M. Shoyab and G. J. Todaro, Arch. Biochem. Biophys. 206:222 (1981).
10. M. Shoyab and G. J. Todaro, Nature 288:451 (1980).
11. G. Carpenter, Life Sci. 24:1691 (1979).
12. V. H. Reynolds, F. H. Boehn, and S. Coren, Surg. Forum 16:108 (1965).
13. J. E. DeLarco and G. Todaro, Proc. Natl. Acad. Sci. USA 75:4001 (1978)

REGULATION OF CYTIDINE UPTAKE IN EHRLICH ASCITES TUMOUR CELLS

Klaus Ring and Ulrich Zabel

Zentrum der Biologischen Chemie, Abteilung für Mikrobiologische Chemie, Universität Frankfurt, Theodor-Stern-Kai 7, D 6000 Frankfurt-70, FRG

INTRODUCTION

Cunningham and Pardee (1969) have described that dialyzed serum is able to stimulate the uptake of uridine in contact-inhibited 3T3 cells. Since then, a large number of studies have been published to describe the effect of hormones, cyclic nucleotides, lectins and viral transformation on the uptake of nucleosides and other nutrients (for reviews see Plagemann and Richey, 1974; Wohlhueter and Plagemann, 1980; Plagemann and Wohlhueter, 1980). In most of these studies, the control mechanism investigated was based on the regulation of de novo formation or breakdown of the proteins presumably involved in nucleoside uptake, i.e. the transmembrane transport catalyzing carrier, and the nucleoside kinase which phosphorylates, and thereby traps, the substrate within the cell.

The present communication deals with a different type of regulation of nucleoside transport, which operates at the level of the membrane by affecting the activity of one of the transport components for cytidine. In contrast to the rather slow adaptional processes based on the regulation of protein turnover, this kind of control allows rapid adaptation of transport to the actual physiological situation.

As stressed by Plagemann and Wohlhueter (1980) and Wohlhueter and Plagemann (1980) in their recent rewies, nucleoside uptake by different types of cultured mammalian cells involves a saturable, nonconcentrative carrier mechanism, operating rapidly enough to establish transmembrane equilibration within a minute. There is, probably , a single carrier of broad specificity capable of transporting all of the natural purine and pyrimidine ribo- and deoxyribonucleosides. The carrier behaves symmetrically with respect to the two sides of the membrane. Having passed the cell membrane, the nucleoside is phosphorylated by a specific kinase. Transport and phosphorylation act in tandem; the latter reaction has been described to govern the rate of nucleoside uptake. Consequently, modulation of uptake in these cells was found to be the result of regulation of the activities of the single kinases responsible for trapping the various nucleosides within the cells.

However, in spite of their consistence, it would be premature to formulate a general scheme for nucleoside uptake in mammalian cells on the basis of these data. There is a considerable number of kinetic data on nucleoside uptake in other cell lines which would not fit into such a scheme, and can only be explained on the basis of the assumption that, in these cases, nucleoside uptake is controlled by the transport step.

In a series of careful kinetic analyses,Eilam and Cabantchik (1976, 1977) presented convincing evidence that in BHK cells the carrier system itself is subjected to regulation. Similar results have been found with Ehrlich ascites tumour cells (Zabel and Ring, 1978; Zabel 1980). In these cells, nucleoside transport is characterized by the following properties, most of which are inconsistent with the data reported by Plagemann and associates: (1) Nucleosides are transported by multiple carrier systems with overlapping substrate specificities; (2) the transmembrane movement is strictly unidirectional; (3) when cells were preloaded with either uridine or cytidine, nucleoside uptake was markedly reduced; (4) in fast growing cells, uridine and cytidine uptake is nonconcentrative; in physiologically older cells, however, 20-30-fold accumulation of free nucleosides has been observed; (5) transport into non-preloaded cells ("trans-zero") is linear with time for more than five minutes. There is no experimental evidence for the assumption that the rate of phosphorylation limits the rate of nucleoside uptake. On the contrary, studies related to control of uptake suggested that the transport components for uridine as well as for cytidine are effectively controlled by different endogeneous and exogeneous factors specifically. One of these systems will be described in this paper.

RESULTS AND DISCUSSION

Fig.1 shows a comparative study of the uptake kinetics of uridine and cytidine in Ehrlich ascites tumour cells. Whereas uridine transport appears to follow conventional Michaelis-Menten type of saturation kinetics, the kinetics of cytidine transport is more complex. At low concentrations, cytidine influx is highly accelerated as the cytidine concentration is increased. At approximately 25 μM, transport operates at maximum rate. As the cytidine concentration is further increased, however, transport is sharply slowed

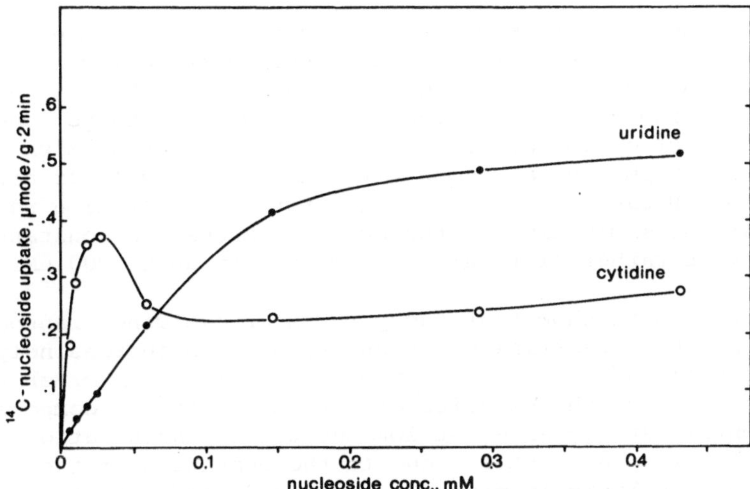

Fig.1. Influx of ^{14}C-labeled cytidine and uridine at different concentrations in the medium (KRP-buffer; 37°C)

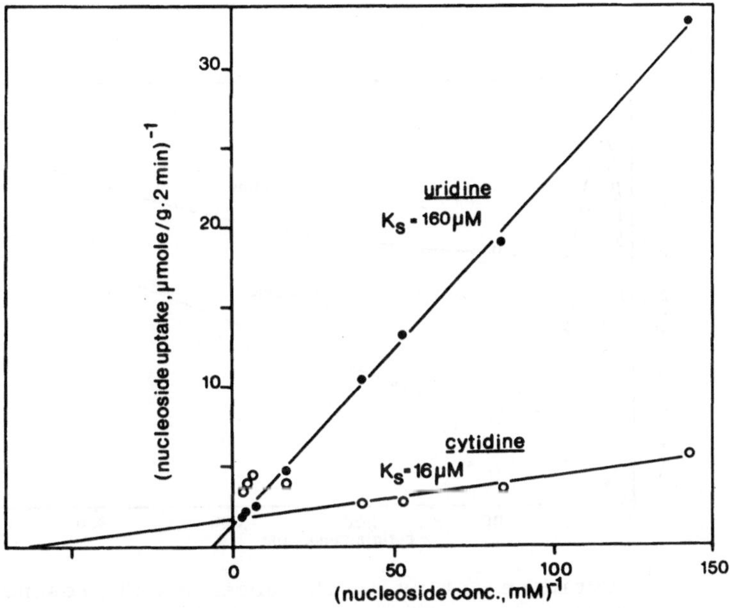

Fig.2. Lineweaver-Burk diagram of the data presented in Fig.1

down. Only at concentrations above 100 μM, cytidine transport increases again.

A Lineweaver-Burk plot of the data reveals a monophasic function for uridine transport. Cytidine transport follows a complex biphasic function (Fig.2) suggesting that this substrate enters the cells via two different saturable catalytic pathways; one with high affinity for cytidine, as expressed by an apparent Michaelis constant of 16 μM, and a second with substantially lower affinity for cytidine. Because of the high substrate concentration required, accurate determination of the apparent Michaelis constant of this pathway is rather difficult; it amounts to about 200 μM.

These data show that the portion of cytidine transport, catalyzed by the low-affinity component, becomes increasingly important as the cytidine level in the medium is raised. In order to obtain a true picture on the kinetics of the high-affinity component, the experimental data obtained at low substrate concentrations have to be corrected for the portions due to the operation of the second pathway. Such analysis is presented in Fig.3, indicating that the activity of the high-affinity component is, in fact, strongly modulated in response to altered extracellular cytidine levels. The increment of influx with increasing extracellular cytidine concentrations is limited to a rather narrow concentration range below 25-30 μM. Above this threshold level, the transport rates become drastically lowered

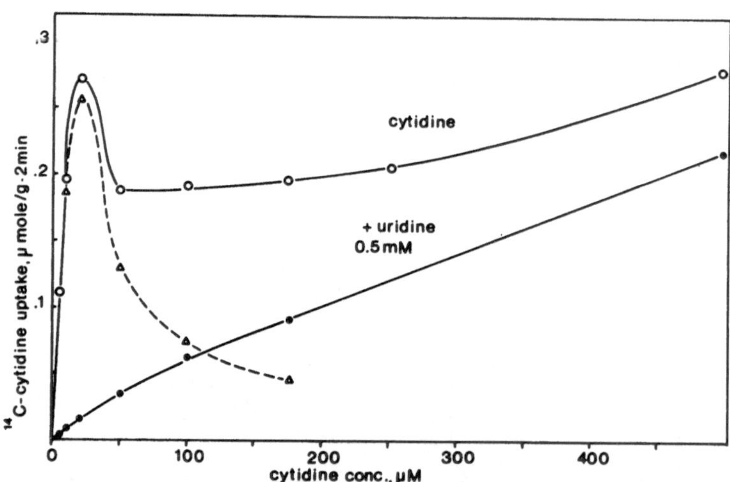

Fig.3. ^{14}C-Cytidine influx in the absence and presence of uridine at different cytidine concentrations in the medium. The dotted line shows the corrected data for transport catalyzed by the high-affinity component

and appear to approach zero at sufficiently high concentrations. This implies that cytidine uptake at higher concentrations is mediated predominantly by the second, low-affinity component.

Transport kinetics of this type have not been described as yet. However, it resembles the kinetics of enzymes that are subject to "substrate inhibition", i.e. which are inhibited by excess of substrate. It is generally accepted that one group among such enzymes are allosteric proteins which are regulated by the substrate via a particular modifier site which is distinct from the catalytic site. If, at high substrate concentrations, the modifier site is occupied by the substrate, the activity of the catalytic site is suppressed.

The apparent similarity between the kinetic properties of the high-affinity transport component for cytidine and the kinetics of substrate-inhibition of enzymes may suggest the involvement of a similar mechanism for cytidine transport. This assumption is further supported by experiments with two different types of inhibitors.

As many other transport systems, also the high-affinity cytidine transport component is sensitive towards inhibitors that specifically react with free sulfhydryl groups. We have used p-chloromercuribenzyl-sulfonate (PCMBS), a hydrophilic molecule which penetrates intact mammalian cell membranes only slowly, and binds reversibly to free sulfhydryl groups. Because of these properties, PCMBS is a suitable tool for studying the function of those sulfhydryl groups that are located at the cell surface.

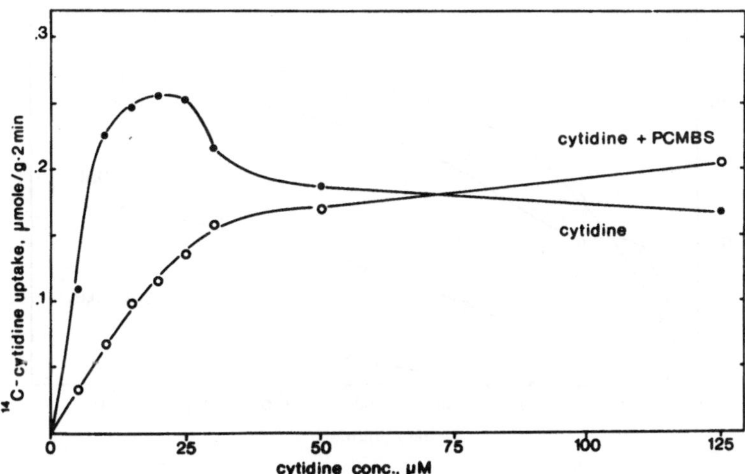

Fig.4. ^{14}C-Cytidine influx in the absence and presence of 1 mM PCMBS

At a concentration of 1 mM, PCMBS effectively inhibits cytidine transport at low extracellular cytidine concentrations; at higher concentrations, i.e. above approximately 70 µM, however, transport appears to be rather unaffected or even stimulated (Fig.4). Detailed kinetic analyses to be presented elsewhere, revealed that the inhibition is of mixed type, comprising a competitive and a non-competitive component. As the PCMBS concentration is lowered to 0.5 mM, the inhibition of transport at low substrate concentrations is diminished, whereas the stimulatory effect at high substrate concentrations becomes more significant (Fig.5). Moreover, the complex two-component kinetics is transformed into a conventional, monophasic Michaelis-Menten type of kinetics.

In the experiment described in Fig.5, a second inhibitor, dinitrophenylthioinosine (DSI) was used, which was synthesized by Fasold and associates (1977). As will be described in a separate communication, DSI binds to the catalytic site of the cytidine transport component competing with cytidine without being transported into the cell itself. Thus, as PCMBS under appropriate conditions, DSI interferes with certain ligands at the cell surface only.

As shown in Fig.5, cytidine influx is inhibited by 47 µM DSI effectively but incompletely. Simultaneous addition of 47 µM DSI

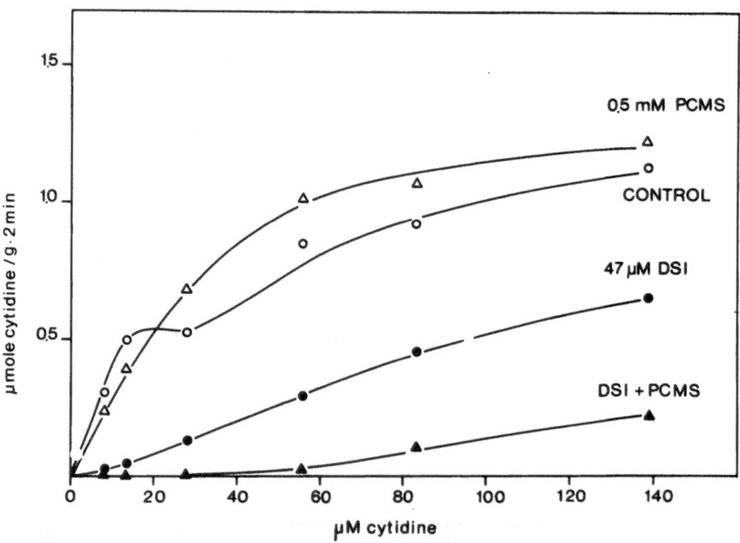

Fig.5. Effect of 0.5 mM PCMBS and 47 µM DSI on ^{14}C-cytidine influx at different cytidine concentrations in the medium

and 0.5 mM PCMBS, however, increases the efficacy of DSI so that, over a wide concentration range, cytidine transport into the cell is entirely inhibited.

These observations strongly support support the assumption that the high-affinity transport component for cytidine is controlled by exogeneous factors. The experimental data available so far are best explained by the following assumptions which are based on the well known kinetic properties of enzymes that are subject to substrate inhibition.

According to this hypothesis, the cytidine transport component is a complex protein with two, functionally different binding sites for cytidine: a transport-catalyzing site with high affinity for cytidine, and a modifier site with substantially lower affinity for cytidine. The substrate specificity of both sites for cytidine, as the main substrate, is not absolute; at appropriately high concentrations, also other ribonucleosides can be accepted as true or pseudosubstrates, as examplified by the inhibitory effect of uridine shown in Fig.3. In the absence of any inhibitor, cytidine at low concentrations predominantly binds to the transport site; consequently, the transport machinery operates with high efficiency. In response to increasing extracellular concentrations, however, cytidine more frequently will occupy also the modifier site, which, in turn, triggers the conversion of the active transport component into its inactive form.

Both sites contain free sulfhydryl groups that are able to react with PCMBS. Again, the sites are characterized by differential affinities: The affinity of the sulfhydryl group in the catalytic site is low for PCMBS, so that rather high PCMBS concentrations are required for substantial inhibition of the transport. In contrast to this, the sulfhydryl group in the modifier site is more affine for PCMBS. Its occupation by PCMBS prevents the binding of the natural substrate, which is a prerequisite for triggering the inactivation of the transport complex. As a result, the activity of the transport site remains unaffected, even at high cytidine concentrations. In other words, the acceleration of influx by PCMBS at high cytidine concentrations is rather the result of de-inhibition than of stimulation. These assumptions imply that under such conditions transport kinetics should follow normal Michaelis-Menten kinetics, which has been demonstrated.

The studies reported here delineate some characteristics of cytidine uptake in Ehrlich ascites tumour cells. Studies with structural analogues of cytidine, e.g. cytosine arabinoside, on the modulation of cytidine uptake in the above system could be important to investigate further the mechanism of cytidine transport. On the other hand, development of specific inhibitors of this transport system could be important in developping compounds of therapeutical interest.

ACKNOWLEDGMENT

We gratefully appreciate the skillfull technical assistence of Mrs.Hella Ehle and Mrs.Beate Foith-Minuth.

This study was supported by the Deutsche Forschungsgemeinschaft.

REFERENCES

Cunningham,D. and Pardee,A., 1969, Transport changes rapidly initiated by serum addition to "contact inhibited" 3T3 cells, Proc.Natl.Acad.Sci.U.S., 64: 1049

Eilam,Y. and Cabantchik,Z., 1976, The mechanism of interaction between high affinity probes and the uridine transport system of mammalian cells, J.Cell.Physiol., 89: 381

Eilam,Y. and Cabantchik,Z., 1977, Nucleoside transport in mammalian cell membranes, II. A specific inhibitory mechanism of high affinity probes, J.Cell.Physiol., 92: 185

Fasold,H., Hulla,F., Ortanderl,F. and Rack,M., 1977, Aromatic thioethers of purine nucleotides, Adv.Enzymol., XLVI: 289

Plagemann,P. and Richey,D., 1974, Transport of nucleosides, nucleic acid bases, choline and glucose by animal cells in culture, Biochim.Biophys.Acta, 344: 263

Plagemann,P. and Wohlhueter,R., 1980, Permeation of nucleosides, nucleic acid bases, and nucleotides in animal cells, in: "Current Topics in Membranes and Transport", F.Bronner and A.Kleinzeller, ed., Academ.Press, New York, 14: 225

Wohlhueter,R. and Plagemann,P., 1980, The roles of transport and phosphorylation in nutrient uptake in cultured animal cells, Internatl.Rev.Cytol., 64: 171

Zabel,U. and Ring,K., 1978, Kinetik der Aufnahme von Pyrimidinnucleosiden in Ehrlich-Maus-Ascites-Tumorzellen, Hoppe-Seyler's Z.Physiol.Chem., 359: 340

Zabel,U., 1980, Nucleosid-Transport bei Ehrlich-Ascites-Tumorzellen, Thesis,University of Frankfurt, FRG.

GLUCOCORTICOID RECEPTORS AND SENSITIVITY IN NORMAL AND NEOPLASTIC HUMAN LYMPHOID TISSUES

Françoise Homo, Sylvie Durant and Dominique Duval

INSERM U7/CNRS LA 318, Dept of Nephrology
Hôpital Necker, 161 rue de Sèvres
75015 Paris, France

INTRODUCTION

It has been known for a long time that glucocorticoids exert many physiological and pharmacological effects on mammalian lymphoid tissues. *In vivo*, administration of corticosteroids or adrenalectomy have been shown to induce marked changes in the size of lymphoid organs as well as in lymphocyte circulation and to alter many immunological reactions (1). *In vitro*, glucocorticoids are generally considered as catabolic agents that induce an inhibition of membrane transports and macromolecules synthesis leading to an arrest of cell growth sometimes accompanied by cell lysis (2). These compounds are also able to modify several immunological functions *in vitro* (3).

These widespread effects on lymphoid cell metabolism and functions constitute the basis for the use of glucocorticoid hormones in the treatment of a wide variety of immunological and inflammatory diseases as well as in the treatment of lymphoid cell neoplasias.

According to the classical mechanism of steroid hormone action, which includes a preliminary step of interaction of the steroid with cytoplasmic receptors (4), numerous studies have been devoted to the question of whether or not the receptor content of lymphoid cells may be representative of the *in vitro* or *in vivo* sensitivity.

This paper presents a critical review of glucocorticoid receptor determination as an index of steroid sensitivity in normal and leukemic human lymphoid tissue.

NORMAL HUMAN LYMPHOID CELLS

Although many studies have attempted to determine the effects of glucocorticoids on human lymphoid tissues, only a few of them have actually related the extent of steroid action either *in vivo* or *in vitro* to the level of glucocorticoid receptors.

Human thymocytes

It was first demonstrated many years ago that administration of corticosteroids can cause an involution of human thymus like in rodents (1, 5). However, Claman postulated that the *in vivo* shrinkage of the infant thymus after hydrocortisone could be explained in terms of growth inhibition rather than by cytolysis (1). Indeed, in contrast to mouse thymocytes which are readily lysed after 6 h incubation in the presence of 10^{-6}M hydrocortisone, human thymocytes are much less susceptible to steroid-induced lysis (6). This resistance of human thymus cells to the cytotoxic action of hydrocortisone was recently confirmed by Galili et al. (7).

Using a whole cell binding assay and [^{3}H]-dexamethasone as tracer, we have determined in thymus cells obtained from children at the time of cardiac surgery the number of glucocorticoid receptors (8). In 9 subjects the average number of specific binding was 3100 sites per cell with an affinity calculated from Scatchard analysis of $K_d \sim 5 \times 10^{-8}$M. Despite this relatively low number of receptors (see below), these human thymocytes were extremely sensitive to the inhibitory effects of glucocorticoids. During a 24 h incubation *in vitro* dexamethasone induced a 70 % decrease in [^{3}H]- uridine incorporation and an almost complete ($\sim$ 95 %) inhibition of [^{3}H]-thymidine incorporation (8). It should be noted that the spontaneous levels of [^{3}H]-thymidine incorporation (i.e. in the absence of steroid) were very high which reflects the high rate of cell division in children thymuses (8).

Normal peripheral lymphocytes (PBL)

One of the most obvious effects of glucocorticoids in human is the lymphopenia which occurs within 4-6 h following steroid administration (9). This depletion of circulating leucocytes appears more pronounced for T cells than for B cells ; and among T cell subpopulations the T_M cells (identified by the presence of an F_C receptor for IgM) are more sensitive to steroid-induced depletion than the T_G cells (identified by an F_C receptor for IgG) (10). This effect which is essentially transient and disappears within 24 h following steroid treatment is probably not due to cell lysis but more likely to a redistribution of the circulating cells among other compartments (9). Similar experiments performed in guinea pigs using ^{51}Cr labeled lymphocytes have shown that most of the cells removed from the circulation after glucocorticoid treatment are transitorily sequestered by the bone marrow (11).

Determinations of in vitro sensitivity to glucosteroids have shown that peripheral lymphocytes are extraordinarily resistant to cytolysis. Schreck demonstrated almost twenty years ago that a 7-10 day incubation of human peripheral lymphocytes in the presence of high amounts of hydrocortisone increases cell survival rather than accelerate the rate of cell death (12). Studies following the effect of glucocorticoids on the incorporation of precursors such as leucine, uridine or thymidine showed only moderate (20-40 %) inhibition of incorporation during 24 h incubation (13, 14), whereas slightly increased inhibition could be observed after a prolonged incubation period (15). In comparison with thymocytes (8), the level of spontaneous incorporation of thymidine in human PBL was very low (almost ten-fold lower) corresponding to the low rate of cell proliferation in the periphery (13).

Measurements of glucocorticoid receptors either by the usual cytosolic assay or by whole cell assay revealed the presence of specific binding sites for glucocorticoid compounds in human peripheral lymphocytes (13-21). In unseparated PBL populations, after monocyte depletion (monocytes have been shown to contain more glucocorticoid receptors than lymphocytes (15, 22)), the number of steroid receptors was in the range 2700 to 6000 binding sites per cell according to different reports. Scatchard plot analysis of the binding curves showed the existence of a single class of binding sites with an affinity ranging from 5.5×10^{-9} to 5×10^{-8}M. Attempts have been made to determine the levels of glucocorticoid receptors in various PBL subpopulations. Lippman and Barr have first demonstrated that purified T lymphocytes prepared by E-rosette formation, and non-T lymphocytes contain the same amount of glucocorticoid receptors (about 3000 sites per cell with an affinity in the range 1.4-1.7×10^{-9}M) (15). Similarly, after separation of PBL subpopulations on an immunoabsorbant column (anti $F(ab')_2$), we failed to show any significant difference between B cells (mean value is 3650 sites per cell) and T + Null cells (mean value is 4700 sites per cell((13). However, a recent report by Distelhorst and Benutto has suggested the existence of different glucocorticoid receptor levels among T cell subpopulations (23).

Up to now there are no indications suggesting that differential effects of glucocorticoids on normal human lymphoid cell metabolism and functions could be accounted for by differences in glucocorticoid receptor levels. Fauci et al. failed to show any detectable differences in intracytoplasmic receptors between T_M and T_G cells, which however do not respond in the same way to in vivo hydrocortisone administration as mentioned above (10).

Normal lymph nodes

Corticosteroids have also been shown to decrease the size of lymph nodes in vivo, but this effect was less marked than that observed on thymus or spleen (3, 24).

Recently, Bloomfield and coworkers have measured the number of glucocorticoid receptors in normal human lymph node cells and determined the in vitro sensitivity of these cells to dexamethasone treatment (25). It appears that dexamethasone (10^{-7}M) induced only a moderate inhibition (20-30 %) of leucine, uridine or thymidine incorporation over a 24 h incubation. The effect of 4 x 10^{-7}M dexamethasone on cell survival was also tested. It was shown that during a 4 day incubation dexamethasone produced only a 20 % decrease of the number of viable cells. The average number of glucocorticoid receptors determined by whole cell assay was 2082 binding sites per cell.

NEOPLASTIC HUMAN LYMPHOID TISSUE

Thymoma

Several indications have led to the suggestion that thymoma cells may represent target tissue for steroid hormones. First, thymus hyperplasia has long been recognized as a pathological form susceptible to regression upon corticoid therapy (26). Then, the presence of specific glucocorticoid receptors was demonstrated in an irradiation-induced mouse thymoma (27). Ranelletti and coworkers have thus studied the level of glucocorticoid receptors in various forms of thymus dysplasia (26, 28). Using a cytosolic assay, they showed that the level of specific receptors was significantly higher in lymphoepithelial thymoma than in pure epithelial form, thymus hyperplasia or even in normal thymus (26, 28, 29). However, despite this high level of glucocorticoid receptors the in vitro sensitivity of thymoma cells to glucosteroids was comparable to that of normal thymus cells either in terms of inhibition of precursor incorporation or in terms of steroid -induced cell lysis (28, 29).

The demonstration of the presence of glucocorticoid receptors in pure epithelial thymoma together with the known influences of epithelial cells on intrathymic lymphocyte maturation suggest that glucocorticoids may exert a dual role on thymus lymphocyte functions either directly and/or through an action on thymus epithelial cells (28, 30).

Acute lymphocytic leukemia (ALL)

It is now widely recognized that ALL is a strongly heterogeneous disease with regard to its clinical aspects, its response to chemotherapy and its prognosis (31). Studies of cell surface markers have shown that in most cases of ALL (70 %) the leukemic cells lack the specific surface membrane markers of B and T lymphocytes and therefore belonged to the so called 'null cell' variety. The great majority of the remaining cases (25-30 %) is of T cell origin (forming rosettes with sheep red blood cells), whereas only a very small proportion (1-2 %) of cases possess B cell characteristics (presence of surface immunoglobulins). More recently, the group of 'null' ALL has been subdivided on the basis of constitutive enzymes and other immunological marker determinations and presumably comprised lymphoid stem cells, pre-B and pre-T cells (32).

This immunological classification of ALL is of practical value, since correlations have been demonstrated between the phenotype of the leukemic cells and the clinical features as well as the prognosis of the disease (31). 'Null cell' ALL can be further subdivided into the common childhood form and the adult form. The childhood form represents the great majority of the cases, responds well to treatment, and is generally associated with a good prognosis. The adult form occurs in patients of over 14, responds poorly to treatment and has a relatively poor prognosis. T cell form is characterized by distinct clinical features: a thymic mediastinal mass, a high leucocyte count in the peripheral blood, a higher incidence in older patients than the common childhood form and presumably a male predominancy. This type of leukemia responds poorly to treatment and has a poor prognosis. Similarly, B cell form of ALL has generally a very poor prognosis.

Glucocorticoids have been used for more than 20 years in the treatment of ALL, first alone and now as a part of combined chemotherapy (31). It appears however that some patients either are at once resistant or cease to be responsive in the course of steroid treatment. Because of the iatrogenic complications of glucocorticoid therapy, it would be of value to select in advance those patients which will respond to treatment.

Over the past 10 years, extensive studies have demonstrated that determinations of estrogen and progestogen receptors in breast cancer represent a major element of the therapeutical management (33). Therefore, similar studies were carried out in leukemia, in an attempt to relate glucocorticoid receptors with the glucocorticoid sensitivity (16, 34). Using a cytosolic assay, Lippman et al. first reported that the cells of 22 patients with previously untreated acute lymphoblastic leukemia contained high levels of glucocorticoid receptors and were, *in vitro*, sensitive to glucocorticoid action as well as *in vivo* (16). Moreover, they showed that among 12 patients previously treated with glucocorticoids, 6 who failed to respond to additional treatment had barely detectable levels of glucocorticoid receptors, whereas 6 who subsequently responded to steroid therapy had levels of receptors similar to those of untreated ALL patients. In addition, glucocorticoids did not induce, *in vitro*, any inhibition of thymidine incorporation in those cells from patients resistant to glucocorticoids and with a very low receptor content.

The potential clinical implication of this correlation between receptor content and the in vitro response to chemotherapy has prompted several groups to undertake comparable studies. However, it soon appeared that the results obtained were less conclusive than previously believed (14, 17-19, 29, 35-38). It turned out that, when assayed by the whole cell method, all the peripheral leucocytes isolated from patients with acute lymphocytic leukemia contained specific receptors for glucocorticoids. The number of binding sites ranged between 1000 sites to more than 20.000 with an average value of about 10.000-15.000 sites, almost 3 to 5-fold greater than measured in normal circulating lymphocytes. Affinities of these receptors for the steroid were comparable to those determined in normal human.

However, attempts to correlate the levels of these receptors to any parameter of in vitro steroid sensitivity such as inhibition of leucine, uridine, thymidine incorporation or even cell lysis were unsuccessful in most of the studies (14, 18, 19, 29, 35, 36).

Antileukemic therapy usually includes in addition to glucocorticoids, two or three other cytostatic agents and it is therefore difficult to ascertain the clinical efficiency of one of these drugs and to relate an in vivo response to the level of steroid receptors. To overcome this difficulty two groups have also attempted to correlate the level of receptors with the short-term response to glucocorticoid therapy (29, 35). Patients were classified as responders or non responders on the basis of a 50 % reduction of the number of circulating blast cells after a 2-4 day treatment with glucocorticoids alone. Iacobelli et al. showed in a small group of patients that cells from responders contained higher levels of receptors than those from non-responders (29). In our study, there was a similar tendency although the difference between the two groups was not statistically significant (35).

On the other hand, several groups have established relationships between the number of glucocorticoid receptors and immunological parameters. It was shown for example that in childhood ALL, there were generally more receptors in the cells from patients with 'null cell' leukemia than in those with T cell form of disease (37, 39, 40). Similarly, there are now some indications that independent of the immunological or cytological type, the amount of glucocorticoid receptors in acute leukemia could be correlated with remission duration after combined chemotherapy (41, 42).

Chronic lymphocytic leukemia (CLL)

Glucosteroids do not represent a major therapeutical agent in the treatment of CLL but are sometimes used when other therapy has failed or in patients with bone marrow cytopenia (31).

Determinations of glucocorticoid receptors have now been done by several groups in approximately 130 patients with CLL (14, 18, 19, 34, 43-47). In two early studies performed by cytosolic assay the receptors were only detected in 1/2 to 1/3 of the patients (34, 44). Again, when assayed by whole cell assay, the presence of glucocorticoid binding sites was detected in all cases. The number of these receptors was on an average similar or even slightly lower than that measured in normal PBL. More recently, Ho et al have demonstrated that the cells from patients who had been treated for 3-5 years with a combination of corticoids and chlorambucil and became resistant to this treatment, contain lower receptor content than cells from newly diagnosed patients (18).

Attempts made to correlate the amount of receptors with _in vitro_ steroid sensitivity were unsuccessful (14, 18, 43). It appears however, that the extent of dexamethasone-induced inhibition of uridine incorporation was generally more important in cells from CLL patients than in normal PBL, and that this _in vitro_ steroid sensitivity increased significantly with the severity of the illness (43). Similarly, it was shown that CLL lymphocytes in contrast to normal PBL were _in vitro_ very sensitive to the cytotoxic action of glucocorticoids (7, 48).

Non Hodgkin's malignant lymphoma (NHML)

Corticosteroids are also included in almost all chemotherapy regimens currently used for the treatment of adult NHML. Recently Bloomfield et al. have attempted to correlate the amount of glucocorticoid receptors in lymphoma cells with their _in vitro_ sensitivity to steroids (25, 49). Investigations were carried out in cell suspensions containing at least 50 % malignant cells. The number of binding sites measured by whole cell assay varied greatly from one patient to another but was on an average higher in cells from patients with NHML than in cells isolated from control lymph nodes (median values are 4110 (n = 42) and 2082 (n = 11) sites per cell respectively). In addition, 'null cell' lymphomas, like in ALL patients, usually contained more receptors than T cell lymphomas. Finally, sequential investigations (at diagnosis and at relapse) showed a decrease of the number of glucocorticoid receptors after treatment.

Despite the high sensitivity of lymphoma cells to the cytotoxic action of glucocorticoids _in vitro_ (almost 50 % of cell lysis after 96 h incubation in the presence of 4×10^{-7}M dexamethasone), no correlation could be found between the level of these glucocorticoid receptors and any parameters of _in vitro_ sensitivity.

On the other hand, the authors have studied in 20 patients the response to therapy with corticosteroids as a single agent. They showed in the group of responders (with a 50 % or more decrease in measurable tumor after corticotherapy) that the number of glucocorticoid receptors was significantly higher than in non-responders (5600 sites versus 3300, $p < 0.01$, $n = 10$, in each group), whereas the dexamethasone-induced inhibition of leucine and uridine incorporation was also more important in the former group.

Sezary syndrome

The Sezary syndrome is characterized by atypical circulating lymphocytes and cutaneous infiltration and may also be improved by steroids (50). Schmidt and Thompson have studied the binding of tritiated dexamethasone in the circulating lymphocytes of seven patients with Sezary syndrome (50). They showed, using a cytosolic assay, the presence of detectable specific receptors in five cases. Although one patient with very high levels of receptors was shown to be dramatically improved by prednisone treatment, whereas one patient with no detectable receptors was refractory to combined chemotherapy, it also appears that two patients with non negligible levels of receptors were not sensitive to combined chemotherapy including glucocorticoids. These authors also proposed that the determination of the activity of the steroid-inducible enzyme glutamine synthetase might represent a useful test for functional glucocorticoid action.

Malignant lymphoid cell lines

Human continuous cell lines represent clonal growth and may thus provide homogeneous material which is difficult to obtain even with very sophisticated separation procedures from patients with hematopoietic malignancies. This advantage led several groups to analyze the expression of specific glucocorticoid receptors in different human leukemic cell lines (19, 51-55). In a recent study Paavonen et al., using a whole cell assay, have reported a wide variation of glucocorticoid receptor content in a panel of phenotypically well defined human leukemic cell lines (54). The number of binding sites ranged from 2200 to 18.100 binding sites per cell, whereas the dissociation constant of the steroid-receptor complexes were very similar in the eleven cell lines ($K_D = 1.5 \times 10^{-8}M$). No clear correlation could be demonstrated between the number of receptors and the phenotypical characteristics of the cells.

In all the studies carried out on human leukemic cell lines of different origin no correlation could be found between the receptor density and the *in vitro* effects of glucocorticoids.

DISCUSSION

The discrepancies seen between the results of various authors may be in part accounted for by differences in the experimental procedures used to determine glucocorticoid receptors and sensitivity of human lymphoid cells but also to the fact that experimental approach was often based on oversimplified assumptions. We will thus review several of these factors.

Selection of the patients

For practical reasons, essentially to collect enough material for receptor determination, the patients to be investigated have been usually selected either on the basis of high white blood cell counts in leukemia or by considering only those patients with more than 50 % of malignant cells in NHML. It should be kept in mind that this selection may well introduce a bias in the interpretation of the results.

Problems of receptor determination

Many of the early studies of receptors were carried out using the classical cytosolic assay, which required cell homogeneization and high speed centrifugation. This method however, presented several disadvantages when compared to the whole cell assay (56). In particular, cytosolic determination leads to a systematic underestimation of the number of binding sites per cell as recently demonstrated by Barrett et al. and Iacobelli and coworkers (33, 57). This could be due to several reasons, such as failure to measure the receptors in the nuclear compartment, loss of binding activity by denaturation or inactivation and even failure to break some cells during homogeneization (47, 55-58). In addition, it was shown that both receptor affinity and specificity may vary according to the type of the assay (57, 59). It should also be mentioned that the use of frozen cells instead of freshly isolated cells may eventually introduce additional variations of the receptor level. Finally, it appears that the whole cell assay is more rapid and requires less material than the cytosolic assay.

Receptor concentration in a given target cell is not constant but varies under the influence of several factors and in particular as a function of the extracellular concentration of steroids. The content of glucocorticoid receptors was shown to be increased following adrenal ablation in heart, liver and thymus (56). Conversely, there is some evidence in favour of a down regulation of these receptors *in vitro* (60). Several authors have also made attempts to demonstrate a similar down regulation of lymphocyte glucocorticoid receptors *in vivo* but these studies remain questionable as they do not take into account other possible steroid action such as sequestration of selected cell population (20). Moreover, it appears that the patients to be investigated should be examined at diagnosis before any corticoid therapy. Indeed, the level of steroid receptors in the cells of previously treated patients, even if they have been withdrawn from steroid treatment 1-2 weeks before, is lower than that of patients at diagnosis (18, 25). Interestingly, sex steroid hormones have also been shown to modulate the glucocorticoid receptor concentration in lymphoid tissue (61, 62), but the role of this modulation awaits more extensive investigations. In addition, glucocorticoid receptor content has been shown to be related to the cell size as well as to the stage of proliferation (19, 63). Several authors demonstrated an increase in receptors following mitogen-induced blast transformation (64, 65), whereas the number of binding sites was shown to increase during the S phase of the cell cycle (63). Therefore, it is necessary to consider that the samples studied represent strongly heterogenous populations containing both normal and malignant cells at different stages of proliferation and in variable proportions from one patient to another. Recently, Bloomfield et al. have attempted to take into account this factor by correcting for the total number of receptors according to the percentage of blast cells (25). This correction, however, which assumes that the non malignant cells have the same number of receptors than the normal lymph-node cells was not clearly validated. It would be, therefore, preferable to separate by any mean available the malignant cells before measuring receptor levels.

Finally, for experimental convenience, most of the studies have been performed using peripheral leucocytes, although it is now well recognized that blood cells do not entirely reflect the proliferative activity of the bone marrow (66, 67). Indeed, comparative studies done in the same patients showed in some cases marked differences between blood and bone marrow samples (67). Bone marrow, which is usually the main focus of the tumoral proliferation may thus represent a more appropriate tissue for the study of glucocorticoid action particularly in leucopenic patients. Furthermore, these possible differences in receptors between blood and bone marrow should be considered when studying the relationship between receptor levels and _in vitro_ or _in vivo_ steroid sensitivity.

Determination of in vitro sensitivity

Studies performed in mouse lymphoma cell lines have demonstrated a fairly good correlation between the level of glucocorticoid receptors and the _in vitro_ action of glucocorticoids (68, 69). Similar investigations were thus carried out in normal and leukemic human lymphoid cells. Several assays have been used to assess the _in vitro_ steroid sensitivity. These include inhibition of glucose transport, inhibition of leucine, uridine or thymidine incorporation, induction of enzyme activity and decrease of cell viability. It appears, however, from these studies that in cell populations with similar number of receptors, corticosteroids may or not induce in vitro an inhibition of cell proliferation and/or cell lysis. Moreover, despite the general assumption that steroid-induced cell death represents a consequence of the catabolic actions of the drug, there was no relationship between the extent of cell lysis and that of metabolic inhibitions as recently emphasized by Young et al. (70). In addition, it was demonstrated in human lymphocytes that the increase of receptors following mitogen treatment was not associated with any significant alteration of steroid inhibitory effects (65).

On the other hand, it should be noted that the inhibitory action of glucocorticoids was more marked in populations characterized by a high level of proliferative activity. It was shown for example that the metabolic effects of glucosteroids are very important in cell populations (human thymocytes, some ALL patients, CLL patients with poor prognosis) characterized by a high proliferative activity on the basis of either high levels of nucleoside incorporation or presence of blood cells in the S phase of the cell cycle.

Kinetic studies of nucleoside incorporation have demonstrated that the in vitro response of a given sample to steroid may vary considerably according to the time of the investigation (67). It thus appears that a single determination of steroid effect on precursor incorporation does not obligatorily represent an adequate indication of the cell sensitivity to the drug.

Finally, recent investigations have shown that the in vitro action of steroids on lymphoid cells may be more complex than previously believed. Smith et al. demonstrated that inhibition of mitogen-induced cell proliferation by steroids represents an indirect action (suppressing the production of a T cell growth factor by a small subset of cells) rather than a direct action at the level of the activated lymphocytes (71).

Evaluation of in vivo sensitivity

Evaluation of the in vivo response to corticoid therapy is even more complex because it represents a combination of different mechanisms including among others inhibition of cell proliferation, and cell lysis but also cell recirculation and sequestration. In addition, it is obvious that in vivo the perturbations caused by steroid treatment on the general hormonal balance, on the vascular functions as well as on the immunological reactions may considerably influence the evolution of the leukemia (31). Therefore, studies of human leukemic cell lines would provide only partial indications on steroid action. Furthermore, in the case of combined chemotherapy it is difficult to evaluate precisely the efficiency of a particular drug. Nevertheless, there are some indications suggesting that the cells of ALL sensitive to short term glucocorticoid therapy tend to contain more receptors than cells of the non responders. These studies however have only be done on a limited number of patients and therefore do not provide definite indications (29, 35).

On the other hand, attempts to relate the number of glucocorticoid receptors to the immunological type of the ALL have shown that in the childhood form of ALL, 'null cell' ALL which are known to have a better response to combined chemotherapy than T cell ALL also contain more binding sites (37, 39, 40). In this regard, the importance of utilizing well characterized hetero-antisera that define T antigens in addition to testing for E-rosettes should be emphasized. Otherwise, many T-derived cells would be classified as 'null cells' on the basis of non E-rosetting (72). Furthermore, Lippman et al. showed that in a given immunological type the induction of remission as well as the length of remission were related to the initial level of receptors (41). Similar findings were also obtained in acute non lymphoid leukemic patients treated with combined chemotherapy but without corticosteroids (73).

It was recently demonstrated that high levels of glucocorticoid receptors were usually associated with the presence of terminal deoxynucleotidyl transferase activity in lymphoid as well as in myeloid keukemic cells (42, 73). Although these results should be confirmed in more extensive studies including careful investigations of immunological markers, they suggest a relationship between the level of receptors and cell differentiation.

Finally, there are several aspects of steroid action which have not been fully explored but may play an important role. Recent findings have shown the existence of a high level of transcortin in patients with lymphatic leukemia and non Hodgkin's lymphoma (74). This abnormality was associated with HLA antigens of the B and C loci. These results, together with the indications suggesting that glucocorticoid receptors and steroid response may be under the control of genetic influence (75, 76), raise the question of the possible role of transcortin in the modulation of glucocorticoid receptors and perhaps cell differentiation ?

On the other hand, several authors have described an increased metabolism of cortisol in malignant lymphoid cells (77-79). It remains to be determined whether this phenomenon may play a role either in the function of glucocorticoid receptors or in the control of cell response ?

1. HN Claman, Corticosteroids and lymphoid cells, New Engl J Med 287: 388 (1972).
2. A Munck and DA Young, Corticosteroids and lymphoid tissue, in: Handbook of Physiology, Section Endocrinology, vol 6 Adrenal Gland, SR Geiger, ed., American Physiological Society, Washington, pp. 231-243 (1975).
3. JF Bach, Corticosteroids, in: The Mode of Action of Immunosuppressive Agents, A Neuberger & EL Tatum, eds., Frontiers of

Biology, vol 41, North Holland Publishing Co, Amsterdam, pp. 21-91 (1975).

4. A Munck and K Leung, Glucocorticoid receptors and mechanisms of action, in: Receptors and Mechanism of Action of Steroid Hormones, Part II, JR Pasqualini, ed., Marcel Dekker, New York, pp. 311-397 (1977).
5. J Caffey and R Sibley, Regrowth and overgrowth of the thymus after atrophy induced by the oral administration of adrenocorticosteroids to human infants, Pediatrics 26: 762 (1960).
6. HN Claman, JW Moorhead and WH Benner, Corticosteroids and lymphoid cells in vitro. I. Hydrocortisone lysis of human, guinea pig and mouse thymus cells, J Lab Clin Med 78: 499 (1971) .
7. U Galili, M Prokocimer and G Izak, The in vitro sensitivity of leukemic and normal leukocytes to hydrocortisone induced cytolysis, Blood 56: 1077 (1980).
8. F Homo and D Duval, Human thymus cells: effects of glucocorticoids in vitro, J Clin Lab Immunol 2: 329 (1979).
9. AS Fauci and DC Dale, The effect of in vivo hydrocortisone on subpopulations of human lymphocytes, J Clin Invest 53: 240 (1974).
10. AS Fauci, T Murakami, DD Brandon, DL Loriaux and MB Lipsett, Mechanisms of corticosteroid action on lymphocyte subpopulations. VI. Lack of correlation between glucocorticosteroid receptors and the differential effects of glucocorticosteroids on T cell subpopulations, Cell Immunol 49: 43 (1980).
11. AS Fauci, Mechanisms of corticosteroid action on lymphocyte subpopulations. I. Redistribution of circulating T and B lymphocytes to the bone marrow, Immunology 28: 669 (1975).
12. R Schreck, Cytotoxicity of adrenal cortex hormones on normal and malignant lymphocytes of man and rat, Proc Soc Exp Biol (NY) 108: 326 (1961).
13. F Homo, D Duval, C Thierry and B Serrou, Human lymphocyte subpopulations: effects of glucocorticoids in vitro, J Steroid Biochem 10: 609 (1979).
14. GR Crabtree, KA Smith and A Munck, Glucocorticoid receptors and in vitro sensitivity of cells from patients with leukemia and lymphoma: a reassessment, in: Glucocorticoid Action and Leukemias, PA Bell and NM Borthwick, eds., Seventh Tenovus Workshop, Alpha Omega Publishing Ltd, Cardiff, pp. 191-204, (1979).
15. ME Lippman and R Barr, Glucocorticoid receptors in purified subpopulations of human peripheral blood lymphocytes, J Immunol 118: 1977 (1977).
16. ME Lippman, RH Halterman, BG Leventhal, S Perry and EB Thompson, Glucocorticoid binding proteins in human acute lymphoblastic leukemic blast cells, J Clin Invest 52: 1715 (1973).
17. P Nanni, C De Giovanni, MC Galli, G Nicoletti, PL Lollini, M Gobbi, S Bartoli and S Grilli, Glucocorticoid binding protein occurrence and glucocorticoid sensitivity in cells from human acute lymphoblastic leukaemia, IRCS Medical Science 8: 624

(1980).

18. AD Ho, W Hunstein and W Schmid, Glucocorticoid receptors and sensitivity in leukemias, Blut 42: 183 (1981).
19. K Kontula, LC Andersson, T Paavonen, G Myllyla, L Teerenhovi and P Vuopio, Glucocorticoid receptors and glucocorticoid sensitivity of human leukemic cells, Int J Cancer 26: 177 (1980).
20. J Schlechte and B Sherman, The glucocorticoid receptor. Regulation by hormone administration, The Endocrine Society, 63rd Annual Meeting, abstract 694 (1981).
21. LA Hansson, SA Gustafsson, J Carlstedt-Duke, G Gahrton, B Högberg and JA Gustafsson, Quantitation of the cytosolic glucocorticoid receptor in human normal and neoplastic leukocytes using isoelectric focusing in polyacrylamide gel, J Steroid Biochem 14: 757 (1981).
22. T Werb, R Foley and A Munck, Interaction of glucocorticoids with macrophages. Identification of glucocorticoid receptors in monocytes and macrophages, J Exp Med 147: 1684 (1978).
23. CW Distelhorst and BM Benutto, Glucocorticoid receptor content of T lymphocytes: evidence for heterogeneity, J Immunol 126: 1630 (1981).
24. J Ahlquist, The adrenal cortex, in: Endocrine Influences on Lymphatic Organs, Immune Responses, Inflammation and Autoimmunity, Almquist and Wiksell International, Stockholm, pp. 18-30 (1976).
25. CD Bloomfield, KA Smith, L Hilde-Brandt, J Zaleskas, KJ Gajl-Peczalska, G Frizzera, BA Peterson, JM Kersey, GR Crabtree and A Munck, The therapeutic utility of glucocorticoid receptor studies in non-Hodgkin's malignant lymphoma, in: Hormones and Cancer, S Iacobelli, ed., Raven Press, New York, pp. 345-359 (1980).
26. FO Ranelletti, M Carmignani, S Iacobelli and P Tonali, Glucocorticoid-binding components in human thymus hyperplasia, Cancer Res 38: 516 (1978).
27. JG Leinen, JL Wittliff, JA Kostyu and RC Brown, Glucocorticoid-binding components in an irradiation-induced thymoma of the C_{57}BL/6J mice, Cancer Res 34: 2779 (1974).
28. FO Ranelletti, S Iacobelli, M Carmignani, G Sica, C Natoli and P Tonali, Glucocorticoid receptors and in vitro corticosensitivity in human thymoma, Cancer Res 40: 2020 (1980).
29. S Iacobelli, P Longo, R Mastrangelo, R Malandrino and FO Ranelletti, Glucocorticoid receptors and steroid sensitivity of acute lymphoblastic leukemia and thymoma, in: Hormones and Cancer, S Iacobelli et al., eds., Raven Press, New York, pp. 371-385 (1980).
30. F Homo, M Papiernik and F Russo-Marie, Steroid modulation of in vitro prostaglandin secretion by human thymic epithelium, J Steroid Biochem (in press).
31. CC BIRD, Clinical classification of leukaemia and lymphoma in relation to glucocorticoid therapy, in: Glucocorticoid Action and Leukaemia, PA Bell and NM Borthwick, eds., 7th Tenovus

Workshop, Alpha Omega Publishing Ltd, Cardiff, pp. 123-142 (1979).

32. KA Foon, RJ Billing, PI Terasaki and MJ Cline, Immunologic classification of acute lymphoblastic leukemia. Implications for normal lymphoid differentiation, Blood 56: 1120 (1980).
33. WL McGuire, An update on estrogen and progesterone receptors in prognosis for primary and advanced breast cancer, in: Hormones and Cancer, S Iacobelli et al., eds., Raven Press, New York, pp. 337-343 (1980).
34. S Galaini, J Minowada, P Silvernail, A Nussbaum, N Kaiser, F Rosen and K Shimaoka, Specific glucocorticoid binding in human hemopoietic cell lines and neoplastic tissue, Cancer Res 33: 2653 (1973).
35. F Homo, D Duval, JL Harousseau, JP Marie and R Zittoun, Heterogeneity of the in vitro responses to glucocorticoids in acute leukemia, Cancer Res 40: 2601 (1980).
36. R Mastrangelo, R Malandrino, R Riccardi, P Longo, FO Ranelleti and S Iacobelli, Clinical implications of glucocorticoid receptor studies in childhood acute lymphoblastic leukemia, Blood 56: 1036 (1980).
37. A Naray, T Revesz, E Walcz, D Schuler and I Horvath, Glucocorticoid receptors in acute leukemia of chidhood, Orv Hetil 121: 3175 (1980).
38. L Danel, P Martin, E Escrich, N Tubiana, D Fière and S Saez, Androgen, estrogen and progestin binding sites in human leukemic cells, Int J Cancer 27: 733 (1981).
39. GS Konior Yarbro, ME Lippman, GE Johnson and BG Leventhal, Glucocorticoid receptors in subpopulations of childhood acute lymphocytic leukemia, Cancer Res 37: 2688 (1977).
40. S Iacobelli, Personal communication.
41. ME Lippman, GS Konior Yarbro and BG Leventhal, Glucocorticoid receptors in normal and leukaemic human leucocytes, in: Glucocorticoid Action and Leukaemia, PA Bell and NM Borthwick, eds., Alpha Omega Publishing Ltd, Cardiff, pp. 175-190 (1979).
42. L Skoog, B Nordenskjöld, A Öst, B Andersson, R Hast, N Giannoulis, S Humla, T Hägerström and P Reizenstein, Glucocorticoid receptor concentrations and terminal transferase activity as indicators of prognosis in acute non-lymphocytic leukaemia, Brit Med J 282: 1826 (1981).
43. F Homo, D Duval, P Meyer, F Belas, P Debré and JL Binet, Chronic lymphatic leukaemia: cellular effects of glucocorticoids in vitro, Brit J Haematol 38: 491 (1978).
44. L Terenius, B Simonsson and K Nilsson, Glucocorticoid receptors, DNA-synthesis, membrane antigens and their relation to disease activity in chronic lymphatic leukemia, J Steroid Biochem 7: 905 (1976).
45. J Stevens, YW Stevens, E Sloan, R Rosenthal and J Rhodes, Nuclear glucocorticoid binding in chronic lymphatic leukemia lymphocytes, Endocrine Res Comm 5: 91 (1978).
46. J Stevens, YW Stevens and RL Rosenthal, Characterization of cy-

tosolic and nuclear glucocorticoid-binding components in human leukemic lymphocytes, Cancer Res 39: 4939 (1979).

47. JC Sloman and PA Bell, Glucocorticoids and myeloid leukaemia, in: Glucocorticoid Action and Leukemia, PA Bell and NM Borthwick, eds., Seventh Tenovus Workshop, pp. 161-169, Alpha Omega Publishing Ltd, Cardiff, pp. 161-169 (1979).

48. R Schreck, Prednisolone sensitivity and cytology of viable lymphocytes as tests for chronic lymphocytic leukemia, J Natl Cancer Inst 33: 837 (1964).

49. CD Bloomfield, KA Smith, BA Peterson, L Hildebrandt, J Zaleskas, KJ Gajl-Peczalska, G Frizzera and A Munck, In vitro glucocorticoid studies for predicting response to glucocorticoid therapy in adults with malignant lymphoma, Lancet 1: 952 (1980).

50. TJ Schmidt and EB Thompson, Glucocorticoid receptors and glutamine synthetase in leukemic Sezary cells, Cancer Res 39: 376 (1979).

51. ME Lippman, S Perry and EB Thompson, Cytoplasmic glucocorticoid-binding proteins in glucocorticoid-unresponsive human and mouse leukemic cell lines, Cancer Res 34: 1572 (1974).

52. MR Norman and EB Thompson, Characterization of a glucocorticoid-sensitive human lymphoid cell line, Cancer Res 37: 3785 (1977).

53. CC Bird, AW Waddell, AMG Robertson, AR Currie, CM Steel and J Evans, Cytoplasmic receptor levels and glucocorticoid response in human lymphoblastoid cell lines, Br J Cancer 33: 700 (1976).

54. T Paavonen, LC Andersson and K Kontula, Lack of correlation between the glucocorticoid receptor density and the in vitro growth-inhibitory effect of dexamethasone in human leukemic cell lines, J Receptor Res 1: 459 (1980).

55. ID Barrett, NS Panesar, CC Bird, AC Abbott, HM Burrow and CM Steel, Human lymphoid cell lines and glucocorticoids: II. Whole cell and cytoplasmic binding properties of lymphoblastoid, leukaemia and lymphoma lines, Diagnostic Histopathology 4: 189 (1981).

56. D Duval and F Homo, Prognostic value of steroid receptor determination in leukemia, Cancer Res 38: 4263 (1978).

57. S Iacobelli, V Natoli, P Longo, FO Ranelletti, G de Rossi, D Pasqualetti, F Mandelli and R Mastrangelo, Glucocorticoid receptor determination in leukemia patients using cytosol and whole cell assays, Cancer Res (in press).

58. EB Thompson, Report on the international union against cancer, Workshop on Steroid Receptors in Leukemia, Cancer Treatment Reports 63: 189 (1979).

59. D Duval and J Simon, Temperature-dependent changes in specificity of glucocorticoid receptors in mouse thymocytes, in: Multiple Molecular Forms of Steroid Hormone Receptors, MK Agarwal, ed., Elsevier/North Holland, Amsterdam, pp. 229-243 (1977).

60. F Svec and M Rudis, Glucocorticoids regulate the glucocorticoid

receptor in the A_tT-20 cell, J Biol Chem 256: 5984 (1981).

61. DB Endres, RJ Milholland and F Rosen, Sex differences in the concentrations of glucocorticoid receptors in rat liver and thymus, J Endocr 80: 21 (1979).

62. P Coulson, D Skafar, S Seaver and J Thornthwaite, Dihydrotestosterone modulation of glucocorticoid receptor levels in thymus and bursa of Fabricius cells in immature chickens, The Endocrine Society, 63rd Annual Meeting, Abstract 692 (1981).

63. GR Crabtree, A Munck and K Smith, Glucocorticoids and lymphocytes. II. Cell cycle-dependent changes in glucocorticoid receptor content, J Immunol 125: 13 (1980).

64. JP Neifeld, ME Lippman and DC Torney, Steroid hormone receptors in normal human lymphocytes. Induction of glucocorticoid receptor activity by phytohemagglutinin, J Biol Chem 252: 2972 (1977).

65. KA Smith, GR Crabtree, SJ Kennedy and AU Munck, Glucocorticoid receptors and glucocorticoid sensitivity of mitogen stimulated and unstimulated human lymphocytes, Nature 267: 523 (1977).

66. AM Mauer and O Fisher, Comparison of the proliferative capacity of acute leukemia cells in bone marrow and blood, Nature 193: 1085 (1962).

67. F Homo, S Durant, D Duval, JP Marie, R Zittoun and JL Harousseau, In vitro hormonal responsiveness of human blood and bone marrow cells in non-lymphocytic leukemia, Leukemia Res 4: 619 (1980).

68. AF Kirkpatrick, RJ Milholland and F Rosen, Stereospecific glucocorticoid binding to subcellular fractions of the sensitive and resistant lymphosarcoma P1798, Nature New Biol 232: 216 (1971).

69. W Rosenau, JD Baxter, GG Rousseau and GM Tomkins, Mechanism of resistance to steroids: glucocorticoid receptor defect in lymphoma cells, Nature New Biol 237: 20 (1972).

70. ML Nicholson and DA Young, An effect of glucocorticoid hormones in vitro on the structural integrity of nuclei in corticoid-sensitive and resistant lines of lymphosarcoma P1798, Cancer Res 38: 3673 (1978).

71. S Gillis, GR Crabtree and KA Smith, Glucocorticoid-induced inhibition of T cell growth factor production. I. The effect on mitogen-induced lymphocyte proliferation, J Immunol 123: 1624 (1979).

72. KA Foon, RJ Billing, PI Terasaki and MJ Cline, Immunologic classification of acute lymphoblastic leukemia implications for normal lymphoid differentiation, Blood 56: 1120 (1980).

73. Y Nakao, S Tsuboi, T Fujita, T Masaoka, S Morikawa and S Watanabe, Glucocorticoid receptors and terminal deoxynucleotidyl transferase activities in leukemic cells, Cancer 47: 1812 (1981).

74. P de Moor and A Louwagie, Association of aberrant transcortin levels with HLA antigens of the B and C loci: high transcortin levels are frequently found in patients with lymphatic leuke-

mia, Hairy cell leukemia, or non-Hodgkin lymphoma, J Clin Endocrinol Metab 51: 868 (1980).
75. B Becker, DH Shin, PF Palmberg and SR Waltman, HLA antigens and corticosteroid response, Science 194: 1427 (1976).
76. MS Butley, RP Erickson and WB Pratt, Hepatic glucocorticoid receptors and the H-2 locus, Nature 275: 136 (1978).
77. AD Forker, RE Bolinger, JH Morris and WE Larson, Metabolism of cortisol-C^{14} by human peripheral leucocyte cultures from leukemic patients, Metabolism 12: 75 (1963)1.
78. JS Jenkins and NH Kemp, Metabolism of cortisol by human leukemic cells, J Clin Endocr 29: 1217 (1969).
79. A Klein, H Kaufmann, S Mannheimer and H Joshua, Cortisol metabolism in lymphocytes from cancer-bearing patients, Metabolism 27: 731 (1978).

IMMUNOCYTOCHEMICAL DETECTION OF GLUCOCORTICOID RECEPTORS IN LYMPHOCYTES AND BREAST TUMOR CELLS

Michael Papamichail, Constantine Ioannidis, Niki Agnanti, John Garas, Nikos Tsawdaroglou, Athena Laventakou, and Constantine E. Sekeris

Hellenic Anticancer Institute, Athens (Greece) and Biological Research Center, National Hellenic Research Foundation, Athens (Greece)

INTRODUCTION

The action of glucocorticoisteroids on target cells is thought to be mediated by the interaction of the steroid with a specific receptor molecule localized in the cytoplasm of the cell, followed by receptor activation and translocation of the steroid-receptor complex into the nucleus (1). The association of the hormone-receptor complex with nuclear structures initiates in a yet unknown way, the events which lead to the expression of the hormonal effects.

To date, all the methods employed to quantitate steroid receptors rely on the ability of these proteins to form complexes with labelled steroids. Most of these methods, however, suffer from a series of limitations, which we have attempted to overcome by an immunocytochemical approach, using an antibody against the glucocorticoid receptor (2). Applying an indirect immunofluorescence technic we have demonstrated the presence of glucocorticoid receptors in the cytoplasm of PHA-stimulated lymphocytes (3) and the translocation of the receptor into the nucleus in the presence of dexamethasone (3). In addition, using an immunoperoxidase method, we have confirmed the presence of glucocorticoid receptors in breast tumors (4) and further demonstrated that most of the malignant breast tumors and intermediate situations (atypical duct or lobular hyperplasia, papillomatosis, etc.) showed positive staining reactions, whereas benign conditions, such as fibrocystic disease simple were negative. These findings suggested that glucocorticoid receptors, as assayed by the immunoperoxidase method, could be useful biological markers to detect early conversion of mormal to hyperplastic tissue and/or malignancy of the mammary gland (4).

EXPERIMENTAL PROCEDURES

Preparation of cytosol and nucleosol (3,5). Lymph cells were disrupted in a glass homogenizer in hypotonic medium with a Teflon pestle and the homogenate contrifuged 1000xg, 3 min. The sediment, representing the crude nuclear pellet, was further purified (see below). The supernatant was submitted to a centrifugation at 105000xg, 60 min to yield the cytosol, which was made 20% in respect to glycerol (v/v). The crude nuclear pellet was then purified in principle according to Chauveau et al., as described in detail in (7). The nuclear pellet thus obtained was washed twice in 50mM tris-HCl, 250 mM sucrose, pH 7.5, containing 25 mM KCl, 2-mercaptoethanol + 10 mM $MgCl_2$ and finally resuspended in 20% glycerol, 50 mM KCl, 10 mM tris-HCl, pH 7.4 and 5 mM 2-mercaptoethanol. The nuclear suspension was then treated with micrococcal nuclease (5 U/OD_{260}) for 20 min at 25° in the presence of 1 mM $CaCl_2$ and the preparation centrifuged 10000xg, 10 min (2). The supernatant obtained represents the nucleosol. As shown by Tsawdaroglou et al., (2) this treatment release 1,5 times more receptor activity from the nuclei than extraction with 0.3 M KCl.

Lymphocyte blast cells and treatment with dexamethasone. Human mononuclear cells were isolated from peripheral blood with Lymphoprep (Nyegaard, Oslo) according to the manufacturer's instructions and subsequently cultured in RPMI 1640 (Gibco Laboratories) (3) and supplements. The cells were incubated at 37° with PHA or dexamethasone (10^{-7}M) when appropriate.

Purification of glucocorticoid receptor and antibody production. The method used was based in principle on that described by Govindan and Sekeris (6) for the purification of glucocorticoid receptors from rat liver, with some modifications (2). In brief, rat thymus cytosol was passed through a column of CH-Sepharose on which 21-dexamethasone sulfonate was linked through a disulfide bond and the receptor-dexamethasone complex was eluted from the column by cleaving the disulfide bond with 2-mercaptoethanol. The hormone-receptor complex was then submitted to DEAE-cellulose chromatography and was resolved into three dexamethasone binding fractions. One of the fractions eluting from the column with 0.13 M_4NH Cl was shown to be composed of mainly one protein with a M_r 45000. This protein was eluted after SDS-acrylamide gel electrophoresis from the gel and used for immunization of rabbits by the method of Vaitukaitis *et al.*, (7). The IgG fraction was then isolated from the rabbit serum (8).

^{3}H-dexamethasone binding assay. The binding of dexamethasone to cytosol obtained from breast tissues or to lymphocyte fractions was performed according to Beato and Feigelson (9).

Immunoperoxidase method. A triple bridge immunoperoxidase method was applied as outlined by Goldenberg et al. (10) in 4-5 μ thick sections from surgical specimens fixed in 10% formol and embedded in paraffin. Staining was considered positive on the basis of the brownish color always by reference to control sections. Further classification was made on the basis of the intensity of staining, irrespective of the presence of non-stained cells in the tissue sections.

Immunofluorescence staining. Smears of cytocentrifuge preparations were fixed in ethanol/acetic acid, 95/5 (v/v). The smears were stained with rabbit IgG against the receptor, followed by incubation with fluorescein-isothiocyanate-conjugated sheep anti-rabbit IgG (3). Preimmune rabbit IgG was always used in parallel as control.

RESULTS

Detection of glucocorticoid receptors in human lymphocytes

The presence and intracellular distribution of glucocorticoid receptors was studied biochemically in freshly prepared human lymphocytes and in PHA-stimulated cells, which, as known, show increased content of glucocorticoid receptors (11,12). In freshly prepared cells glucocorticoid receptor content, particularly in the nucleus, is low (Figure 1). In 24 hours PHA-treated cells, an increase in glucocorticoid receptors is evident both in the nucleus and in the cytoplasm, either expressed as binding activity per mg protein or on a per cell basis (4 fold and 2 fold increase, respectively) (Figure 1).

Very similar findings were obtained using the indirect immunofluorescence method. In unstimulated cells, staining is restricted to a faint ring surrounding the nucleus, corresponding to the cytoplasm (3). Twentyfour hours after the addition of PHA, the intensity of fluorescence increases both in the cytoplasm and in the nucleus (Figure 2a). On the third day, nuclear fluorescence is faint, contrasting to the bright cytoplasmic staining (Figure 2b).

The movement of the glucocorticoid receptor from the cytoplasm into the nucleus in 24 hours blasts has obviously taken place in the absence of glucocorticoids as steroids possibly present in the serum of the culture medium have been eliminated by charcoal treatment of the serum, which we routinely performed.

Under these conditions an almost complete disappearance of cytoplasmic antigenic sites is observed in more than 90% of the cells, with concomittant, although faint, appearance of fluorescence in the nucleus, supporting previous biochemical findings (2).

Detection of glucocorticoid receptors in breast tumors by the immunoperoxidase staining

Glucocorticoid receptors were assayed by the immunoperoxidase method in various malignant and benign tumors, such as infiltrating duct and lobular carcinomas, fibrocystic disease simple and complex and fibroadenoma with duct hyperplasia (4). The results are shown in Figure 3-6 and Table I.

A typical case of infiltrated duct carcinoma is shown in Figure 3 stained by the immunoperoxidase method. Sections stained with the antibody to the receptor were always compared with those treated with preimmune rabbit gamma globulin (Figure 4). Positivity was graded according to the intensity of stained cells, independent of the relations of stained to unstained cells. Lobular carcinoma, as well as other histological types of breast tumors, showed similar staining reactions.

In contrast to breast carcinomas all the preparations from fibrocystic disease simple were negative by the immunoperoxidase method (Table I).

We have examined 23 cases with varying degree of mammary hyperplasia, including fibrocystic disease complex and fibroadenomas. Of these, 20 cases were found positive with the immunoperoxidase method (Table I). In Figure 5 some hyperplastic areas are shown which show positive reaction with the immunoperoxidase method and in Figure 6 the respective control section treated with preimmune gamma globulin.

In some carcinomas certain malignant cells were more positive than others in the same section. We could not find a correlation between the intensity of staining and the histological grade of the tumor, the lymphatic infiltration of the tumor, the lymph node metastasis or the associated fibrocystic disease.

Using the ^{3}H-dexamethasone binding assay and considering amounts of receptor above 5 fmol/mg protein as positive only 40% of the carcinomas and 21% of fibrocystic disease simple were positive (data not shown) whereas with the immunoperoxidase method almost all carcinomas were positive and all cases of fibrocystic disease were negative (one case was considered ±). Frequently, the values obtained by the biochemical assay could not be correlated with the intensity of the immunoperoxidase staining reaction.

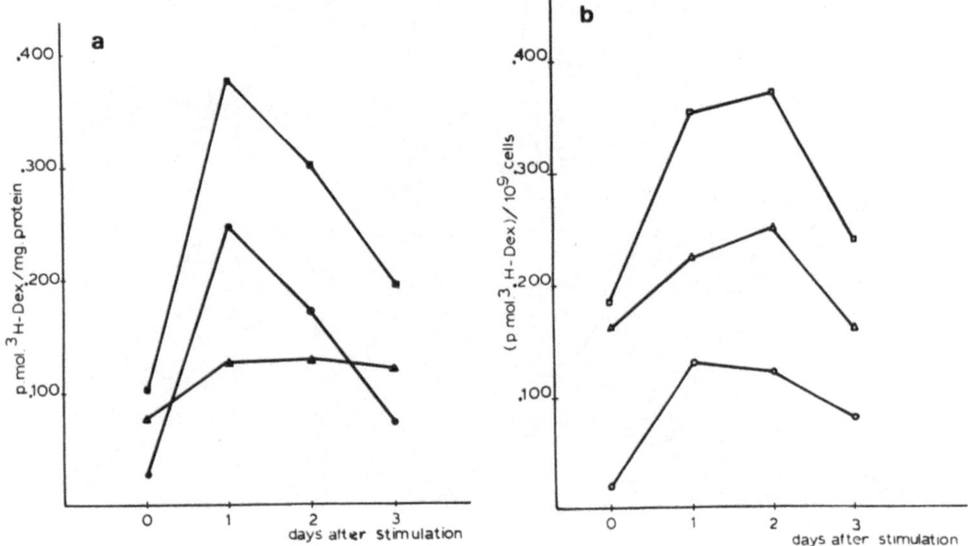

Fig. 1. Distribution of ^{3}H-dexamethasone binding activity between cytosol and nucleosol of PHA-stimulated human lymphocytes.

(a) ^{3}H-dexamethasone binding activity expressed as pmol/mg ● protein activity in nucleosol ▲ activity in cytosol ■ total activity

(b) ^{3}H-dexamethasone binding activity expressed as pmol/10^{9} cells ○ activity in nucleosol △ activity in cytosol □ total activity (Data from Ref. 5).

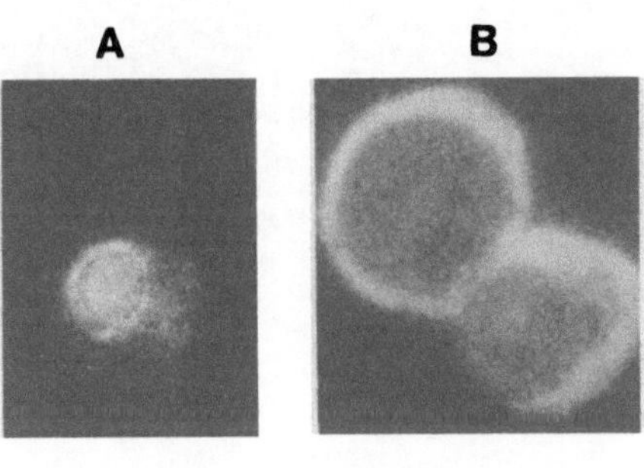

Fig. 2. Intracellular distribution of glucocorticoid receptor in PHA-stimulated human lymphocytes. Human mononuclear cells were treated with PHA for (A) 24 h and (B) 72 h. Smears of the cytocentrifuged preparations were submitted to immunofluorescence staining. X1300 (Data from Ref. 5).

In addition to the translocation of the receptor in the absence of hormone ("automatic") the receptor can be induced to translocate by dexamethasone treatment of the PHA-induced blasts, under conditions known to promote the entrance of the hormone into the nucleus.

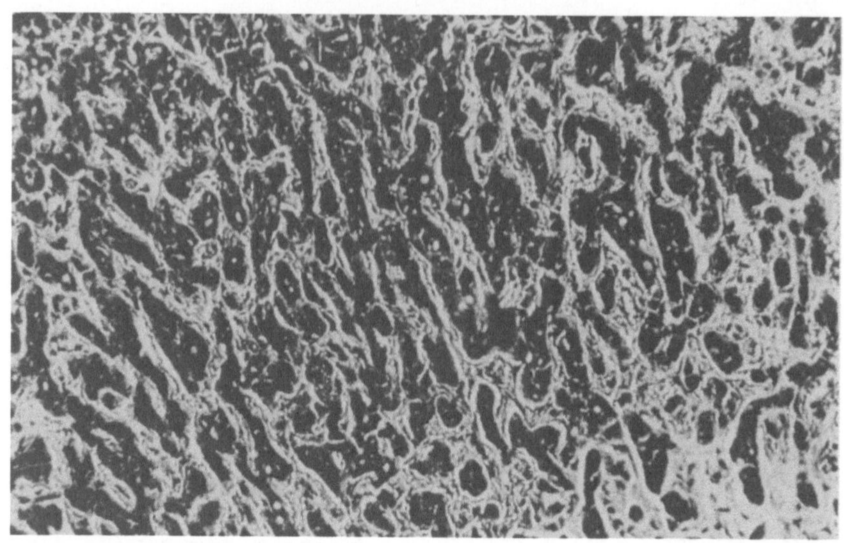

Fig. 3. Infiltrated duct carcinoma stained by the immunoperoxidase method. First layer is antibody to the glucocorticoid receptor (x1000) Data from (4).

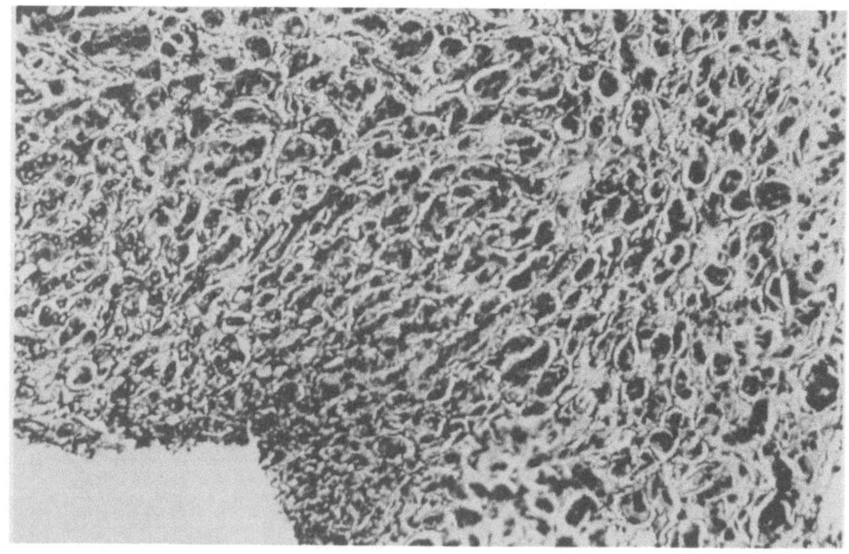

Fig. 4. Infiltrated duct carcinoma stained by the immunoperoxidase method. First layer is preimmune rabbit gamma globulin (x1000) Data from (4).

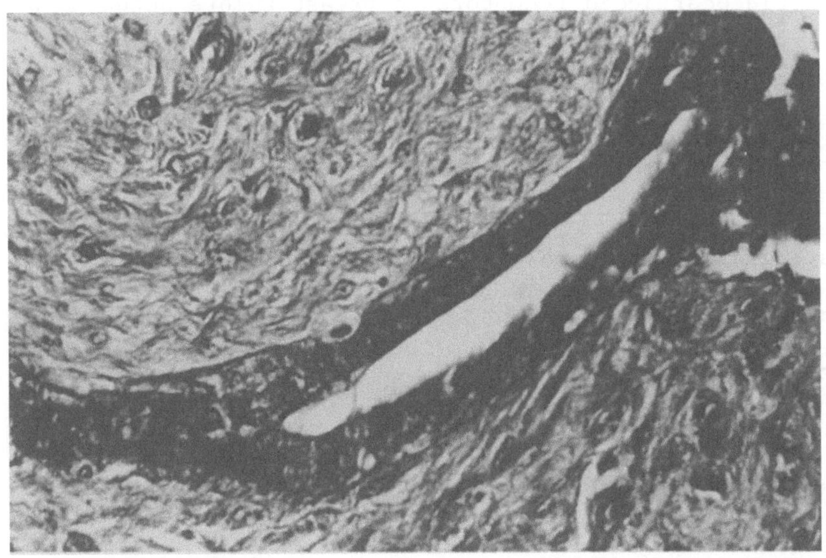

Fig. 5. Hyperplastic areas of breast stained by the immunoperoxidase method. First layer is antibody to the glucocorticoid receptor (x1000) Data from (4).

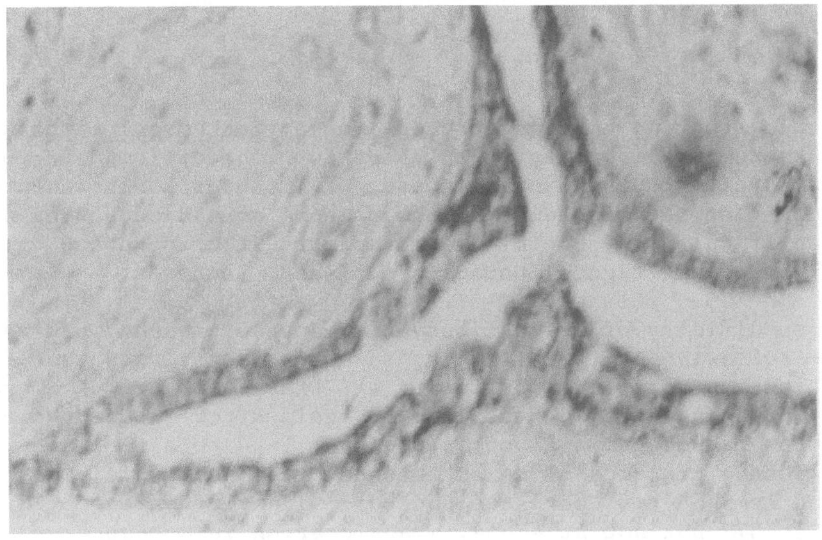

Fig. 6. Hyperplastic areas of breast stained by the immunoperoxidase method. First layer is preimmune rabbit gamma globulin (x1000) Data from (4).

Table I. Glucocorticoid receptors in breast tumors detected by the immunoperoxidase method

Material	No. of cases	Intensity of immunoperoxidase staining				
		-	±	+	++	+++
Infiltrating duct carcinoma	31		1	2	8	20
Intraductal carcinoma	8		1	4	2	1
Infiltrat. lobular carcinoma	9			1	2	6
Lobular carcinoma in situ	5			1	4	
Fibrocystic disease simple	14	13	1			
Fibrocystic disease complex (high risk group)	11		1	2	2	6
Fibroadenoma with duct hyperplasia	12		2	1	7	2

Data from (4)

DISCUSSION

In this study we have introduced immunocytochemical methods made possible by the availability of antibody to glucocorticoid receptor as an alternative to the biochemical assay of glucocorticoid receptors. It is known that the biochemical methods have some disadvantages, mainly the difficulty in determining receptors either occupied by endogenous hormone or tightly bound to nuclear structures.

The specificity of the antibody used in the present work was based on the following criteris: 1. The receptor preparation used for immunization was purified by affinity and DEAD-cellulose chromatography and eluted as a single band from SDS-gels after electrophoresis 2. Preimmune rabbit gamma globulin does not lead to a positive reaction 3. In the presence of dexamethasone cytoplasmic staining of PHA-stimulated lymphocytes decreases and concomitantly increases in the nucleus, at 37° but not at 4° 4. Translocation is not observed with steroids which bind to the receptor but do not activate it (e.g. progesterone) 5. The antibody inhibits binding of ^{3}H-dexamethasone to rat thymocyte cytosol and precipitates already fromed ^{3}H-dexamethasone receptor complexes (2). This is not seen with preimmune gamma blobulin.

In this study both an immunofluorescence and an immunoperoxidase technic were applied (3,4,5). The immunofluorescence method was used mainly on isolated cell populations, i.e. human peripheral lymphocytes. In resting lymphocytes only a faint fluorescence is observed, whereas in PHA-induced blasts, which have increased amounts of receptor (11,12), a bright fluorescence is evident. An interesting observation in this series of experiments was that 24 hours after PHA-stimulation glucocorticoid receptor, initially restricted to the cytoplasm, translocates into the nucleus in the absence of hormone, suggesting that "activation" of the receptor can be induced by agents other than the steroid hormone (5).

The immunoperoxidase method was applied for the detection of glucocorticoid receptors in breast tumors (4). Most breast cancers and intermediate conditions, such as atypical duct and lobular hyperplasias, were positive. In contrast, breast tissue obtained from fibrocystic disease simple was negative for glucocorticoid receptors. With the biochemical assay, only 40% of the breast cancers showed more than 5 f moles dexamethasone bound per mg protein, which was arbitrarily considered as denoting positivity. The discrepancy between the immunocytochemical and the biochemical findings could be due to the fact that few malignant foci in a given cancer specimen will probably give false negative results with the biochemical assay. Furthermore, some of the breast cancers were found to be composed of a heterogeneous population of cancer cells, differing in intensity of staining. Finally, as already mentioned, the biochemical method measures only the available, easily solubilized receptor sites, not occupied by endogenous steroid. We feel, therefore, that the immunoperoxidase method offers certain advantages over the biochemical method.

One interesting point concerns some intermediate situations, such as fibroadenoma with duct hyperplasia and fibrocystic disease complex, where a positive reaction with the immunoperoxidase method was observed. This finding may have clinical application in the selection of patients with high risk in developing breast cancer.

In conclusion,our results demonstrate the usefulness of immunocytochemical methods in detecting functionally active glucocorticoid receptors in different cell types as well as in studying some of the biological functions of the receptor at the cellular level. The detection of glucocorticoid receptor by the immunoperoxidase method in malignant breast tumors and some hyperplastic conditions, in contrast to the negative findings in benign conditions, suggest that glucocorticoid receptors could be useful biological markers to detect early conversion of normal to hyperplastic tissue and/or malignancy of the mammary gland.

REFERENCES

1. Raspe, G., Advances in the Biosciences (1971) Vieweg Verlag and Pergamon Press, Braunschweig and Oxford.
2. Tsawdaroglou, N., Govindan, M.V., Schmid, W., and Sekeris, C.E., Europ. J. Biochem., 114 (1981), 305.
3. Papamichail, M., Tsokos, G., Tsawdaroglou, N., and Sekeris, C.E., Exp. Cell Res., 125 (1980), 490.
4. Ioannidis, C., Papamichail, M., Agnanti, N., Garas, J., Tsawdaroglou, N., and Sekeris, C.E., Int. J. Canc., 29 (1982), 147.
5. Papamichail, M., Ioannidis, C., Tsawdaroglou, N., and Sekeris, C.E., Exp. Cell Res., 133 (1981), 461.
6. Govindan, M.V. and Sekeris, C.E., Europ. J. Biochem. 89 (1978), 95.
7. Vaitukaitis, J., Robbins, T.B., Nieschlag, E., and Ross, G.T., J. Clin. Endocr. 33 (1971), 988.
8. Livingston, D.M., in Methods in Enzymology, Vol. 34 (1974), 723, (N.O. Kaplan and S.P. Colowick, eds.), Academic Press, New York.
9. Beato, M. and Feigelson, P., J. Biol. Chem. 247 (1972), 7890.
10. Goldenberg, D.M., Sharkey, R.M., and Primus, F.J., J. Nat. Canc. Inst. 57 (1976), 11.
11. Smith, K.A., Crabtree, G.R., Kennedy, S.J., and Munck, A.U., Nature 267, (1977), 523.
12. Neifeld, J.P., Lippman, M.E., and Tormey, D.C., J. Biol. Chem. 254, (1977), 2972.

OESTROGEN AND PROGESTIN RECEPTORS AS MARKERS FOR THE BEHAVIOUR OF HUMAN BREAST TUMOURS

R. J. B. King

Hormone Biochemistry Department
Imperial Cancer Research Fund
P. O. Box 123, Lincoln's Inn Fields
London WC2A 3PX, U.K.

INTRODUCTION

It has now been established that most actions of steroid hormones are mediated by intracellular protein receptors. The steroid combines with an extranuclear receptor and the complex translocates to the nucleus where further biochemical changes are initiated. The presence of this receptor machinery is characteristic of cells capable of responding to that steroid whereas the absence of detectable receptor indicates hormone insensitivity[1]. These principles have been applied clinically to determine the hormone sensitivity of tumours of the endometrium[2,3], prostate[4,5], kidney[6], white blood cells[7,8] and breast[9]. It has also been suggested that oestrogen receptor (RE) analysis may be useful for tumours like malignant melanoma[10], pancreas[11], ovary[12] and colorectal cancer[13]. It should be stressed that the RE levels are very low in the latter types of tumour and their prognostic significance has not been proven. Most data are available for breast cancer and this article will be devoted to that topic.

The main initial treatment for primary breast cancer is surgery. Depending on the stage of the disease at first presentation, recurrence can occur at times varying from a few months to many years[14] and frequently occurs in sites like brain, liver or bone that make further surgery impossible; medical treatment is thus prescribed. Various forms of hormone therapy was, until the advent of receptor assays and of chemotherapy, the routine treatment. Premenopausal women were usually ovariectomized to remove the source of mitogenic oestrogenic steroids; post-menopausal women were given a number of different hormonal treatments such as hypophysectomy, adrenalectomy, androgens, high doses of oestrogens, antioestrogens, glucocorticoids or progestins. With the exception of the ablative procedures, the hormone regimes were well-tolerated, especially with antioestrogens like Tamoxifen which is now the main form of additive treatment. Unfortunately, only about one third of patients respond to hormone treatment and it is important to distinguish the responsive from the unresponsive tumours. The latter group can then be directed immediately to chemotherapy which cannot be given routinely to all patients because of multiple side-effects.

METHODOLOGY

Receptors for steroid hormones are high molecular weight proteins that reversibly bind steroids with Kd's varying from 10^{-10}M (oestradiol) to 10^{-8}M (progesterone)[15]. They readily aggregate and debate proceeds as to the biological significance of the various physical forms of receptor[16,17]. There are about 10^4 receptor molecules/responsive cell which is equivalent to about 100 f mol/mg tissue protein which, in theory, could be detected with a specific ligand labelled with either tritium or a fluorescent probe. In practice, most data have been obtained with tritiated steroids although recent interest and controversy have been directed at both fluorescent labelled steroids and the use of unlabelled steroids subsequently recognised by antibodies to that steroid. The main reason that so much interest has been directed at fluorescent steroids and antibodies to steroids which can then be monitored by an enzyme-labelled second antibody is the urgent requirement for a histochemical method of detecting receptors. Unfortunately, none of the methods thus far published satisfy the criteria required of an assay that purports to detect receptors rather than a compound(s) indirectly related to the receptor[18-21]. Thus, with the possible exception of the very new methods utilising antibody to the receptor itself, autoradiography is the only method capable of detecting receptors histologically[22]. However, autoradiography is not practical at a clinical level because of its relative insensitivity and the long exposure times required (up to six months). Very recently, antibodies have been raised against several receptor proteins[23-26] which should have a large impact on methodology. The importance of these antibodies cannot be underestimated but because of the current paucity of data, they will not be discussed here.

Most assays used in clinical practice measure receptor present in the cytosol fraction after a period of labelling at 0-4° with ^{3}H ligand. Because of the high affinity of oestradiol for its receptor, such assay conditions only quantitate receptor unoccupied by endogenous oestradiol, a point of importance for premenopausal samples[27]. On the other hand, progesterone has a low affinity for its receptor (RP) so endogenous steroid exchanges with added ligand even at low temperatures[27].

As steroids bind to many molecules apart from their specific receptors, the latter must be distinguished from the former. All of the methods utilise the high affinity and specificity of the receptor protein as a means of distinguishing them from the low affinity and specificity of the non-specific sites. It is this feature of the assay design that proves most variation amongst the published methods. The majority of assays use dextran-coated charcoal to separate specifically bound ligand from the free and weakly associated material although several other methods have been utilised[28-31]. It is usual to include tubes containing an excess of unlabelled compound (diethylstilboestrol for RE, R5020 for RP) that will compete with the ^{3}H ligand for specific sites and thus correct for residual non-specific binding. The other major methodological variable is the use of a single, saturating concentration of ^{3}H ligand as opposed to multiple concentration assays[32]. The latter are more demanding of time and material but do provide an estimate of binding affinity which the single concentration methods do not. Because of these methodological variations, a number of quality control programmes have been set up to compare results obtained by the different methods[33,34].

Table 1. Objective remissions in advanced breast cancer

Hormone Therapy	RE+	RE-
Castration	44/87 *	1/68 *
Adrenalectomy	89/171	6/65
Hypophysectomy	28/58	0/17
Oestrogens	47/84	0/35
Androgens	14/84	0/22
Aminoglutethimide	19/38	1/7
Antioestrogens	44/89	3/42
Total	285/519 = 55%	11/256 = 4%

Correct diagnosis in <u>68%</u> of cases

* No. patients responding/no. patients treated

Data from Consensus Meeting, 1979. Articles published in ref. 9

CLINICAL RESULTS

Advanced disease

The prognostic value of RE analysis in advanced breast cancer is now of proven clinical value (Table 1). RE negative (RE-) tumours rarely respond to any form of endocrine therapy whereas over half of the RE positive (RE+) tumours respond to such treatments. These data posed the question of why some RE+ tumours did not respond. Many reasons have been advanced[31,35] and it is probable that multiple causes exist. I will deal with what I think are the more important ones.

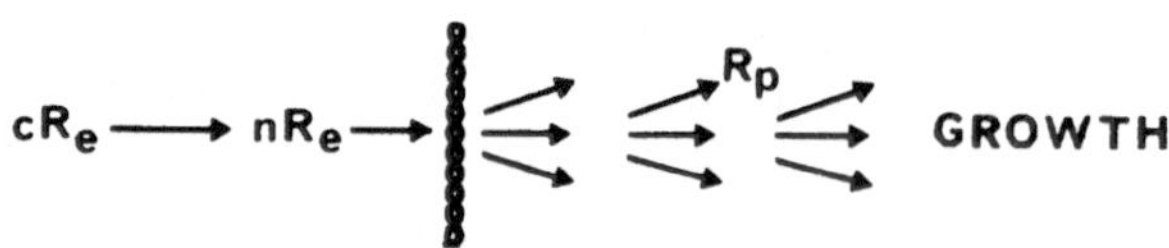

Fig. 1. Model for oestrogen-stimulated growth of breast tumour cells. cRe = cytosol oestradiol receptor; nRe = nuclear oestradiol receptor; Rp = progesterone receptor; [chromatin symbol] = chromatin.

Progesterone receptor. The experimental basis for using RE is depicted in Fig. 1. By analysing only the first component (RE) of the chain of events leading to cell division, one might be missing more distal defects in the reaction sequence. It was therefore suggested that an oestrogen-sensitive product like RP be also assayed[36]. This has proven to be useful (Table 2). The response rate decreases in the order RE+ RP+ > RE+ RP- > RE- RP-. The small number of RE- RP+ tumours are difficult to categorise in clinical terms. Clearly, identification of the poorly responsive

RE+ RP- tumours is helpful but we now need to know why approximately one fifth of such tumours do respond to treatment. It has been suggested that the responders are predominantly postmenopausal and that such tumours are in fact RE+ RP+ but, because of the low oestrogenic environment RP has not been stimulated to detectable levels[37,38].

Nuclear oestradiol receptor. The logic behind the assay of this receptor is analogous to that for RP and the limited clinical data point to its potential use[39,40].

Tumour heterogeneity. Several histological types of breast cancer exist although somewhat surprisingly all but one (medullary carcinoma with lymphoid infiltration) show similar distributions of receptor[41,42]. However, more subtle differences probably exist that are not evident at the histological level. If a tumour is composed of variable numbers of responsive and unresponsive cells then the cell proportions could determine its biological behaviour[43]. Logically, one might suppose that such cellular heterogeneity would be reflected in the quantity of receptor per unit weight of tumour. The observed wide range of receptor values is consistent with this view but cannot be so interpreted at the present time because we cannot disprove the alternative explanation that some responsive cells have more receptor than others. The figure of 100 f mol RE/mg protein mentioned above is an approximation derived from oestrogen-sensitive organs like rat uterus. With human breast tumours, we have no data to say how much or how little RE is required for oestrogen-mediated growth. Only two oestrogen-sensitive, human breast cancer cell culture

Table 2. Objective remissions in advanced breast cancer

	E+ P+	E+ P-	E- P-	E- P+
Remissions †	87/113	33/121	12/111	6/13 *
	77%	27%	11%	46%

Correct diagnosis in 78% of cases (E- P+ not included)

* No. patients responding/no. patients treated

Data from Consensus Meeting, 1979. Articles published in ref. 9

† Type of treatment not specified

systems are available, one of which (MCF-7 cells) shows variable proliferative responses to oestradiol[44]. These systems have not therefore been helpful in determining the relationship between receptor numbers and proliferation response.

Cut-off points. Regardless of whether variability in receptor numbers reflect heterogeneity in cell types or receptors per cell, the relationship between receptor numbers and clinical response is an important one. Data from different centres are generally in agreement that response rate increases with RE content but there is no consensus as to the best cut-off value for defining patients with a > 90% probability of failing endocrine therapy[9]. The majority of laboratories agree that tumours with low but detectable levels of RE (< 10 f mol/mg protein) do not respond to therapy.

Definition of response. Most definitions are similar to that published by the UICC that requires a decrease of at least 50% in the product of two tumour diameters before being classified as a response[45]. This requires that about 90% of the tumour cells be killed so one could get an appreciable endocrine effect in an RE+ tumour that would be classified as a non-response.

Histological features. The differentiation state (histological grade) of the tumour has a marked effect on both RE and RP phenotype[37,46]. As Grade 1 tumours (well-differentiated) have a better prognosis than the more anaplastic Grade 3 neoplasms[47] it is important to determine whether or not histological grade and receptor phenotype are independent variables. Preliminary data suggest some overlap but that both variables provide useful information[48].

The elastic tissue content (elastosis) of the tumour also has prognostic value and tumours with appreciable elastic tissue tend to be RE+[49]. Limited data indicate that RE+ tumours with high elastosis respond more frequently to endocrine therapy than do those with no elastosis[50].

Chemotherapy. It is now accepted dogma that RE- tumours should receive chemotherapy rather than endocrine therapy. It was therefore of special interest when reports appeared suggesting that RE- tumours responded to chemotherapy more frequently than the RE+ tumours[51]. However, the consensus view is now that both phenotypes respond equally well to chemotherapy[9].

Early disease

Receptor analysis on primary tumours has predictive application in at least three areas, response of subsequent metastases, response to adjuvant endocrine therapy and disease-free interval.

Subsequent metastases. Tissue is available for analysis from all except the smallest primary tumours whereas subsequent metastases are often inaccessible. Thus it is reasonable to obtain and store information on the primary tumour for later use. This philosophy is dependent on the constancy of tumour phenotype over long periods. Changes do occur with estimates of those changes varying from 10-30% (Table 3). This implies that the predictive value of the primary analysis will be less efficient than that of the metastasis and this appears to be the case (Table 4). Responses were largely confined to the RE+ tumours although the predictions were less accurate than for the advanced disease (compare Tables 1 and 4). The RE-tumours were equally correct in the early and advanced groups. The main clinical conclusion from these data is that the primary tumour does provide useful information but that where practical a sample of the metastasis should also be analysed.

Response to adjuvant endocrine therapy. Tumours which, at first presentation have already spread to the lymph nodes are increasingly being treated immediately after surgery with

Table 3. Analysis of tumours from the same patient at different times

Receptor Phenotype	% of total tumours	
	RE	RP
+ → +	53	38
- → -	20	33
+ → -	17	19
- → +	10	10
% constant phenotype	73	71
no. analyses	104	42

King et al., unpublished data

Table 4. RE analysis of primary tumours. Response of subsequent metastases

Group	Reference	RE+	RE-
Blamey	52	13/30 *	5/27 *
De Sombre	53	9/13	1/21
King	54	24/77	2/56
Osborne	55	22/44	1/6
Total		68/164	9/110
		41%	8%

Correct diagnosis in 62% of cases

* No. patients responding/no. patients treated

adjuvant chemotherapy and/or endocrine therapy. Receptor status has a use in predicting response to the endocrine arm of such treatments[56,57,58] (Fig. 2).

Disease-free interval and site of subsequent metastasis. RE-tumours recur more rapidly than the RE+ ones[52,55], a result that is probably related to the higher ^3H thymidine labelling index of the former phenotype[59,60]. Interestingly, the RE+ tumours tend to reappear in bone whereas the RE- ones recur more frequently in liver and brain[61,62].

CONCLUSIONS

RE is not a tumour marker in the sense that it is unique to a tumour but it is a very useful index of tumour behaviour and as such is widely used in determining treatment of breast cancer. RE- tumours have an unacceptably low probability of responding to any form of hormone therapy and can therefore be immediately treated by chemotherapy whereas RE+ tumours should be treated by hormonal means. RP analysis is also helpful in this respect. Knowledge of receptor phenotype is also useful for determining adjuvant treatments in early breast cancer.

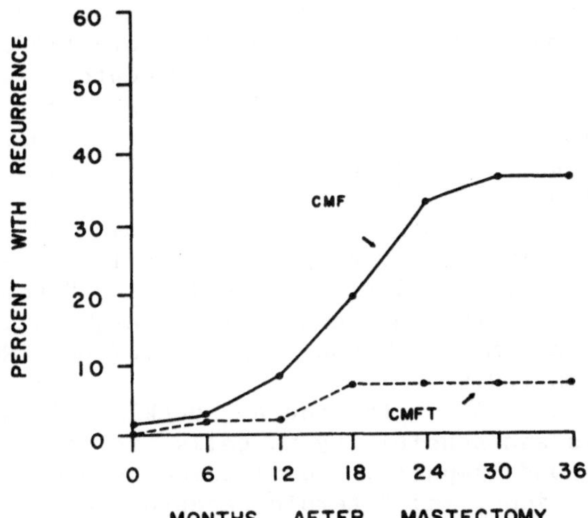

Fig. 2. Life table plot of recurrence rates for RE+, Stage II breast tumours. CMF = chemotherapy with cyclophosphamide + methotrexate + fluorouracil; CMFT = CMF + the antioestrogen, Tamoxifen. Significance of difference between CMF and CMFT, p = 0.0176. Reproduced from Hubay et al.[56]

REFERENCES

1. R.J.B. King and W.I.P. Mainwaring, "Steroid-Cell Interactions," Butterworths, London (1974).
2. P.C.M. Young and C.E. Ehrlich, Progesterone receptors in human endometrial cancer, in: "Steroid Receptors and the Management of Cancer," Vol. 1, E.B. Thompson and M.E. Lippman, eds., pp.135-159, CRC Press, Boca Raton (1979).
3. T.J. Benraad, L.G. Friberg, A.J.M. Koenders, and S. Kullander, Do estrogen and progesterone receptors (E_2R and PR) in metastasizing endometrial cancers predict the response to gestagen therapy? Acta Obstet. Gynecol. Scand. 59:155-159 (1980).
4. B.G. Mobbs, Steroid hormone receptors in normal and neoplastic hormone-sensitive tissues, in: "Influences of Hormones in Tumor Development," Vol. 1, J.A. Kellen and R. Hilf, eds., pp.11-53, CRC Press, Boca Raton (1979).
5. P. Ekman, E. Dahlberg, J.-A. Gustafsson, B. Hogberg, A. Pousette, and M. Snochowski, Present and future clinical value of steroid receptor assays in human prostatic carcinoma, in: "Hormones and Cancer. Progress in Cancer Research and Therapy," Vol.14,

S. Iacobelli, R.J.B. King, H.R. Lindner, and M.E. Lippman, eds., pp.361-370, Raven Press, New York (1980).

6. G. Concolino, Renal cancer: steroid receptors as a biochemical basis for endocrine therapy, in: "Steroid Receptors and the Management of Cancer," Vol. 1, E.B. Thompson and M.E. Lippman, eds., pp.173-195, CRC Press, Boca Raton (1979).

7. M.E. Lippman, G.K. Yarbro, B.G. Leventhal, and E.B. Thompson, Clinical correlations of glucocorticoid receptors in human acute lymphoblastic leukemia, in: "Steroid Receptors and the Management of Cancer," Vol. 1, E.B. Thompson and M.E. Lippman, eds., pp.67-80, CRC Press, Boca Raton (1979).

8. G.R. Crabtree, K.A. Smith, and A. Munck, Glucocorticoid receptors and sensitivity in cells from patients with leukemia and lymphoma. Results and methodologic considerations, in: "Steroid Receptors and the Management of Cancer," Vol. 1, E.B. Thompson and M.E. Lippman, eds., pp.81-97, CRC Press, Boca Raton (1979).

9. Cancer 46:(12) Suppl. (1980).

10. R.I. Fisher and M.E. Lippman, Steroid receptors in malignant melanoma, in: "Steroid Receptors and the Management of Cancer," Vol.1, E.B. Thompson and M.E. Lippman, eds., pp.197-204, CRC Press, Boca Raton (1979).

11. B. Greenway, M.J. Iqbal, P.J. Johnson, and R. Williams, Oestrogen receptor proteins in malignant and fetal pancreas, Brit. Med. J. 283:751-753 (1981).

12. O. Janne, A. Kauppila, P. Syrjala, and R. Vihko, Comparison of cytosol estrogen and progestin receptor status in malignant and benign tumors and tumor-like lesions of human ovary, Int. J. Cancer 25:175-179 (1980).

13. R.E. Leake, L. Laing, K.C. Kalman, and F.R. MacBeth, Estrogen receptors and antioestrogen therapy in selected human solid tumours, Cancer Treatment Rep. 64:797-799 (1980).

14. S.J. Cutler, The prognosis of treated breast cancer, in: "Prognostic Factors in Breast Cancer," A.P.M. Forrest and P.B. Kunkler, eds., pp.20-31, Livingstone, Edinburgh (1968)

15. J.H. Clark and E.J. Peck, "Female Sex Steroids," Springer-Verlag, Berlin (1979).

16. J.L. Wittliff, Steroid receptor interactions in human breast carcinoma, Cancer, 46:2953-2960 (1980).

17. J.H. Clark and E.J. Peck, Characteristics of cytoplasmic and nuclear receptor forms, in: "Female Sex Steroids", pp.46-57, Springer-Verlag, Berlin (1979).

18. L.P. Pertschuk, E.H. Tobin, P. Tanapat, E. Gaetjens, A.C. Carter, N.D. Bloom, R.J. Macchia, and K.B. Eisenberg, Histochemical analyses of steroid hormone receptors in breast and prostatic carcinoma, J. Histochem. Cytochem. 28:799-810 (1980).

19. D.R. Zehr, P.G. Satyaswaroop, D.M. Sheehan, Nonspecific staining in the immunolocalization of estrogen receptors, J. Steroid Biochem. 14:613-617 (1981).
20. G.C. Chamness, W.D. Mercer, and W.L. McGuire, Are histochemical methods for estrogen receptor valid? J. Histochem. Cytochem. 28:792-798 (1980).
21. I. Nenci, G. Fabris, A. Marzola, and E. Marchetti, Hormone receptor cytochemistry in human breast cancer, in: "Hormones and Cancer. Progress in Cancer Research and Therapy," Vol.14, S. Iacobelli, R.J.B. King, H.R. Lindner, and M.E. Lippman, eds., pp.227-239, Raven Press, New York (1980).
22. W.E. Stumpf and M. Sar, Autoradiographic localization of estrogen, androgen, progestin, and glucocorticosteroid in "target tissues" and "nontarget tissues," in: "Receptors and Mechanism of Action of Steroid Hormones," Vol.1, J.R. Pasqualini, ed., pp.41-84, Marcel Dekker, New York (1976).
23. G.L. Greene, C. Nolan, J.P. Engler, and E.V. Jensen, Monoclonal antibodies to human estrogen receptor. Proc. Natl. Acad. Sci. USA, 77:5115-5119 (1980).
24. A.I. Coffer and R.J.B. King, Antibodies to estradiol receptor from human myometrium. J. Steroid Biochem. 14: 1229-1235.
25. F. Logeat, M.T.Vu Hai, and E. Milgrom, Antibodies to rabbit progesterone receptor: crossreaction with human receptor, Proc. Natl. Acad. Sci. USA, 78:1426-1430 (1981).
26. M.V. Govindan and C.E. Sekeris, Purification of two dexamethasone-binding proteins from rat liver cytosol. Eur. J. Biochem. 89:95-104 (1978).
27. J.H. Clark and E.J. Peck, Receptor states and ^{3}H-steroid exchange, in: "Female Sex Steroids," pp.22-28, Springer-Verlag, Berlin (1979).
28. G.C. Chamness and W.L. McGuire, Steroid receptor assay methods in human breast cancer, in: "Steroid Receptors and the Management of Cancer," Vol. 1, E.B. Thompson and M.E. Lippman, eds., pp.3-30, CRC Press, Boca Raton (1979).
29. J.H. Clark and E.J. Peck, Methods of receptor assay, in: "Female Sex Steroids," pp.28-31, Springer-Verlag, Berlin, (1979).
30. P.W. Jungblut, S. Hughes, A. Hughes, and R.K. Wagner, Evaluation of various methods for the assay of cytoplasmic oestrogen receptors in extracts of calf uteri and human breast cancers, Acta endocrinol. 70:185-195 (1972).
31. R.J.B. King, Clinical relevance of steroid-receptor measurements in tumours, Cancer Treatment Rev. 2:253-273 (1976).

32. R.J.B. King, S. Redgrave, J.L. Hayward, R.R. Millis, and R.D. Rubens, The measurement of receptors for oestradiol and progesterone in human breast tumours, in: "Steroid Receptor Assays in Human Breast Tumours: Methodological and Clinical Aspects,"R.J.B. King, ed.,pp.55-73, Alpha Omega, Cardiff (1979).
33. R.J.B. King, D.M. Barnes, R.A. Hawkins, R.E. Leake, P.V. Maynard, R.M. Millis, and M.M. Roberts, Measurement of oestrogen receptors by five institutions on common tissue samples, in: "Steroid Receptor Assays in Human Breast Tumours Methodological and Clinical Aspects,"R.J.B. King, ed., pp.7-15, Alpha Omega, Cardiff (1979).
34. A.J.M. Koenders, J. Juarts, T. Hendreks, and T.J. Benraad, The Nederlands interlaboratory quality control programme of receptor assays in breast cancer, in: "Estrogen Receptor Assays in Breast Cancer. Laboratory Discrepancies in Quality Assurance,"G.A. Safferty, E.R. Nash, and D. Caghtley, eds., pp.69-82, Masson, New York (1980).
35. M.E. Lippman and E.B. Thompson, Pitfalls in the interpretation of steroid receptor assays in clinical medicine, in: "Steroid Receptors and the Management of Cancer," Vol. 1, E.B. Thompson and M.E. Lippman, eds., pp.235-238, CRC Press, Boca Raton (1979).
36. W.L. McGuire and K.B. Horwitz, A role for progesterone in breast cancer, Ann. N.Y. Acad. Sci. 286:90-100 (1977).
37. R.J.B. King, Analysis of estradiol and progesterone receptors in early and advanced breast tumors, Cancer, 46:2818-2822 (1980).
38. G.A. Degenshein, N. Bloom, and E. Tobin, The value of progesterone receptor assays in the management of advanced breast cancer, Cancer, 46:27892794 (1980).
39. R.E. Leake, L. Laing, and D.C. Smith, A role for nuclear oestrogen receptors in prediction of therapy regime for breast cancer patients, in:"Steroid Receptor Assays in Human Breast Tumours: Methodological and Clinical Aspects," R.J.B. King, ed., pp.73-85, Alpha Omega, Cardiff (1979).
40. L.G. Skinner, D.M. Barnes, and G.G. Ribeiro, The clinical value of multiple steroid receptor assays in breast cancer management, Cancer, 46:2939-2945 (1980).
41. P.P. Rosen, C.J. Menendez-Botet, J.S. Nisselbaum, J.A.Urban, V. Miké, A. Fracchia, and M.K. Schwartz, Pathological review of breast lesions analyzed for estrogen receptor protein, Cancer Res. 35:3187-3194 (1975).
42. R.R. Millis, Correlation of hormone receptors with pathological features in human breast cancer, Cancer, 46:2869-2871 (1980).
43. W.L. McGuire and G.C. Chamness, Studies on the estrogen receptor in breast cancer, in: "Receptors for Reproductive

Hormones", B.W. O'Malley and A.R. Means, eds., pp.113-136, Plenum Press, New York (1973).

44. M. Lippman, Hormonal regulation of human breast cancer cells in vitro, in: "Hormones and Breast Cancer. Banbury Report," Vol.8, M. C. Pike, P. K. Siiteri and C.W. Welsch, eds., pp.171-181. Cold Spring Harbor (1981).

45. J.L. Hayward, P.P. Carbone, J.-C. Heuson, S. Kumaoka, A. Segaloff, and R.D. Rubens, Assessment of response to therapy in advanced breast cancer, Europ.J.Cancer 13: 89-94 (1977).

46. P.V. Maynard, C.J. Davies, R.W. Blamey, C.W. Elson, J. Johnson, and K. Griffiths, Relationship between oestrogen-receptor content and histological grade in human primary breast tumours, Br.J.Cancer 38:745-748 (1978).

47. H.J.G. Bloom and W.W. Richardson, Histological grading and prognosis in breast cancer, Br.J.Cancer 11:359-377 (1957).

48. R.J.B. King, J.L. Hayward, J.R.W. Masters, R.R. Millis, and R.D. Rubens, Steroid receptor assays as prognostic aids in the treatment of breast cancer, in: "Steroid Receptors and Hormone-Dependent Neoplasia, J.L.Wittliff and O. Dapunt, eds., pp.249-256, Masson, New York (1980).

49. J.R.W. Masters, R.A. Hawkins, K. Sanster, W. Hawkins, I.I. Smith, A.A. Shivas, M.M. Roberts, and A.P.M. Forrest, Oestrogen receptors, cellularity, elastosis and menstrual status in human breast cancer, Europ.J.Cancer, 14:303-307 (1978).

50. J.R.W. Masters, R.R. Millis, R.J.B. King, and R.D.Rubens, Elastosis and response to endocrine therapy in human breast cancer, Br.J.Cancer, 39:536-539 (1979).

51. M.E. Lippman, J.C. Allegra, E.B. Thompson, R. Simon, A. Barlock, L. Green, K.K. Huff, H.M.T. Do, S.C. Aitken, and R. Warren, The relationship between estrogen receptors and response rate to cytotoxic chemotherapy in metastatic breast cancer, New Engl.J.Med. 298:1223-1228 (1978).

52. R.W. Blamey, H.M. Bishop, J.R.S. Blake, P.J. Doyle, C.W. Elston, J.L. Haybittle, R.I. Nicholson, and K.Griffiths, Relationship between primary breast tumor receptor status and patient survival, Cancer 46:2765-2774 (1980).

53. E.R. DeSombre and E.V. Jensen, Estrophilin assays in breast cancer: quantitative features and application to the mastectomy specimen, Cancer 46:2783-2788 (1980).

54. R.J.B. King, J.F. Stewart, R.D. Rubens, and J.L. Hayward, unpublished results.

55. C.K. Osborne, M.G. Yochmowitz, W.A. Knight, and W.L.McGuire, The value of estrogen and progesterone receptors in the treatment of breast cancer, Cancer, 46:2884-2888 (1980).

56. C.A. Hubay, O.H. Pearson, J.S. Marshall, R.S. Rhodes, S.M. Debanne, E.G. Mansour, R.E. Hermann, J.C. Jones, W.J. Flynn, C. Eckert, W.L. McGuire and 27 participating investigators, Adjuvant chemotherapy, antiestrogen therapy and immunotherapy for stage II breast cancer, in: "Breast Cancer - Experimental and Clinical Aspects," H.T. Mouridsen and T. Palshof, eds., pp.189-195, Pergamon Press, Oxford (1980).

57. C.A. Hubay, O.H. Pearson, J.S. Marshall, R.S. Rhodes, S.M. Debanne, J. Rosenblatt, E.G. Mansour, R.E. Hermann, J.C. Jones, W.J. Flynn, C. Eckert, W.L. McGuire and 27 participating investigators, Adjuvant chemotherapy, antiestrogen therapy and immunotherapy for stage II breast cancer: 45-month follow-up of a prospective, randomized clinical trial, Cancer, 46:2805-2808 (1980).

58. B. Fisher, C. Redmond, A. Brown, N. Wolmark, J. Wittliff, E.R. Fisher, D. Plotkin, D. Bowman, S. Sachs, J. Wolter, R. Frelick, R. Desser, N. LiCalzi, P. Geggie, T. Campbell, E.G. Elias, D. Prager, P. Koontz, H. Volk, N. Dimitrov, B. Gardner, H. Lerner, H. Shibata, and other NSABP investigators, Treatment of primary breast cancer with chemotherapy and Tamoxifen, New Engl.J.Med. 305:1-6 (1981).

59. J.S. Meyer and J.Y. Lee, Relationships of S-phase fraction of breast carcinoma in relapse to duration of remission, estrogen receptor content, therapeutic responsiveness, and duration of survival, Cancer Res. 40:1890-1896 (1980).

60. R. Silvestrini, M.G. Daidone, and G. DiFronzo, Relationship between proliferative activity and estrogen receptors in breast cancer, Cancer, 44:665-670 (1979).

61. J.F. Stewart, R.J.B. King, S.A. Sexton, R.R. Millis, R.D. Rubens, and J.L. Hayward, Oestrogen receptors, sites of metastatic disease and survival in recurrent breast cancer, Europ.J.Cancer, 17:449-453 (1981).

62. F.C. Campbell, R.W. Blamey, C.W. Elston, R.I. Nicholson, K. Griffiths, and J.L. Haybittle, Oestrogen-receptor status and sites of metastasis in breast cancer, Br.J.Cancer, 44:456-460 (1981).

ANDROGEN RECEPTOR ASSAY IN HUMAN PROSTATE CANCER

Peter Ekman, John Trachtenberg, and Patrick C. Walsh

Brady Urological Institute
The Johns Hopkins Hospital
Baltimore, Maryland 21205, USA

INTRODUCTION

Because intracellular hormone receptors appear to be necessary for the hormonal dependence of target organs, theoretically steroid receptor measurements should be of value in predicting the response to endocrine therapy for tumors in these target tissues. With this generally accepted concept in mind, much effort has been spent to develop assay methods for steroid receptors in different human cancers. Breast cancer has been extensively studied for more than 10 years and a clear relationship has been demonstrated between the presence of cytosol estrogen receptors and response to endocrine manipulations.[1] The clinical correlation can be improved by also taking into account the presence of progestin receptors and measurement of the nuclear compartment.[2]

Although little controversy exists concerning the value of such measurements in selecting appropriate therapy, their value in improving survival needs to be further elucidated. Other human malignancies that have been analyzed for steroid receptor content include cancers of the male and female urogenital system and leukemia.

Prostatic cancer is very common in elderly men and one of the two most common causes of cancer death in men in the western world. The hormone dependency of the prostate and of prostatic malignancies has been well known for over 40 years. Prostatic cancer is usually first diagnosed in elderly men in declining condition; approximately half of them have already developed wide-spread disease. Therefore, radical prostatectomy is only employed in a small fraction of the cases for treatment of prostatic cancer. These circumstances also limit the application of radiotherapy. Moreover, radiotherapy only occasionally leads to definite cure. New cytotoxic drugs are continuously under trial against prostatic cancer; so far, however, no generally accepted chemotherapeutic regimen exists. Hence, endocrine therapy remains a major tool in controlling the growth of prostatic carcinoma.

Endocrine therapy aims at reducing circulating androgen levels, which is achieved either by castration or estrogen treatment. The effect and duration of response to the two modes of treatment is comparable; estrogen therapy has a higher incidence of unwanted side effects, especially feminization and risk of cardiovascular complications.[3]

Notwithstanding the therapeutic shortcomings described above, it is of major interest to explore the value of steroid receptor assays in human prostate cancer. Lack of steroid receptors in a cancer specimen should favor the use of alternative therapeutic approach saving the patient from the negative side effects of endocrine treatment. The presence of steroid receptors may indicate a more favorable prognosis (compare below[4]) and, therefore, treatment may be withheld until symptoms demand interaction. Furthermore, steroid receptor assay may increase our knowledge of etiological factors influencing the pathogenesis of prostatic carcinoma.

It is the purpose of this paper to review the progress in androgen receptor assays of human prostate cancer.

MATERIALS

Animal Models

Most information about the androgen receptor stems from extensive studies of the rat ventral prostate. Studies of a spontaneous prostatic cancer the rat, the Dunning tumor, has been of particular interest since a hormone sensitive line of the tumor appeared to be receptor-positive whereas a hormone insensitive line of the tumor lacked androgen receptor[5].

The canine prostate has been used primarily for the studies of benign prostatic hyperplasia (BPH). The role of estrogens in the induction of hyperplasia, apparently via an estrogen receptor mediated increase in nuclear androgen receptor content, has been clearly demonstrated[6].

However, no suitable animal model exists for studies of the spontaneously occurring human prostatic diseases. The Dunning tumor lacks many characteristics of the human cancer. Canine BPH differs from human BPH and an estrogen-androgen inter-relationship has so far not been proven. Attempts to grow human prostate cancer in cell cultures or nude mice have met with little success and, therefore, so far only the human tumor itself can be used for a clinically oriented study.

Tissue Sampling

Whereas benigh prostatic hyperplasia is readily available from routine operations, the availability of carcinomatous tissue is limited. Some cancer tissue may be obtained from radical prostatectomy specimens. The best and most homogenous cancer tissue is obtained by removal of soft tissue metastases. However, the major information has to be achieved from needle biopsies, whereby only 0.1 to 0.2 grams of tissue are available. Electroresected specimens can only be used for qualitative studies since heat denaturation makes receptor quantitative unreliable[7] and the periurethral carcinomatous tissue is frequently contaminated with benign elements. Also, normal tissue is difficult to obtain. Occasionally normal tissue can be obtained during cystectomy for bladder cancer and from cadaver prostates at the time of kidney removal for transplantation.

Tissue Storage

Usually the specimens are stored in liquid nitrogen (-96°C) until used. Pilot studies have shown an androgen receptor loss of 20-30% from the freezing procedure, but the receptor thereafter seems to be stable for at least six months when stored as bulk tissue.[8] When stored as powdered tissue, considerable loss occurs within 2-3 weeks and in cytosol preparation the receptor may be totally lost within even less time of storage.[9]

METHODS

Pitfalls

Compared with estrogen receptor studies in breast cancer, androgen receptor studies of the prostate have proven far more complicated. Many pitfalls have hampered the development of reliable assay methods. One major problem has been tissue contamination with testosterone-estradiol binding globulin (TeBG) which binds testosterone and dihydrotestosterone (DHT) with similar affinity as the receptor. Moreover, most androgen receptors are endogenously bound necessitating the use of an exchange assay. This is further complicated by the slow dissociation of the steroid receptor complex at low temperature and the instability of the receptor at higher temperatures. The use of natural ligands, such as DHT, is also limited due to their rapid metabolism even at low temperature.[10]

Ligands

Therefore, a major step forward was made by the introduction of synthetic steroids. These do not bind to TeBG, they bind to the receptor with even higher affinity than the natural ligands, and they are resistant to metabolic conversion allowing long time exchange incubation. For androgen receptor measurement (^{3}H)-labeled methyltrienolone (R1881) (17-β hydroxy-17-α methyl-4,9,11-estratrien-3one[11]) is used. Similarly R2858[12] and R5020[13] have been used for estrogen and progesterone receptor studies respectively. (^{3}H)-dexamethasone has long been used for assays of the glucocorticoid receptor.[14]

A major disadvantage using R1881 for androgen receptor studies is its lack of specificity. R1881 also binds to the progesterone receptor which is present in most specimens of human BPH.[8,15] It is also present within some samples containing prostatic cancer and therefore, R1881 can not be used alone for quantitation of the androgen receptor binding. It has been suggested that incubation for 20 hours at 15°C should destroy the progesterone receptor (which is less stabilized by endogenously bound steroids) but leave the androgen receptor (which is endogensouly bound and thereby more stable) intact.[16] This has only proven partly true.[17,18] However, the binding to the progestin receptor can be totally blocked by adding a 500-fold excess of triamcinolone acetonide to the incubation medium[19]. At present the most secure and reproducible androgen receptor assay under stable conditions seems to be performed over night at 4°C using R1881 as ligand in the presence of a 500-1000 fold excess of triamcinolone acetonide.[17]

Receptor Quantitation

One of the most reliable methods to make quantitative receptor analysis still remains calculations by Scatchard plots[20] which give information on maximum number of binding sites (B_{max}), dissociation constant (K_d) and also permits statistical quality control thereby avoiding false positive results. A full Scatchard plot, however, needs approximately 1 gram of tissue and therefore, when cancer biopsies weighing 0.1-0.2 grams are analyzed, the reliability is decreased and the number of false negative specimens may increase.[7] To minimize the tissue need, single point assays have been introduced.[21,22,23] The results have been diverging and they have proven poor specificity and reproducibility. Recently, however, a triplicate hydroxyl apatite single point assay has been developed with high reliability permitting receptor assay of cytosol as well as nuclei in small cancer biopsies.[9] When performing receptor measurements with single point assays, their reliability must be carefully checked with e.g. Scatchard analysis.

Receptor Stabilization

Glycerol in 10-30% saturation seems to be of great importance for stabilization of the receptor.[24] Further improvement in stabilization has been shown when sodium molybdate and phenyl methyl sulfonyl fluoride (PMSF) is added to the cytosol.[25] Indeed, a 3-fold increase of receptor content was reported.[9] It has also been shown that separation of free from bound ligand with dextran coated charcoal (DCC) might give false low receptor values when the protein concentration is below 1mg[9] which may often be the case when dealing with cancer biopsies. This may be avoided by using hydroxyl apatite for the separation step[9,26].

Sucrose density gradient centrifugation is also widely used for receptor quantitation and further characterization. Trachtenberg et al have shown that the use of a vertical rotor system to ensure short centrifugation time enhances the reliability of the human prostatic androgen receptor assay.[9]

RESULTS

Reproducible methods now exist for androgen receptor measurements in the human prostate. The receptor has been carefully investigated and characterized in cytosol preparations[8,10,16,27,28] as well as in the nuclear compartment.[16,27,28] Binding data and other characteristics vary slightly in different laboratories. In cytosol, the mean value for B_{max} is between 150-500 fmol/mg DNA and the K_d 0.8 to 1.8 nM.[8,28] In nuclei the B_{max} seems to be higher as does the K_d.[28] On sucrose density gradient the steroid receptor complex sediments as an 8 S peak.[27] When comparing normal and hyperplastic human prostate there does not seem to be any difference either in the cytosol[8,28] or in the nuclei[28] in androgen receptor content.

The receptor profile of prostatic cancer seems more heterogeneous. In soft tissue metastases of human prostate cancer Ekman et al found measurable androgen receptor in 7 of 8 cases with B_{max} varying from over 8000 to 20 fM/mg DNA[18](Table I). In punch needle

Table I. Steroid receptor profiles in cytosol of prostatic cancer metastases B_{max} fmoles/ mg DNA

Patient number	R1881	R5020	R2858	Dexamethasone
I	8,150	402	n.d.*	1,350
II	3,440	92	n.d.	2,270
III	2,870	n.d.	n.d.	n.d.
IV	1,790	n.d.	n.d.	254
V	1,319	54	n.d.	401
VI	674	n.d.	n.d.	942
VII	21	n.d.	n.d.	n.d.
VIII	n.d.	n.d.	n.d.	n.d.-

*n.d. - not detectable (slope of Scatchard plot not significantly ($p<0.05$) different from zero)

Table II. Relation between receptor content and response to endocrine therapy.

R1881-binding	Number of patients responding to endocrine therapy	Total number of patients given endocrine therapy	Percent responders
"Positive"	18	21	86
"Negative"	3	8	37

biopsies of primary prostatic carcinoma androgen receptor was measurable by Scatchard analysis in 75% of the cases.[7,29] Eighty-six percent of the androgen "receptor-positive" cases did respond to endocrine therapy for >3 months compared to 37% of the androgen "receptor-negative" cases[18](Table II). There was, however, no clear relation between B_{max} values and duration of response to therapy. It is possible that a better prognostic index is achieved from androgen receptor assay of the nuclear compartment[4](Table III). Also presence of other steroid receptors may appear to be a favorable prognostic sign (Table I).

However, the terms "receptor-positive" and "receptor-negative" are probably somewhat misleading. With more sensitive techniques all prostatic cancers may turn out to be "receptor-positive". Therefore, the term "receptor-poor" and "receptor-rich" may be more adequate. In the studies mentioned above a critical level for B_{max} of the androgen receptor with regard to duration of response seemed to be around 110 fM/mg DNA.[4,29]

On the other hand, also may patients with tumors containing more than 110 fM/mg DNA only experienced a short response (<1 year) to endocrine treatment.[4] Recent investigations indicate that the

Table III. Duration of response to endocrine therapy compared to receptor content.

Preparation	B_{max} fmol/mg DNA	n	Duration of response ≤1 year	>1 year
Cytosol*	<110	17	12	5
	>110	22	9	13
Nuclei	<110	7	7	0
	>110	16	5	11

*Cytosol data are taken from 2 different studies[4,34]. Please observe that B_{max} data of cytosol are assayed without incubation in sodium molybdate or PMSF which would have given higher data (cf text).

salt resistant nuclear receptor is more closely related to biological activity[30] and may therefore serve as a more valuable parameter for clinical use.

DISCUSSION

Throughout the last years advances in steroid receptor research has led to the development of reproducible methods for androgen receptor assays in the human prostate. The techniques are under continuous improvement and recently a microassay method has been introduced allowing nuclear and cytosol receptor assays in needle biopsy specimens.[4] The figures on relation between receptor content and response to therapy plus prognosis are good enough to encourage further attempts to optimize and simplify the methods.

However, prostatic cancer usually is a most heterogeneous tumor, often with areas of varying differentiation intermingled. Similarly it can be expected that hormone sensitive and insensitive cells are mixed in the same cancer. It is reasonable to believe that a "receptor-rich" tumor is dominated by hormone sensitive cells and a "receptor-poor" primarily of hormone insensitive cells. Present assays and also biopsy techniques do not permit us to estimate the relative distribution of these important parameters, however. Possibly progress towards this goal will not come until immunological methods, allowing cell per cell assays, have been developed.[18]

By using anti-steroid antibodies attempts have been made to specify receptor binding.[31,32] The results, however, have so far been disappointing with little specificity and reproducibility.[33] It seems as if no major success will be gained until the receptor itself has been purified and monoclonal antibodies developed.

Hence, much further research needs to be performed before androgen receptor assay of human prostatic carcinoma can be introduced into clinical routine as a means to select patients expected to benefit from endocrine therapy.

Acknowledgement - This work was supported by grant 1 F05 TW03029-01 from NIH and 502206286 from the Swedish Medical Research Council.

REFERENCES

1. E.V. Jensen, B.E. Block, S. Smith, K. Kyser, and E.R. DeSombre, Estrogen receptors and breast cancer response to adrenalectomy, National Cancer Institute Monograph-Prediction of Responses in Cancer Therapy, 34:55 (1971).
2. K.B. Horwitz and W.L. McGuire, Estrogen control of progesterone receptor in human breast cancer: correlation with nuclear processing of estrogen receptor, J. Biol. Chem. 253:2223, (1978).

3. Veterans Administrative Cooperative Urological Research Group, Treatment and survival of patients with cancer of the prostate, Surg. Gynecol. Obstet. 124: 1011 (1967).
4. J. Trachtenberg and P.C. Walsh, Correlation of prostatic nuclear androgen receptor content with duration of response and survival following hormonal therapy in advanced prostatic cancer, J. Urol. in press (1982).
5. F.S. Markland, R.T. Chopp, M.D. Gosgrove, and E.B. Howard Characterization of steroid hormone receptors in the Dunning R-3327 rat prostatic adenocarcinoma, Cancer Res. 38:2818 (1978).
6. J. Trachtenberg, L.L. Hicks, and P.C. Walsh, Androgen and estrogen receptor content in spontaneous and experimentally induced canine prostatic hyperplasia, J. Clin. Invest. 65: 1051 (1980).
7. P. Ekman, Clinical significance of steroid receptor assay in the human prostate. In: Steroid receptors, metabolism and prostatic cancer, F.H. Schroder and H.J. deVoogt, eds, Excerpta Medica, Amsterdam, 208 (1980).
8. P. Ekman, M. Snochowski, E. Dahlberg, D. Bression, B. Hogberg and J.A. Gustafsson, Steroid receptor contents in cytosol from "normal" and hyperplastic human prostates, J. Clin. Endocrinol. Metab. 49: 205 (1979).
9. J. Trachtenberg, L.L. Hicks, and P.C. Walsh, Methods for the determination of androgen receptor content in human prostatic tissue, Invest. Urol. 18: 349 (1981).
10. M. Snochowski, A. Pousette, P. Ekman, D. Bression, L. Andersson, B. Hogberg, and J.A. Gustafsson, Characterization and measurement of the androgen receptor in human benign prostatic hyperplasia and prostatic carcinoma, J. Clin. Endocrinol. Metab. 45: 920 (1977).
11. C. Bonne and J.P. Raynaud, Assay of androgen binding sites by exchange with MT (R1881), Steroids 27:497 (1976).
12. J.P. Raynaud, M.M. Bouton, D. Ballet-Bourquin, D. Philibert, C. Tournemine,and G. Azadien-Boulanger, Comparative study of estrogen action, Mol. Pharmacol. 9:520 (1973).
13. D. Philibert and J.P. Raynaud, Progesterone binding in the immature mouse and rat uterus, Steroids 22:89 (1973).
14. J.D. Baxter and G.M. Tomkins, The relationship between glucocorticoid binding and tyrosine aminotransferase induction in hepatoma tissue culture cells, Proc. Natl. Acad. Sci. U.S.A. 65: 709 (1970).
15. J. Asselin, F. Labrie, Y. Gourdeau, C. Bonne, and J.P. Raynaud, Binding of (^{3}H) methyltrienolone in rat prostate and human benign prostatic hypertrophy (BPH), Steroids 28:449 (1976).
16. S.A. Shain, R.W. Boesel, D.L. Lamm, and M.M. Radwin, Characterization of unoccupied (R) and occupied (RA) androgen binding components of the hyperplastic human prostate, Steroids 31:541 (1978).
17. L.L. Hicks, and P. C. Walsh, A microassay for the measurement of androgen receptors in human prostatic tissue, Steroids 33:389 (1979).
18. P. Ekman, E. Dahlberg, J.A. Gustafsson, B. Hogberg, A. Pousette,

and M. Snochowski, Present and future clinical value of steroid receptor assay in human prostatic carcinoma. In: Hormones and Cancer, S. Iacobelli et al, eds, Raven Press, New York, p.361 (1980).

19. T. Zava, B. Landrum, K.B. Horwitz, and W.L. McGuire, Androgen receptor assay with (^{3}H) methyltrienolone (R1881) in the presence of progesterone receptors, Endocrinology 104:1007 (1979).

20. G. Scatchard, The attraction of proteins for small molecules and ions, Ann. N.Y. Acad. Sci. 41:660 (1949).

21. R.K. Wagner, K.H. Schulze, and P.W. Jungblut, Estrogen and androgen receptor in human prostate and prostatic tumor tissue, Acta Endocrinol. Suppl. 193: 52 (1975).

22. H.J. deVoogt, and P. Dingjan, Steroid receptors in human prostatic cancer, Urol. Res. 6:151 (1978).

23. N. Bashirelahi, J.H. O'Toole, and J.D. Young, A specific 17-β estradiol receptor in human benign prostatic hypertrophy, Biochem. Med. 15:254 (1976).

24. P.D. Feil, S.R. Glaser, D.O. Toft, and B.W. O'Malley, Progesterone binding in the mouse and rat uterus, Endocrinology 91: 738 (1972).

25. C.B. Lazier, and A.J. Haggarty, Estradiol binding in cockrel liver cytosol, Biochem. J. 180:347 (1979).

26. R.E. Garola, and W.L. McGuire, A hydroxylapatite micromethod for measuring estrogen receptors in human breat cancer, Cancer Res. 38:2216 (1978).

27. M. Menon, C.E. Taninis, L.L. Hicks, E.F. Hawkins, M.G. Mc Loughlin, and P.C. Walsh, Characterization of the binding of a potent synthetic androgen; methyltrienolone, to human tissues, J. Clin. Invest. 61:150 (1978).

28. J. Trachtenberg, P. Bujnovszky, and P.C. Walsh, Androgen receptor content of normal and hyperplastic human prostate, J. Clin. Endocrinol. Metab. (in press) (1981).

29. P. Ekman, M. Snochowski, A. Zetterberg, B. Hogberg, and J.A. Gustafsson, Steroid receptor content in human prostatic carcinoma and response to endocrine therapy, Cancer 44:1173 (1979).

30. J.H. Clark and E.J. Peck Jr., Nuclear retention of receptor-oestrogen complex and nuclear acceptor sites, Nature 260:635 (1976).

31. E. Castaneda, and S. Liao, The use of antisteroid antibodies in the characterization of steroid receptors, J. Biol. Chem. 250: 883 (1975).

32. L.P.Pertschuk, E.H. Tobin, P. Tanapat, E. Gaetjens, A.C. Carter, N.D. Bloom, R.J. Macchia, and K.B. Eisenberg, Histochemical analyses of steroid hormone receptors in breast cancer and prostatic carcinoma, J. Histochem. Cytochem. 28:799 (1980).

33. G.C. Chamness, W.D. Mercer, and W.L. McGuire, Are histochemical methods for estrogen receptor valid?. J. Histochem. Cytochem. 28: 792 (1980).

34. P. Ekman, K.Svennerus, A. Zetterberg and J.A. Gustafsson, Cytophotometric DNA analyses and steroid receptor content in human prostatic carcinoma,Scand. J. Urol. Nephrol. (in press) (1981).

MODULATION OF 5-FLUOROURACIL TOXICITY VIA ESTROGEN RECEPTOR

Chris Benz and Ed Cadman

Departments of Medicine and Pharmacology

Yale School of Medicine, New Haven, CT 06510

SUMMARY

Modulation of 5-fluorouracil (FUra) metabolism and toxicity by methotrexate (MTX) and other antimetabolites occurs in many cultured tumor cell lines, including the estrogen receptor positive human mammary carcinoma, 47-DN. The growth rate of this cell line depends on exogenously administered insulin and estradiol, and can be reversibly inhibited by the antiestrogen, tamoxifen (TAM).

47-DN cell-cycle kinetics are altered by doses of TAM which suppress the synthesis of estrogen and progesterone receptors. In cloning assays, TAM is synergistic with FUra and sequentially combined MTX→FUra; in biochemical assays, TAM inhibits FUra intracellular accumulation and incorporation into RNA. This unique form of drug modulation may represent a form of "complementary inhibition," and supports clinical trials in breast cancer suggesting that TAM + FUra-containing chemotherapy is superior to chemotherapy alone.

INTRODUCTION

More than half of all women diagnosed with breast cancer will ultimately require systemic treatment with either chemo- or hormonal therapy. The use of combination chemotherapy in both adjuvant treatment and management of advanced breast cancer has improved response rates over single agent therapy, but it is still unclear whether combined chemo-hormonal therapy produces better response rates than chemotherapy alone[1]. We have used a hormone sensitive human mammary carcinoma cell line, 47-DN to study the in vitro interaction of two antimetabolites, MTX and FUra, with the antiestrogen, TAM.

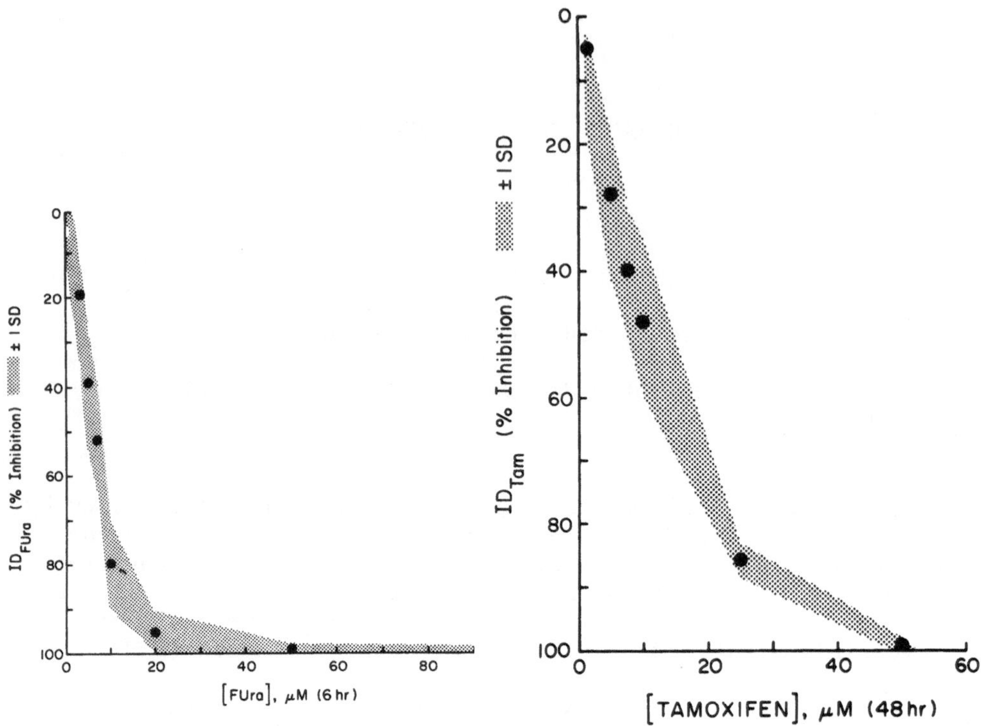

Fig.1. Clonal growth inhibition of 47-DN exposed to FUra for 6 hr

Fig.2. Clonal growth inhibition of 47-DN exposed to TAM for 48 hr.

RESULTS

In a monolayer cloning assay FUra and TAM inhibit 47-DN clonal growth in dose-dependent fashions (Figures 1 and 2). We have previously shown by this cloning assay that in vitro pretreatment of 47-DN cells with $\geq$ 12 hrs of MTX synergistically enhances FUra toxicity.[2]

Biochemical studies have shown that synergistic enhancement of FUra toxicity by MTX and other inhibitors of de novo purine synthesis occurs by increasing intracellular FUra accumulation, mediated by a build-up of 5-phosphoribosyl-1-pyrophosphate (PRPP) pools.[3,4] Table 1 shows the enhancement of FUra accumulation in 47-DN cells pretreated with MTX and other antimetabolites known to enhance FUra toxicity. Figure 3 shows that treatment of 47-DN with various concentrations and exposure intervals of MTX produces

Table 1. Enhancement of Intracellular Accumulation of FUra (100μM) in Pretreated 47-DN

Pretreatment (24 hr)	% Control
Methotrexate (10μM)	310
Methylmercaptopurine riboside (0.1μM)	178
Azaserine (10μM)	195
6-Diazo-5-Oxo-Norleucine (10μM)	246
L-Alanosine (10μM)	136

Control = 2.26 ± .61 nmol/10^6 cells/6 h.

a positive correlation between PRPP pools and rate of intracellular FUra accumulation. After a 24 hr exposure to 10μM MTX, PRPP pools are maximally enhanced 8-fold and FUra accumulation is enhanced 5-fold. The increase in FUra toxicity following MTX pretreatment may result from increased intracellular formation of FdUMP, which inhibits synthesis of thymidylate and DNA; or FUTP, which incorporates into RNA, interfering with transcription and translation of the genetic message. The intracellular nucleotide derivatives of FUra can be measured by high performance liquid chromatography (HPLC) of cell extracts prepared from treated tumor cells.[3] In MTX pretreated 47-DN cells, there is a 5-fold increase in FUTP, but no net increase in total (protein-complexed and freely soluble) FdUMP, consistant with the notion that MTX enhances the RNA-directed toxicity of FUra.

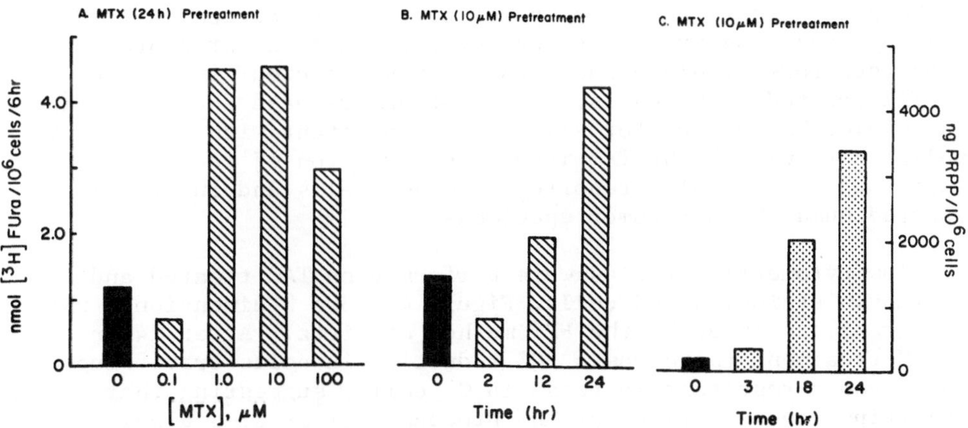

Fig.3. The effect of MTX concentrations (A) and intervals (B) on intracellular accumulation of 100μM [6-^{3}H] FUra (20Ci/mmol) and pools of PRPP (C) in 47-DN.

Table 2. Effect of Exogenous Insulin (I), Estradiol (E) and Tamoxifen (TAM) on 47-DN Doubling Time

Condition*	HR.	P.
+I, +E	26	<.01
-I, -E	34	
+I, -E, +TAM	37	
+I, +E, +TAM	26	

* +I = insulin 0.2 IU/ml; +E = estradiol 1nM; +TAM = tamoxifen 10nM; -I = no detectable endogenous insulin; -E = less than 10pM endogenous estradiol

The 47-DN cell line is maintained in culture by addition of both insulin (0.2 IU/ml) and estradiol (1nM) to the culture media, which is also supplemented with 10% fetal calf serum.[5] Radioimmunoassay of estradiol content indicates that cells grown in the absence of exogenously added estradiol are exposed to $\leq$ 10 pmol endogenous estradiol from the addition of serum to the media (no attempt was made to completely remove estradiol by dextrancoated charcoal adsorption). Table 2 shows that depriving these cells of exogenous insulin and estradiol significantly reduces 47-DN doubling time. Although decreasing estradiol content alone has no significant effect on 47-DN growth rate, the addition of tamoxifen inhibits cell growth and this inhibition can be reversed by increasing the estradiol content of the media.

We attempted to show the specificity of TAM action on 47-DN cells by measuring its effect on estrogen (ER) and progesterone (PgR) receptors. Table 3 shows ER and PgR measured‡ in control and TAM treated 47-DN cells. TAM treatment reduced ERc 5-fold and PgR 4-fold. These receptor values are consistant with previously published results.[6] The TAM induced suppression of receptors is consistant with results reported in MCF-7 cells and in freshly resected human breast tumor specimens.[7,8]

Flow cytometric studies were performed on TAM treated and acriflavin stained 47-DN cells. Figure 4 shows 2-dimensional DNA histograms, analyzed by the FPi method of Zietz.[9] After 24 hr of TAM treatment there was a 50% reduction in cells traversing S-phase, and a resultant increase in G_1 cells, suggesting that growth inhibiting concentrations of TAM produce a G_1 or G_1-S block.

Since cycle-arresting agents can antagonize cycle-active antimetabolites such as FUra and MTX, we expected to observe in cloning

‡Performed by A. Eisenfeld, Yale School of Medicine, using a modification of the methods of Horwitz and McGuire.[6,7]

Table 3. 47-DN Estrogen (ER_c + ER_n) and Progesterone (PgR) Receptors

47-DN	ER_c (fmol/mg protein)	ER_n (fmol/mgDNA)	PgR (fmol/mg protein)
untreated	32	282	1,366
TAM(10μMx72hr)	6	-	349

ER_c and ER_n = cytoplasmic and nuclear estrogen receptors detectable at 0°C (unfilled) + 30°C (filled).

experiments a less than additive effect between TAM pretreatment and administered MTX→FUra (18 hr MTX pretreatment followed by 6 hr FUra). On the contrary, Table 4 shows greater than additive cytotoxicity in TAM pretreated cells exposed to FUra or combined MTX→FUra. The index value, I.C.E., is a ratio between observed and expected results assuming independent mechanisms of action between TAM, FUra and MTX. An I.C.E. value of 1 ± .2 indicates an additive effect; <1 or >1 indicates synergism or antagonism. No significant synergism was observed between TAM + MTX; however, TAM + Fura or TAM + MTX→FUra had I.C.E. values of 0.48 and 0.33, indicating significant synergism.

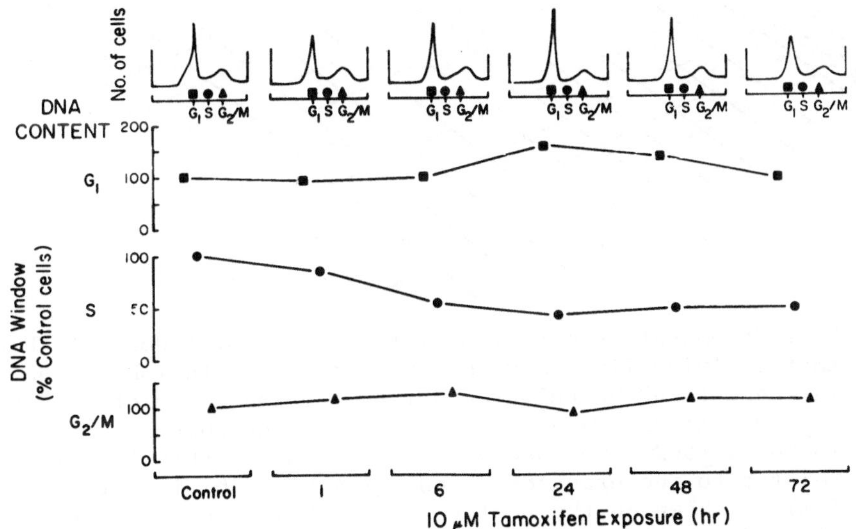

Fig.4. DNA histogram analysis of acriflavin stained 47-DN cells treated with 10μM tamoxifen for up to 72 hr.

Table 4. Clonal Growth of 47-DN Cells Treated with MTX, FUra, and TAM (10μM x 72 hr)

Condition	Mean % Clonal Growth (±10%)		*I.C.E.(±.2)
	-TAM	+TAM	
Control	100%	58%	
MTX(0.5μM,24 hr)	47%	22%	.81
FUra(5μM,6 hr)	93%	26%	.48
MTX→FUra	11%	2%	.33

* Independent Combined Effect:ratio of observed clonal growth to expected clonal growth.eg. FUra+TAM=(.26)/(.93)(.58) = .48
If I.C.E. = 1.0, additive; <1.0 synergistic; >1.0, antagonistic.

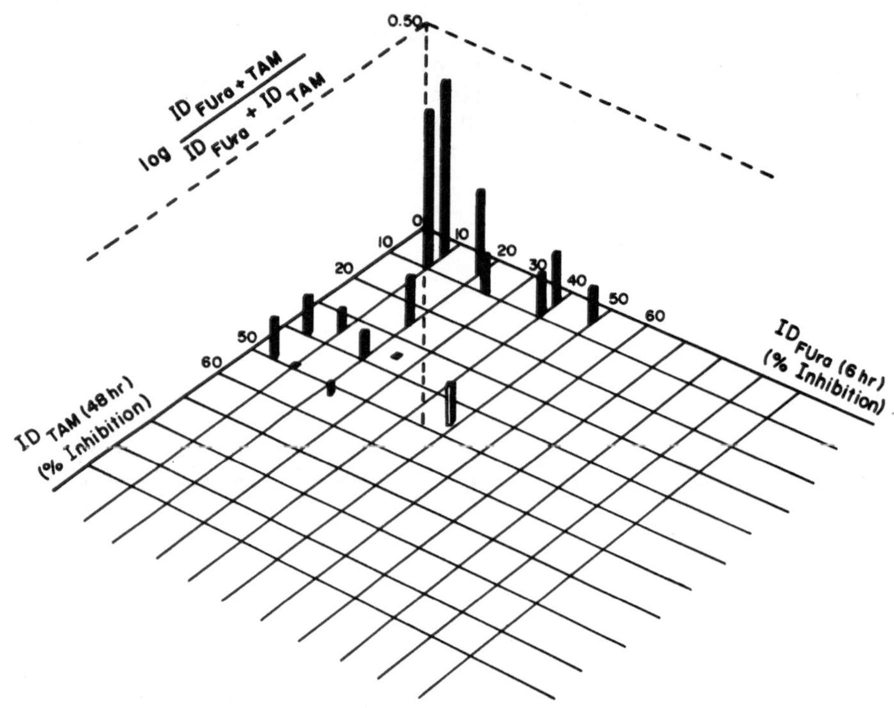

Fig.5. Dose-dependent interaction of FUra and TAM. A dose of FUra that individually produces ≤50% clonal growth inhibition is added to 47-DN cultures during the last 6 hr of a 48 hr exposure to TAM, given at a concentration which also inhibits ≤50%. The observed inhibition (ID FUra + TAM) is related to the expected clonal growth inhibition (ID Fura+ID TAM) as shown on the vertical axis. Values on the horizontal plane denote additive effects, those above the plane (solid bars) represent synergistic effects and those below the plane (empty bars) represent antagonistic effects.

Table 5. Distribution of Soluble FUra Metabolites in TAM Treated 47-DN

	% FUTP	% FUDP	%FUDP-glu	%FUMP	%FUra/FUrd
control	76(±4)	6(±1)	11(±4)	7(±3)	1 (±1)
TAM (>1μM,≥24hr)	70(±3)	6(±2)	11(±1)	10(±3)	3 (±1)

Toxicity of the TAM-FUra combination was experimentally determined over many concentrations, and the synergistic interaction of these two drugs was examined graphically. If doses of FUra (1-7.5μM) representing ID_{FUra} 5-50% inhibition are combined with similar doses of TAM (1-10μM), the measured combined inhibition, $ID_{TAM + FUra}$, could range from 0-100%. Plotting the logarithm of the ratio between the observed $ID_{TAM + FUra}$ and the expected sum of $ID_{TAM} + ID_{FUra}$ results in the 3-coordinate graph of Figure 5. Values on the horizontal plane represent additive effects, those above and below the plane represent synergistic and antagonistic effects. As can be seen, the synergistic interaction between TAM and FUra occurs at many different concentrations, but appears most pronounced at drug exposures which individually kill fewer cells, that is, when ID_{TAM} or ID_{FUra}=5-20%.

The biochemical influence of TAM on FUra metabolism is shown in Tables 5 and 6. TAM treatment does not significantly alter soluble FUra metabolites. But unlike other drugs which enhance FUra toxicity by increasing FUra metabolism, TAM actually decreases intracellular accumulation and RNA incorporation of FUra by 20-60%, in a pattern which may be inversely correlated with TAM toxicity.

DISCUSSION

Tamoxifen appears to be a unique modulator of FUra, enhancing FUra toxicity without increasing its intracellular accumulation. Notably, TAM is synergistic with the sequenced combination of MTX→FUra but not with MTX alone. Since MTX enhances the RNA-directed toxicity of FUra,[10] these results suggest that TAM's interaction with FUra is mediated by the template-directed activities of both drugs. Sartorelli first introduced the concept of "complementary inhibition" to describe synergistic effects observed between drugs that bind and inactivate DNA and RNA (such as actinomycin D and mitomycin C) and others such as FUra which alter the biosynthetic pathways leading to formation of these macromolecules.[11,12] TAM and other antiestrogens are known to bind to nuclear chromatin before exerting their cytotoxic effects.[7,13,14] Actinomycin D has been shown to mimic some antiestrogen effects on receptor processing.[13] Thus it is indeed possible that the TAM-FUra synergistic interaction is another example of complementary inhibition. This concept lacks a more detailed mechanistic explanation; but it is

Table 6. Intracellular Accumulation and RNA Incorporation of FUra In TAM Treated 47-DN

TAM Exposure		FUra Accumulation	RNA Incorporation
μM	hr	% control*	% control*
*	*	(3.7nmol/10^6cell/6hr)	(451pmol/10^6cells/6hr)
.01	24	100	95
.10	24	100	96
1	24	36	47
10	24	63	72
10	48	68	87
10	72	87	76

conceivable that TAM modulates RNA polymerases and/or nucleases, enhancing FUra toxicity by selectively directing FUra incorporation into critical species of RNA.

The TAM-FUra interaction is dose dependent with the greatest degree of synergism observed at individual drug concentrations which, by themselves, are minimally toxic (TAM$\leq$4μM, FUra$\leq$3μM). Pharmacokinetic studies suggest that these doses are clinically achievable during standard therapy.[15,16] Thus, in vitro studies attempting to understand the biochemical basis of TAM-FUra synergism would best be performed at these lower doses.

We feel that this unique form of FUra modulation observed in vitro supports the results of clinical trials suggesting that TAM + FUra-containing chemotherapy is superior to chemotherapy alone in the treatment of ER positive breast cancer.[17,18] Further studies are also necessary to determine whether the cellular phenomenon of chemo-hormonal interaction can be exploited more generally in the design of better cancer treatment protocols.

ACKNOWLEDGEMENTS

We wish to thank Terrence Wu, John Gwin, Joan Gesmonde, and Hillary Raeffer for their skillful technical assistance. This work was supported by the following N.C.I. grants: CA-24187, CA-27130, CA-08341 and grant CH-145 from the American Cancer Society.

REFERENCES

1. L.G. Kardinal, W.L. Donegan and J.S. Spratt, eds. Chemotherapy, in: "Cancer of the Breast," W.B. Saunders Company, Philadelphia. 405-447 (1979).
2. C. Benz, M. Schoenberg, M. Choti, and E. Cadman. Schedule dependent cytotoxicity of methotrexate and 5-fluorouracil in human colon and breast tumor cell lines. J. Clin. Invest. 66: 1162-1165 (1980).

3. C. Benz, and E. Cadman. Modulation of 5-fluorouracil metabolism and cytotoxicity by antimetabolite pretreatment in human colorectal adenocarcinoma, HCT-8. Cancer Res.41:994-999(1981).
4. E. Cadman, C. Benz, R. Heimer, and J. O'Shaughnessy. Effect of de novo purine synthesis inhibitors on 5-fluorouracil metabolism and cytotoxicity. Biochem. Pharmac.30:2469-2472(1981).
5. I. Keydar, L. Chen, S. Karby, F. Weiss, J. Delarea, M. Radu, S. Chartcik, H. Brenner. Establishment and characterization of a cell line of human breast carcinoma origin. Eur. J. Cancer 15: 659-679 (1979).
6. K. Horwitz, D. Zava, A. Thilagar, E. Jensen, and W. McGuire. Steroid receptor analysis of nine human breast cancer cell lines. Cancer Res. 38: 2434-2437 (1978).
7. K. Horwitz and W. McGuire. Nuclear mechanisms of estrogen action:Effects of estradiol and antiestrogens on estrogen receptors and nuclear receptor processing. J. Biol. Chem. 253: 8185-8191 (1978).
8. N. Waseda, Y. Kato, H. Imura, and M. Kurata. Effects of Tamoxifen on estrogen and progesterone receptors in human breast cancer. Cancer Res. 41: 1984-1988 (1981).
9. S. Zietz. FPi analysis-theoretical outline of a new method to analyze time sequences of DNA histograms. Cell Tissue Kinet. 13: 461-471 (1980).
10. E. Cadman, R. Heimer and C. Benz. The influence of methotrexate pretreatment on 5-fluorouracil metabolism in L1210 cells. J. Biol. Cehm. 256: 1695-1704 (1981).
11. A. Sartorelli. Combination chemotherapy with actinomycin D and ribonuclease: an example of complementary inhibition. Nature, 203: 877-878 (1964).
12. A. Sartorelli and B. Booth. The synergistic antineoplastic activity of combinations of mitomycins with either 6-thioguanine of 5-fluorouracil. Cancer Res. 25: 1393-1499 (1965).
13. K. Horwitz and W. McGuire. Studies on mechanisms of estrogen and antiestrogen action in human breast cancer. Recent Results Cancer Res. 71: 45-58 (1980).
14. L. Baudendistel and T. Ruh. Antiestrogen action: differential nuclear retention and extractability of the estrogen receptor. Steroids 28: 223-237 (1976).
15. C. Fabian, L. Sternson, and M. Barnett. Clinical pharmacology of tamoxifen in patients with breast cancer: comparison of traditional and loading dose schedules. Cancer Treat Rep. 64: 765-773 (1980).
16. J. Speyer, J. Collins, R. Dedrick, M. Brennan, A. Buckpitt, H. Londer, V. DeVita, Jr., and C. Myers. Phase I and Pharmacological studies of 5-fluorouracil administered intraperitoneally. Cancer Res. 40: 567-572 (1980).

17. H. Mouridsen, T. Palshof, E. Engelsman, and R. Sylvester. CMF versus CMF plus tamoxifen in advanced breast cancer in post-menopausal women: An EORTC trial, *in*: "Breast Cancer - Experimental and Clinical Aspects," H.T. Mouridsen and T. Palshof, eds., Pergamon Press, Oxford. 119-123 (1980).
18. B. Fisher, C. Redmond, A. Brown, N. Wolmark, J. Wittliff, E. Fisher, D. Plotkin, D. Bowman, S. Sachs, J. Wolter, R. Frelick, R. Desser, N. LiCalzi, P. Geggie, T. Campbell, E. Elias, D. Prager, P. Koontz, H. Volk, N. Dimitrov, B. Gardner, H. Lerner, H. Shibata, and other NSABP investigators. Treatment of primary breast cancer with chemotherapy and tamoxifen. N.E.J.M. 305: 1-6 (1981).

T-LYMPHOCYTE TOLERANCE IN RNA TUMOR VIRUS ONCOGENESIS: A MODEL FOR THE CLONAL ABORTION HYPOTHESIS

Luigi Chieco-Bianchi, Dino Collavo, Giovanni Biasi, Paola Zanovello, and Franca Ronchese

Laboratory of Oncology
University of Padova, Italy

I. INTRODUCTION

Retrovirus genes are contained in the chromosomal DNA of most vertebrates and may be transmitted vertically from parent to offspring through the germ line, as well as horizontally as infectious virus particles.

Most retroviruses have been causally associated with tumor induction in the host animal. These oncogenic viruses are usually divided into two main groups; viruses that induce rapid tumor development and transform target cells in culture (*rapidly transforming viruses*), and viruses that induce neoplasia with a prolonged latency (i.e. 2 to 12 months) and do not cause observable effects in tissue cultures despite adequate virus replication (*slowly transforming viruses*). The rapidly transforming viruses contain a gene coding for a protein which is considered responsible for the transformation. On the other hand, no transforming protein has been detected yet for the slowly transforming viruses, and therefore the mechanism responsible for their oncogenic potential is still unclear (1). In this regard two interesting hypotheses have been recently proposed. The *receptor-mediated* hypothesis (2), which is specifically concerned with lymphatic leukemia induction, postulates that neoplastic transformation results from the continuous stimulation of antigen reactive T cells by viral antigens that interact with specific cell surface receptors. The *promoter insertion* hypothesis (3) has a wider application and suggests that through its long terminal repeat sequences, the integrated provirus could be able to induce the expression of an adjacent, potentially oncogenic, cellular gene whose transcriptional activation could lead to neoplastic transformation.

However, both hypotheses are based on the prerequisite that high and long lasting levels of virus production are needed to produce transformation of appropriate target cells. Host failure to rid the body of replicating virus would thus greatly increase the risk of developing a neoplastic disease.

During the past years we have been studying the genetic, virologic and immunologic factors which influence type C retrovirus oncogenesis in the mouse. Among other significant findings, we have observed that immunological tolerance against viral antigens involving the T lymphocyte population may be of critical importance for the appearance and growth of the induced tumors. In this paper more recent data are presented, which further confirm the existence of immunological tolerance in mice chronically infected with murine leukemia virus (MuLV), and possibly throw light on the operational mechanism underlying the tolerance. To illustrate some characteristic features of the experimental system used, a short outline of previous work will also be presented.

II. BIOLOGICAL PROPERTIES OF THE M-MuLV/M-MuSV TUMOR SYSTEM

With the aim of investigating the host immune response to MuLV, we have performed in the past an extensive series of experiments making use of the Moloney-murine sarcoma virus (M-MuSV) tumor system. This experimental model has been proven highly reproducible and its most advantageous aspects are a short latent period of 5 to 10 days preceding tumor appearance, and spontaneous regression of the established primary tumor. Even if M-MuSV belongs to the acute transforming virus class, its defectiveness for replication requires the helper activity of slowly transforming MuLV in order to synthesize the envelope components necessary for infection. Consequently, M-MuSV particles, as well as virus infected cells, possess the antigenic specificities of associated helper virus, and the vigorous host immune response evoked by the tumor tissue is directed mostly against the MuLV-coded antigens (4). Thus, tumor induction by M-MuSV may be considered a sort of amplifying system as far as immune reactivity to MuLV is considered.

Because of its non-clonal growth, it has been suggested that M-MuSV-induced tumors are sustained by a high constant rate of virus replication with continuous recruitment of newly infected transformed cells (5). Therefore, when virus synthesis is slowed down by host immune reaction, prevention as well as tumor regression may result. Conversely, lack of regression leading to fatal progressive tumor growth is usually found in newborn and very aged mice, or in mice which have been immunologically depressed by a variety of treatments (6).

The phenomenon of spontaneous tumor regression depends mostly on the T lymphocyte population, since M-MuSV injection in T lymphocyte-deficient (7) or in nude mice (8) produces tumors that grow until the host's death. In addition, extensive studies have indicated that in lymphoid organs from M-MuSV immune mice cytotoxic T lymphocytes (CTL) are generated that exert in vitro a strong specific cytotoxicity against target cells bearing the relevant viral antigens, detectable by a short term ^{51}Cr release test (9,10).

We have also observed that M-MuSV oncogenicity may be greatly enhanced by previous infection of newborn mice with MuLV. Thus, mice characterized by early expression of endogenous MuLV (e.g. AKR, C58) or mice injected at birth with exogenous Graffi- or passage A Gross-MuLV, when challenged as young adults with M-MuSV had a remarkably higher percentage of fatally growing tumors (11). Since the MuSV recovered from the progressing sarcoma possessed an antigenicity and a host range identical to that of pre-infecting helper MuLV, it was evident that a phenotypic mixing phenomenon had occurred. Accordingly, the possibility was advanced that progressing tumors could be sustained by the new MuSV pseudotype, formed in vivo with endogenous or pre-existing exogenous MuLV, against which the host was poorly immunoresponsive (11).

A similar enhancement of M-MuSV oncogenesis was observed when mice were neonatally injected with Moloney (M)-MuLV, the natural helper for M-MuSV. As shown in Table 1, a high percentage of dually virus infected mice had sarcoma at the injection site, an incidence which was comparable to that observed in mice inoculated with the M-MuSV only. However, only a few dually infected mice regressed their tumor, while the great majority of mice ultimately died with a large tumorous mass. In addition, some mice died with sarcoma and concomitant generalized leukemia. In the control groups, only 8 percent of the mice injected with M-MuSV died with progressing tumor, and 79 percent of the mice receiving M-MuLV at birth developed lymphatic leukemia or localized thymic lymphoma at 4 to 6 months of age.

Thus, by injecting the M-MuLV/M-MuSV virus complex either tumor *progressor* or *regressor* mice could be obtained, depending on whether the host animals had been infected with M-MuLV at birth (*carrier* mice) or not. We then utilized this simple tumor system to investigate the difference in the immune reactivity existing between M-MuSV tumor regressor and progressor mice at the cellular level. Inbred mice of the C57BL/6 strain were used throughout the following studies.

First, the cytotoxic activity of spleen cells from normal mice injected with M-MuSV (*regressors*) was compared to that of mice dually infected with M-MuLV and M-MuSV (*progressors*). In agreement with previous results (12), it was observed (Table 2) that spleen

Table 1

Tumor Induction in C57BL/6 Mice Injected at Birth With M-MuLV and as Adults With M-MuSV.[a]

Treatment	Total No. mice	% mice with sarcoma[b]	% mice dead with sarcoma[c]	% mice dead with lymphoma[b]
M-MuLV	63	0	0	79
M-MuSV	60	85	8	0
M-MuLV+M-MuSV	58	81	89	22

a - M-MuLV ($5x10^8$ PFU/ml on SC-1/XC cells) was injected subcutaneously in newborn mice at a dose of 0.05 ml; M-MuSV ($1-5x10^6$ FFU/ml on 3T3FL cells) was injected intramuscularly in 6-8 week old mice at a dose of 0.05 ml.
b - Percentage of mice calculated from total number of mice.
c - Percentage of mice calculated from number of mice with sarcoma.

cells from tumor progressor mice, tested either directly or after in vitro restimulation with MBL-2 cells (a C57BL/6 transplantable leukemia cell line originally induced by M-MuLV), did not lyse the MBL-2 target cells, whereas those from regressor mice gave high cytotoxicity.

The failure to generate virus specific CTL could not be ascribed to a nonspecific immune depression caused by M-MuLV neonatal infection, since we have noted that spleen cells from M-MuLV carrier mice were able to generate alloreactive CTL (12).

It is known that mice injected at birth with MuLV express viral antigens, that are detectable on the cell surface early in the infection by serological methods (13). We have also observed that in M-MuLV carrier mice viral antigens could be revealed on the surface of lymphoid cells by using virus-immune CTL (12). A further experiment was then carried out to study the M-MuLV antigen expression on different cell populations from young adult M-MuLV carrier mice. From Table 3 it is evident that target structures for virus-immune CTL are widely represented on different T cell subsets, as well as on B lymphocytes and macrophages, since all these targets were efficiently lysed at levels comparable to those of MBL-2 leukemia targets. No lysis was noted when different

Table 2

Cytotoxic Activity of Spleen Cells
From M-MuSV Tumor Regressor or Progressor Mice[a]

Effector cell donors[b]	Stimulating cells	Target cells	% specific ^{51}Cr release at effector to target ratio of:		
			50:1	25:1	12:1
Regressors	none	MBL-2	38	19	10
Progressors	none	MBL-2	0	1	1
Regressors	MBL-2	MBL-2	54	39	24
Progressors	MBL-2	MBL-2	3	0	1

a - The cytotoxic activity was evaluated directly or after cocultivation with MBL-2 stimulating cells pretreated with mitomycin C.

b - Mice injected with M-MuLV 14-18 days previously were splenectomized and used as donors of effector cells. All mice injected with M-MuSV only eventually regressed the tumor. All mice infected at birth with M-MuLV, and as adults with M-MuSV, ultimately died with progressing tumor.

cell subpopulations obtained from normal donors were used (data not shown). This finding, together with the observation that viral antigens detected by immune CTL could be revealed on spleen and thymus cells as soon as 10 days after M-MuLV neonatal injection (12), strongly suggests that appearance of viral antigens on different lymphoid cell populations is causally related to the failure of specific CTL activity in M-MuLV carrier or M-MuSV tumor progressor mice.

III. MECHANISM OF IMMUNOLOGICAL TOLERANCE TO M-MuLV INDUCED ANTIGENS

The conventional concept of immunological tolerance as originally proposed by Burnet and Fenner more than 30 years ago implies that specific lymphocytes encountering an antigen may receive a negative signal resulting in their physical elimination (14). This possibility was particularly advanced for animals developing, during perinatal life, a state of immunological unresponsiveness to their own body constituents or to some infectious, self-replicating agents.

Table 3
Lysis of Different Cell Populations From M-MuLV Carrier Mice By Immune Cytotoxic T Lymphocytes Restimulated in Vitro With MBL-2 Leukemia Cells

Target cells	% specific ^{51}Cr release at effector to target ratio of:		
	45:1	15:1	5:1
Thymocytes[a]	46	28	17
Spleen T lymphocytes (unselected)[a]	30	20	15
Spleen T lymphocytes Lyt 1^+ [b]	33	20	17
Spleen T lymphocytes Lyt 2^+ [c]	33	29	18
Spleen B lymphocytes[d]	38	24	12
Macrophages[e]	41	37	31
MBL-2	50	32	27

a - Con A-stimulated blast cells
b - Con A-stimulated blast cells pretreated with anti-Lyt 2.2 monoclonal antibody (NEN, Dreieich, Germany) and C'.
c - Con A-stimulated blast cells pretreated with anti-Lyt 1.2 monoclonal antibody (NEN, Dreieich, Germany) and C'.
d - LPS-stimulated blast cells;
e - Adherent peritoneal exudate cells.

However, this concept has been challenged by other models of unresponsiveness as they were experimentally developed (15,16). We have undertaken experiments to investigate which of the commonly proposed mechanisms is at work in our M-MuLV/M-MuSV tumor system.

A - Lymphoid cells physically present but incapable of response to antigen

Exposure of immunocompetent cells to high concentrations of specific antigen, particularly in the case of multivalent antigens including antigen-antibody complexes, could lead to saturation of the antigen receptors and block subsequent antigen-specific stimulation. Although it has been reported that irreversible inactivation of B cells may be induced in some experimental situations by such a receptor blockade mechanism, in other cases the unresponsiveness can be reversed by enzymatic treatment of cell membrane to remove the blockading antigens (17). In rats neonatally infected with Gross-MuLV, Myburg, and Mitchihson (18) have reported data suggesting that activation of lymphoid cells unresponsive to virus-specific antigens could be obtained by pretreatment of effector cells with trypsin. Accordingly, we performed an experiment to verify the effect of trypsinization on spleen cells of M-MuSV tumor progressor mice. As illustrated in Table 4, trypsinization of spleen cells from progressor or carrier mice did not modify their incapability to exert a cytotoxic activity when directly assayed on MBL-2 target cells. These data confirm our previous results (12) showing that preincubation of effector cells for 14-24 hours in medium alone, in order to shed virus-antibody complexes possibly blocking antigen receptors from the cell membrane, was ineffective in reversing the unresponsive state.

B - <u>Active suppression by regulatory T cells</u>

Antigen-specific suppressor T cells have been demonstrated to mediate immunological unresponsiveness in many experimental systems (19). The existence of active tolerance has been also hypothesized in the M-MuLV/M-MuSV tumor system on the basis of results reported by Plater et al. (20) who observed that tumor progressor mice possess specific suppressor T cells. On the other hand, negative results have been obtained in repeated attempts carried out in our laboratory to ascertain whether suppressor T cells play a role in determining specific unresponsiveness of CTL in M-MuLV carrier or in M-MuSV tumor progressor mice. In fact, no suppression was observed on the in vivo protective effect conferred by immune spleen cells from M-MuSV regressors transferred together with progressor spleen cells into T cell-deficient mice subsequently challenged with M-MuSV (21).

To further investigate the possibility that active suppression by regulatory T cells could be at work in M-MuSV tumor progressor mice, the following experiment was set up. Virus-specific CTL generation was studied by cocultivating M-MuSV immune spleen cells with different doses of unblocked stimulator cells, obtained from spleen of progressor mice. Since variable numbers of living cells were recovered from the cultures, the results of this experiment (reported in Table 5) are expressed as lytic units (LU) per culture. It is clear that the capacity of spleen cells from M-MuSV progressor mice to induce CTL generation increased by increasing their dose in culture, and at 1:1 effector/stimulator cell ratio it was even higher than that seen at the optimal dose ($5x10^6$) of MBL-2 leukemia cells. Pretreatment of stimulating cells with anti-Thy 1.2 serum

Table 4

Effect of Enzyme Pretreatment on Cytotoxic Activity of Spleen Cells From M-MuSV Tumor Progressor of Regressor Mice and From M-MuLV Carrier Mice

Effector Cell donors[a]	Trypsinization[b]	% specific ^{51}Cr release at effector/target cell ratio of: 100:1	50:1	25:1
Regressors	-	30	26	17
	+	38	21	16
Progressors	-	3	2	2
	+	4	3	3
Carriers	-	2	1	1
	+	3	2	2

a - See Table 2
b - The spleen cell suspensions were incubated for 40 min. at 37°C in medium containing 0.125% trypsin; they were then washed twice, resuspended and assayed for cytotoxic activity on target MBL-2 leukemia cells.

and C' to eliminate T suppressor cells possibly present, paradoxically reduced the stimulatory activity; this reduction could be partially ascribed to the decrease in the stimulatory cell number due to preincubation with antiserum and C' that caused 32 percent cell lysis.

These findings support the conclusion that suppressor T cells are not responsible for the failure of M-MuLV carrier or M-MuSV tumor progressor mice to generate virus-specific CTL.

C - Functional clonal deletion

To obtain direct evidence that T cell tolerance in the M-MuLV/ M-MuSV tumor system is due to a central defect - i.e. elimination of specific cell clones - virus-specific CTL precursor frequency was evaluated by making use of a limiting dilution microculture

Table 5

Virus-Specific CTL Generation Following Stimulation of M-MuSV Immune Spleen Cells With Unblocked Spleen Cells From M-MuSV Tumor Progressor Mice[a]

Stimulating cells[b]	Stimulating cell pretreatment[c]	Stimulating cell dose ($x10^{-6}$)	LU per culture[d]
progressor spleen cells	none	1	4.5
		2	8.3
		5	12.5
		20	96.0
progressor spleen cells	anti-Thy 1.2	2	2.9
		5	4.7
		20	8.2
MBL-2 cells	none	5	42.0

a - $20x10^6$ spleen cells, obtained 14 days after M-MuSV injection and restimulated in vitro, were tested against MBL-2 leukemia cells.

b - Unblocked spleen cells from progressor mice or mitomycin C blocked MBL-2 cells.

c - Stimulator cells were treated with serum and C' before adding in cultures.

d - LU = Lytic units; 1 LU = No of CTL/culture able to give 50% lysis of 10^4 target cells.

assay according to Brunner et al. (22). Briefly, limiting numbers of spleen cells were plated in round-bottomed microwells together with $3x10^4$ mitomycin C blocked MBL-2 leukemia cells and $1x10^6$ mytomycin C blocked syngeneic spleen cells as a feeder layer (virus-specific CTL-precursor evaluation), or with $1x10^6$ allogeneic spleen cells (allogeneic CTL-precursor evaluation). Secondary mixed leukocyte culture supernate was also added to the cultures as a source of T cell growth factor (TCGF). After 7 days each microculture was assayed for cytotoxicity on appropriate target cells

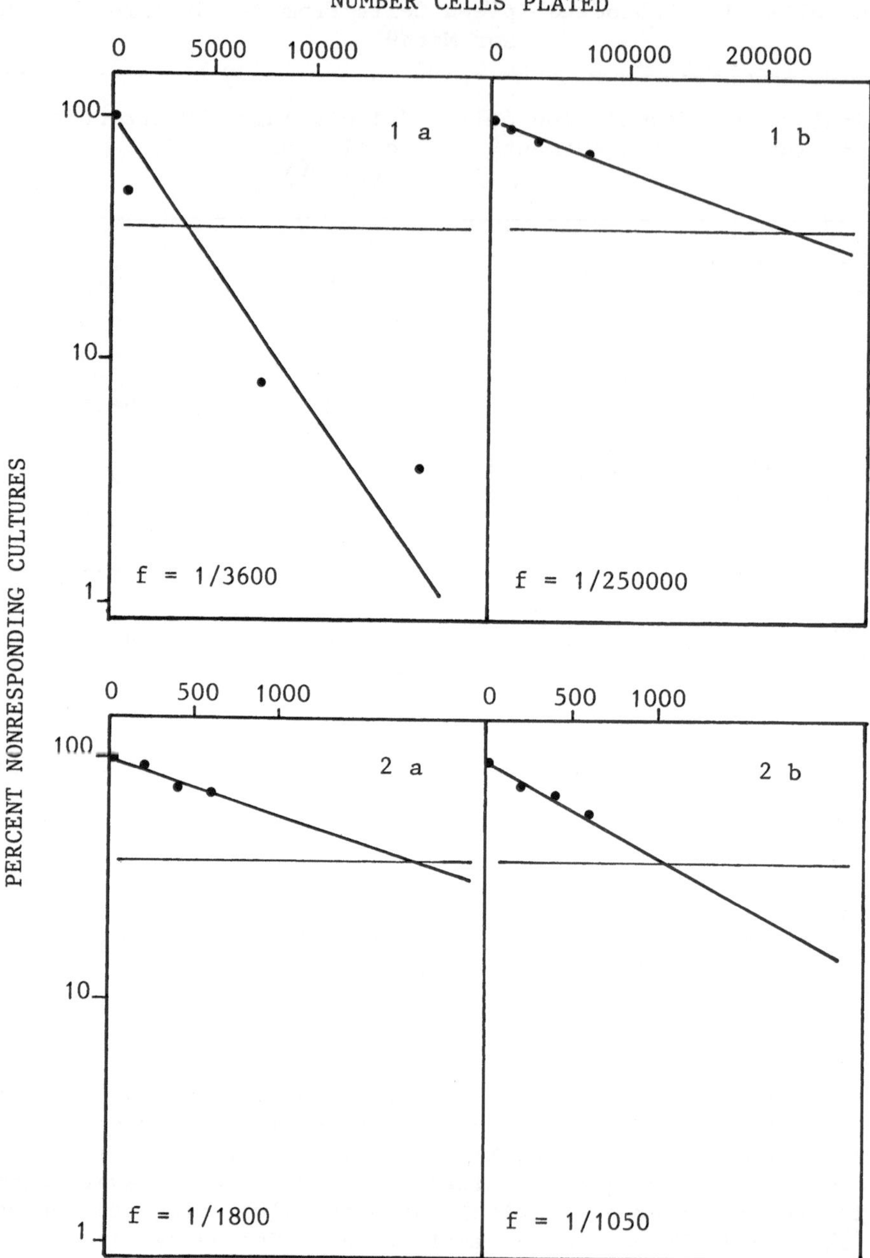
NUMBER CELLS PLATED
PERCENT NONRESPONDING CULTURES
0
5000
10000
0
100000
200000
100
10
1
1 a
1 b
f = 1/3600
f = 1/250000
0
500
1000
0
500
1000
100
10
1
2 a
2 b
f = 1/1800
f = 1/1050

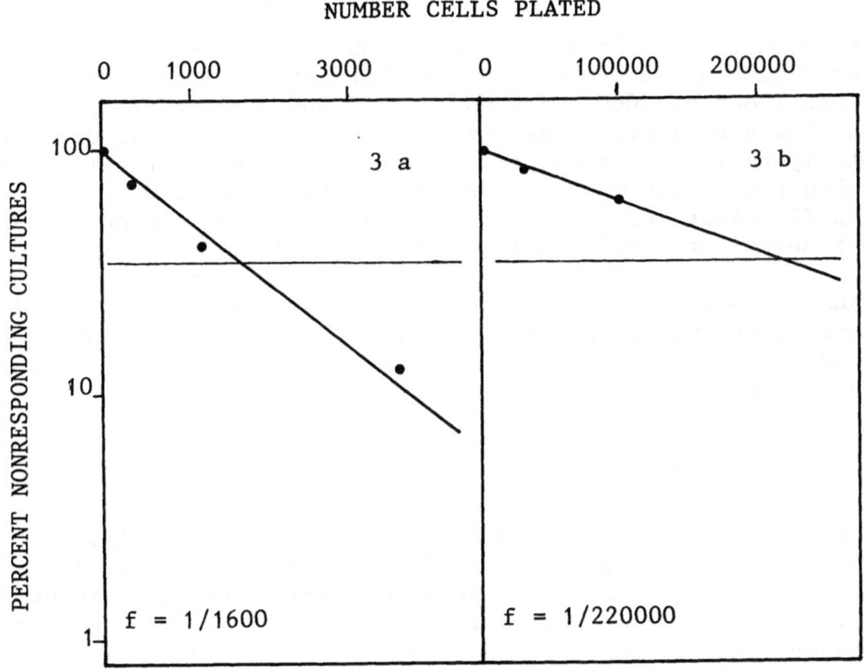

Fig.1 - Minimal estimate of the frequency (f) of CTL precursors specific for MBL-2 leukemia cells in normal (1a) or carrier (1b) mice.

Fig.2 - Minimal estimate of the frequency (f) of CTL precursors specific for P815 allogeneic cells in normal (2a) or carrier (2b) mice.

Fig.3 - Minimal estimate of the frequency (f) of CTL precursors specific for MBL-2 leukemia cells in M-MuSV tumor regressor (3a) or progressor (3b) mice.

and the cultures were considered positive (responding) when the ^{51}Cr release exceeded the mean spontaneous release, in the absence of responding cells, by at least 3 SD. The minimal estimate of CTL precursor frequency was then calculated according to the Poisson distribution by analysis of the relationship between the percentage of non-responding cultures and the dose of responding cells plated.

As shown in Fig. 1a and b, the frequency of virus-specific CTL precursors (as determined by the slope of the linear regression curve) was 1 out of 3600 and 1 out of 250.000 cells plated, for spleens from normal or, respectively, carrier mice. This difference was not imputable to a non-specific unresponsiveness of carrier mice since when the frequency of alloreactive CTL precursors against P815 (DBA/2) tumor cell target was calculated it was found comparable for normal or carrier spleen (Fig. 2a and b).

Finally, when spleen cells from regressor and progressor mice were tested, the virus-specific CTL precursor frequencies observed, were 1 out of 1600 and 1 out of 220.000 cells plated, respectively (Fig. 3a and b).

IV. DISCUSSION

For many years it has been known that mice perinatally infected with slowly transforming leukemia viruses, while retaining infectious virus throughout life, mount a deficient antibody response and fail to produce transplantation resistance towards syngeneic MuLV-induced leukemia cells even after repeated immunizations (23, 24). The most obvious interpretation for these findings was in terms of immunologic tolerance, as originally defined by Burnet and Fenner (14). This conclusion has been reappraised, however, following recent acquisitions regarding the complexity of immunoregulatory mechanisms; consequently, alternative models of tolerance have been developed.

In previous studies, we were able to demonstrate that mice neonatally injected with M-MuLV were devoid of virus-specific CTL, albeit still active in producing antibodies against viral reverse transcriptase (25). In more recent experiments (12) as well as in the present studies, using the M-MuLV/M-MuSV tumor system, the existence of a tolerant state involving the CTL population has been confirmed. Mice injected at birth with M-MuLV and challenged as adults with M-MuSV were unable to regress the induced sarcoma, nor could they generate virus-specific CTL. Investigations to elucidate the possible mechanism responsible for the CTL tolerance indicated that blockade of antigen receptors or suppressor T cells are not involved. On the other hand, through the use of the limiting dilution microculture assay, evidence has been obtained for a profound deficiency in virus-specific CTL precursors, both in M-MuLV carrier and in M-MuSV tumor progressor mice.

The functional deletion of maturing cells during a stage of lymphocyte differentiation in which the cells are highly sensitive to tolerigenesis following contact with antigen has been termed *clonal abortion* by Nossal and Pike (26). This mechanism, originally demonstrated for developing B cells in the adult bone-marrow (26) and in the neonatal spleen (27), has been recently proposed also for the T cells which are tolerant to antigens of the major histocompatibility complex (28). In the case of MuLV-induced antigens, it is clear that T cell tolerance is preferentially established in mice which are infected during the first two weeks of life (29); moreover, a few days after infection of newborn but not of adult mice, viral antigens detectable by CTL are widely expressed among the various lymphoid cell populations, including maturing thymocytes (12 and unpublished results). Therefore, it seems likely that the virtual deletion of virus-immune CTL precursors observed in the present studies might represent an early event, thus satisfying the requirement for the *clonal abortion* mechanism of tolerance.

Preliminary results indicate that in M-MuLV carrier mice, besides CTL precursor deletion and absence of suppressor T cells, there is also a very low frequency of virus-specific helper T cell precursors. These findings, on the whole, do not support the possibility that the development of T cell lymphomas in neonatally MuLV-infected mice is the consequence of a chronic stimulation of virus-specific T cells by viral antigens (2,30,31). Therefore, our data substantiate the alternative hypothesis that the specific lack of immune reactivity in these mice allows a persistent T cell infection by M-MuLV, that in turn may induce neoplastic transformation at a given hot spot of target cell genome, or by facilitating the generation of recombinant viruses that function ultimately as leukemogenic agents.

V. SUMMARY

To investigate the phenomenon of immunological tolerance to Moloney murine leukemia virus (M-MuLV)-induced antigens, mice injected at birth with M-MuLV were subsequently challenged when adults with M-MuSV. In contrast with normal mice injected with M-MuSV only, the dually infected mice failed to regress the induced sarcomas and died with fatally growing sarcoma. Spleen cells from M-MuSV tumor progressor mice, unlike those from regressors, were unable to generate virus-specific cytotoxic T lymphocytes (CTL) in vitro. This lack of immune reactivity does not appear due to: a) non-specific immune depression of T cell reactivity; b) blockade of T cell receptors; and c) presence of specific suppressor of T cells.

Since it was found that maturing T cells are highly susceptible to virus infection and a strong reduction in the frequency of virus-specific CTL precursors in M-MuLV carrier and M-MuSV tumor progressor mice was observed by a limiting dilution assay, it seems likely that a *clonal abortion* of virus-specific T cell precursors represents the basic mechanism for the tolerant state. The significance of tolerance to viral antigens in reference to neoplastic cell transformation by retroviruses is discussed.

ACKNOWLEDGMENTS

This work was supported by grants from Consiglio Nazionale delle Ricerche, progetto finalizzato Controllo della Crescita Neoplastica N°80.01514.96 and 80.01519.96, Roma, and Associazione italiana per le Ricerche sul Cancro, Milano. The technical assistance of Mr. S. Mezzalira, Ms. G. Miazzo, L. Canova and P. Segato is gratefully acknowledged.

REFERENCES

1. P. Duesberg, and K. Bister, Transforming genes of retroviruses: definition, specificity and relation to cellular DNA, in: "Cancer: Achievements, Challenges and Prospects for the 1980s," (J.H. Burchenal and H.F. Oettgen eds.), Vol. 1, p. 111, Grune and Stratton, New York (1981).

2. M.S. McGrath, E. Pillemer, D. Kooistra, and I.L. Weissman, The role of MuLV receptors on T-lymphoma cells in lymphoma cell proliferation, in "Contemporary Topics in Immunobiology," (N.L. Warner, ed.), Vol. 11, p. 157, Plenum Press, New York/ London (1980).

3. W. Hayward, B. Neel, and S. Astrin, Activation of a cellular oncogene by promoter insertion in ALV-induced lymphoid leukosis, Nature (London), 290:475 (1981).

4. L. Chieco-Bianchi, and D. Collavo, Some illustrative systems of viral carcinogenesis: The leukemia-sarcoma virus complex in the mouse, in: "Scientific Foundations of Oncology" (T. Symington and R.L. Carter eds.), p. 388, William Heinemann Medical Books Ltd, London (1977).

5. L. Chieco-Bianchi, A. Colombatti, D. Collavo, F. Sendo, T. Aoki, and P.J. Fischinger, Tumor induction by murine sarcoma virus in AKR and C58 mice: reduction of tumor regression associated with appearance of Gross leukemia virus pseudotypes, J. Exp. Med., 140:1162 (1974).

6. J.P. Levy, and J.C. Leclerc, The murine sarcoma virus induced tumors: exception or general model in tumor immunology? Adv. Cancer Res., 24:1 (1977).

7. D. Collavo, A. Colombatti, L. Chieco-Bianchi, and A.J.S. Davies, T-lymphocyte requirement for MSV tumor prevention or regression, Nature, (London) 249:169 (1974).

8. O. Stutman, Delayed tumour appearance and absence of regression in nude mice infected with murine sarcoma virus, Nature (London), 253:142 (1975).

9. F. Plata, J.C. Cerottini, and K.T. Brunner. Primary and secondary in vitro generation of cytolytic T lymphocytes in the murine sarcoma virus system. Eur. J. Immunol., 5:227 (1975).

10. D. Collavo, A. Parenti, G. Biasi, L. Chieco-Bianchi, and A. Colombatti, Secondary in vitro generation of cytolytic T lymphocytes (CTLs) in the murine sarcoma virus system. Virus-specific CTL induction across the H-2 barrier, J. Natl. Cancer Inst., 61:885 (1978).

11. L. Chieco-Bianchi, D. Collavo, A. Colombatti, G. Biasi, A. Parenti, E. D'Andrea, and A. De Rossi, Multifactorial control of M-MSV tumor induction and progression, in: "Tumor-associated Antigens and their Specific Immune Response" (F. Spreafico and R. Arnon, eds.), p. 71, Academic Press, London (1979).

12. D. Collavo, P. Zanovello, G. Biasi, and L. Chieco-Bianchi, T lymphocyte tolerance and early appearance of virus induced cell surface antigens in Moloney-murine leukemia virus neonatally injected mice, J. Immunol., 126:187 (1981).

13. W.H. Burns, and A.C. Allison, Virus infections and the immune responses they elicit, in: "The Antigens" (M. Sela, ed.), Vol. 3, p. 479, Academic Press, New York (1975).

14. F.M. Burnet, and F. Fenner, The Production of Antibodies, Mac Millan, Melbourne (1949).

15. D.E. Parks, and W.O. Weigle, Current perspectives on the cellular mechanisms of immunological tolerance, Clin. Exper. Immunol., 39:257 (1980).

16. G.J.V. Nossal, and B.L. Pike, Antibody receptor diversity and diversity of signals, in: "Progress in Immunology 80," (M. Fougereau and J. Dausset, eds.), Vol. 1, p. 136, Academic Press, London (1980).

17. C. Fernandez, and G. Möller, Irreversible immunological tolerance to thymus-independent antigens is restricted to the clone of B-cells having both Ig and PBA receptors for the tolerogen, Scand. J. Immunol., 7:137 (1978).

18. J.A. Myburgh and N.A. Mitchison, Suppressor mechanism in neonatally acquired tolerance to a Gross virus-induced lymphoma in rats. Transplantation 22:236 (1976).

19. I. Kamo and H. Friedman, Immunosuppression and the role of suppressive factors in cancer, Adv. Cancer Res., 25:271 (1977).

20. C.P. Plater, P. Debré, and J.C. Leclerc, T cell-mediated immunity to oncorna-virus-induced tumor. III. Specific and non specific suppression in tumor-bearing mice, Eur. J. Immunol. 11:39 (1981).

21. D. Collavo, F. Ronchese, P. Zanovello, G. Biasi, and L. Chieco-Bianchi, T cell tolerance in Moloney-murine leukemia virus (M-MuLV) carrier mice low cytotoxic T lymphocyte precursor frequency and absence of suppressor T cells in carrier mice with Moloney-murine sarcoma (M-MSV)-induced tumors, J. Immunol., 28:774 (1982).

22. K.T. Brunner, H.R. MacDonald, and J.C. Cerottini, Antigenic specificity of the cytolytic T lymphocyte (CTL) response to murine sarcoma virus-induced tumors. II. Analysis of the clonal progeny of CTL precursors stimulated in vitro with syngeneic tumor cells, J. Immunol., 124:1627 (1980).

23. G. Klein and E. Klein, Immunological tolerance of neonatally infected mice to the Moloney leukemia virus. Nature (London) 209:163 (1966).

24. L. Chieco-Bianchi, L. Fiore-Donati, G. Tridente, and N. Pennelli, Graffi virus and immunological tolerance in leukaemogenesis, Nature (London), 214:1227 (1967).

25. L. Chieco-Bianchi, F. Sendo, T. Aoki, and O.L. Barrera, Immunological tolerance to antigens associated with murine leukemia viruses: T-cell unresponsiveness? J. Natl. Cancer Inst., 52: 1345 (1974).

26. G.J.V. Nossal and B.L. Pike, Evidence for the clonal abortion theory of B-lymphocyte tolerance, J. Exp. Med., 141:904 (1975).

27. E.S. Metcalf and N.R. Keinman, In vitro tolerance induction of neonatal murine B cells, J. Exp. Med. 143:1327 (1976).

28. G.J.V. Nossal and B.L. Pike, Functional clonal deletion in immunological tolerance to major histocompatibility complex antigens, Proc. Natl. Acad. Sci. USA, 78:3844 (1981).

29. L. Chieco-Bianchi, L. Fiore-Donati, D. Collavo, and G. Tridente, Immunological problems in virus-induced leukemia in the mouse, in "Immunity and Tolerance in Oncogenesis (L. Severi, ed.), p. 599, Division of Cancer Research, Perugia (1970).

30. J. Lee, I. Horak, and J. Ihle, Mechanisms in T cell leukemogenesis. II. T cell responses of preleukemic BALB/c mice to Moloney leukemia virus antigens, J. Immunol., 126:715 (1981).

31. J. Lee and J. Ihle, Chronic immune stimulation is required for Moloney leukaemia virus-induced lymphomas, Nature (London), 289:407, 1981.

TRANSFORMING GENES OF AVIAN RETROVIRUSES AND THEIR RELATION TO CELLULAR PROTOTYPES*

P. Duesberg+, T. Robins+, W.-H. Lee+, C. Garon++, T. Papas++, and K. Bister+

+Department of Molecular Biology, University of California, Berkeley, California 94720
++Tumor Virus Genetics Laboratory, National Cancer Institute, Bethesda, Maryland 20205

ABSTRACT

The relationship between two types of retroviral onc genes and cellular structural homologs termed proto-onc genes was studied. The type I Rous sarcoma virus (RSV) src gene, which is unrelated to essential virion genes, was found to have a complete structural homolog in cloned chicken DNA based on fingerprinting RNA-DNA hybrids. By the same techniques only the specific part (mcv) of the type II MC29 virus onc gene, which is a hybrid that also includes part (Δ) of the structural gag gene of retroviruses (Δgag-mcv), was found to have a structural homolog in the cell. Hence, the onc gene of MC29 does not have a complete homolog in the cell. Both onc-related cellular loci are not linked to any other virion sequences. Presumed host markers of certain viral src genes, said to be experimentally transduced from the cell, were not detected in the proto src-locus. The cellular mcv-locus was found to be interrupted by one sequence of non-homology relative to the viral counterpart; the src-locus is known to be interrupted by six. We deduce that there is a close qualitative sequence-homology between the virion gene-unrelated sequences of viral onc genes and cellular proto-onc genes. However, functional homology between viral onc genes and proto-onc loci cannot be deduced due to the different arrangements of onc-related sequences in viruses and cells and to scattered single nucleotide differences in their

*Essentially the same manuscript was presented at the International Symposium for Comparative Leukemia Research in Los Angeles, California, August 31 - September 5, 1981.

primary structures and due to the lack of Δ gag in cellular prototypes of hybrid onc genes, such as Δgag mcv. Considering the genetic structures of RSV and MC29 and those of the corresponding cellular DNA loci, it follows that the generation of viruses like RSV and MC29 by transduction of cellular sequences into the genome of a retrovirus must have involved rare, illegitimate recombinations and specific deletions.

INTRODUCTION

The hallmark of the transforming onc genes of acutely-transforming retroviruses such as Rous sarcoma virus (RSV) and avian myelocytomatosis (MC29) virus, is a onc-specific, coding RNA sequence that is unrelated to virion genes which are essential for virus replication[1,2,3]. Since the identification of src, the onc gene of RSV, in 1970 by deletion[3,4] and subsequent recombination analysis, over a dozen different onc-specific sequences have been identified in various oncogenic retroviruses. Seven of these belong to oncogenic viruses of the avian tumor virus group (Fig. 1)[1]. The relative abundance of different onc genes in the avian tumor virus group compared to known onc genes in other taxonomic groups[3] may either reflect the fact that the avian group of viruses has been studied more extensively than others or that chicken are more permissive for oncogenic retroviruses than other animal species. To date no such onc genes have been discovered in other taxonomic groups of viruses with oncogenic properties where oncogenesis appears to be an indirect consequence of genetic elements necessary for virus structure or replication[3].

Based on analyses of the genetic structures of oncogenic avian tumor viruses and the products they encode, we have recently distinguished two types of onc genes[1,5]. The coding sequence of type I consists entirely of specific sequences. The original example is the src gene of RSV[4,6] which encodes a 60 kd protein (Fig. 1)[1,7]. The onc gene of avian myeloblastosis virus is another example in the avian tumor virus group[8]. The coding sequence of type II onc genes is a hybrid consisting of a specific sequence and of elements of essential virion genes typically including a partial (Δ)gag gene. The original example is the onc gene of MC29 in which both Δ gag and a specific sequence, termed mcv, function as one genetic unit that encodes a 110 kd gag-related, probable transforming protein[3,9,10] (Fig. 1). The hybrid onc genes of Fujinami sarcoma virus[11] and other avian sarcoma viruses[1] and of avian erythroblastosis virus[1,12] are other examples in the avian tumor virus group (Fig. 1).

In contrast to type I onc genes, the type II onc genes of a given subgroup of oncogenic avian tumor viruses, which are defined

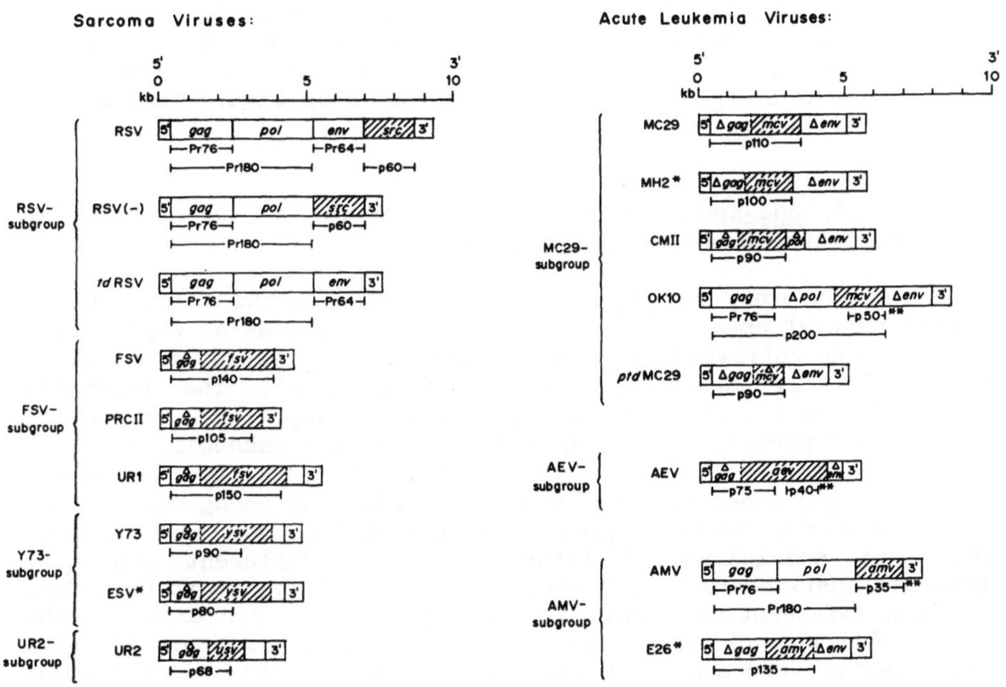

Fig. 1. Genetic structures of oncogenic retroviruses of the avian leukosis/sarcoma group: Boxes indicate the size of viral RNAs in kilobases (kb) and segments within boxes indicate map locations in kb of complete or partial (Δ) complements of the three essential virion genes gag, pol and env of the onc-specific sequences and of the non-coding regulatory sequences at the 5' and 3' end of viral RNAs. Dotted lines are used to indicate that borders between genetic elements are uncertain. Based on onc gene-specific RNA sequences (hatched boxes), four subgroups of sarcoma viruses and three subgroups of acute leukemia viruses can be distinguished. The three-letter code for onc-specific RNA sequences extends the one used previously by the authors. src represents the onc-specific sequence of the RSV-subgroup of avian sarcoma viruses; fsv that of the Fujinami-subgroup, ysv that of the Y73-subgroup, and usv that of the UR2-subgroup of avian sarcoma viruses; mcv that of the MC29-subgroup, aev that of the avian erythroblastosis virus subgroup and amv that of the AMV-subgroup of acute leukemia viruses. Lines and numbers under the boxes

symbolize the complexities in kilodaltons (kd) of the precursors (Pr) for viral structural proteins and of the transformation-specific polyproteins (p). For some viruses (*) complete genetic maps are not yet available, and some protein products (**) have only been identified in cell-free translation assays. This figure is from Reference 1.

by related, onc-specific sequences[1,3], appear to be variable in size and sequence complexity (Fig. 1)[3].

The discovery of onc genes in retroviruses has allowed us to answer several specific questions: (i) It provides an explanation for why some retroviruses, namely those with onc genes, are acutely and inevitably oncogenic, while others, namely the lymphatic leukemia viruses, are only rarely oncogenic and, if so, only after long latent periods consistent with an indirect mechanism of transformation[1,3]. (ii) Further, the existence of completely different onc genes with the same or overlapping oncogenic spectra within the same taxonomic group of viruses argues that there must be different mechanisms of transformation. Different onc gene products must interact with different cellular targets to transform a given differentiated cell into a tumor cell. For example, the four subgroups of avian sarcoma viruses (Fig. 1) and the two subgroups of sarcomagenic acute leukemia viruses (Fig. 1) all cause sarcomas in the animal with totally different transforming genes[3]. Likewise do all three subgroups of acute leukemia viruses cause erythroblastosis with different onc genes[3] (Fig. 1). The argument that different onc genes encode structurally different, but functionally identical, transforming proteins fails to explain why distinct onc genes behave differently in different cells. For example, acute leukemia viruses (but not sarcoma viruses) cause carcinomas and leukemias, in addition to sarcomas[1,3,13].

The identification of onc genes in retroviruses, however, has also raised critical questions about the origin of these genes and their relevance to natural cancers. Since retroviruses with onc genes are rarely found in natural cancers, they do not appear to play a significant role in natural carcinogenesis[14]. Several retroviruses with onc genes have been isolated from animals that developed tumors after inoculation with lymphatic leukemia viruses that do not contain onc genes as, for example, Harvey[15], Kirsten[16], and Moloney[17] sarcoma viruses and the murine Abelson[18] acute leukemia and the MC29 virus-related OK10 acute leukemia virus[19]. The rare, spontaneous, or lymphatic leukemia virus-induced, occurrence of retroviruses with onc genes and the complete lack of evidence for an epidemic, horizontal spread of retroviruses with acute onc genes suggest that retroviral onc genes must exist in cells in a latent form or may be generated from cellular genetic from cellular genetic elements. These questions and the questions of how viruses with related onc genes (like MC29 and OK10 or Harvey and Kirsten virus) appear in seemingly independent spontaneous cancers and whether lymphatic leukemia viruses are involved in the emergence of viral oncogenes were first addressed by the oncogene hypothesis of Huebner and Todaro[20]. The oncogene hypothesis postulates that viral onc genes are present in normal cells and may cause cancer if induced by carcinogens or other oncogenic agents. Subsequently, it was hypothesized that cellular genetic elements with a potential of becoming viral onc genes, termed proto-viruses (and later proto-onc genes)[21], can be transduced by retroviruses and can evolve into viral onc genes[22,23].

An experimental test of the oncogene hypothesis became possible with the identification of onc gene-specific sequences initially in RSV[4,6], then in Kirsten and Harvey sarcoma viruses[24,25,26,27] and later in many other avian and mammalian retroviruses[1,3]. The first direct evidence of sequence homology between cellular DNA and onc-specific sequences was obtained in the cases of Kirsten and Harvey sarcoma viruses[25,26,27] and later also with those of Moloney sarcoma[28], Rous sarcoma[29], MC29[30,31] and other viruses. These results lend indirect support to the oncogene and transduction hypotheses. However, these experiments did not determine whether viral onc genes (referred to as the quantitative model of oncogene hypothesis) or structural relatives of viral onc genes with possibly different functions (referred to as the qualitative model of oncogene hypothesis) were detected in normal cells.

In the case that viral onc genes have direct counterparts in normal cells, as postulated by the quantitative model, cellular transformation by viruses with onc genes is thought to be due to enhanced gene dosage[21]. In accord with the quantitative model, it was proposed that viruses without viral onc genes cause cancer by integrating proviral DNA adjacent to, and consequently promoting the expression of, cellular genes related to viral onc genes[32]. Specifically, it was concluded that the "downstream promotion" of a cellular MC29-related proto-mcv sequence is the cause for lymphomas in chicken infected by lymphatic leukemia viruses[32]. In the murine system, the quantitative model derives support from experiments which showed transformation of 3T3 fibroblasts with cellular sequences homologous to the specific-sequence of Moloney sarcoma virus (MSV) after ligation with terminal sequences of MSV[33] or with cellular sequences homologous to the transforming region of Harvey MSV again after ligation with viral terminal sequences[34]. However, with regard to the relevance of these experiments for the quantitative model, one would have to know whether viral and related cellular sequences encode functionally similar transforming proteins. In addition, it would be critical to know whether only viral or also cellular promoters could induce onc-related sequences to transform cells in order to exclude the possibility that besides promoter functions other functions are encoded in the viral terminal sequences. (It would also be interesting to know whether viral promoters could also transform cells by inducing cellular sequences unrelated to viral onc genes.) The quantitative model appears to be most directly supported in the avian system by the claim that partial src deletion mutants of RSV which lack over 75% of src[35,36,37,38] including one src-terminus[36,39,40] can reproducibly transduce src from the cells of infected chicken to regenerate complete RSVs. Such RSVs have been termed recovered (r)RSVs[35]. Proof for the transductional origin of the src genes of rRSVs has been based on presumably host-derived oligonucleotide and peptide src markers[36,37,38,40].

In an effort to distinguish between the quantitative and the qualitative model, we have compared here a prototype of each of the two classes of viral onc genes to DNA of their cellular prototypes: the src-gene of a Rous strain that was reportedly transduced from the cell[37,38] and the type II hybrid Δgag-mcv gene of MC29.

RESULTS

a. The complexities of the cellular src-related locus and viral src are about the same, but the cellular locus is not linked to viral sequences outside src and lacks markers reported as characteristic of transduced src genes.

We have recently identified about 20 RNase T_1-resistant oligonucleotides in the src genes of 10 strains of Rous sarcoma virus including two reportedly transduced froom the cell[41]. The purpose of this study was to determine the degree of variability among src genes of different viral strains as a basis for the identification of markers of transduction in src genes thought to be transduced from the cell[37,38,40]. The study concluded that src genes of all RSV strains tested, including two reportedly transduced from the cell, are completely allelic differing only in scattered single base variations[41].

Here we describe the relationship between a viral src gene reportedly transduced from the cell and a cloned cellular DNA sequence homolog in terms of sequence complexity and sequence arrangement. Fig. 2C shows the fingerprint of about 20 unique src oligonucleotides of rRSV 14-2, with a src gene presumably transduced from the cell, which was isolated by Vigne et al.[37,38,42]. This fingerprint was obtained from rRSV 14-2 (^{32}P)RNA hybridized by a src-specific cDNA. This src-specific cDNA was prepared by annealing rRSV 14-2 cDNA with unlabelled RNA of an isogenic src-deletion mutant which lacked the 1.6 kb src gene of RSV and about 200 additional nucleotides mapping adjacent to the 5' end

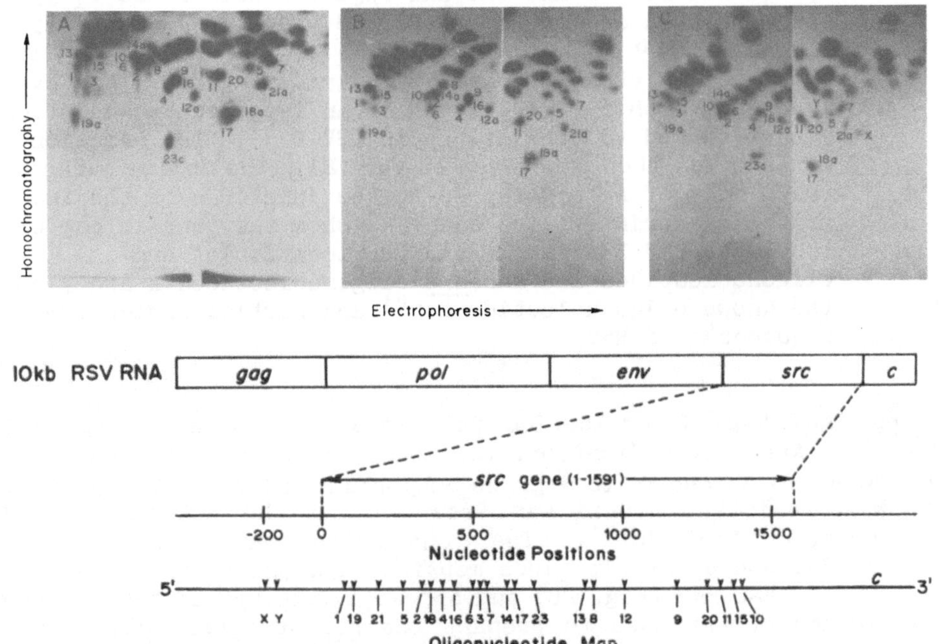

Fig. 2. Src-specific oligonucleotides of a RSV strain termed rRSV 14-2, hybridized by DNA of proto-src, the cellular src-related locus. (A) rRSV 14-2 (^{32}P)RNA (0.25 μg or 1 x 10^6 cpm) was hybridized for 8 hr at 40°C in 10 μl 70% formamide, 0.3 M NaCl, 0.03 M Na-citrate, and 10 mM Na-phosphate at pH 7 with about 20 μg of src-related DNA of the chicken cell cloned in a lambda phage CS3 that had been degraded for 12 min at 95-100°C in 0.3 N NaOH. The reaction mixture was then incubated in 200 μl 0.3 M NaCl 0.03 M Na-citrate pH 7.0 for 30 min at 40°C with RNase T_1 at 50 units/ml and the hybrid was prepared and fingerprinted as described[41]. To enhance resolution of oligonucleotides, electrophoresis was on cellulose acetate strips at pH 2.5 over a distance of 30 cm. The strip was then cut and each half was chromatographed on a 15 x 30 cm commercial DEAE-cellulose thin layer plate[41]. The letters associated with oligonucleotide numbers identify the rRSV 14-2 specific alleles of variable src oligonucleotides[41]. (B) A rRSV 14-2 (^{32}P) RNA- λCS3 DNA hybrid was prepared as for (A). The reaction mixture was then incubated in 200 μl 0.3 M NaCl, 0.03 M Na-citrate pH 7.0 with RNases A (25 μg/ml), T_1 (50 units/ml) and T_2(10 units/ml) at 40°C for 30 min and the hybrid isolated by gel exclusion chromatography. After phenol extraction, the

hybrid was melted and the RNA digested with RNase T_1 and fingerprinted as for (A). (C) rRSV 14-2 (^{32}P)RNA (0.25 μg or 1 x 10^6 cpm) was hybridized for 2 hr as for (A) with src-specific cDNA, prepared by hybridizing 0.5 μg rRSV 14-2 cDNA with 5 to 10 μg RNA of an isogenic src-deletion mutant termed td Schmidt-Ruppin RSV-D2[41]. The reaction mixture was then processed as for (B). The splice marks in the middle of A to C represent the junctures of the two half-fingerprints. The diagrams show the genetic map of the 10 kb RSV RNA genome and the location of src-oligonucleotides and non-src oligonucleotides x and y on the known oligonucleotide map[41] and partial nucleotide sequence[43] of RSV.

of src (see Fig. 2 nucleotide positions -200 to about 1600 of src[43])[41]. After hybridization, the reaction mixture was incubated with RNases A, T_1 and T_2 to degrade unhybridized RNA. Subsequently, the RNase-resistant hybrid was isolated and the RNA melted and fingerprinted (Fig. 2C)[41]. The oligonucleotides x and y derive from the 200 non-src nucleotides mapping adjacent to the 5' end of src in rRSV 14-2 RNA (Fig. 2C) and appear in the hybrid because the src-deletion mutant used to prepare the src-specific cDNA lacked these sequences[41]. The genetic map of RSV has been described (Fig. 1)[1,3]. The nucleotide positions of src are those from Czernilofsky et al.[43] and the oligonucleotide map is that from Lee et al.[41]. The composition of src oligonucleotides and oligonucleotides x (4U, 3C, G, C, 2AU, A_2C) and y (2U, C, AC, AU, A_2G) has been described elsewhere[41]. The genetic structure of RSV (5' gag-pol-env-src-c 3') [1,3] and the location of the above oligonucleotides within the known oligonucleotide map[41] and nucleotide sequence of RSV[43] have been determined and are diagrammed in Fig. 1.

To determine whether proto-src, the src-related locus of the chicken[29,44], contains all sequences represented by the src oligonucleotides of rRSV 14-2, we have used hybridization of viral RNA with cellular src-related DNA cloned in lambda phage. The particular clone was a gift of G. Cooper and has been termed λ CS3[44]. Our sequence comparisons were conducted under two conditions of stringency: a relaxed condition in which the resulting hybrid was only digested with RNase T_1, a condition which would preserve all completely and partially base-paired T_1-resistant oligonucleotides hybridized by λCS3, and a stringent condition using RNases cleaving mismatched regions within T_1-oligonucleotides, which would eliminate T_1-oligonucleotides that incompletely base-paired by DNA. A fingerprint of T_1-oligonucleotides of a rRSV 14-2 (^{32}P)RNA-λ CS3 DNA hybrid treated with RNase T_1 is shown in Fig. 2A. The fingerprint contained all src-oligonucleotides shown in Fig. 2C. However, it lacked all known, non-src virion oligonucleotides of rRSV[41] even oligonucleotides x and y from the adjacent non-coding 5' region as well as all known oligonucleotides of the adjacent 3' c-region[41](Fig. 2).

Next, the RNA of a rRSV 14-2 RNA-λCS3 DNA hybrid, which had been incubated with RNases A, T_1 and T_2 to degrade mismatched sequences, was fingerprinted. As can be seen in Fig. 2B, the hybrid lacked src oligonucleotide 23c. By contrast, a hybrid formed with src-specific cDNA contained oligonucleotide 23c after the same treatment with the same three RNases (Fig. 2C).

We can conclude that (i) the complexities of the viral src and the cellular src-related locus are about the same and (ii) that the cellular src-related locus is not linked to any viral sequences outside src including probable non-coding sequences mapping directly adjacent to src. This extends the results of others[44,45] that src-related loci of the chicken are not linked to other coding or non-coding sequences of endogenous retroviruses. (iii) Further, we conclude that the 23c allele of src oligonucleotide 23 has no identical counterpart in the chicken src locus of λCS3. Since all rRSVs from one laboratory[37,38] and some of those from another[36,40] shared only one non-parental oligonucleotide, i.e., the 23c allele of oligonucleotide 23 that was not found in other RSV strains[41], it has been argued that this oligonucleotide is a marker of cell-derived src sequences[37,38,40]. (Other RSV strains contain different alleles of oligonucleotide 23, e.g., oligonucleotide 23a or 23b[41]). Since this marker is absent from λCS3, it appears unlikely that this src marker was directly transduced from this cellular src-related locus. Nevertheless, it is possible that src-related loci of chicken may differ in scattered base changes and that other chicken may contain a proto-src with 23c. However, if one concedes that cellular src-related loci vary in single bases, one can no longer use single base changes as markers of src transduction since src genes, like other viral genes, are also subject to spontaneous point mutations[41].

b. Structural relationship between a normal chicken DNA locus and the onc gene of MC29.

In order to compare sequence-relationship between a representative type II, hybrid onc gene and a cellular counterpart, we have analyzed here the homology between MC29 RNA and proto-mcv, the mcv-related chicken locus[30] cloned in lambda phage. A lambda phage containing the mcv-related locus, termed λproto-mcv3, was selected by screening a chicken library with molecularly cloned proviral MC29 DNA as a hybridization probe[46]. We have used the same library of chicken DNA that had been used to select the above src-related locus[47].

The chicken DNA of our λproto-mcv3 recombinant phage measured about 17 kb. To locate the MC29-related sequence in the recombinant phage, a heteroduplex was prepared with proviral MC29 DNA prepared from a recombinant lambda phage[48]. The cloned fragment of MC29 DNA extended from the 5' end of the viral genome to a restriction endonuclease Eco RI site at about 2.5 kb from its 3' end. It also included about 4.5 kb of quail cell DNA mapping adjacent to the 5' end of MC29 DNA[48]. The genetic structure of MC29 has been determined previously to read 5' Δgag-mcv-Δenv-c 3'[10,13,49] and is schematically represented in Fig.4. It can be seen in Fig. 3 that the MC29-related sequence of the chicken was located in two discontinuous regions of 0.9 and 0.7 kb respectively and was interrupted relative to viral DNA by a 1 kb region of non-homology (see arrow in Fig. 4). Based on the heteroduplex and the known structure of the cloned fragment of MC29 DNA[48], the cellular 0.9 kb region was identified as co-linear with the 5' half and the 0.7 kb region as co-linear with the 3' half of most or all of the mcv sequence.

In order to determine whether the cellular MC29-related DNA contains the complete onc gene of MC29 or only part of it, the RNA of a RNase T_1-resistant hybrid formed between MC29 RNA and λproto-mcv3 DNA was fingerprinted. As shown in Fig. 4A, this hybrid contained all eight MC29-specific oligonucleotides defined previously, i.e., nos. 1, 3, 6, 7b, 8b, 15, 26, and 120 but no other MC29 oligonucleotides. Treatment of the hybrid with RNases T_1, A and T_2 prior to fingerprint analysis virtually eliminated oligonucleotides nos. 3 and 7b (Fig. 4B). Under the same conditions of digestion none of the eight MC29-specific oligonucleotides are eliminated from a MC29-specific cDNA hybrid[10] or from a MC29 RNA hybrid formed with an endonuclease Bam HI-resistant fragment of proviral MC29 DNA which was cloned in pBR322 (Fig. 4C). This fragment includes the 3' part of Δ gag, all of mcv and the 5' end of Δenv[48]. It can be seen in Fig. 4C that all mcv oligonucleotides were recovered at near equimolar ratios. The molar recovery of some gag and env-related oligonucleotides was higher since the hybrid was formed with excess DNA and hence also includes gag and env oligonucleotides of the helper virus. The gag oligonucleotide 20a has not been identified previously in MC29, but has been analyzed in CMII under the label 17b[5] and in OK10 under the label 20a[5]. The map location of all MC29 oligonucleotides described here on the MC29 RNA genome is diagrammed in Fig. 4; these include the eight mcv oligonucleotides in addition to the gag oligonucleotides nos. 20a and 13 and the env oligonucleotides nos. 14a, 7a and 2.

Based on homology with MC29-specific oligonucleotides and on the size of heteroduplexed sequences, we conclude that the complexities of the primary sequence of the cellular proto-mcv locus and of the viral mcv are about the same. However, no

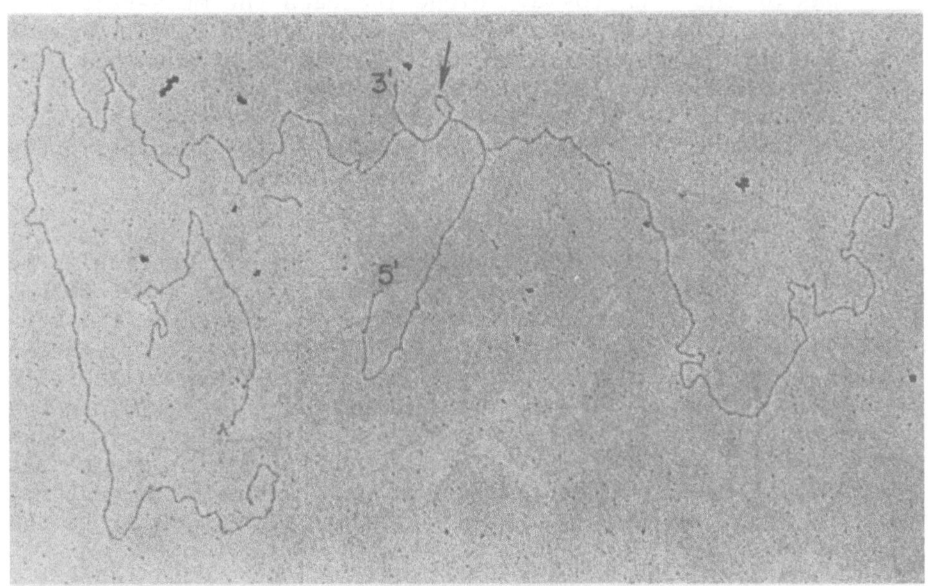

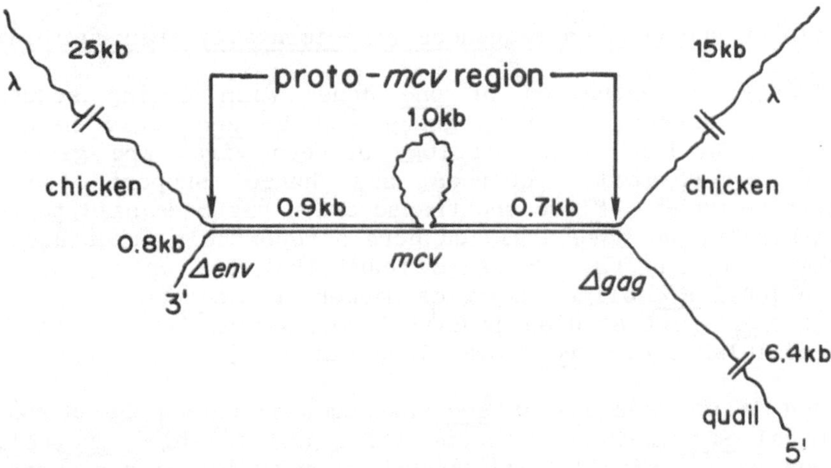

Fig. 3. Electronmicrograph of a heteroduplex formed between a fragment of molecularly cloned MC29 DNA and proto-mcv, the cellular MC29-related locus of the chicken cloned in lambda phage. Procedures for heteroduplex formation and analysis have been described[48]. The MC29 DNA used was a restriction endonuclease Eco RI-resistant DNA fragment that extends from the 5' end and includes about 4.5 kb quail cell DNA adjacent to the 5' end of the viral DNA[48].

DNA of the λ proto-mcv3 clone includes the MC29-related locus flanked by about 6 to 7 kb chicken DNA at either side and then by the two arms of the lambda phage vector[46]. The arrow marks the 1 kb sequence of nonhomology that interrupts the MC29-related sequence of λ proto-mcv3. The diagram reports length measurements of the respective DNA regions of the heteroduplex in kilobases.

oligonucleotides of virion sequences outside mcv, in particular no gag-related oligonucleotides, were detected in the MC29 RNA-λ proto-mcv3 DNA hybrids analyzed here (Fig. 4), although the mcv-related sequence of λ proto-mcv3 is flanked by 7 kb of chicken DNA on either side (Fig. 3). This conclusion is consistent with unpublished evidence of Sheiness et al. (quoted in ref. 30) that a 10 kb restriction enzyme-resistant fragment of proto-mcv DNA lacks sequences related to essential virion genes. It would appear that the cellular locus analyzed here does not contain an entire, structural homolog of the Δgag-mcv gene which is thought to be the onc gene of MC29[1,10,13,49].

DISCUSSION

Are cellular src-related sequences experimentally transducible?

Clearly, transduction of onc genes with coding sequences unrelated to essential virion genes such as src would be direct evidence for a functional homology between viral src genes and cellular src-related sequences and hence support for the quantitative model. Since src transduction has been said to occur experimentally, we have analyzed here a reportedly transduced src gene for host markers. We have found that the only presumably transductional src oligonucleotide marker of the reportedly transduced src gene studied by us[41,42] and others[37,38] is not found in the cellular proto-src locus cloned in λ CS3.

Even if we assume that src-transduction cannot be proven by biochemical src markers, because viral and cellular src-related sequences are too similar, the transduction frequencies reported in the above system cannot be reconciled with the evidence that the src-related locus is not linked to any other viral sequences[41,44,45] (Fig. 2). It has been reported that, in this system, src deletions that lack either the 5' end[36,40] or the 3' end[37,39] as well as deletions that retain both ends[36,40] nevertheless transduce src at the same high rate of over 50% within two months[35,38]. This appears surprising since illegitimate recombination with proto-src would be required for src transduction by deletions lacking one end of src while the more efficient and plausible homologous recombination would suffice to regenerate src from deletions with two

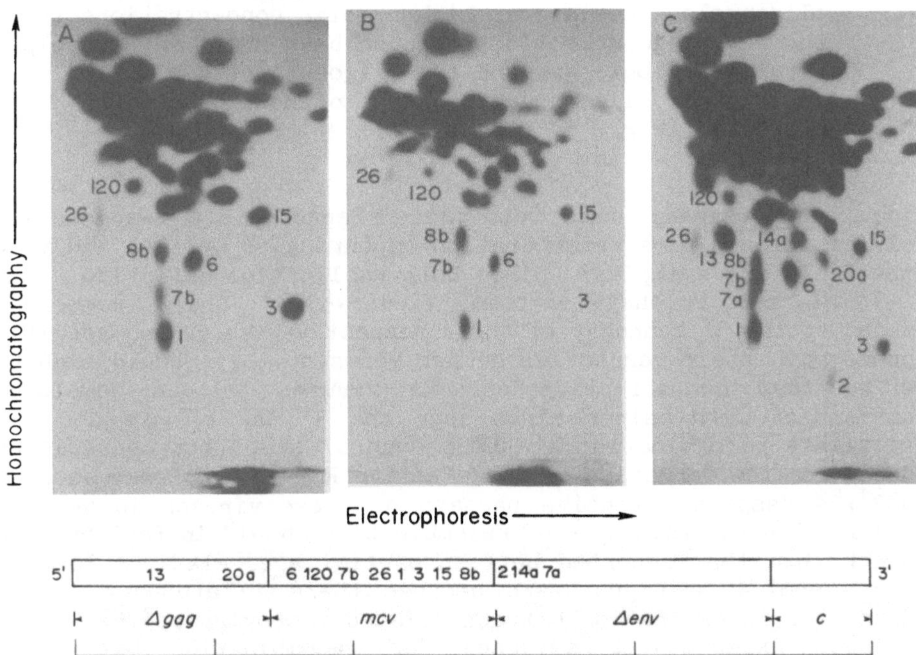

Fig. 4. The MC29 oligonucleotides hybridized by DNA of proto-mcv and by cloned proviral MC29 DNA. (A) MC29(ring neck pheasant [RPV]) (^{32}P)RNA (0.25 μg or 1 x 10^6 cpm) was hybridized with 25 μg of alkali-degraded λproto-mcv3 DNA as described for Fig. 2. After treatment of the reaction mixture with RNase T_1 as described for Fig. 2, the RNA of the hybrid was fingerprinted as described[5,10,49]. The MC29-specific (mcv) oligonucleotides are fingerprinted and numbered as in previous publications which also describe their compositions in terms of RNase A-resistant fragments[10,13,49]. The genetic structure of the MC29 genome (5' Δgag-mcv- Δenv-c 3') is schematically represented at the bottom of the figure. The order of all oligonucleotides identified here on the known viral oligonucleotide map[10,13,49] is recorded in the diagram. (B) The mcv oligonucleotides from a hybrid formed as in (A) but isolated after treating the reaction mixture with RNases A, T_1 and T_2 as detailed in Fig. 2. (C) The mcv oligonucleotides and some gag- and env-related oligonucleotides from a RNase A, T_1 and T_2-resistant hybrid formed with a restriction endonuclease Bam HI-resistant fragment of molecularly cloned MC29 DNA. The hybrid was prepared as for (B). The MC29 DNA fragment included most of the Δ gag, all of mcv and about 0.3 kb of

Δ env of the MC29 genome[48]. Some Δgag and Δenv-related oligonucleotides are at higher molar concentrations than the mcv oligonucleotides because both MC29 and RPV helper virus contribute gag and env oligonucleotides. The genetic and oligonucleotide maps of MC29 have been described (Fig. 1, ref. 1,13).

residual src termini and proto-src. Reproducible transduction involving illegitimate recombination appears also unlikely in view of the following examples: Since both replication-defective RSV(-)[3,50] and Moloney sarcoma virus[3,24,33] share sequences adjacent to the 3' boundary of their respective src or onc-specific sequence with their respective helper viruses, they should each be able to form non-defective sarcoma viruses by one homologus recombination with helper virus near the 3' end of src and one illegitimate recombination at the 5' end of src. Yet there is no evidence for the formation of nondefective RSV[51] or Moloney sarcoma virus[51,52] despite extensive passage of these viruses in animals and cell culture. This type of recombination should in fact be more frequent than that reported between partial src deletions lacking one src-terminus and proto-src because there is abundant direct evidence for recombination between related retroviruses[3,49,53] but no direct experimental evidence for recombination between a retrovirus and a cell. Furthermore, in viral recombination no introns need to be removed from viral src.

Hence, it cannot be excluded that cross-reactivation between the predominant src-deletion and nonoverlapping src-deletions possibly present as minor components in the stocks of src-deletion mutants used to generate recovered RSVs or reversion of minor variants containing complete but mutated src genes rather than transduction was the origin of the src genes of recovered RSVs[41,42]. This is suggested because long-term persistence of stable heterozygotes has been described previously in retroviruses subjected to extensive biological cloning even under selective conditions, non-permissible for the host range of one of two viral components[52,53] and because the partial src-deletion mutants used to generate rRSVs were not molecularly cloned[35,38,39].

On the Origin of Oncogenic Viruses.

Although our evidence casts doubt on the idea that the specific src sequence of recovered RSVs originated recently by experimental transduction from the cell, the close relationship between src and the cellular src-related sequences argues that such an event occurred at one time in the evolution of RSV. Likewise does the similarity between mcv and proto-mcv argue for a cellular origin of viral mcv.

However, in view of our previous results that onc-specific sequences are located within viral RNA genomes at very specific sites[1,2,3,13] (Fig. 1), transduction of these sequences must be a complex process for the following three reasons:

(i) Since the src and mcv-related cellular loci lack any detectable linkage to other virion sequences, transduction must involve double illegitimate recombination.

(ii) Moreover, the six introns (relative to src) of the proto-src locus[44] and the one intron (relative to mcv) of proto-mcv (Fig. 3) would have to be deleted in order to make these sequences co-linear with the known viral onc genes. (This is also true for the cellular homologs of other viral onc genes[1].) It is conceivable that this is accomplished by a splicing mechanism during or after transcription. However, since the src-related mRNA of normal cells measures about 3.5 kb but viral src mRNA only about 2 kb[54] and since the MC29-related mRNA of normal cells measures about 2.8 kb[30] but the mcv-sequence of MC29 only about 1.6 kb[10], it cannot be assumed that cellular splicing removes the intervening sequences of non-homology upon transcription, until this is directly demonstrated. We recognize that normal cells contain a protein that is similar to the viral src gene product[21]. However, it remains to be demonstrated that this protein is indeed translated from the cellular 3.5 kb RNA and that the protein is potentially oncogenic.

(iii) Finally, in the case of type II onc genes, such as the Δgag-mcv gene of MC29, specific deletions of virion genes of the transducing retrovirus would have to occur in order to form the Δgag hybrid onc genes[10](cf. Figs. 1 and 3).

A hypothesis which postulates that onc genes evolved from cellular prototypes in a multiphase process via endogenous, defective retrovirus-like intermediates has been described by us recently[1,3].

Qualitative or Quantitative Model?

Due to a total lack of direct evidence at this time for the function of cellular src- and mcv-related DNA, it is difficult to assess whether onc-specific sequences and their cellular, structural homologs are also functionally related as postulated by the quantitative model[21] or different as postulated by the qualitative model.

If the cellular onc-related sequences do represent functional homologs of viral onc genes and are cellular transforming genes as postulated by the quantitative model, it follows that all normal chicken cells must regularly prevent expression of potential cancer genes over a critical level . These potential cancer genes would correspond to at least about 0.01% of the genetic information of the chicken. This is estimated as follows. Chicken cellular DNA sequences related to eight known viral onc genes, e.g., the seven of the avian tumor virus group[1](Fig. 1) and one of reticuloendotheliosis virus group[55,56] together represent at least 15 kb. In the cases where the respective cellular loci have been analyzed, e.g., src[44], mcv[46](Figs.3 and 4), aev (Bishop, J.M., personal communication), amv[57], rel (for reticuloendotheliosis-specific sequence)[55], the onc-related cellular DNA sequences are separated by one (mcv, amv) or more (src, aev, rel) intervening sequences unrelated to viral counterparts. Hence, the total contiguous, chicken cellular DNA regions flanked by and including onc-related DNA represent at least about 100 kb, i.e., about 0.01% of the haploid chicken genome of 1 x 10^6 kb. This number would go up if more onc genes were discovered and if chicken DNA would contain sequences related to onc genes of mammalian retroviruses as has been suggested in one case[58]. It appears then that a prediction of the quantitative model would be that a substantial number of normal cellular genes are potential transforming genes.

We have detected minor differences between the primary structures of two viral onc-specific sequences and cellular sequence counterparts: For example, the src oligonucleotide marker, 23c, had an allelic but no identical counterpart in the cellular src-related locus of the chicken cloned in λ CS3. Likewise did mcv oligonucleotides nos. 3 and 7b of MC29 have allelic but no identical counterparts in λ proto-mcv3. Since the permissible range of sequence variation that does not affect onc gene function is not known, we cannot deduce whether these qualitative differences could explain the apparent functional difference between viral and cellular sequences.

Further, it is uncertain at this time whether the sequences that interrupt the src and mcv-related sequences of the cell are indeed noncoding introns. It is possible that these sequences together with onc-related sequences encode products that are qualitatively different from viral transforming proteins.

However, a clear qualitative difference was detected between the Δ gag-mcv gene of MC29 and the cellular homolog of MC29 which lacks Δ gag altogether. Hence, proto-mcv cannot represent a structural counterpart of the onc gene of MC29, although the role of Δ gag in transforming function of the viral Δgag-mcv protein remains to be determined. It could either be a functionally essential element of the onc gene of MC29 or else a necessary vector element for MC29 RNA to be packaged by helper viral coat proteins or both[1]. However, the second alternative fails to explain why the Δ gag element of MC29 is expressed as a Δ gag-mcv hybrid protein. If the 1 kb coding sequence of Δ gag were functionally irrelevant, it could be spliced out prior to expression of mcv[1]. (The notion that the viral Δgag-mcv gene and the cellular prototype are structurally different is nevertheless compatible with the hypothesis of Hayward et al.[32] that expression of the cellular proto-mcv locus without Δ gag is the cause of lymphatic leukemia. Clearly, a lymphatic leukemia is qualitatively different from the acute leukemias, carcinomas and sarcomas caused by MC29[3].)

In conclusion, our evidence supports the qualitative model in the case of type II onc genes of MC29, since only a part of the hybrid onc gene is found in uninfected cells, but does not at this time distinguish between the two models in the case of type I onc genes, like src. To distinguish further between the two models it would be necessary to determine whether the gene products of cellular, structural homologs of viral onc sequences are also functional homologs of viral onc genes; and whether experimental transduction of cellular onc-related sequences by molecularly cloned retroviruses without functional onc genes generates oncogenic viruses.

ACKNOWLEDGMENTS

We are grateful to Mike Botchan and associates for assistance with DNA cloning and Mike Kriegler for review of the manuscript. This work was supported by NIH Research Grant CA 11426 from the National Cancer Institute.

REFERENCES

1. K. Bister and P. H. Duesberg, Genetic structure and transforming genes of avian retroviruses, in: "Advances in Viral Oncology," G. Klein, ed., Raven Press, New York, in press, (1982).
2. P. H. Duesberg and K. Bister, in: "Cancer: Achievements, Challenges and Prospects for the 1980's," J. Burchenal and J. Oettgen, ed., Grune and Stratton, New York (1981).
3. P. H. Duesberg, Transforming genes of retroviruses, Cold Spring Harbor Symp. Quant. Biol., 44:13-29 (1980).
4. P. H. Duesberg and P. K. Vogt, Differences between the ribonucleic acids of transforming and nontransforming avian tumor viruses, Proc. Natl. Acad. Sci., 67:1673-1680 (1970).
5. K. Bister, G. Ramsay, M. J. Hayman, and P. H. Duesberg, OK10, an avian acute leukemia virus of the MC29 subgroup with a unique genetic structure, Proc. Natl. Acad. Sci., 77:7142-7146 (1980).

6. M. M.-C. Lai, P. H. Duesberg, J. Horst, and P. K. Vogt, Avian tumor virus RNA: A comparison of three sarcoma viruses and their transformation-defective derivatives by oligonucleotide fingerprinting and DNA-RNA hybridization, Proc. Natl. Acad. Sci., 70:2266-2270 (1973).
7. J. Brugge and R. J. Erikson, Identification of a transformation-specific antigen induced by an avian sarcoma virus, Nature, 269:346-348 (1977).
8. P. H. Duesberg, K. Bister, and C. Moscovici, Genetic structure of avian myeloblastosis virus released from transformed myeloblasts as a defective virus particle, Proc. Natl. Acad. Sci., 77:5120-5124 (1980).
9. K. Bister, M. J. Hayman, and R. K. Vogt, Defectiveness of avian myelocytomatosis virus MC29: Isolation of long-term nonproducer cultures and analysis of virus-specific polypeptide synthesis, Virology, 82:431-448 (1977).
10. P. Mellon, A. Pawson, K. Bister, G. S. Martin, and P. H. Duesberg, Specific RNA sequences and gene products of MC29 avian acute leukemia virus, Proc. Natl. Acad. Sci., 75:5874-5878 (1978).
11. W.-H. Lee, K. Bister, A. Pawson, T. Robins, C. Moscovici, and P. H. Duesberg, Fujinami sarcoma virus: An avian RNA tumor virus with a unique transforming gene, Proc. Natl. Acad. Sci., 77:2018-2022 (1980).
12. K. Bister and P.H. Duesberg, Structure and specific sequences of avian erythroblastosis virus RNA: Evidence for multiple classes of transforming genes among avian tumor viruses, Proc. Natl. Acad. Sci., 76:5023-5027 (1979).
13. K. Bister and P. H. Duesberg, Genetic structure of avian acute leukemia viruses, Cold Spring Harbor Symp. Quant. Biol., 44:801-822 (1980).
14. L. Gross, "Oncogenic Viruses," Pergamon Press, New York (1970).
15. J. J. Harvey, An unidentified virus which causes the rapid production of tumors in mice, Nature, 284:1104-1105 (1964).
16. W. M. Kirsten and L.A. Mayer, Morphologic responses to a murine erythroblastosis virus, J. Nat. Cancer Inst., 39:311-335 (1967).
17. J. B. Moloney, A virus-induced rhabdomyosarcoma of mice, in: "Conference on Murine Leukemia," Nat. Cancer Inst. Monograph No. 22, U. S. Public Health Service, Bethesda, Maryland (1966).
18. H. T. Abelson and L. S. Rabstein, Lymphosarcoma: virus-induced thymine-independent disease in mice, Cancer Res., 30:2213-2222 (1970).

19. N. Oker-Blom, H. Westermark, and S. Rosengard, Effect of - 1-adamantanamine hydrochloride (Amantadine) on chicken leukosis, in: "Progress in Antimicrobial and Anticancer Chemotherapy Vol.2," University Press, Baltimore, (1970).
20. R. J. Huebner and G. J. Todaro, Oncogenes of RNA tumor viruses as determinants of cancer, Proc. Natl. Acad. Sci., 64:1087-1094 (1969).
21. J. M. Bishop, Enemies within: the genesis of retrovirus oncogenes, Cell, 23:5-6 (1981).
22. H. M. Temin, The protovirus hypothesis: Speculations on the significance of RNA directed DNA synthesis for normal development and for carcinogenesis, J. Nat. Canc. Inst., 46:3-7 (1971).
23. H. M. Temin, Origin of retroviruses from cellular moveable genetic elements, Cell, 21:599-600 (1980).
24. J. Maisel, V. Klement, M. M. C. Lai, W. Ostertag, and P. H. Duesberg, Ribonucleic acid components of murine sarcoma and leukemia viruses, Proc. Natl. Acad. Sci., 70:3536-3540 (1973).
25. E. M. Scolnick, E. Rands, D. Williams, and W. P. Parks, Studies on the nucleic acid sequences of Kirsten sarcoma virus: A model for the formation of a mammalian RNA-containing, sarcoma virus, J. Virol., 12:456-463 (1973).
26. E. M. Scolnick and W. P. Parks, Harvey sarcoma virus: A second murine type C sarcoma virus with rat genetic information, J. Virol., 13:1211-1219 (1974).
27. N. Tsuchida, R. V. Gilden, and M. Hatanaka, Sarcoma virus-related RNA sequences in normal rat cells, Proc. Natl. Acad. Sci., 71:4503-4507 (1974).
28. A. D. Frankel and P. J. Fischinger, Nucleotide sequences in mouse DNA and RNA specific for Moloney sarcoma virus, Proc. Natl. Acad. Sci., 73:3705-3709 (1976).
29. D. Stehelin, H. E. Varmus, J. M. Bishop, and P. K. Vogt, DNA related to the transforming gene(s) of avian sarcoma viruses is present in normal avian DNA, Nature. 260:170-173 (1976).
30. D. Sheiness, S. M. Hughes, H. E. Varmus, E. Stubblefield, and J. M. Bishop, The vertebrate homolog of the putative transforming gene of avian myelocytomatosis virus: Characteristics of the DNA locus and its RNA transcript, Virology, 105:415-424 (1980).

31. D. Stehelin, S. Saule, M. Roussel, A. Sergeant, C. Lagrou, C. Rommens, and M. B. Raes, Three new types of viraloncogenes in defective avian leukemia viruses: I. Specific nucleotide sequences of cellular origin correlate with specific transformation, Cold Spring Harbor Symp. Quant. Biol., 44:1214-1223 (1980).

32. W. S. Hayward, G. B. Neel, and S. M. Astrin, Activation of a cellular onc gene by promoter insertion in ALV-induced lymphoid leukosis, Nature, 290:475-480 (1981).

33. M. Oskarsson, W. C. McClements, D. G. Blair, J. V. Maizel, and G. F. Vande Woude, Properties of a normal mouse cell DNA sequence (sarc) homologous to the src sequence of Moloney sarcoma virus, Science, 207:1222-1224 (1980) .

34. D. DeFeo, M. A. Gonda, H. A. Young, E. J. Chang, D. R. Lowy, E. M. Scolnick, and R. W. Ellis, Analysis of two divergent rat genomic clones homologous to the transforming gene of Harvey murine sarcoma virus, Proc. Natl. Acad. Sci., 78:3328-3332 (1981).

35. H. Hanafusa, C. C. Halpern, D. C. Buchhagen, and S. Kawai, Recovery of avian sarcoma virus from tumors induced by transformation-defective mutants, J. Exp. Med., 146:1735-1747 (1977).

36. H. Hanafusa, L.-H. Wang, T. Hanafusa, S. M. Anderson, R. E. Karess, and W. S. Hayward, The nature and origins of the transforming gene of avian sarcoma viruses, In: "Animal Virus Genetics," B. Fields, R. Jaenisch, and C. F. Fox, eds., ICN-UCLA Symposia on Molecular and Cellular Biology, Academic Press, New York, (1980).

37. R. Vigne, J. C. Neil, M. L. Breitman, and P. K. Vogt, Recovered src genes are polymorphic and contain host markers, Virology, 105:71-85 (1980).

38. R. Vigne, M. Breitman, C. Moscovici, and P. K. Vogt, Restitution of fibroblast-transforming ability in src-deletion mutants of avian sarcoma virus during animal passage, Virology, 93:413-426 (1979).

39. M. M.-C. Lai, S. S. F. Hu, and P. K. Vogt, Occurrence of partial src deletion and substitution of the src gene in the RNA genome of avian sarcoma virus, Proc. Natl. Acad. Sci., 74:4781-4785 (1977).

40. L.-H. Wang, C. C. Halpern, M. Nadel, and H. Hanafusa, Recombination between viral and cellular sequences generates transforming sarcoma virus, Proc. Natl. Acad. Sci., 75:5812-5816 (1978).

41. W.-H. Lee, M. Nunn, and P. H. Duesberg, Src genes of ten Rous sarcoma virus strains, including two reportedly transduced from the cell, are completely allelic; Putative markers of transduction are not detected, J. Virol., 39:758-776 (1981).
42. T. Robins, and P. H. Duesberg, Specific RNA sequences of Rous sarcoma virus (RSV) recovered from tumors induced by transformation-defective RSV deletion mutants, Virology, 93:427-434 (1979).
43. A. P. Czernilofsky, A. D. Levinson, H. E. Varmus, and J. M. Bishop, Nucleotide sequences of an avian sarcoma virus oncogene (src) and proposed amino acid sequence for gene product, Nature, 287:198-203 (1980).
44. D. Shalloway, A. D. Zelentz, and G. M. Cooper, Molecular cloning and characterization of the chicken gene homologous to the transforming gene of Rous sarcoma virus, Cell, 24:531-541 (1981).
45. S. J. Hughes, F. Stubblefield, F. Payvar, J. D. Engel, J. G. Dodgson, D. Spector, B. Cordell, R. T. Schimke, and H. Varmus, Gene localization by chromosome fractionation: Globin genes are on at least two chromosomes and three estrogen-inducible genes are on three chromosomes, Proc. Natl. Acad. Sci., 76:1348-1352 (1979).
46. T. Robins, K. Bister, C. Garon, T. Papas, and P. H. Duesberg, Structural relationship between a normal chicken DNA locus and the transforming gene of the avian acute leukemia virus MC29, J. Virol., in press, (1982).
47. J. G. Dodgson, Strommer and J. A. Engel, Isolation of the chicken -globin gene and a linked embryonic - like globin gene from a chicken DNA recombinant library, Cell, 17:879-887 (1979).
48. J. A. Lautenberger, R. A. Schulz, C. F. Garon, P. H. Tsichlis, and T. S. Papas, Molecular cloning of avian myeloblastosis virus (MC29) transforming sequences, Proc. Natl. Acad. Sci., 78:1518-1522 (1981).
49. P. H. Duesberg, K. Bister, and C. Moscovici, Avian acute leukemia virus MC29: Conserved and variable RNA sequences and recombination with helper virus, Virology, 99:121-134 (1979).
50. L.-H. Wang, P. H. Duesberg, S. Kawai, and H. Hanafusa, Location of envelope-specific and sarcoma-specific oligonucleotides in RNA of Schmidt-Ruppin rous sarcoma virus, Proc. Natl. Acad. Sci., 73:447-451 (1976).
51. J. Tooze, "The Molecular Biology of Tumor Viruses," Cold Spring Harbor Press, New York (1973).

41. W.-H. Lee, M. Nunn, and P. H. Duesberg, Src genes of ten Rous sarcoma virus strains, including two reportedly transduced from the cell, are completely allelic; Putative markers of transduction are not detected, J. Virol., 39:758-776 (1981).
42. T. Robins, and P. H. Duesberg, Specific RNA sequences of Rous sarcoma virus (RSV) recovered from tumors induced by transformation-defective RSV deletion mutants, Virology, 93:427-434 (1979).
43. A. P. Czernilofsky, A. D. Levinson, H. E. Varmus, and J. M. Bishop, Nucleotide sequences of an avian sarcoma virus oncogene (src) and proposed amino acid sequence for gene product, Nature, 287:198-203 (1980).
44. D. Shalloway, A. D. Zelentz, and G. M. Cooper, Molecular cloning and characterization of the chicken gene homologous to the transforming gene of Rous sarcoma virus, Cell, 24:531-541 (1981).
45. S. J. Hughes, F. Stubblefield, F. Payvar, J. D. Engel, J. G. Dodgson, D. Spector, B. Cordell, R. T. Schimke, and H. Varmus, Gene localization by chromosome fractionation: Globin genes are on at least two chromosomes and three estrogen-inducible genes are on three chromosomes, Proc. Natl. Acad. Sci., 76:1348-1352 (1979).
46. T. Robins, K. Bister, C. Garon, T. Papas, and P. H. Duesberg, Structural relationship between a normal chicken DNA locus and the transforming gene of the avian acute leukemia virus MC29, J. Virol., in press, (1982).
47. J. G. Dodgson, Strommer and J. A. Engel, Isolation of the chicken -globin gene and a linked embryonic -like globin gene from a chicken DNA recombinant library, Cell, 17:879-887 (1979).
48. J. A. Lautenberger, R. A. Schulz, C. F. Garon, P. H. Tsichlis, and T. S. Papas, Molecular cloning of avian myeloblastosis virus (MC29) transforming sequences, Proc. Natl. Acad. Sci., 78:1518-1522 (1981).
49. P. H. Duesberg, K. Bister, and C. Moscovici, Avian acute leukemia virus MC29: Conserved and variable RNA sequences and recombination with helper virus, Virology, 99:121-134 (1979).
50. L.-H. Wang, P. H. Duesberg, S. Kawai, and H. Hanafusa, Location of envelope-specific and sarcoma-specific oligonucleotides in RNA of Schmidt-Ruppin rous sarcoma virus, Proc. Natl. Acad. Sci., 73:447-451 (1976).
51. J. Tooze, "The Molecular Biology of Tumor Viruses," Cold Spring Harbor Press, New York (1973).

52. J. Maisel, D. Dina, and P. H. Duesberg, Murine sarcoma viruses: the helper independence reported for a Moloney variant is unconfirmed; distinct strains differ in the size of their RNAs, Virology, 76:295-312 (1977).
53. P. H. Tsichlis, K. F. Conklin, and J. M. Coffin, Role of the c-region in relative growth rates of endogenous and exogenous avian oncoviruses, Proc. Natl. Acad. Sci., 77:536-540 (1980).
54. S. Y. Wang, W. S. Hayward, and H. Hanafusa, Genetic variation in the RNA transcripts of endogenous virus genes in uninfected chicken cells, J. Virol., 24:64-73 (1977).
55. I. S. Y. Chen, T. W. Mak, J. J. O'Rear, and H. M. Temin, Characterization of reticuloendotheliosis virus strain T [REV-T(REV-A)] DNA and the isolation of a novel variant of REV-T by molecular cloning, J. Virol., in press (1981).
56. S. S. F. Hu, M. M. C. Lai, T. C. Wong, R. S. Cohen, and M. Sevoian, Avian reticuloendotheliosis virus: Characterization of the genome structure by heteroduplex mapping, J. Virol., 37:899-907 (1981).
57. B. Perbal, and M. Baluda, The avian myeloblastosis virus transforming gene is related to unique chicken DNA regions separated by at least one intervening sequence, J. Virol., in press (1981).
58. S. P. Goff, E. Gilboa, O. N. Witte, and D. Baltimore, Structure of the Abelson murine leukemia virus genome and the homologus cellular gene: Studies with cloned viral DNA, Cell, 22:777-785 (1980).

NUCLEOTIDE SEQUENCE ANALYSIS OF INTEGRATED AVIAN MYELOBLASTOSIS VIRUS (AMV) LONG TERMINAL REPEAT (LTR) AND THEIR HOST AND VIRAL JUNCTIONS: STRUCTURAL SIMILARITIES TO TRANSPOSABLE ELEMENTS

T. S. Papas*, K. E. Rushlow*, J. A. Lautenberger*, K. P. Samuel*, M. A. Baluda† and E. P. Reddy+

*Laboratory of Molecular Oncology, and +Laboratory of Cellular and Molecular Biology, National Cancer Institute, Bethesda, MD; †UCLA School of Medicine and Molecular Biology Institute, Los Angeles, CA

INTRODUCTION

The nucleotide sequence of the integrated avian myeloblastosis virus (AMV) long terminal repeat (LTR) has been determined. The sequence is 385 bp long and is present at both ends of the viral DNA. The cell-virus junctions at each end consist of a six base pair direct repeat of cell DNA next to the inverted repeat of viral DNA. The LTR also contains promoter-like sequences, a mRNA capping site, and polyadenylation signals. Several features of this LTR suggest a structural and functional similarity with sequences of transposable and other genetic elements. Comparison of these sequences with LTRs of other avian retroviruses indicates that there is a great variation in the 3' unique sequence (U_3) while the 5' specific sequences are highly conserved.

Avian myeloblastosis virus (AMV) transforms only specific target cells: It induces acute myeloblastosis in chickens and transforms only granulocyte precursors *in vitro* (1). The genomic structure of AMV and its helper-associated virus (MAV) has been determined by molecular cloning of both viruses followed by restriction enzyme analysis and heteroduplex and R-looping electron microscopic mapping (2,3,4). These studies revealed that both genomes share about a 5 kb region of homology from the 5' end of the genomes. AMV and MAV RNAs direct the synthesis of the two precursor polypeptides Pr76^{gag} and Pr$^{gag\text{-}pol}$ (5). Furthermore, Duesberg et al. (6) have found that leukemic cells transformed by AMV without concomittant infection by helper viruses produce noninfectious particles and synthesize these two precursor polypeptides.

Following infection, the RNA genome of avian retroviruses is copied into circular and linear double-stranded DNA forms (7). One or both of these forms integrate into the cellular genome and is transcribed by the host into viral mRNAs and progeny viral genomes. The proviral DNA is integrated colinearly with respect to the free viral DNA with long terminal repeats (LTR) at the 5' and 3' termini of the proviral genome (10).

A prerequisite to understanding the function of the viral LTRs is to determine their nucleotide sequence. In this report we present the complete nucleotide sequence of the LTR as well as adjacent host sequences of an integrated provirus of AMV.

MATERIALS AND METHODS

DNA Isolation

The 2.0 kbp DNA fragment containing the 3'-terminal sequences of AMV proviral DNA from the λ11A1-1 clone of Souza et al. (3) was subcloned into the EcoRI site of pBR325. The fragment was isolated from an EcoRI digest of the recombinant plasmid and purified from a preparative agarose gel using the procedure described by Reddy et. al. (9).

End Labelling of DNA

DNA fragments used in restriction endonuclease mapping or chemical sequencing were labelled at their 5'-termini using [γ-^{32}P]ATP (Amersham; 2000 Ci/mmole) and T4 polynucleotide kinase (P-L Biochemicals; Cat. No. 0734). Prior to labelling, the 5'-ends were dephosphorylated with bacterial alkaline phosphatase (P-L Biochemicals; Cat. No. 0976). Treatment of the DNA with phosphatase and labelling of the 5'-termini were carried out according to the procedure of Maxam and Gilbert (10).

Restriction Endonuclease Mapping

Restriction endonucleases HindIII, XhoI, XbaI, BglII, and HinfI were purchased from New England Biolabs. The 2.0 kbp DNA fragment labelled with [^{32}P] at its 5'-ends was digested with XhoI to yield two fragments of approximately 900 bp and 1100 bp in length. The fragments were separated by gel electrophoresis on 5% (wt/vol) acrylamide, located by autoradiography, and purified from the gel using previously described methods (10). The individual DNA fragments were mapped by the method of Smith and Birnstiel (11).

DNA Sequencing

Sequence determination of DNA fragments labeled with [^{32}P] at a unique 5'-terminus was carried out by the chemical modification method of Maxam and Gilbert (10). The five base-specific reactions utilized were G, G+A, A>C, C+T, and C. Bases 1-150 from the 5'-labeled end were identified by fractionation of the chemical cleavage products on 10-20% acrylamide-8M urea gels. For sequence information beyond 150 bases, the products were fractionated on sequencing gels 80 cm in length and containing 6% acrylamide.

RESULTS

Organization of the EcoRI fragment containing the LTR sequences

AMV specific sequences were selected from a library of λ-chicken recombinant phage constructed with DNA from leukemic myeloblasts producing AMV-B (4). The clone described in this study contained the complete AMV provirus and a substantial amount of cellular sequences. As shown in Fig. 1, the clone contains four EcoRI sites, two of which are within the proviral sequences. The three fragments delimited by the EcoRI sites were subcloned in the plasmid vector pBR325. One of these subclones contains the left-hand LTR and adjoining host sequences, while a second contains the right-hand LTR and adjoining cellular sequences. The restriction map of the

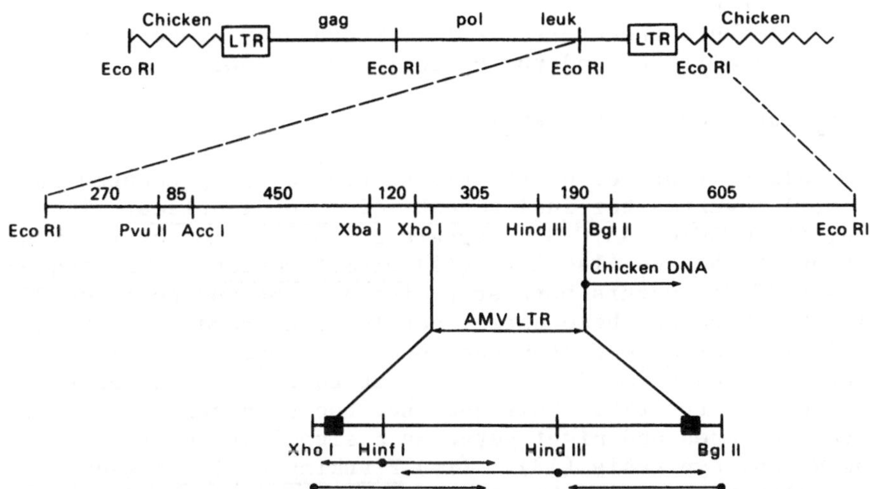

Fig. 1. Strategy for determining the nucleotide sequence of the 5'-LTR of AMV. The restriction sites were determined by the partial restriction mapping technique of Smith and Birnstiel (11). Fragments generated by XbaI, HinfI, HindIII and BglII were used for sequence analysis.

EcoRI fragment carrying the complete right-hand LTR and host flanking sequences is shown in Fig. 1. The fragment is oriented from left to right, 5' to 3', with respect to the viral RNA. We chose to initially determine the nucleotide sequence of the right-hand LTR since it is adjacent to the transforming gene of the virus.

Sequence strategy

Fig. 1 provides a summary of the strategy employed to determine the nucleotide sequence of the right-hand LTR. After digestion of DNA samples with appropriate restriction enzymes, the fragments to be sequenced were isolated from agarose or polyacrylamide gels and sequenced in either the 5'—→3' or 3'—→5' direction as described in Methods.

Nucleotide sequence

We have sequenced the right LTR and adjacent host sequences and have also determined the cell proviral junction sequences of the left LTR (Fig. 2). The LTR boundaries were determined by comparison of our AMV sequences to the sequences of the terminal repeats of other avian viruses (12,13) as well as to the AMV "strong stop" DNA sequence (14). The AMV LTR is 385 base pairs long and contains the following regions: the 3'-specific sequence (U_3) which extends from position 42 to 327, and the 5' derived sequence (RU_5), from positions 328 to 426. The latter sequence, the so-called "strong stop," is the first region transcribed by the reverse transcriptase. This region contains the R sequence of 21 bp which represents the short terminal redundancy present at the ends of the genomic RNA and the U_5 region of 75 bp.

Notable features of the sequences

The salient features of the LTR sequences are presented in Fig. 3. (i) Inverted terminal repeats of 13 nucleotides. The sequence, 5'-TGTAGTCTTAATC-3', appears at the termini of the LTR at positions 42-54 and 414-426. (ii) Direct repeats. The sequence, 5'-ACCAAATAAGG-3', occurs both at position 74-84 and position 99-109. The function of these direct duplications remains to be determined. (iii) Plus strong stop sequences. A sequence of 11 consecutive purines occurs immediately adjacent to the 5'- end of the inverted repeat of the LTR. This sequence has been observed in other cloned retroviruses and might serve as a signal or initiation point for plus strand synthesis (13). (iv) Flanking cellular sequences. A short duplication of six bp host sequences was observed flanking the termini of each LTR. This feature is common to procaryotic transposable elements where the number of duplicated nucleotides appears to be variable and characteristic of the elements (8,12). Examination of the host flanking sequences of right and left LTR indicate that both share a six base duplication which is not tandem.

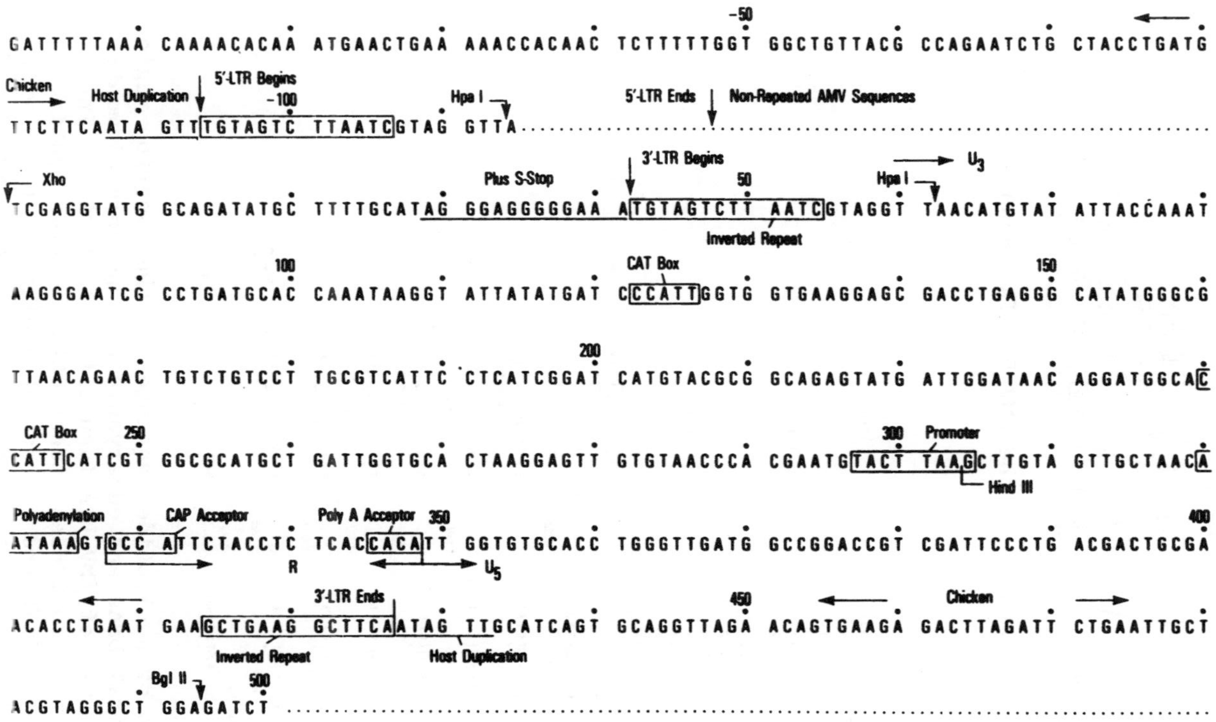

```
                                                     -50
GATTTTTAAA CAAAACACAA ATGAACTGAA AAACCACAAC TCTTTTTGGT GGCTGTTACG CCAGAATCTG CTACCTGATG

Chicken   Host Duplication | 5'-LTR Begins -100   Hpa I         5'-LTR Ends | Non-Repeated AMV Sequences
TTCTTCAATA GTT[TGTAGTC TTAATC]GTAG GTTA..................................................

 Xho                        Plus S-Stop   | 3'-LTR Begins  50        Hpa I  → U3
TCGAGGTATG GCAGATATGC TTTTGCATAG GGAGGGGGAA A[TGTAGTCTT AATC]GTAGGT TAACATGTAT ATTACCAAAT
                                                  Inverted Repeat
                     100                        CAT Box                       150
AAGGGAATCG CCTGATGCAC CAAATAAGGT ATTATATGAT C[CCATT]GGTG GTGAAGGAGC GACCTGAGGG CATATGGGCG

                                          200
TTAACAGAAC TGTCTGTCCT TGCGTCATTC CTCATCGGAT CATGTACGCG GCAGAGTATG ATTGGATAAC AGGATGGCA[C

CAT Box  250                                                   300 Promoter
CATT]CATCGT GGCGCATGCT GATTGGTGCA CTAAGGAGTT GTGTAACCCA CGAATG[TACT TAAG]CTTGTA GTTGCTAAC[A
                                                                    Hind III
Polyadenylation  CAP Acceptor  Poly A Acceptor 350                                       400
ATAAA]GT[GCC A]TTCTACCTC TCAC[CACA]TT GGTGTGCACC TGGGTTGATG GCCGGACCGT CGATTCCCTG ACGACTGCGA
                         R               U5
                          3'-LTR Ends           450              Chicken
ACACCTGAAT GAA[GCTGAAG GCTTCA]ATAG TTGCATCAGT GCAGGTTAGA ACAGTGAAGA GACTTAGATT CTGAATTGCT
              Inverted Repeat     Host Duplication
      Bgl II   500
ACGTAGGGCT GGAGATCT ....................................................................
```

Fig. 2. The nucleotide sequence of AMV LTR and its host and viral junctions. The nucleotide sequence includes the 5'-LTR junction point and flanking chicken sequences. The 3'-LTR nucleotide sequence is completely included as well as its junction with flanking chicken sequences. From left to right (5' to 3' with respect to orientation of viral RNA) the following sequences are identified: 13 base pair inverted repeat (box), promoter-like sequences, 5' end of viral RNA; possible polyadenylation site, possible polyA acceptor site, CAT-boxes; the restriction endonuclease sites, HpaI, XhoI, BglII, the six base duplicated host sequence A-T-A-G-T-T.

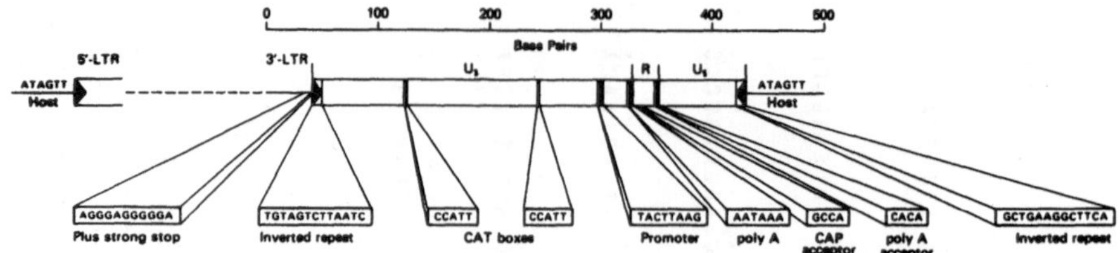

Fig. 3. Salient features of the AMV 5'-LTR sequences.

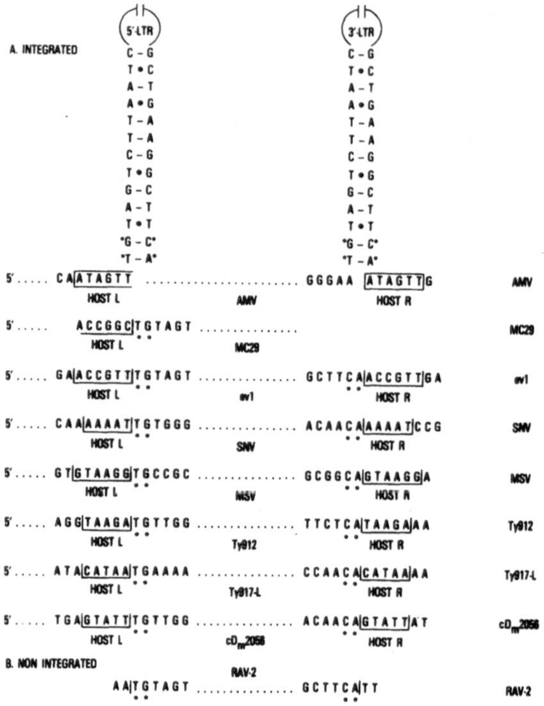

A. INTEGRATED

5'-LTR: C – G, T • C, A – T, A • G, T – A, T – A, C – G, T • G, G – C, A – T, T • T, *G – C*, *T – A*

3'-LTR: C – G, T • C, A – T, A • G, T – A, T – A, C – G, T • G, G – C, A – T, T • T, *G – C*, *T – A*

5'..... CA|ATAGTT .... GGGAA |ATAGTT|G — AMV (HOST L, AMV, HOST R)

5'..... ACCGGC|TGTAGT .... — MC29 (HOST L, MC29)

5'..... GA|ACCGTT|TGTAGT .... GCTTCA|ACCGTT|GA — ev1 (HOST L, ev1, HOST R)

5'..... CAA|AAAAT|TGTGGG .... ACAACA|AAAAT|CCG — SNV (HOST L, SNV, HOST R)

5'..... GT|GTAAGG|TGCCGC .... GCGGCA|GTAAGG|A — MSV (HOST L, MSV, HOST R)

5'..... AGG|TAAGA|TGTTGG .... TTCTCA|TAAGA|AA — Ty912 (HOST L, Ty912, HOST R)

5'..... ATA|CATAA|TGAAAA .... CCAACA|CATAA|AA — Ty917-L (HOST L, Ty917-L, HOST R)

5'..... TGA|GTATT|TGTTGG .... ACAACA|GTATT|AT — cD_m2056 (HOST L, cD_m2056, HOST R)

B. NON INTEGRATED

RAV-2

AA|TGTAGT .... GCTTCA|TT — RAV-2

Fig. 4. Comparison of nucleotide sequences in the junction region of integrated proviral DNAs and eucaryotic transposable elements. The stem loop structure of AMV LTR identifies similarities in organization between retroviruses, eucaryotic and procaryotic transposable elements. LTR and transposable elements begin with TG and end with CA (asterisk is placed on these nucleotides). The identical base host sequences bracketing the integrated elements are underlined. SNV, evl, Ty912, Ty917-L, cDm2056 sequences are taken from reference (12). Non-integrated proviral sequence (RAV-2) has four additional bases two at each end of the provirus. These four bases are lost during integration.

Signals for regulation of transcription

Analysis of the LTR sequence identified several sequences which may function in the regulation of transcription (Fig. 3). These include: (i) Transcription initiation signal. A promoter-like sequence (15), 5'-TACTTAAG-3', was found at position 297-304. This A-T rich sequence precedes by 21 nucleotides the GCCA sequence most likely to be the RNA capping site. (ii) Polyadenylation signal. The sequence, 5' AATAAA 3', is usually found 10-30 nucleotides upstream from the dinucleotide CA, the preferred site for polyadenylation (16). For the AMV LTR, the sequence, 5'-AATAAA-3', (325-320) occurs 20 nucleotides upstream of the CA dinucleotide. (iii) Termination signals. The presence of termination sequences in the right LTR may serve to terminate the provirus expression.

The correlation between the position of the promoter-like sequence and the initiation site for mRNA synthesis in the LTR suggests that the right-hand LTR may serve as a promoter for the mRNA extending into the adjacent host sequences. This argument is further strengthened by the presence of the sequence 5'-CCATT-3' at position 240-244, 88 bases upstream from the 5'-cap structure. An analogous sequence has been shown to occur 77 ± 10 bp upstream from the 5' end of the mRNA capping site of most eukaryotic structural genes (17).

Comparison of the AMV LTR to the LTRs from other RNA tumor viruses and procaryotic transposable elements

The external junction joining LTR with host sequences can be best visualized by pairing the LTR elements at their inverted repeat sequences (Fig. 4). The stem loop structure makes it convenient to identify similarities in organization between the retroviruses eucaryotic and some procaryotic transposable elements. The AMV LTR exhibits a six-base host sequence (ATAGTT) that is found at the site of integration flanking the termini of the proviruses. In addition, AMV proviral sequences also integrate into A-T rich DNA, and AMV, as well as transposable elements begin with the dinucleotide T-G and finish with C-A.

It has been previously observed that integration of proviruses is followed by a two-base pair loss at each point of integration. In the case of RSV, a T-T dinucleotide which is present in the right terminus of the unintegrated genome is lost during integration (18).

The DNA sequence of the AMV LTR region was compared to the sequences of other avian LTRs by the two dimensional dot matrix homology program of Maizel (19,20) and by the program of Queen and Korn (21) (Fig. 5). This analysis showed that while there was considerable homology between the U_5 and R regions of AMV and ev1, RAV, and RSV, most of the U_3 region is quite different. Short regions of similarity in the U_3 include the leftmost invert repeat and the "promoter" sequence and the polyA signal. The later sequences are in the rightmost portion of the U_3 region. The sequences just to the left of the AMV 3'-LTR contains a polypurine stretch that is closely related to similar stretches in the evl and RSV sequences. These sequences are thought to be involved in the binding of the primer for plus strand DNA synthesis (20). Each of these four LTRs have a 21 base R region. The AMV R region shares 17 bases with the evl and RAV-2 R regions and 18 bases with the RSV R region. The U_5 regions of AMV, RAV-2 and RSV each contain 80 bases. AMV shares 75 of these with RAV-2 and 76 with RSV. While the U5 of evl is two bases shorter, 70 of these bases are shared with the AMV LTR sequence. No homology in any region was found between the AMV LTR and the SNV LTR sequence (25). This was not surprising since previous studies have found that SNV along with REV are not related to other avian viruses but are related to some mammalian retroviruses (26).

DISCUSSION

The genomic structure of AMV is unique among acute leukemia viruses because it contains the complete sequence of the "gag" proteins and most of the polymerase. Evidence for this comes from restriction enzyme analysis of cloned DNAs of both AMV and helper associated MAV. In addition, both AMV and MAV RNAs direct the synthesis in an *in vitro* translation system of both the "gag" precursor of 76000 daltons and the "gag-pol" precursor of 180,000 daltons.

The AMV LTR is 385 bp long and therefore is slightly larger than the LTRs of other avian retroviruses (12) but considerably smaller than the LTR of the mammalian virus MSV (8). The AMV LTR sequence is flanked by an inverted repeat which is also characteristic of other LTRs and of the movable transposable elements. The AMV sequence contains a unique 3' (U_3) region and a 5' (RU_5) region closely related to that of other avian retroviruses. The highly conserved 5' specific sequences suggest that they are structurally and functionally important. This 5' specific region is the first sequence to be transcribed by the reverse transcriptase and may be therefore essential in the synthesis of the provirus. There is great variation in the U_3 region of viral LTRs but even within this poorly conserved region, several small and highly conserved sequences can be identified. The sequence, 5'-TACTTAAG-3', tentatively identified as a promoter is identical with sequences which have been shown to direct the synthesis of distinct species of mRNA in the RSV system (27). Upstream from this promoter site, and still within the U_3 region, is the sequence, 5'-CCATT-3', which is present in a wide variety of eucaryotic regulatory regions. The polyadenylation signal, 5'-AATAAA-3', upstream from the capping site, 5'-GCCA-3', is also present in the U_3 region. The R sequence of the AMV LTR is similar to RSV with only four base mismatches in the 21 nucleotides. The presence of the identical R sequences at both the 5' and 3' ends of the RNA genome allows synthesis of the complementary DNA beyond the 5' end of the template. The regions which flank the LTR sequence at the 3' end of the left LTR and at the 5' end of the right LTR represent sequences involved with DNA synthesis priming of minus and plus strands respectively (13,24). The 11 base polypurine sequence, 5'-AGGGAGGGGCA-3', represents a portion of the sequence necessary for priming of the strong stop DNA, although the mechanism by which this priming function occurs remains to be elucidated. This sequence seems to be conserved in all RNA tumor viruses with very little variation.

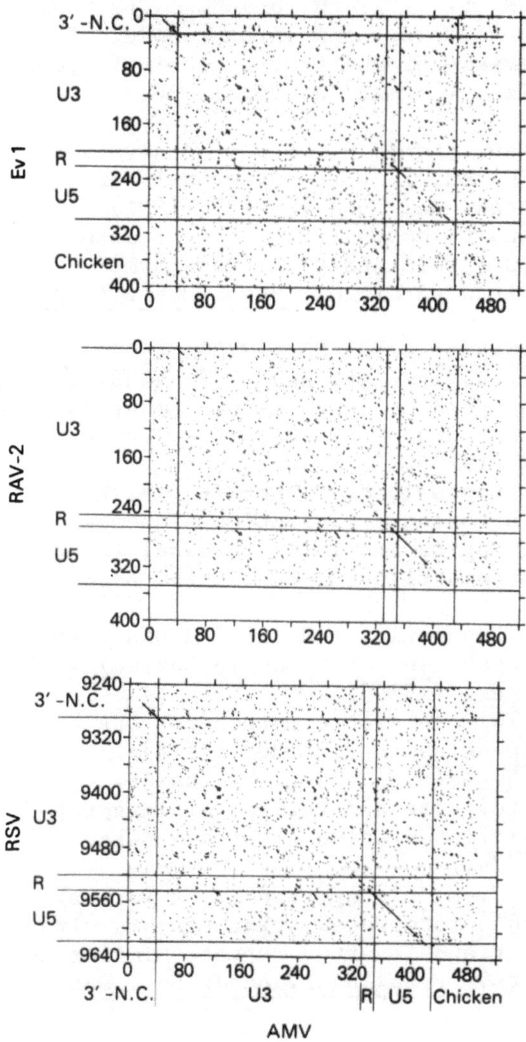

Figure 5. The DNA sequence of the AMV 3'-LTR was compared to that of the LTRs of ev1 (12) RAV-2 (22) RSV (23) by the two dimensional dot matrix homology diagram of Maizel (19,20). The plots were generated by a program in BASIC running on a Tektronix 4052 computer. The numbering system of the AMV LTR starts at the XhoI site in the 3' noncoding region while that used for RAV-2 and ev-1 assign begins at the start of these sequences as published.

REFERENCES

1. H. Beug, A. von Kirchbach, G. Doderlein, J.-F. Conscience, J.-F., and T. Graf. Chicken hematopoietic cells transformed by seven strains of defective avian leukemia viruses display three distinct phenotypes of differentiation. Cell 18:375 (1979).
2. L. M. Souza and M. A. Baluda. Identification of the avian myeloblastsosis virus genome. I. Identification of restriction endonuclease fragments associated with acute myeloblastic leukemia. J. Virol. 36:317 (1980).
3. L. M. Souza, M. C. Komaromy, and M. A. Baluda. Identification of a proviral genome associated with avian myeloblastic leukemia. Proc. Natl. Acad. Sci. USA 77:3004 (1980).
4. L. M. Souza, M. J. Briskin, R. L. Hillyard, and M. A. Baluda. Identification of the avian myeloblastosis virus genome. II. Restriction endonuclease analysis of DNA from λ proviral recombinants and leukemic myeloblast clones. J. Virol. 36:325 (1980).
5. R. A. Schulz, J. G. Chirikjian, and T. S. Papas. Analysis of avian myeloblastosis viral RNA and in vitro sysnthesis of proviral DNA. Proc. Natl. Acad. Sci. USA 78:2057 (1981).
6. P. M. Duesberg, K. Bister, and C. Moscovici. Genetic structure of avian myeloblastsosis virus, released from transformed myeloblasts as a defective virus particle. Proc. Natl. Acad. Sci. USA 77:5120 (1980).
7. J. M. Bishop. Annu. Rev. Biochem. 47:35-88 (1978).
8. R. Dhar, W. L. McClements, L. W. Enquist, and G. F. Vande Woude. Nucleotide sequences of integrated Moloney sarcoma provirus long terminal repeats and their host and viral junctions. Proc. Natl. Acad. Sci. USA 77:3937 (1980).
9. P. E. Reddy, M. J. Smith, E. Canaani, K. C. Robbins, S. R. Tronick, S. Zain, and S. A. Aaronson. Nucleotide sequence analysis of the transforming region and large terminal redundancies of Moloney murine sarcoma virus. Proc. Natl. Acad. Sci. USA 77:5234 (1981).
10. A. M. Maxam and W. Gilbert. Sequencing end-labeled DNA with base-specific chemical cleavages, in: "Methods in Enzymology," K. Moldave and L. Grossman, eds. (1980).
11. H. O. Smith and M. A. Birnstiel. A simple method for DNA restriction mapping. Nucleic Acids Res. 3:2387 (1976).
12. F. Hishinuma, P. J. DeBone, S. Astrin, and A. M. Skalka. Nucleotide sequence of acceptor site and termini of integrated avian endogenous provirus evl: integration creates a 6 bp. Cell 23:155 (1981).
13. R. Swanstrom, W. J. DeLorbe, J. M. Bishop, and H. E. Varmus. Nucleotide sequence of cloned unintegrated avian sarcoma virus DNA: viral DNA contains direct and inverted repeats similar to those in transposable elements. Proc. Natl. Acad. Sci. USA 78:124 (1981).

14. E. Stoll, M. A. Billeter, A. Palmenberg, and C. Weissman. Avian myeloblastosis virus RNA is terminally redundant: Implications for the mechanism of retrovirus replication. Cell 12:57 (1977).
15. D. Pribnow. Nucleotide sequence of an RNA polymerase binding site at an early T7 promoter. Proc. Natl. Acad. Sci. USA 72: 784 (1975).
16. N. J. Proudfoot and G. G. Brownlee. 3' Non-coding region sequences in eukaryotic messenger RNA. Nature (London) 263:211 (1976).
17. A. Efstratiadis, J. W. Posakony, T. Maniatis, R. M. Lawn, C. O'Connell, R. A. Spritz, J. K. DeRiel, B. G. Forget, S. M. Weissman, J. L. Slightom, A. E. Blechl, O. Smithies, F. E. Baralle, C. C. Shoulders, N. J. Proudfoot. The structure and evolution of the human β-globin gene family. Cell 21:653 (1980).
18. S. H. Hughes, A. Mutshchler, J. M. Bishop, and H. E. Varmus. A Rous sarcoma virus provirus is flanked by short direct repeats of a cellular DNA sequence present in only one copy prior to integration. Proc. Natl. Acad. Sci. USA 78:4299 (1981).
19. D. A. Konkel, J. V. Maizel, and P. Leder. The evolution and sequence comparison of two recently diverged mouse chromosomal β-globin genes. Cell 18:865 (1979).
20. J. V. Maizel. Proc. Natl. Acad. Sci. USA, in press.
21. C. L. Queen and L. J. Korn. Computer analysis of nucleic acids and proteins, in: "Methods in Enzymology," K. Moldave and L. Grossman, eds. (1980).
22. G. Ju and A. M. Skalka. Nucleotide sequence analysis of the long terminal repeat (LTR) of avian retroviruses: structural similarities with transposable elements. Cell 22:379 (1980).
23. D. Schwartz, R. Tizard, and W. Gilbert. Personal communication.
24. E. Gilboa, S. W. Mitra, S. Goff, and D. Baltimore. A detailed model of reverse transcription and tests of crucial aspects. Cell 18:93 (1979).
25. K. Shimotohno, S. Mizutani, and H. M. Temin. Sequence of retrovirus provirus resembles that of bacterial transposable elements. Nature 285:550 (1980).
26. M. Barbacid, E. Hunter, and S. A. Aaronson. Avian reticuloendotheliosis viruses: Evolutionary linkage with mammalian type C retroviruses. J. Virol. 30:508 (1979).
27. T. Yamamoto, B. deCrombrugghe, and I. Pastan. Identification of a functional promoter in the long terminal repeat of a Rous sarcoma virus. Cell 22:787 (1980).

QUANTITATIVE CHANGES OF SOME CELLULAR POLYPEPTIDES IN C3H MOUSE CELLS FOLLOWING TRANSFORMATION BY MOLONEY SARCOMA VIRUS

Jes Forchhammer

The Fibiger Laboratory
Ndr. Frihavnsgade 70
DK 2100 Copenhagen, Denmark

Keywords: Transformation, tropomyosin, ts-mutant of murine sarcoma virus, quantitation of proteins, two-dimensional gel electrophoresis.

INTRODUCTION

Over the past few years changes have been reported in proteins synthesized by transformed cells when compared to normal cells using various modifications of two dimensional polyacrylamide electrophoresis (2D-gel). It can be observed that the relative amount of some proteins differs in transformed cells when compared to untransformed cells. In this report the question of turnover of some of these proteins will be reported. This was studied by pulse-chase experiments of tissue cultures of mouse embryo fibroblast, 10T½-cl8 infected with Moloney sarcoma virus and a mutant thereof MSV-CP27 at permissive and nonpermissive conditions. It was observed that some proteins are preferentially synthesized in the transformed state and turnover rapidly with halflives of 0.5-3 hours. For most of the proteins that show a quantitative difference, pulse chase experiments gave no indication of differences in the rate of turnover. Although most of the polypeptides that change in quantities during transformation were unidentified, four were shown to be related to tropomyosin by immuno precipitation. Two of these seem to turnover faster in transformed cells than in untransformed cells.

RESULTS

Pulse-chase Experiments in Tissue Cultures Labelled with [^{35}S]-methionine

The quantity of any protein in a cell reflects the rate of synthesis and the rate of degradation or modification. The aim of this study was to shed light on changes in some of these processes with regard to some of the changes observed in proteins in transformed cells. In this study Moloney sarcoma virus (MSV), and a ts-mutant of the murine src gene MSV-CP27 was used. As described previously virus are produced at both 32°C and 39°C (Forchhammer and Turnock, 1978), and it is possible to study the function of the mutated gene which is expressed at 32°C but is greatly reduced in function at 39°C.

Acidic and neutral cellular proteins were separated by 2D-gel electrophoresis (O'Farrell, 1975) using a modification of ampholyte composition during the isoelectric focusing (IEF) covering the pH range 8.2-3.8. The proteins were located by fluorography and special attention was paid to polypeptides where changes in intensity following transformation was observed as previously reported by a number of groups (Strand and August, 1978; Forchhammer and Klarlund, 1979; Bravo and Celis, 1980; Brzeski and Ege, 1980). Some of these changes were detected in the present study, however in many cases it has been difficult to identify some of the polypeptides studied in the different systems because of technical differences (see discussion). The turnover of polypeptides in growing cells was studied labelling cultures with [^{35}S]-methionine for 1 hour in a medium containing 1.5μCi/ml methionine, followed by removal of the radioactive medium and replacement with a chase medium containing 30 μg/ml

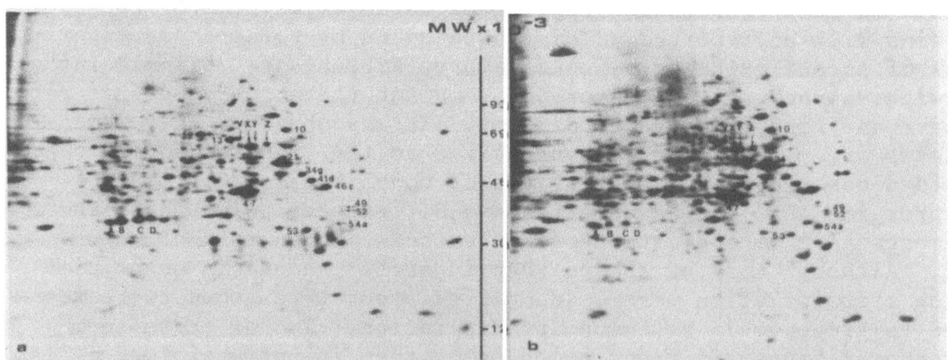

Fig.1. Fluorograms of [^{35}S]-methionine labelled polypeptides from 10T½ cells. The figure represents polypeptides labelled with [^{35}S]-methionine for 1 hour at 36° from a culture infected with MSV and fully transformed a), or the uninfected controls b).

methionine. The results of some experiments are shown in Fig.1.

Polypeptides are numbered in accordance to the numbering system for murine kidney fibroblasts (Fey et al., 1981). Only polypeptides to be discussed here are indicated plus a few prominent marker polypeptides. The relative coordinates of proteins are sometimes given in brackets (MW:pI) to facilitate the location of proteins. Special attention was paid to polypeptides in three regions between IEF 10 (76.000 : 5.0) and IEF 13 (72.000 :5.5), around IEF 50h (39.000 : 6.9) and between IEF 51f (40.000 : 4.8) and IEF 55 (35.000 : 4.8). The kinetics of changes during a chase in some of these polypeptides are shown in Fig. 4, 5 and 6. Other changes will be mentioned but some have been difficult to quantitate owing to overlapping peptide spots.

When the same type of experiments were performed using cells infected with a ts-mutant which is temperature sensitive for transformation MSV-CP27 (Forchhammer and Turnock, 1978) complete transformation is seen at 32° and the polypeptides labelled in a 1 hour pulse is shown in Fig. 2 a. At 39° no transformation can be observed and Fig. 2 b shows the polypeptides synthesized at this non-permissive temperature. Uninfected control cultures were also labelled at 32° and 39° and showed results similar to those obtained at 36° C. Striking similarity between the two transformed cell cultures can be observed (Fig.1a and Fig.2a) as well as between the polypeptides from the uninfected control cultures (Fig. 1b), and the polypeptides from the culture infected with the ts-CP27 grown at the non-permissive temperature of 39°C (Fig. 2b).

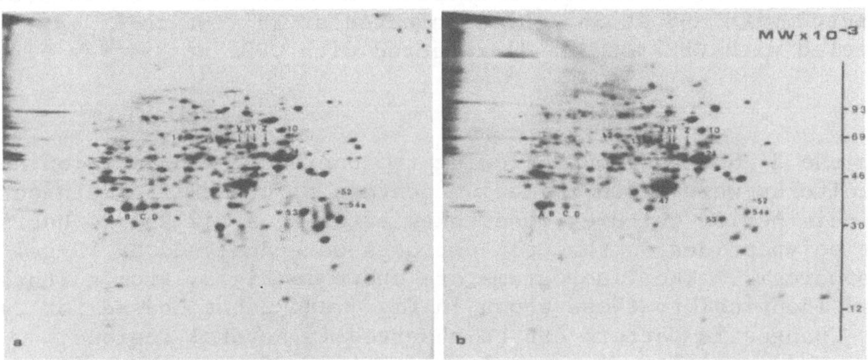

Fig.2. Fluorograms of polypeptides of cells infected with the ts-mutant MSV-CP27. Parallel cultures of 10T$\frac{1}{2}$ cells were infected with CP27 with identical inocula of virus (moi:0.2 FFU/cell +0.3 XCPFU/cell). After 4 days of growth at 32° cells were completely transformed a), but retained normal morphology at 39°b). All cultures were labelled for 1 hour at the appropriate temperatures.

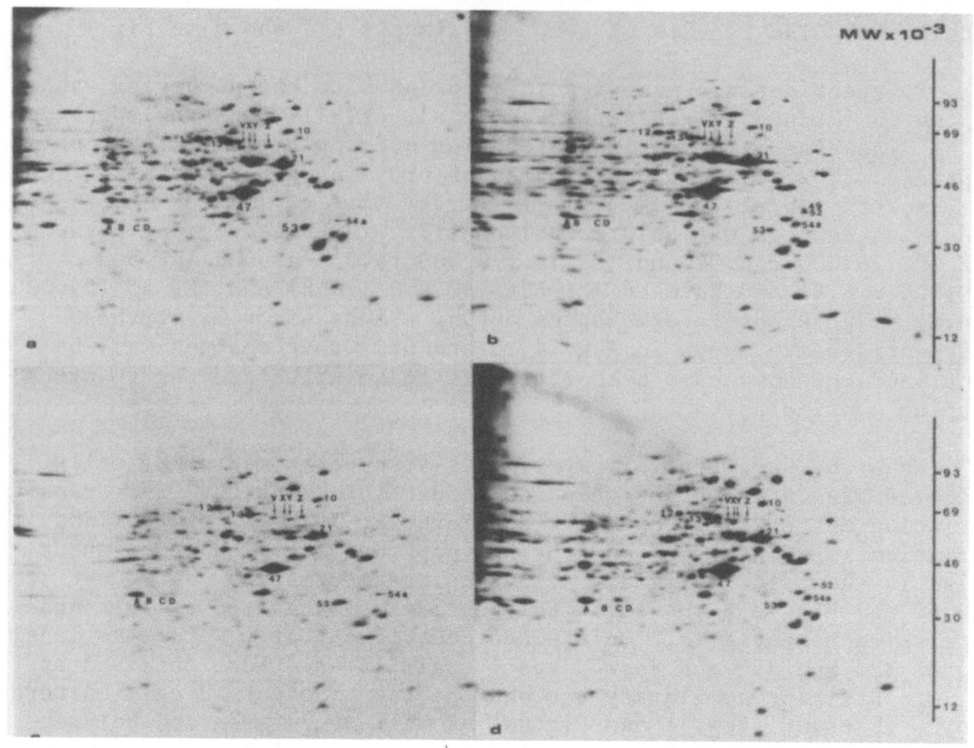

Fig.3. Fluorograms of polypeptides from pulse-chase experiments. Cultures labelled for 1 hour with $[^{35}S]$-methionine and chased for 24 hours with non-radioactive methionine at unchanged temperature. a) Infected with MSV at 36° b) uninfected at 36° (as in Fig.1.) c) infected with CP27 at 32^0 d) infected with CP27 at 39° (as Fig.2).

In the 24 hour period following the one hour $[^{35}S]$-methionine pulse cultures were grown in medium containing unlabelled methionine. During this period cultures were taken after 2, 5, 12 and 24 hours and the polypeptides in the cell extracts were analysed by 2D-gel electrophoresis. The fluorograms are shown in Fig.3. from cultures labelled identical to those shown in Fig.1 and 2 but chased for 24 hours. Changes in pattern can be observed in several regions. Thus the radioactivity of several polypeptides are now much less intense in the transformed cultures for instance B, C and D as well as for a series of vertical streaks in the right side of the gels of polypeptides with MW 30-40,000 and pI 5.7 to 4.7. The relative amounts of the proteins 49, 52 and 54a are also reduced in the transformed cultures and different proportions can be observed in polypeptides V,X,Y and Z.

Quantitation of the Changes Observed

Quantitation of autoradiograms was performed by cutting out individual spots from gels and counting the quantity of $[^{35}S]$-methionine although it is difficult for polypeptides located within close proximity. Approximately 100 polypeptides were cut out per IEF gel and the remaining gel was reexposed to assure the accuracy of the cutting. Finally the total gel was cut into pieces and counted to estimate the counts in proteins dissociated during IEF, leaving out the counts at the origin of the gel used for the first dimension(10-20% of the total cpm). After correction for $[^{35}S]$ decay individual polypeptides can be expressed as a fraction of the total radioactivity in dissociated polypeptides. This has been calculated and results for several proteins are shown in Fig. 4, 5 and 6.

It can be seen that the counts in β- and γ actin(IEF 47)constitutes a slightly increasing fraction of the total counts present in polypeptides during the chase (Fig.4a) most likely reflecting that approximate 1/4 of the labelled polypeptides turn over during 24 hours of growth both in transformed and untransformed cells. Some proteins turnover more rapidly in transformed than in untransformed cells especially a series of polypeptides around IEF50h. The polypeptides B, C and D disappear with halflives of ½-2 hours (Fig. 4b and 5). It is possible that A corresponds to IEF 50 h and C corresponds to IEF 53 but it may also be due to superimposition of the proteins; a question which can only be answered after peptide mapping. Changes during the chase can also be observed in the polypeptides IEF 49 and 52, which decrease in relative amounts in transformed cultures when compared to untransformed growing cultures. These polypeptides were shown to be related to tropomyosin by immuno precipitation , see later (Fig.7). Other changes have also been observed including the polypeptides indicated with V, X, Y and Z in Fig. 1-3. Both IEF V and Z contain a doublet of polypeptides with a very small difference in MW only allowing the doublets to be cut out and counted together. The general trend for IEF V and Z in transformed cells is, that they are preserved during a chase, whereas IEF Y is greatly reduced during the chase (compare Fig. 1a to Fig. 3a and Fig. 2a to Fig. 3c). In contrast untransformed cells maintain IEF Y as the most prominent of these polypeptides during a chase, (compare Fig. 1b to 3b and Fig. 2b to Fig. 3d). Counting IEF Y has been complicated by polypeptides located in the vicinity, since all the pulse labelled cultures exhibit a labile partly overlapping polypeptide IEF X in amounts similar to IEF Y.

Identification by immunoprecipitation of some of the polypeptides reduced in amounts in transformed cells

It has previously been reported that a protein (33,000 : 4.8)

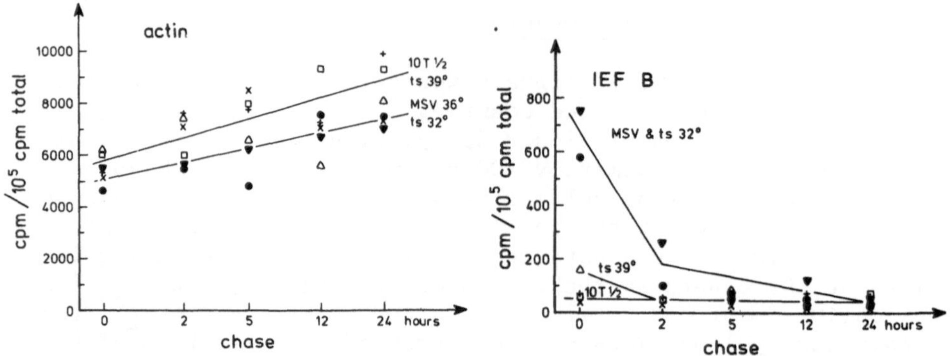

Fig.4. Relative amounts of [^{35}S]-methionine present in individual polypeptides at various times during the 24 hour chase. a) α+γ actin (IEF 47) b) IEF B. Open symbols represent untransformed cultures i.e. uninfected at 32° +; at 36° □ ;at 39° x and ts-CP27 infected at 39° Δ . Black symbols represent transformed cultures i.e. MSV infected at 36° ● and ts-CP27 infected at 32° ▼.

was reduced in transformed cells as well as in tumors induced by MSV and the ts-mutant CP27 (Forchhammer and Klarlund, 1979). At the same time Paulin et al. (1979) have reported, that the amounts of tropomyosin in a teratocarcinoma cell line PCC3/A/1 was reduced in a tumorigenic form of this cell line when compared to a flat, non tumorigenic form of PCC 3/A/1. To study a relation between these observations the proteins in this area were identified by immuno precipitation with anti tropomyosin antibody. A purified IgG (1.5 mg/ml) from antisera against chicken gizzard tropomyosin was obtained from D. Louvard (EMBL, Heidelberg). As shown in Fig. 7 four polypeptides (IEF 49, 52, 54a and 55) was specifically precipitated with this IgG. In addition some actin (IEF 47) and vimentin (IEF 22) was non-specifically precipitated in all samples including the controls where either IgG or protein A was omitted. Precipitation of extracts from MSV transformed cells show the same

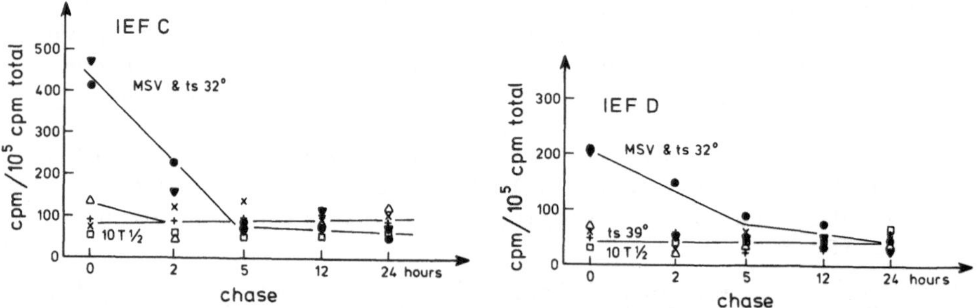

Fig.5. Relative amounts of [^{35}S]-methionine present in polypeptides during a 24 hour chase a) IEF C and b) IEF D. Symbols are as in Fig.4.

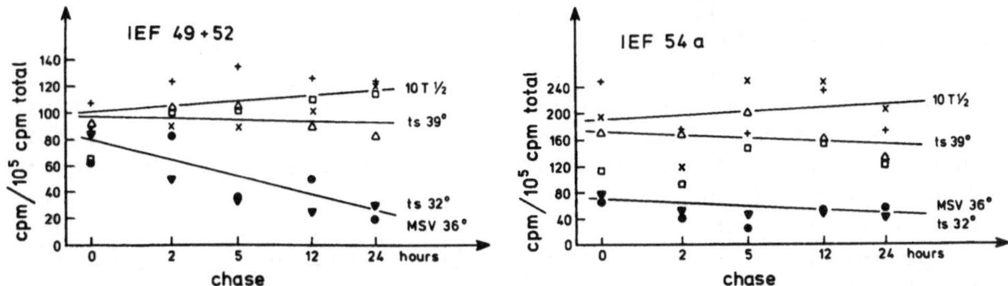

Fig.6. Relative amounts of [^{35}S]-methionine present in individual polypeptides related to tropomyosin during the chase. a) IEF 49 + IEF 52 and b) IEF 54a. Symbols are as in Fig. 4.

proteins in amounts lowered with a factor two to four. This identification is in agreement with analysis of the polypeptides IEF 49, 52 , 54a and 55 performed with partial peptide digestion using protease V-8 (Bravo et al., 1981).

Pulse-chase experiments shown in Fig. 6 furthermore indicate that the proteins IEF 49 + 52 presumably two forms of β-tropomyosin are synthesized but are more labile in transformed cells; the reduced amount of protein 54a presumably α-tropomyosin is observed already after 1 hour of pulse with [^{35}S]-methionine indicating a lower rate of synthesis of this protein in transformed cells. Experiments performed with the ts-mutant MSV-CP27 show that these differences can only be observed at the permissive temperature 32°C and not at 39°C (Fig. 6) indicating an association of this effect with the transforming function.

DISCUSSION

Over the past few years separation of proteins by 2D-gel electrophoresis has been used by several groups to study cellular proteins in transformed cell lines (Strand and August, 1978; Forchhammer and Klarlund, 1979; Bravo and Celis, 1980; Brzeski and Ege, 1980). Several modifications of the original technique proposed by O'Farrell (1975) have been used which make it difficult to identify polypeptides analysed with different modifications of the technique as recently discussed (Clark, 1981),and in this paper the numbering system proposed by Fey et al. (1981) has been adopted. A multitude of changes in quantities of polypeptides has been observed in transformed cells some of which may reflect the pleiotropic effects of transformation, whereas other may be related to the cloned origin of the cell lines studied. Brzeski and Ege (1980) have reported differences in 269 polypeptides in a variety of established transformed cell lines. By grouping the cell lines they were able to reduce the number of common changes to be around 3o polypeptides.

In this study, to avoid problems of clonal variations, infected and transformed cells were used. Analysis took place 3-5 days after the infection when the total cell population appeared transformed. Similarly changes were unlikely.to be due to differences in growth rate, since all cultures had 4-10 fold surface area available for growth. Another indication of good growth comes from the fact that similar amounts of the vimentin degradation polypeptides was seen in all cultures used for the pulse-chase experiments. The level of these proteins IEF 34g, 41d and 46c (indicated in Fig. 1a) have been linked to the mitotic - and G2-phase of the growth cycle (Bravo et al, 1981). Finally the changes in polypeptides discussed here seem not to be related to viral structural proteins present in infected cells, since immuno precipitation with goat anti sera against p30 or gp70 in experiments performed in parallel to those in Fig. 7 did not precipitate any of the polypeptides mentioned. It therefore

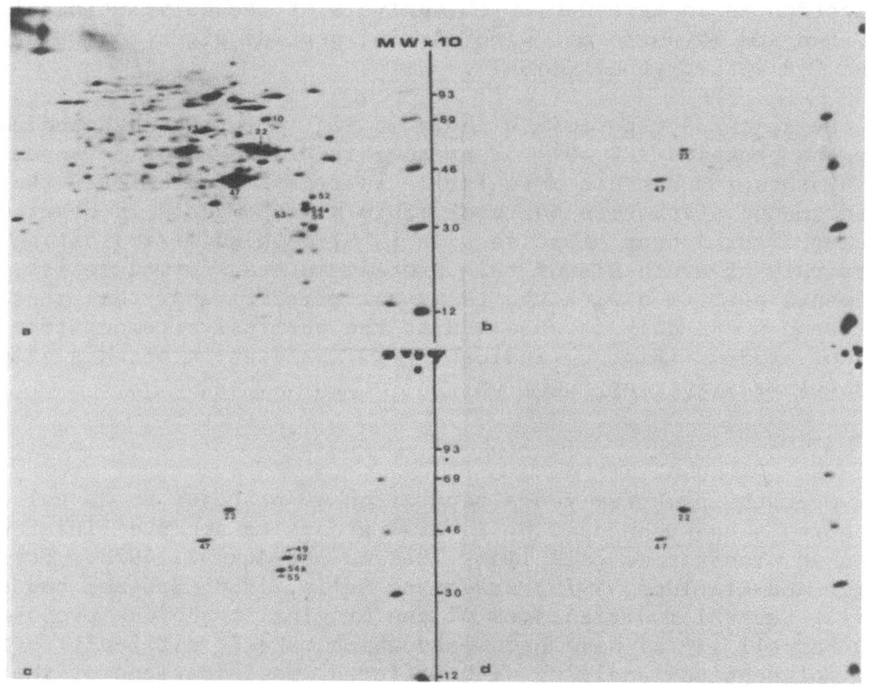

Fig.7. IEF 49, 52, 54a and 55 are related to tropomyosin. Polypeptides immuno precipitated (IP) from $[^{35}S]$-labelled proteins of 10T½ cells. a) polypeptides in the 100.000 g supernatant used for IP. b) IP using normal rabbit serum plus sepharose coupled protein A (S-PIA). c) IP using affinity purified IgG from rabbit anti chicken gizzard tropomyosin plus S-PIA. d) IP using the same IgG plus sepharose not coupled to protein A. MW markers are included as in Fig. 1.

seems likely to assume that the changes in cellular polypeptides are related to the function of the murine sarcoma gene. This supposition is supported by the experiments with the ts-mutant giving results at 32° closely similar to wild type infection with MSV at 36°. At 39° the non-permissive temperature the polypeptides synthesized are equal to those synthesized in uninfected cells.

Attention has here been focused on three sets of polypeptides. Two of these are indicated with capital letters on Fig. 1-3, and one more which was difficult to quantitate will be mentioned only in broader terms. The group which has been designated B, C and D consists of polypeptides with MW around 38,000 and pI = 6.8 - 6.5. Their synthesis was greatly increased in transformed cells and they have halflives of ½-2 hours as indicated in Fig. 4b and Fig. 5. The proteins encoded by the murine src gene has recently been identified by hybrid-arrest translation in vitro to consist of four polypeptides the two larger of these being 43K and 40-37K (Papkoff et al. 1980, Cremer et al. 1981, Klarlund and Andersson unpublished). Whether polypeptides B, C or D are related to the murine src gene product has presently not been answered and will await peptide mapping. Polypeptide A seems in contrast to the other to be relatively stable independently of transformation and seem to correspond in location to IEF 50 h. In the same size range a series of 7-8 vertical streaks can be observed in the acidic part of gels from pulse labelled transformed cultures (Fig. 1a and Fig. 2a). These have MW between 30,000 and 40,000 and pI ranging from 5.7 to 4.7. They can only be observed in the extracts from pulse labelled transformed cultures. Thus it is also possible that the murine src gene product could be found among these polypeptides which turn over rapidly and possess multiple charge classes.

Finally changes in polypeptides around IEF 54a has been reported several times in relation to transformation (Forchhammer and Klarlund, 1979; Bravo and Celis, 1980; Brzeski and Ege, 1980) and also in relation to tumorigenesis in mice (Paulin et al. 1979, Forchhammer and Klarlund, 1979). These polypeptides have been shown to be associated with the cytoskeleton (Bravo and Celis, 1980). Identification of the murine polypeptides IEF 49, 52, 54a and 55 as being related to tropomyosin has been indicated by peptide analysis (Fey et al. 1981). This concur with the results presented in Fig.7 where IgG specific for chicken gizzard tropomyosin was used in immuno precipitations to identify the spots. The kinetics of labelling and chase of IEF 49 + 52 (Fig. 6a) indicates, that these two polypeptides are more labile in transformed cells. The reduction in IEF 54a seems to indicate reduced synthesis since the labelling of this polypeptide stayed at the same relative level throughout the chase (Fig. 6b). Reduced amounts of these tropomyosinrelated polypeptides seem to be a general phenomenon observed in many different transformed cell lines. This raises the possibility that the dissociation of microfilaments in transformed cells (Edelman and Yahara, 1976) is caused by the reduction of one or several forms of tropomyosin rather than the small reduction of β- and γ actin (IEF 47) observed in Fig. 4a.

Besides the polypeptides mentioned here other differences can be observed as previously mentioned and observed by others. (Strand and August, 1978; Forchhammer and Klarlund, 1979; Bravo and Celis, 1980; Brzeski and Ege, 1980). The complex functions of several of these proteins may lead to the transformed phenotype of cells and the oncogene potential of transformed cells. This oncogenic potential has been shown for MSV (Aaronson and Rowe, 1970), and is temperature sensitive in the ts-mutant MSV-CP27 used in this study (Forchhammer and Klarlund, 1979, Klarlund and Forchhammer, 1980). All the changes in polypeptides described here has been temperature sensitive with this mutant, indicating that a functional murine src gene is necessary in the regulation of the polypeptides studied, some of which may be of importance for tumorigenesis.

METHODS AND MATERIALS

The cell line 10T½-cl 8 from C3H mice, the virus M-MSV (MuLV) and the ts-mutant M-MSV (MuLV)-CP27 have previously been described as well as labelling of cells with ^{35}S -methionine in tissue culture (Forchhammer and Turnock, 1978). Cultures were washed in PBS + 1 mM PMSF at 0°, scraped off, and centrifuged at 130 g for 30 sec. The unfractionated cell extract was dissolved in lysis buffer "new A" pH 5-7. 2D-gel electrophoresis was performed as described by O'Farrell (1975) with slight modifications (Forchhamer and Klarlund, 1979).

Immuno precipitatio of cell proteins with protein A-sepharose CL-4B. Cells in N2TET (20 mM Tris, pH 7.6 0.1 M NaCl, 1 mM EDTA, 0.2% TRITON X 100) were sonicated for 20 sec. and clarified by ultracentrifugation at 100,000 g for 30 min. The supernatants (S-100) were used for immunocomplex formation, sedimented after incubating 60 µl cell extract (5 x 10^{5}cpm) with protein A-sepharose CL-4B (Pharmacia, Uppsala, Sweden) for 20 min. at 4°C. The absorbed immune complexes were washed three times with 600 µl N2TET containing 1 mM PMSF and once with 600 µl of N2TE contining 1 mM PMSF. The sediment was collected at 2.000 g for 1.5 min. and resuspended in 60 µl of lysis buffer "New A" containing 2.3% SDS and solubilized by heating in a waterbath at 70°C for 3 min. before electrophoresis.

ACKNOWLEDGMENT: These studies were supported by the Danish Cancer Society. I would like to thank Helle Jensen, Rita Nielsen and Birgette Rask for their skillful praticipation in the experiments. The numbering of proteins has been discussed in details with Rodrigo Bravo and Julio Celis, whom I would like to thank for many useful discussions. The IgG from rabbit antisera against chicken gizzard tropomyosin was kindly supplied by David Louvard, EMBL, Heidelberg.

REFERENCES

Aaronson, S.A. and Rowe, W.P., 1970, Nonproducer clones of murine sarcoma virus transformed Balb/3T3 cells, Virology, 42:9-19.

Bravo, R. and Celis, J.E., 1980, Gene expression in normal and virally transformed mouse 3T3B and hamster BHK21 cells, Exp. Cell Res., 127: 249-260.

Bravo, R., Fey, S.J., Bellatin, J., Larsen, P.M., Arevalo, J. and Celis, J.E., 1981, Identification of a nuclear and of a cytoplasmic polypeptide whose relative proportions are sensitive to changes in the rate of cell proliferation, Exp.Cell Res. In press.

Bravo, R., Fey, S.J., Larsen, P.M. and Celis, J.E, 1981, Modificacations of vimentin polypeptides during mitosis. Cold. Spr. Harbor Symp. Quant. Biol. 46, In press.

Brzeski, H. and Ege, T., 1980, Changes in polypeptide pattern in ASV-transformed rat cells are correlated with the degree of morphological transformation, Cell, 22:513-522.

Clark, B.F.C., 1981, Towards a total human protein map, Nature, 292: 491-492.

Cremer, K., Reddy, E.P. and Aaronson, S.A., 1981, Translational products of Moloney murine sarcoma virus RNA: Identification of proteins encoded by the murine sarcoma virus src gene, J.Virol.38: 704-711.

Edelman, G.M. and Yahara, I., 1976, Temperature-sensitive changes in surface modulating assemblies of fibroblasts transformed by mutants of Rous sarcoma virus, P.N.A.S. USA, 73: 2o47-2051.

Fey, S.J., Bravo, R., Larsen, P.M., Bellatin, J. and Celis, J.E., 1981, ^{35}S -methionine labelled polypeptides from secondary mouse kidney fibroblasts: Coordinates and one dimensional peptide maps of some major polypeptides, Cell Biology International Reports, 5: 491-500.

Forchhammer, J. and Turnock, G., 1978, Glycoproteins from murine C-type virus are more acidic in virus derived from transformed cells than from non-transformed cells, Virology, 88: 177-182.

Forchhammer, J. and Klarlund, J., 1979, Changes in proteins from transformed cultures and tumors induced by sarcoma virus, in "Advances in Medical Oncology, Research and Education", vol.1: 51-60, P.G. Margison, ed., Pergamon Press, Oxford.

Klarlund, J. and Forchhammer, J., 1980, Temperature-sensitive tumorigenicity of cells transformed by a mutant of Moloney sarcoma virus, P.N.A.S.USA, 77:1501-1505.

O'Farrell, P.H., 1975, High resolution two-dimensional electrophoresis of proteins, J. Biol. Chem. 250: 4007-4021.

Papkoff,J.,Hunter,T. and Beemon,K.,1980,In vitro translation of virion RNA from Moloney murine sarcoma virus,Virology,101:91-103.

Paulin,D.,Perreau,J.,Jakob,H.,Jacob,F.and Yaniv,M.,1979, Tropomyosin synthesis accompanies formation of actin filaments in embryonal carcinoma cells induced to differentiate by hezamethylene bisacetamide, P.N.A.S. USA. 76 : 1891-1895.

Strand,M. and August,J.T., 1978, Cell, 13, 399-408.

TUMOR ANTIGENS AND TRANSFORMATION-RELATED PROTEINS ASSOCIATED WITH FELINE RETROVIRUSES

M. Essex and A.P. Chen

Department of Microbiology, Harvard University School of Public Health, Boston, Massachusetts 02115 U.S.A.

INTRODUCTION

The oncogenic feline retroviruses that cause leukemia, lymphoma, and fibrosarcoma have been defined in detail. The feline leukemia virus (FeLV) a replication-competent agent, causes leukemia, lymphoma, and aplastic anemia, as well as a syndrome of immunosuppression which predisposes cats to the development of numerous other infectious diseases (Essex, 1975). FeLV will not transform fibroblasts. The tumors induced by FeLV occur after a prolonged induction period (Francis, et al., 1979b), and they are monoclonal (Mullins, personal communication). The feline sarcoma virus (FeSV), which is defective for replication, can only be transmitted from cell to cell in the presence of FeLV. FeSV will transform cultured fibroblasts with a very high degree of efficiency. When FeSV is inoculated into kittens, it usually causes multicentric fibrosarcomas. These tumors are polyclonal in origin and they occur after a very brief latent period (Snyder and Theilen, 1969).

FeLV and FeSV have genome structures that are typical of many other retroviruses of avian, murine, and primate origin (Kurth, et al., 1979). The genome of FeLV, about 8.5 kilobases in length, has long terminal repeat or LTR sequences at both ends, and three genes, all of which encode for products essential for replication. The *gag* gene, which is closest to the 5' end of the RNA, represents a polyprotein which undergoes post-translational cleavage. The peptides represented in this polyportein, designated by their molecular weight in thousands, occur in the order p15-p12-p30-p10 from the 5' end of the genome. The other two proteins made by the FeLV genome are reverse transcriptase and the gp70-p15e envelope protein complex.

FeSV, expecially the ST and GA strains, has also been completely mapped and compared to FeLV (Sherr, et al., 1980). The same LTR sequence is found at both ends of FeSV, which is only five kilobases in length. Essentially all of the *pol* gene has been deleted, as well as a large portion of the *env* gene and a smaller portion of the *gag* gene. In the deleted middle region of the genome a new 1.5 kilobase sequence designated *fes* (Coffin, et al., 1981) has been inserted. This stretch of nucleotides represents information that has been rescued from genetic information present in all normal cat cells. As represented in the cell genome, however, the nucleotides found in *fes* cover a span of more than four kilobases, due to the presence of three introns which separate the four stretches which come together to form this gene in the virion (Franchini, et al., 1981). To date, only one protein product has been identified as encoded by FeSV, the polyprotein designated *gag-fes* (Stephenson, et al., 1977b).

This polyprotein, which has autophosphorylating kinase activity, (Barbacid, et al., 1980; Van de Ven, et al, 1980), is approximately 85,000 daltons in the case of ST-FeSV, and about 110,000 daltons for the related GA strain. *Fes* and other comparable *"onc"* genes derived from the cell apparently represent highly conserved sequences. One sush gene insert found in transforming avian retrovirus, the Fujinami sarcoma virus, has been reported to be distantly related to *fes* (Shibuya, et al., 1980; Barbacid, et al., 1981).

ONCORNAVIRUS ASSOCIATED CELL MEMBRANE ANTIGEN

When adult cats are inoculated with FeSV they usually do not develop progressively growing tumors. Instead, they either fail to develop any detectable tumors or develop tumors which regress after a brief growth phase. Such cats are classified as "regressors". Conversely, when young kittens are inoculated with FeSV they usually develop tumors that continue to grow locally and metastasize to kill the animal. Those that respond in this way are classified as progressors. Although the "regressor" vs "progressor" response is usually associated primarily with age at the time of inoculation, other factors such as the dose of FeSV obtained and perhaps the strain of virus are also important (Essex, et al., 1971a,b).

The mechanism of resistance in regressor cats seems to be the immune response. Cats classified as regressors have high levels of serum antibodies which react with a surface antigen or antigen complex designated the feline oncornavirus-associated cell membrane antigen (FOCMA) at the membrane of feline leukemia cells, fibrosarcoma cells, and FeSV-transformed fibroblasts (Essex, 1975; Hardy, et al., 1977). Cats classified as progressors usually lack detectable levels of serum antibodies to FOCMA, as do cats that have not been exposed to FeLV or FeSV.

Although fibrosarcomas associated with FeSV occur periodically in outbred domestic cats, leukemias and lymphomas occur more frequently, and 65-80% of the cases occur in cats that are viremic with FeLV (Francis, et al., 1979a). Cats with clinical leukemia or lymphoma do not have high titers of antibodies to FOCMA whether they are viremic or not (Essex, et al., 1975a). Serologically, such cats present the same picutre as FeSV-inoculated "progressor" animals.

FeLV is efficiently transmitted via saliva among cats that reside together (Francis, et al., 1977). In multiple cat households where one or more FeLV-excretor animals are present, most or all of the other exposed animals become either transiently or persistently viremic (Essex, et al., 1975b,c). Some of the exposed healthy animals develop high FOCMA antibody titers, while others fail to develop significant titers of such antibodies. Those cats that fail to develop high levels of antibodies following such exposures have a greatly increased risk for developing leukemia or lymphoma, regardless of whether they become persistently viremic (Essex, et al., 1975a,c).

Antibodies directed to the FOCMA complex are distinct from those directed to the structural proteins of the virion (Essex, et al., 1977; Stephenson, et al., 1977a). However, the virion proteins also occur at the cell surface, and antibodies to either FOCMA or the major virion proteins are lytic for virus producer cells in the presence of complement. This action can be demonstrated both *in vitro* and *in vivo* (Grant, et al., 1978; Noronha, et al., 1978; Cotter, et al., 1980). Since virion proteins are present on FeLV-producer non-leukemic cells as well as on producer leukemic cells, while FOCMA is present on both producer and virus-negative leukemia cells, we have postulated that an immune response to virus proteins in the absence of a response to FOCMA could result in the development of virus negative leukemias by immunoselection (Essex, et al., 1980). Virus-negative leukemias occur at greatly increased rates in cats that are exposed to FeLV in multiple cat households (Hardy, et al., 1980; Francis, et al., 1981).

Once cats develop high titers of antibodies to FOCMA following natural exposure to FeLV, they usually maintain these high antibody titers for prolonged periods, even in the absence of apparent re-exposure to FeLV. The persistence of such high antibody titers in non-viremic cats that are not re-exposed to FeLV suggests that such animals have FOCMA-positive cells somewhere in their body. Such cells would then expand in number to immunogenic doses as antibody titers decrease, and the resulting antibody response would eliminate many of the FOCMA-positive cells.

The nature of the FOCMA target antigens that are distinct from the virion structural proteins has been partially elucidated. One class of protein that fits this category is the *fes* portion of the *gag-fes* polyprotein encoded by the ST and GA strains of FeSV (Stephenson, et al., 1977b; Sherr, et al., 1978a,b). A second class of protein that fits criteria for FOCMA-type activity is the p65, p68 molecules found in FeSV-nonproducer transformed fibroblasts as well as lymphoma cells (Snyder, et al, 1978; Worley and Essex, 1980). We recently described a third class of antigen, designated NCP^{105} which is also distinct from the virion structural proteins (Chen, et al., 1982b). Although antibodies to this antigen are also found in FeLV-exposed healthy cats this protein does not fit the FOCMA pattern since it is found in normal uninfected cells as well as FeSV-transformed cells. Characteristics of these antigens are discussed in the following sections, and summarized in Table 1.

GAG-FES

The demonstration of a *gag-fes* protein which has FOCMA-type activity was made possible by the development of well-characterized FeSV-transformed nonproducer cells (Sliski, et al., 1977). The *gag-fes* protein contained the virus portions representing the 5' end of the *gag* gene, p15, p12, and a portion but not all of p30, as well as the non-*gag* portion equivalent to *fes*. This molecule was first demonstrated in whole cell homogenates taken from FeSV-transformed nonproducer cells, as well as in FeSV virus rescued by helper retroviruses of non-feline origin (Stephenson, et al., 1977b; Sherr, et al., 1978a). The *gag-fes* antigen identified in this was also designated FOCMA-S (Sherr, et al., 1978a,b). It was estimated to be about 85,000 daltons in the case of the protein encoded by ST-FeSV and about 110,000 daltons in the case of the protein encoded by GA-FeSV.

Gag-fes was characteristic of most of the polyproteins described for other defective retroviruses which could transform cultured cells such as the avian acute leukemia viruses and the Abelson murine leukemia virus (Witte, et al., 1978; Hayman, 1981). One property that all these proteins had in common was the ability to phosphorylate tyrosine. In some instances this was a kinase activity that acted on other substrates, while in other cases it appeared to be limited to autophosphorylation. *Gag-fes* appears to localize in the cytoplasmic membrane of the cell. When FeSV-transformed cells are metabolically labelled with ^{35}S methionine and subjected to immunoprecipitation, essentially all of the label was found in the cell membrane, while none was found in the cytosol, nucleus, or mitochondria-lysosome fractions (Essex, et al., 1980).

The *gag-fes* protein was known to be present in nonproducer cells of non-feline origin that were transformed by FeSV *in vitro* (Stephenson, et al., 1977b). In recent studies we found that the same *gag-fes* proteins were also present in FeSV-transformed cells of feline origin, and present in cells that produce FeLV as well as nonproducer cells (Chen, et al., 1980, 1981). *Gag-fes* was not present in cat cells transformed by chemical carcinogens such as dimethylbenzanthracene (Rhim, et al., 1981). Also, FOCMA-positive feline lymphoma cells, both FeLV-positive and FeLV-negative, lack this protein. In fact, attempts to introduce FeSV into lymphoid cell lines have been unsuccessful.

To determine if cats could mount an antibody response to the *fes* portion of the *gag-fes* molecule, animals were inoculated with their own biopsied cells following transformation of the cells *in vitro* with FeSV (Chen, et al., 1980, 1981). The cats were inoculated with a limited number of transformed cells as young adults so that they experienced a "regressor" type of response. The resulting antisera were then absorbed extensively with FeLV structural proteins. Such antisera, confirmed to be negative for reactivity with all virion structural proteins, still gave the typical FOCMA-type membrane immunofluorescense reaction on both transformed fibroblasts and lymphoma cells. Antisera prepared in this manner would also successfully immunoprecipitate the *gag-fes* protein, obviously on the basis of reactivity with the *fes* determinant (Chen, et al., 1981). Perhaps of even greater interest, high titered anti-FOCMA sera taken from cats that had only been exposed to FeLV would also react specifically with *fes*. This suggested that FeLV must somehow activate *fes*-related sequences (c-*fes*) in some of the cells of infected cats, at least those cats that mount high anti-FOCMA responses. Some anti-FOCMA serum samples with anti-*fes* activity were from healthy cats that had persistent viremia (Essex, et al., 1975b). Presumably in many of these cases adsorption of all antibodies to the FeLV structural proteins has occurred *in vivo*. In the limited number of samples examined from both FeSV-inoculated cats and FeLV-exposed animals, the anti-FOCMA response coincided with the anti-*fes* response. While cats with high anti-FOCMA titers had detectable antibodies to *fes*, those with low anti-FOCMA titers or no detectable anti-FOCMA lacked antibodies to *fes*.

In another series of experiments, tumor cells taken from cats with FeSV-induced tumors were examined to determine if they expressed the same *gag-fes* polyproteins as cells transformed with FeSV *in vitro* (Chen, et al., 1981, 1982a). With the exception of the Rous sarcoma virus transforming protein $pp60^{src}$, which is different from the other "*onc*" proteins in that it is translated separately from any virus structural gene products, little or no information is available about whether or not the "*onc*" gene products found in cells transformed *in vitro* are also present in tumor cells that arose *de novo in vivo*. To address this question, cells from progressively growing fibrosarcomas were biopsied, prepared as single-cell suspensions, and allowed to metabolize in culture media at 37°C for periods that were sufficient to allow uptake of ^{35}S methionine (Chen, et al., 1982a).

TABLE 1. NON-VIRON ANTIGENS DETECTED USING ANTISERUM FROM CATS EXPOSED TO FELINE RETROVIRUSES

Category	FOCMA	GAG-FES	p65	p68	NCP105
Method of detection	MIF,AMC[a]	MIF,IPT,RIA	MIF, ^{125}I	MIF, ^{125}I	IPT
Presence of antigen in:					
FeSV-transformed nonproducer feline cells	+	+	+	–	+
FeSV-transformed nonproducer mink cells	+	+	+	–	+
Normal uninfected cat cells	–	–	–	–	+
Normal uninfected mink cells	–	–	–	–	+
FeLV-infected non-transformed cat cells	–	–	–	–	+
Carcinogen-transformed cat cells	–	–	ND	ND	+
Normal cat lymphoid cells	–	–	–	–	+
FeLV-producer cat lymphoma cells	+	–	–	+	+
Nonproducer cat lymphoma cells	+	–	–	+	+
Presence of antibodies in:					
FeSV regressor cats	+	+	+	+	–
FeSV progressor cats	–	–	–	–	–

Category	FOCMA	GAG-FES	p65	p68	NCP^{105}
Lymphoma bearing cats	–	–	–	–	–
FeLV-exposed healthy cats (high FOCMA antibody levels)	+	+	+	+	–
FeLV-exposed healthy cats (low or negative FOCMA antibody levels)	±	–	–	–	+
Healthy unexposed cats	–	–	–	–	–
Major site in cell	Membrane	Membrane	Membrane	Membrane	ND
Size (daltons x 1,000)	ND	85(ST) 110(GA)	65	68	105
Phosphorylated	ND	Yes	ND	ND	ND
Protein kinase activity	ND	Yes	ND	ND	ND

[a]Abbreviations: MIF=membrane immunofluorescence, AMC=antibody mediated cytotoxicity, IPT=immunoprecipitation, ND=not done, ^{125}I=antibody binding of iodinated protein.

Fibrosarcomas induced with both ST-FeSV and GA-FeSV were examined. Although some were examined at *in vitro* culture intervals of one to two days or less, others were examined after 20 or more subcultivations to determine if the *gag-fes* proteins were expressed under all such conditions. In every instance tumor cells from animals with ST-FeSV-induced tumors expressed the characteristic polyprotein of 85,000 daltons and tumor cells from animals with GA-FeSV-induced tumors expressed the characteristic polyprotein of 110,000 daltons. The *gag-fes* polyprotein found in the tumor cells was compared to the analogous protein found in cells transformed *in vitro* by FeSV using two-dimensional tryptic peptide mapping. No differences could be detected. Additionally, the *gag-fes* proteins found in tumor cells had the same protein kinase activity found for the protein detected in cells transformed *in vitro*.

Tumor cells taken from cats with various types of spontaneous tumors were also examined for the presence of *gag-fes* polyproteins (Chen, et al., 1982a). These included a melanoma, a neurofibrosarcoma, a chondrosarcoma, an osteosarcoma, and a fibrosarcoma; all taken from cats that were negative for FeLV and FeSV. None of these tumor cells had transformation-related polyproteins that were characteristic of ST-FeSV or GA-FeSV. The possibility that such tumor cells express *fes* or a portion of *fes* in a protein of a different structure could not be eliminated by these studies.

Finally, since different "*gag-onc*" polyproteins identified in association with the avian leukemia viruses have been postulated to mimick differentiation proteins that might function in a given restricted lineage of hematopoietic cells (Graf, et al., 1980), we decided to check tumor cells taken from FeSV-induced tumors orginating from different embyonic germ layers (Chen, et al., 1980, 1981). To do this, we took advantage of the ability of FeSV to induce melanomas when inoculated by the intradermal or intraocular routes (Niederkorn, et al., 1981). Melanomas reportedly arise from neuroectodermal cells whereas both fibrosarcomas and lymphomas arise from mesodermal cells. In this study cells from FeSV-induced melanomas were found to have the same *gag-fes* polyprotein as cells from fibrosarcomas or cells transformed by the same strain of FeSV *in vitro* (Chen, et al., 1981).

OTHER PROTEINS THAT REACT WITH FOCMA ANTIBODIES

Using cell surface radioiodination, Snyder et al. (1978, 1980) identified two non-glycosylated proteins that lacked any reactivity with p15, p12, p30, or any of the other virion structural proteins. Both proteins reacted with antisera to FOCMA. One, detected on FeSV-transformed fibroblasts was estimated to be 65,000 daltons, and the second, detected on feline lymphoma cells, was estimated to be about 70,000 daltons. Both were also biochemically distinct by peptide mapping from the virion envelope gp70 protein (Snyder, et al., 1980). Neither p65 or p70 was present in normal uninfected mink or cat cells.

In a separate set of experiments, starting with cell membrane fractions using gel filtration and ion-exchange chromatography, p65 was purified from the same FeSV-transformed mink nonproducer fibroblasts (Worley and Essex, 1980, 1982a,b). p65 identified in this manner regularly reacted with high-titered anti-FOCMA cat sera, but not with antisera from cats that lacked FOCMA antibody activity. Using immunoaffinity columns that contained antisera raised in rabbits to the purified p65, a FOCMA-reactive molecule of about 68,000 daltons was identified in 3 M KCL extracts of feline lymphoma cells (Worley and Essex, 1982a,b). This protein, designated p68, was also non-glycosylated and distinct from the virus structural proteins. It reacted with the appropriate spectrum of cat anti-FOCMA sera, and was presumably similar or identical to the p70 protein described by Snyder et al. (1978, 1980).

Recently we detected another protein, designated NCP^{105}, which could be immunoprecipitated with antisera obtained from healthy cats that had prior exposure to FeLV (Chen, et al., 1980, 1982b). Antibodies to this protein were not present in cats that lacked any exposure to FeLV, but such antibodies seemed to be lacking in cats that had high antibody titers to FOCMA and reactivity for the *fes* portion of *gag-fes*. The NCP^{105} protein was however not limited to FeSV-tranformed cells or lymphoma cells. NCP^{105} could be detected in a wide variety of normal and malignant rapidly dividing cells originating from cats as well as a wide variety of other species including mink, dog, and ox (Chen, et al., 1982b). NCP^{105} clearly lacked the *gag*-encoded determinants present in *gag-fes*.

Since NCP^{105} appeared to be functioning at an immunogenic level only in cats that were exposed to FeLV and evolutionarily conserved in the sense of expression across specied barriers, the possibility that this protein might be encoded by a cellular form of an *onc* gene such as *c-fes* was considered. A recent report by Barbacid, et al., (1980) in fact described a normal cell protein that had a wide species distribution, protein kinase activity, serologic cross-reactivity with *fes*, and was very similar in size to NCP^{105}. The NCP described by Barbacid and his colleagues was found to have serologic cross-reactivity with the v-*fes* protein when hyperimmune goat v-*fes* antiserum was used and thus was interpreted as the protein product of c-*fes*.

However, when comparing the tryptic peptide map of NCP^{105} to that of *gag-fes*, no evidence was found for the sharing of common peptides by these two proteins.

SUMMARY

The oncogenic feline retroviruses occur in two distinct categories. The feline sarcoma viruses (FeSV), which are defective for replication, cause transformation of cultured cells and fibrosarcomas or other tumors after a very brief latent period. The viral genomes lack portions or all of the nucleotides found in the three genes normally present in replication competent retroviruses such as the feline leukemia virus (FeLV). Instead, an insertion sequence *onc* or *fes* gene, recently acquired from the cell genome, is present in FeSV. FeLV is a typical helper virus, competent for replication, and lacking any *onc* gene. FeLV causes lymphoma or leukemia after a prolonged latent period.

Cats that undergo a "regressor" response following exposure to FeSV develop high titers of humoral antibodies which react with the feline oncornavirus associated cell membrane antigen (FOMCA) which is on both lymphoma cells and transformed fibroblasts. Cats with lymphoma or "progressor" fibrosarcoma lack detectable levels of such antibodies.

Three classes of non-virion antigens have been described that react with antibodies found in cats exposed to FeLV or FeSV. The first, *gag-fes*, is a polyprotein that represents the only gene product of FeSV that has been recognized thus far. It is transformation-specific and it possesses protein kinase activity. It is associated with the malignant phenotype both *in vitro* and *in vivo* and even in tumors that originate in different embyonic germ layers, if in fact the tumor was induced by FeSV. The second class of antigen is the p65, p68 series which is found as p65 in FeSV-transformed fibroblasts and as p68 in lymphoma cells. This protein also occurs at the cell membrane. The third class of antigen is designated NCP^{105}. Although antibodies to this protein only appear in cats exposed to FeLV, and the protein is similar in size to *gag-fes*, it is distinct from *gag-fes* by tryptic peptide mapping.

ACKNOWLEDGMENTS

Work done in the laboratories of the authors was supported by NIH grant CA-13885 and CA-18216 and American Cancer Society grant PDT-36.

REFERENCES

Barbacid, M., Beemon, K., and Devare, S.G., 1980, Origin and functional properties of the major gene product of the Snyder-Theilen strain of feline sarcoma virus, Proc. Natl. Acad. Sci. 77:5158.

Barbacid, M., Breitman, M.L., Lauver, A.V., Long, L.K., and Vogt, P.K., 1981, The transformation-specific proteins of avian (Fujinami and PRC-11) and feline (Snyder-Theilen and Gardner-Arnstein) sarcoma viruses are immunologically related, J. Virol. 110:441.

Chen, A.P., Essex, M., Mikami, T., Albert, D., Niederkorn, J.Y., and Shadduck, J.P., 1980, The expression of transformation-related proteins in cat cells, in: Feline Leukemia Virus (W.D. Hardy, Jr., M. Essex, and A.J. McClelland, eds.) pp. 441-456, Elsevier/North Holland, New York.

Chen, A.P., Essex, M., Shadduck, J.A., Niederkorn, J.Y., and Albert, D., 1981, Retrovirus-encoded transformation specific polyproteins: Expression coordinated with malignant phenotype in cells from different germ layers, Proc. Natl. Acad. Sci. 78:3915.

Chen, A.P., Essex, M., Kelliher, M., de Noronha, F., Shadduck, J.A., Niederkorn, J.Y., and Albert, D., 1982a, Expression of feline sarcoma virus-specific transformation related proteins in tumor cells, Submitted for publication.

Chen, A.P., Mikami, T., and Essex, M., 1982b, Infection with feline leukemia virus associated with induction of humoral immune response to a normal cell protein, Submitted for publication.

Coffin, J.M., Varmus, H.E., Bishop, J.M., Essex, M., Hardy, W.D.Jr., Martin, G.S., Rosenberg, N.W., Scolnick, E.M., Weinberg, R.A., and Vogt, P.K., 1981, A proposal for naming host cell-derived inserts in retrovirus genomes, J. Virol. 40:953.

Cotter, S.M., Essex, M., McLane, M.F., Grant, C.K., and Hardy, W.D. Jr., 1980, Passive immunotherapy in naturally occurring feline mediastinal lymphoma in: Feline Leukemia Virus (W.D. Hardy, Jr., M. Essex, and A.J. McClelland, eds.) pp. 219-226, Elsevier/North Holland, New York.

Essex, M., 1975, Horizontally and vertically transmitted oncornaviruses of cats, Advan. Cancer Res. 21:175.

Essex, M., Klein, G., Snyder, S.P., and Harrold, J.B., 1971a, Antibody to feline oncornavirus-associated cell membrane antigen in neonatal cats, Int. J. Cancer 8:384.

Essex, M., Klein, G., Snyder, S.P., and Harrold, J.B., 1971b, Feline sarcoma virus (FSV) induced tumors: Correlations between humoral antibody and tumor regression, Nature 223:195.

Essex, M., Cotter, S.M., Hardy, W.D. Jr., Hess, P., Jarrett, W., Jarrett, O., Mackey, L., Laird, H., Perryman, L., Olsen, R.G., and Yohn, D.S., 1975a, Feline oncornavirus associated cell membrane antigen. IV. Antibody titers in cats with naturally occuring leukemia, lymphoma, and other diseases, J. Natl. Cancer Inst. 55:463.

Essex, M., Jakowski, R.M., Hardy, W.D. Jr., Cotter, S.M., Hess, P. and Sliski, A., 1975b, Feline oncornavirus-associated cell membrane antigen. III. Antibody titers in cats from leukemia cluster household, J. Natl. Cancer Inst. 54:637.

Essex, M., Sliski, A., Cotter, S.M., Jakowski, R.M., and Hardy, W.D. Jr., 1975c, Immunosurveillance of naturally occurring feline leukemia, Science 190:790.

Essex, M., Stephenson, J.R., Hardy, W.D. Jr., Cotter, S.M., and Aaronson, S.A., 1977, Leukemia, lymphoma, and fibrosarcoma of cats as models for similar diseases of man, Cold Spring Harbor Proc. on Cell Proliferation 4:1197.

Essex, M., Sliski, A.H., Worley, M., Grant, C.K., Snyder, H. Jr., Hardy, W.D. Jr., and Chen, L.B., 1980, Significance of the feline oncornavirus associated cell membrane antigen (FOCMA) in the natural history of feline leukemia, Cold Spring Harbor Proc. on Cell Proliferation 7:589.

Franchini, G., Even, J., Sherr, C.J., and Wong-Staal, F., 1981, *Onc* sequences (v-*fes*) of Snyder-Theilen feline sarcoma virus are derived from non-contiguous regions of a cat cellular gene (c-*fes*), Nature 290:154.

Francis, D.P., Essex, M., and Hardy, W.D. Jr., 1977, Excretion of feline leukemia virus by naturally infected pet cats, Nature 269:252.

Francis, D.P., Cotter, S.M., Hardy, W.D. Jr., and Essex, M., 1979a, Comparison of virus-positive and virus-negative cases of feline leukemia and lymphoma, Cancer Res. 39:3866.

Francis, D.P., Essex, M., Cotter, S.M., Jakowski, R.M., and Hardy, W.D. Jr., 1979b, Feline leukemia virus infections: The significance of chronic viremia, Leukemia Res. 3:435.

Francis, D.P., Essex, M., Cotter, S.M., Gutensohn, N., Jakowski, R.M., and Hardy, W.D. Jr., 1981, Epidemiologic association between virus-negative feline leukemia and the horizontally transmitted feline leukemia virus, Cancer Letters 12:37.

Graf, T., Beug, H., and Hayman, M.J., 1980, Target cell specificity of defective avian leukemia virus: hematopoietic target cells for a given virus type can be infected but not transformed by strains of a different type, Proc. Natl. Acad. Sci. 77:389.

Grant, C.K., Essex, M., Pedersen, N.C., Hardy, W.D. Jr., Stephenson, J.R., Cotter, S.M., and Theilen, G.H., 1978a, Lysis of feline lymphoma cells by complement-dependent antibodies in feline leukemia virus contact cats. Correlation of lysis and antibodies to feline oncornavirus-associated cell membrane antigen, J. Natl. Cancer Inst. 60:161.

Hardy, W.D. Jr., Zuckerman, E.E., MacEwen, E.G., Hayes, A.A., and Essex, M., 1977, A feline leukemia virus- and sarcoma virus-induced tumor-specific antigen, Nature 270:249.

Hardy, W.D. Jr., McClelland, A.J., Zuckerman, E.E., Snyder, H.W. Jr., MacEwen, E.G., Francis, D., and Essex, M., 1980, Development of virus nonproducer lymphyosarcomas in pet cats exposed to feline leukemia virus, Nature 288:90.

Hayman, M.J., 1981, Transforming proteins of avian retroviruses, J. Gen. Virol. 52:1.

Kurth, R., Fenyo, E.M., Klein, E., and Essex, M., 1979, Cell surface antigens induced by RNA tumor viruses, Nature 279:197.

Niederkorn, J.Y., Shadduck, J.A., Albert, D., and Essex, M., 1981, Serum antibodies against feline oncornavirus-associated cell membrane antigen in cats bearing virally-induced uveal melanomas, Invest. Ophthalmol. 20:598.

Noronha, F. de, Schafer, W., Essex, M., and Bolognesi, D.P., 1978, Influence of antisera to oncornavirus glycoprotein (gp71) on *in vivo* infections, Virol. 85:617.

Rhim, J.S., Koh, K.S., Chen, A., and Essex, M., 1981, Characterization of clones of tumorigenic feline cells transformed by a chemical carcinogen, Int. J. Cancer 28:51.

Sherr, C.J., Fedele, L.A., Oskarsson, M., Maizel, J., and Vande Woude, G., 1980, Molecular cloning of Snyder-Theilen feline leukemia and sarcoma viruses: Comparative studies of feline sarcoma virus with its natural helper virus and with Moloney murine sarcoma virus, J. Virol. 34:200.

Sherr, C.J., Sen, A., Todaro, G.J., Sliski, A., and Essex, M., 1978a, Pseudotypes of feline sarcoma virus contain an 85,000 dalton protein with feline oncornavirus-associated cell membrane antigen (FOCMA) activity, Proc. Natl. Acad. Sci. 75:1505.

Sherr, C.J., Todaro, G.J., Sliski, A., and Essex, M., 1978b, Characterization of a feline sarcoma virus-coded antigen (FOCMA-S) by radioimmunosassay, Proc. Natl. Acad. Sci. 75:4489.

Shibuya, M., Hanafusa, T., Hanafusa, H., and Stephenson, J.R., 1980, Hemology exists among the transforming sequences of avian and feline sarcoma viruses, Proc. Natl. Acad. Sci. 77:6536.

Sliski, A.H., Essex, M., Meyer, C., and Todaro, G., 1977, Feline oncornavirus-associated cell membrane antigen: Expression in transformed nonproducer mink cells, Science 196:1336.

Snyder, H.W. Jr., Hardy, W.D. Jr., Zuckerman, E.E., and Fleissner, E., 1978, Characterization of a tumor-specific antigen on the surface of feline lymphosarcoma cells, Nature 275:656.

Snyder, H.W. Jr., Phillips, K.J., Hardy, W.D., Jr., Zuckerman, E.E., Essex, M., Sliski, A.H., and Rhim, J., 1980, Isolation and characterization of proteins carrying the feline oncornavirus-associated cell membrane antigen (FOCMA), Cold Spring Harbor Symp. Quant. Biol. 44:787.

Snyder, S.P., and Theilen, G.H., 1969, Transmissible feline fibrosarcoma, Nature 221:1074.

Stephenson, J.W., Essex, M., Hino, S., Aaronson, S.A., and Hardy, W.D. Jr., 1977a, Feline oncornavirus-associated cell-membrane antigen (FOCMA): Distinction between FOCMA and the major virion glycoprotein, Proc. Natl. Acad. Sci. 74:1219.

Stephenson, J.R., Kahn, A.S., Sliski, A.H., and Essex, M., 1979b, Feline oncornavirus-associated cell membrane antigen (FOCMA): Identification of an immunologically cross-reactive feline sarcoma virus-coded protein, Proc. Natl. Acad. Sci. 74:5608.

Van de Ven, W.J.M., Reynolds, F.H., and Stephenson, J.R., 1980, The nonstructural components of polyproteins encoded by the replication-defective mammalian transforming retroviruses are phosphorylated and have associated protein kinase activity, Virol. 101:185.

Witte, O.N., Rosenberg, N., Paskind, M., Shields, A., and Baltimore, D., 1978, Identification of an Abelson murine leukemia virus-encoded protein present in transformed fibroblast and lymphoid cells, Proc. Natl. Acad. Sci. 75:2488.

Worley, M., and Essex, M., 1980, Identification of membrane proteins associated with transformation-related antigens shared by feline lymphoma cells and feline sarcoma virus-transformed fibroblasts, in: Feline Leukemia Virus (W.D. Hardy, Jr., M. Essex, and A.J. McClelland, eds.) pp. 431-440, Elsevier/North Holland, New York.

Worley, M., and Essex, M., 1982a, Isolation and characterization of a transformation-related antigen expressed on feline sarcoma virus transformed fibroblasts, Submitted for publication.

Worley, M., and Essex, M., 1982b, Purification of a transformation-specific antigen from feline lymphoma cells that shares antigenic determinants with membrane proteins of feline sarcoma virus transformed fibroblasts, Submitted for publication.

LEUKEMOGENESIS BY BOVINE LEUKEMIA VIRUS

J. Deschamps[(1)], R. Kettmann[(1)(2)], G. Marbaix[(1)] and A. Burny[(1)(2)]

(1) Department of Molecular Biology, University of Brussels 67, rue des Chevaux, 1640 Rhode St-Genèse, Belgium
(2) Faculty of Agronomy, 5800 Gembloux, Belgium

THE VIRUS

Bovine leukemia virus (BLV) is a type C retrovirus which is the causative agent of a B cell lymphocytic leukemia horizontally transmitted in cattle (1, 2, 3). This virus was proven to be totally exogenous to its target animal as well as to sheep, goat, mouse, cat, chicken and man (4). It appears to be unrelated, as judged by nucleic acid hybridization and immunological properties, to any other retrovirus tested up to now. However, a slight relatedness between the amino terminal sequence of the major internal protein (p24) of BLV and the corresponding polypeptide (p30) of a murine type C virus has recently been uncovered (5). This suggests the existence of a remote common ancestor for both viruses.

BLV genetic organization is similar to that of the avian and mammalian type C retroviruses, namely : cap GAG-POL-ENV-?-poly(A). In agreement with the fact that BLV is a long latency leukemia virus, its genome does not seem to contain any cellular gene similar to the "onc" genes of the sarcoma (6) or acute leukemia viruses (7). However, the BLV genome contains an unidentified sequence about 2 kb long in its 3' half (8). The question of the possible role of this viral information in the onset of leukemia remains unanswered so far.

THE DISEASE

BLV leukemogenesis is characterized by the occurrence of distinct possible animal responses to the viral infection. Primary BLV infection is detected by the presence in the blood of the animals of a high level of antibodies raised against the envelope glycoprotein (gp51) and the major core protein (p24) of the virus (Ab^+ Animals). Infected animals may or may not develop persistent lymphocytosis (PL). This stage of the disease is characterized by an increased number of circulating B lymphocytes. Ab^+ or PL animals may develop lymphoid tumors after a latency period variable in length (from a few months to several years) (3). A recent study showed that no cellular information was present in the cloned complete tumor-derived proviral DNA. This differs from the AKR-MuLV system (9) where a recombinational event between a slow-acting leukemia provirus and a cellular DNA sequence generates an oncogenic viral strain.

HOST-VIRUS INTERACTION AT THE DNA LEVEL

In the PL and tumor phase of the disease, BLV infection is always characterized by the presence of one or a few copies of the proviral information integrated in the genome of the host cell. However, a striking difference was noticed between the pattern of provirus integration in the PL and tumor phase of the disease (10). The infected lymphocytes from animals in PL are polyclonal whereas the tumor cells are monoclonal with respect to the BLV integration sites. This fact was interpreted as meaning that the BLV integration site might be crucial for tumorisation. One possible molecular mechanism for BLV oncogenesis is the insertion of a viral promoter upstream to a cellular gene involved in the control of the normal B cell cycle, as was shown to be the case in avian leukosis virus - induced tumors (11, 12).

However, the restriction patterns of tumor cell DNA from different animals, when using the same endonuclease, reveal that the integration sites seem to be different (Fig. 1). Further more, recent experiments using labelled cloned tumor DNA fragments containing 3' viral - cellular junctions (Fig. 2) confirm that many possible sites may accomodate a BLV provirus (Fig. 3). These data argue against the existence of a unique bovine "onc" gene activated by insertion of a proximate upstream BLV promoter. Moreover they suggest that integration of the provirus takes place into one single chromosome of a given pair (see Fig. 3).

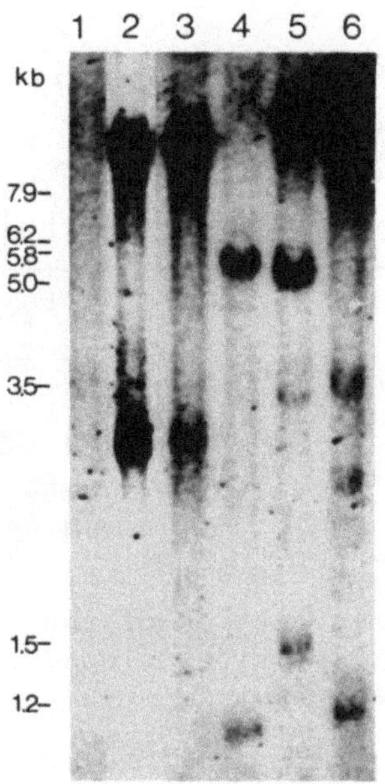

FIG. 1. DNA hybridization using nick-translated cloned BLV-DNA as a probe.

10 µg each of normal bovine leukocyte DNA (animal 94, lane 1) and bovine tumor DNA (animal 3168, lanes 2, 3; animal 1345, lane 4; animal 119, lane 5; animal 3261, lane 6) were digested to completion by Eco RI and electrophorezed on a 0,8 % agarose gel. The Southern blots of the DNA fragments were soaked in the prehybridization mixture at 65° C and hybridized for 24 h with 5 x 10^6 cpm per ml of ^{32}P nick-translated cloned BLV DNA (specific activity : 2 x 10^8 cpm/µg) except that the nitrocellulose strip corresponding to lane 2 which was hybridized for 24 h with 5 x 10^6 cpm of 3' enriched BLV ^{32}P cDNA per ml (specific activity : 3 x 10^8 cmp/µg). The last whashings were performed in 15 mM NaCl/1,5 mM Na citrate. Filters were exposed at - 70° C to preflashed Kodak-X omat - R film in presence of Siemens "special" intensifying screens for 3 days. Fragments length is given in kilobases (kb). Only fragments larger than 0,5 kb were detected on those gels. Eco RI λDNA as well as Hae III digested Ø x 174 DNA were used as molecular weight markers.

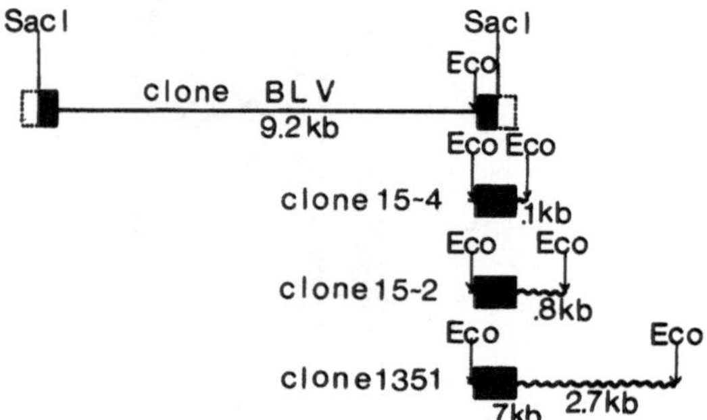

FIG. 2. Schematical representation of the cloned integrated BLV genome, and of the various cloned 3' viral - cellular junction DNA fragments.

Dark box, LTR DNA, empty box, LTR DNA not present in the cloned fragments; — viral DNA, cellular DNA; 15-4, 15, 2, 1351, numerotation of the tumors which the probes have been cloned from.

10 μg each of normal bovine leukocyte DNA (lane 1) and bovine tumor DNA (animal 1351, lane 2; animal 104, lane 3; animal 106, lane 4; animal 15, lane 5; animal 3202, lane 6; animal 3261, lane 7 and animal 79-2, lane 8) were digested to completion by Eco RI and electrophorezed on a 0,8 % agarose gel. The Southern blot of the DNA fragments was treated as described in Fig. 1 except that $0,5 \times 10^6$ cpm per ml of nick-translated cloned 1351 BLV DNA (see Fig. 1) (specific activity : 3×10^7 cpm/μg) was used as a probe. Autoradiography was from a 5 days exposure. The LTR moiety of the probe reveals the viral fragments present in the 8 tumor cell DNAs tested (two per integrated proviral copy). The viral patterns were strictly

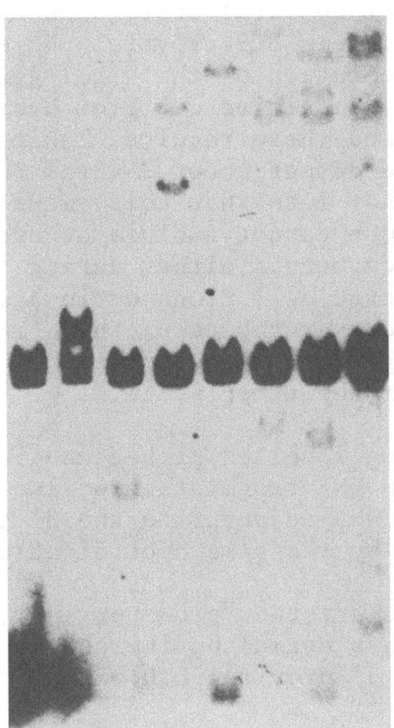

FIG. 3.
DNA hybridization using cloned 1351 DNA as a probe.

identical (except in the intensity of the bands) with those obtained when BLV cDNA or cloned BLV DNA were used as probes (data not shown). In all DNAs tested including the control leukocyte DNA, the cellular moiety of the probe strongly hybridized to a same 2-9 kb fragment (not revealed by a viral probe (lanes 1-8)). In addition the cloned DNA hybridized to a 3-4 kb fragment in the homologous 1351 DNA (lane 2), this fragment being also revealed by a viral probe. Summing the intensities of both the hybridizing bands in this latter case gave a result roughly similar to the darkness of the unique band found in the control DNA. Knowing that all tumor cells harbor the BLV provirus (data not shown), these results suggest that integration of the provirus can take place at many different possible sites into one single chromosome of a given pair.

EXPRESSION OF VIRAL AND 3'- ADJACENT CELLULAR INFORMATION IN TUMOR CELLS

Previous liquid hybridization experiments between ^{3}H - BLV cDNA and total RNA from tumors and circulating leukocytes of leukemic animals have shown that no viral expression occurred in these cells. Using a cloned complete BLV proviral DNA which constitutes a more homogenous representation of the viral genome than BLV cDNA and thanks to the very sensitive dot blot hybridization technique (13), we have confirmed these results. Taking into account the fact that the method should detect about 2 viral RNA molecules per cell, one could estimate that less than this amount of BLV RNA is present per infected cell. One cannot exclude however that transcription of BLV information occurs either during a short period in the development of the infected cell clone or in a minor subpopulation of infected cells. Whatever the case, these results suggest that a high level of proviral genome expression is not required for induction and/or maintenance of the tumor state.

In other experiments, using labelled probes consisting of cloned 3' viral - cellular junction DNA fragments, we also showed that no transcription of tumor cell DNA adjacent to the 3' LTR of the BLV provirus could be detected by the same hybridization method.

This fact again argues against the "promoter insertion" model for tumorisation (11), in which a normal bovine gene would be transcribed, starting from the 3' proviral LTR.

One possible mechanism for oncogenesis by BLV would be an undefined alteration of the chromatin structure caused by BLV integration. This alteration would lead to a modification in the rate of expression of a physically remote cellular gene important for regulation of the B cell cycle. To test this hypothesis, it would be help ful to use large cloned cellular sequences originating from both sides of a given integrated proviral genome in order to test the relatedness between the chromosomal areas which have accomodated the BLV provirus. It would also be interesting to estimate the level of transcriptional activity of these regions as compared to their normal, non tumorous counterpart.

SUMMARY

Previously published data concerning the B cell lymphocytic leukemia horizontally transmitted in cattle through BLV infection led to the conclusion that the provirus integration is a necessary event to switch on the tumorigenic process. The hereabove results prove that this insertion can take place at many different possible genomic sites and does not give rise to detectable viral RNA production nor to 3' adjacent cellular DNA transcription.

Consequently, tumorisation by BLV does not seem to result from the insertion of a viral promoter initiating the downstream transcription of a bovine "onc" gene.

REFERENCES

1. Callahan, R., M.M. Lieber, G.J. Todaro, D.C. Graves and Ferrer, 1976. Bovine leukemia virus genes in the DNA of leukemic cattle. Science, 192, 1005.

2. Kettmann, R., D. Portetelle, M. Mammerickx, Y. Cleuter, D. Dekegel, M. Galoux, J. Ghysdael, A. Burny and H. Chantrenne, 1976. Bovine leukemia virus : An exogenous RNA oncogenic virus. Proc. Natl. Acad. Sci. USA, 73, 1014.

3. Burny, A., F. Bex, H. Chantrenne, Y. Cleuter, D. Dekegel, J. Ghysdael, R. Kettmann, M. Leclercq, J. Leunen, M. Mammerickx, and D. Portetelle, 1978. Bovine Leukemia virus involvement in enzootic bovine leukosis. Adv. Cancer Res., 28, 251.

4. Deschamps, J., R. Kettmann and A. Burny, 1981. Experiments using cloned complete tumor derived bovine leukemia virus information prove that the virus is totally exogenous to its target animal species. J. Virol., in press.

5. Oroszlan, S., T.D. Copeland, L.E. Henderson, J.R. Stephenson and R.V. Gilden, 1979. Amino-terminal sequence of bovine leukemia virus major internal protein. Homology with mammalian type C virus p30 structural proteins. Proc. Natl. Acad. Sci. USA, 76, 2996.

6. Erikson, R.L., 1980. Avian sarcoma viruses: molecular biology. In : "Viral oncology" (Klein G., ed.) pp. 39-53. Raven Press, New York, NY.

7. Roussel, M., S. Saule, C. Lagrou, C. Rommens, H. Beug, T. Graf and D. Stehelin, 1979. Three new types of viral oncogene of cellular origin specific for haematopoietic cell transformation? Nature, 281.

8. Ghysdael, J., R. Kettmann and A. Burny, 1979. Translation of Bovine Leukemia Virus virion RNA in heterologous protein synthesizing systems. J. Virol., 29, 1087.

9. Hartley, J.W., N.K. Wolford, L.J. Old and W.P. Rowe, 1977. A new class of murine leukemia viruses associated with the development of spontaneous lymphomas. Proc. Natl. Acad. Sci. USA, 74, 789, 792.

10. Kettmann, R., Y. Cleuter, M. Mammerickx, M. Meunier-Rotival, G. Bernardi, A. Burny and H. Chantrenne, 1980. Genomic integration of bovine leukemia provirus : comparison of persistent lymphocytosis with lymph node tumor form of enzootic bovine leukosis. Proc. Natl. Acad. Sci. USA, 77, 2577-2581.

11. Hayward, W.S., B.G. Neel and S.M. Astrin, 1981. Activation of a cellular onc gene by promoter insertion in ALV- induced lymphoid leukosis. Nature, 290, 475-480.

12. Neel, B.G., W.S. Hayward, H.L. Robinson, J. Fang and S. Astrin, 1981. Avian leukosis virus- induced tumors have common proviral integration sites and synthesize discrete new RNAS : oncogenesis by promoter insertion. Cell 23, 323-334.

13. Thomas, P., 1980. Hybridization of denaturated RNA and small DNA fragments transferred to nitrocellulose. Proc. Natl. Acad. Sci. USA, 77, 5201-5205.

THE CLONING AND ANALYSES OF HUMAN CELLULAR GENES HOMOLOGOUS TO RETROVIRAL *ONC* GENES

Flossie Wong-Staal, Eric Westin, Genoveffa Franchini, Edward Gelmann, Riccardo Dalla Favera, Vittorio Manzari and Robert C. Gallo

Laboratory of Tumor Cell Biology
National Cancer Institute
Bethesda, MD 20205

INTRODUCTION

Type-C retroviruses are associated with certain forms of naturally occurring leukemias and lymphomas of many species, including humans, as recently evidenced. These viruses generally do not transform cells directly *in vitro* and apparently do not contain a specific transforming gene. In contrast, a more unusual class of retroviruses is the acutely transforming viruses[1,2]. They cause diseases rapidly *in vivo*, have the capacity to transform appropriate target cells *in vitro*, and contain genomes which are usually defective for replication and include a specific transforming (*v-onc*) gene.

All *v-onc* genes are derived from normal cellular genes of their host species of origin[1]. Although the function(s) of these cellular gene products are not yet clearly defined, there is evidence that the viral and cellular counterparts are similar functionally, and that viral induced transformation is sometimes correlated with enhanced level of expression of these genes. Furthermore, the strong conservation of cellular *onc* genes suggests an essential role for these genes in normal cellular processes. It is also reasonable to speculate that the *onc* genes may play a role in neoplasias regardless of whether viruses are involved etiologically.

Our laboratory has recently cloned the genomes of simian sarcoma virus (SSV), the only acutely transforming primate retrovirus, and its helper associated virus (SSAV)[3]. Using cloned SSV and FeSV[4] as probes, we had further identified and cloned the human DNA sequences homologous to the transforming genes of these two retroviruses. We will describe analyses of these human onc genes, and experiments to study expression of these and other onc genes in human hematopoietic cells.

The Transforming Gene of Simian Sarcoma Virus (v-sis).

Closed circular SSV and SSAV viral DNA intermediates were cleaved with a one-cut enzyme and ligated to phage vector arms to generate clones of complete, permuted SSV and SSAV genomes[3], which were compared by restriction enzyme and heteroduplex mapping. The SSAV DNA genome is a 9.0 Kb molecule with a long terminal repeat unit (LTR) at both ends. In the permuted SSAV genome, the two LTR units in the middle of the cloned DNA can be recognized by the tandem array of a constellation of restriction sites; Pst I, Sst I, Kpn I (Fig. 1) and Sma I (not shown). The distance between two adjacent sites of the same enzyme (except for Sst I which cuts twice in the LTR) is 550 bp, which marks the size of one LTR unit. Two SSV clones from viral DNA intermediates were extensively analyzed[3,5]. When compared to SSAV the two clones share three regions of deletion and one substitution: a 0.2 Kb deletion near the beginning of the gag gene, a 1.9 Kb deletion probably comprising most of the pol gene and a 1.5 Kb deletion in the env gene where a substitution of 1.1-1.2 Kb of SSV-specific (v-sis) sequences is found. In addition, one clone (C14) lacks one of the two LTRs and the other (C60) has an inversion of one LTR and 1.0 Kb of adjacent sequences (Fig. 1). C14 transforms mouse fibroblasts in vitro[3]. Labeled SSAV failed to hybridize to SSV DNA fragments within the v-sis substitution[3,6]. This suggests that v-sis is SSV specific and is the transforming gene of this virus. Comparison to v-sis to other viral onc genes showed no detectable homology[7].

Hybridization as assayed by S_1 nuclease resistance of a single-strand M13 recombinant phage containing v-sis sequences to DNA from different primate and non-primate species showed that v-sis has highest homology to woolly monkey DNA[7]. Since SSV was originally isolated from a pet woolly monkey which cohabited with a pet gibbon ape, and since SSAV is highly homologous to GaLV isolates[8] we concluded that SSV arose from a recombination between a retrovirus transmitted from the gibbon to the woolly monkey and a cellular gene of the woolly monkey during the life time of that animal.

Although the genetic origin of v-sis can be traced to the woolly monkey whose tumor yielded SSV/SSAV, similar genes can be found in other vertebrate species including man. When DNA from chimpanzeee, human, gibbon ape, woolly monkey, dog and chicken were digested with Bam HI and blot hybridized to labeled pBR322-SSV plasmid, two fragments (~8.6 and 1.9 Kb) were found in DNA from all four primates, one band of 4.3 Kb was detected in chicken DNA

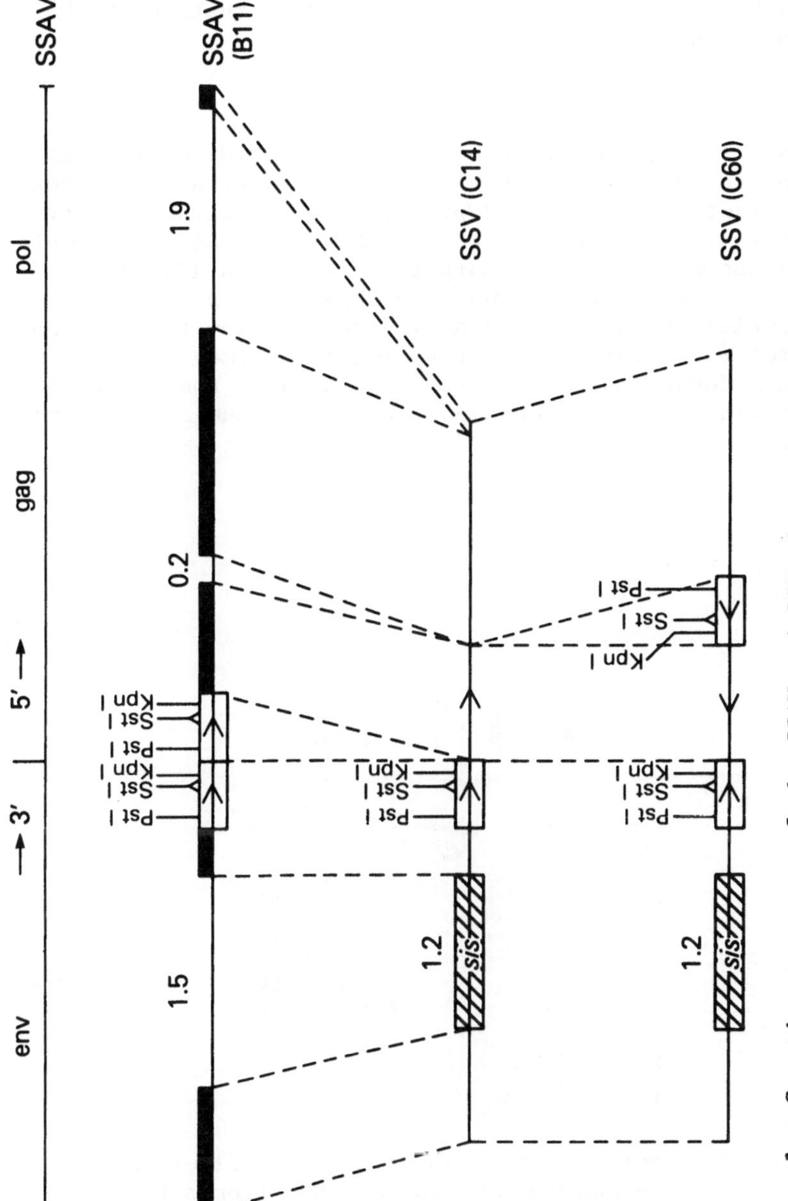

Fig. 1. Genetic structure of the SSAV and SSV clones derived from unintegrated viral DNA. Dark bars on the SSAV genome indicate regions conserved in SSV clones. The dotted lines connect corresponding regions of SSAV (B11) and the two SSV (C60 and C14) clones.

and three bands of 7.3, 3.9 and 2.1 Kb were detected in dog DNA[9]. These bands were obtained using moderately stringent conditions for hybrid formation (3XSSC, 50% formamide, 37°) and detection (1XSSC, 0.5% SDS, 60°) suggesting a strong conservation of the homologous sequences from chicken to man.

The Human Gene Homolog of v-sis.

When SSV and SSAV DNA probes were blot hybridized to various human DNA samples digested with different enzymes, discrete DNA bands were detected with the SSV probe, but none with the SSAV probe[9]. The most frequent patterns of the SSV-related fragments observed are shown in Fig. 2. Both EcoRI and Hind III gave a single high molecular weight band (>20 Kbp), suggesting there is one or few copies of the v-sis gene in man. Inside this region, cleavage sites for several other restriction enzymes (Bam HI, Xba I, Bgl II) are found. Survey of DNA from a large number of individuals revealed a rarer second genotype (our unpublished data).

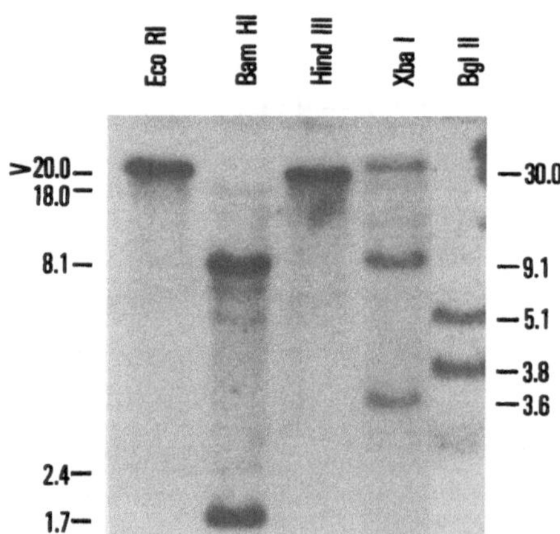

Fig. 2. Restriction enzyme digestion patterns of the human c-sis locus. DNA from a human promyelocytic cell line HL60 (24) was digested with the indicated enzymes and blot hybridized to the [^{32}P]-labeled plasmid containing the SSV insert.

A clone of the human c-sis gene was isolated from a recombinant phage library[10]. The DNA insert of this clone (L33) has no internal EcoRI site, and contains all the coding sequences of v-sis. Two techniques were used to locate the regions of homology: restriction endonuclease mapping and heteroduplex formation between L33 and an SSV clone. Both analyses revealed that the 1.2 Kb of v-sis homologous sequences in L33 span a region of 12Kb and are interrupted by four non-homologous regions. The genetic structure of L33 and a representative heteroduplex molecule are shown in Figs. 3 and 4. The presence of intervening sequences has been shown in many eukaryotic cellular genes, and in all but one cellular onc gene analyzed so far[11]. It is not clear whether the intervening sequences non-homologous to v-onc are also non-coding for the cellular gene products. At least in the case of Rous sarcoma virus, the cellular and viral products are almost identical structurally and functionally[12], even though the cellular src gene has many introns[13]. Therefore, it is likely that the non-homologous sequences of c-sis will be processed in the corresponding mRNA, although this remains to be proven.

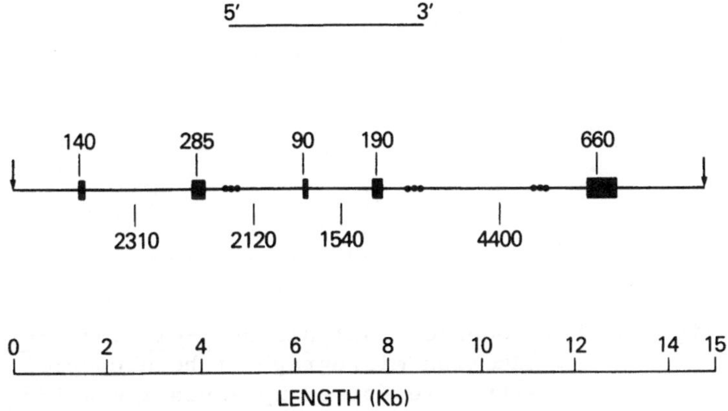

Fig. 3. A schematic representation of the organization of c-sis sequences in L33. The lengths of the v-sis homologous regions (black boxes) and intervening sequences (solid lines) are derived from both restriction enzyme mapping and heteroduplex measurements. Charon 4A DNA sequences are not shown. ●●● indicate the approximate position of repeated sequences of the human Alu family.

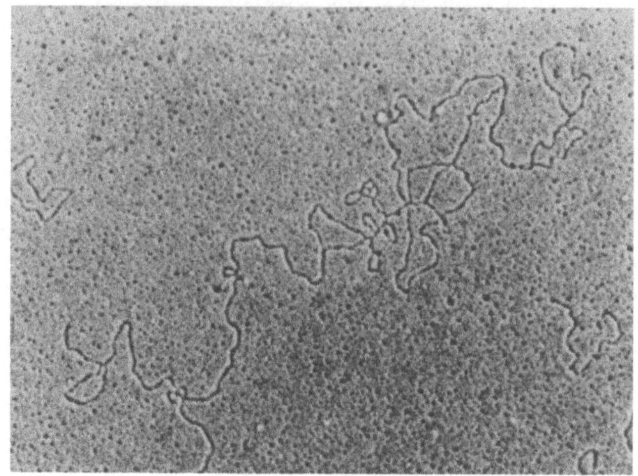

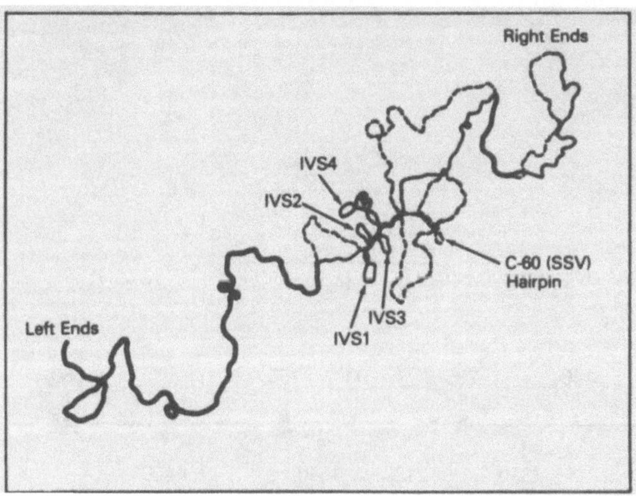

Fig. 4. Heteroduplex analysis of c-sis and v-sis.
a. Electron micrograph of heteroduplex structure formed by annealing λ-L33 DNA and λ C60 (SSV) DNA.
b. An interpretive drawing of the heteroduplex structure. IV = intervening sequences numbered in the 5'-->3' direction of c-sis and SSV DNA.

The Human Gene Homolog of the Transforming Gene of Feline Sarcoma Virus (v-fes).

Feline sarcoma viruses (FeSV) were isolated from fibrosarcomas of domestic cats[14-16]. They can induce fibrosarcomas *in vivo* and transform fibroblasts *in vitro*. Three isolates of FeSV have been characterized; the Snyder-Theilen (ST) strain, the Gardner-Arnstein (GA) strain and the McDonough-Sarma (SM) strain. ST and GA FeSV have apparently acquired the same transforming gene (*fes*) from cats[17]. Several strains of avian sarcoma virus (Fujinami, PRC II) have also acquired a homologous transforming gene from chickens[18]. Restriction enzyme mapping of the cat cellular *fes* locus revealed good correspondence of restriction enzyme sites between *v-fes* and homologous sequences in *c-fes*[19]. However, although *v-fes* is only 1.4 Kb in size, *c-fes* spans about 4.5 Kb and contains at least three intervening sequences[19].

We used molecularly cloned ST-FeSV DNA as probe in Southern blot hybridization and detected a unique *c-fes* locus in DNA from chicken to man[9]. Human DNA digested with EcoRI yielded a single 14 Kb band. In order to further analyze the structure of the human *c-fes* gene, a human DNA library was screened. Three positive clones were found to be overlapping with each other. Together they constitute >20 Kb of DNA sequences inclusive of the entire 14 Kb EcoRI fragment detected in total human DNA. The clones were further mapped and oriented with respect to *v-fes* using probes derived from the complete or specific regions of *v-fes*. Fig. 5 presents a map of the human *c-fes* locus. Like the cat *c-fes* locus, human *c-fes* is more complex than *v-fes* and contains three discernible intervening sequences.

Expression of Human onc Gene Homologs in Hematopoietic Cells.

Accumulating evidence indicates that all viral *onc* genes are derived from phylogenetically conserved cellular genes and that human DNA is likely to contain counterparts of all the *onc* genes. In addition to SSV and FeSV, molecularly cloned genomes of Abelson-MuLV, Harvey-MSV, Balb MSV, MC29, and AMV have been reported to detect homologous genetic loci in man[9,20-22]. The identification of human *onc* gene homologs obviously raises the question whether they are functional genetic elements and whether they play a role in normal or neoplastic cell growth. We have examined a wide spectrum of human hematopoietic cells derived from leukemic and normal individuals for the expression of *onc* gene homologues of six transforming retroviruses.

Leukemic cells are currently best viewed as equivalent to clones of normal cells "frozen" at various stages of hematopoiesis. During the past decade, a number of permanent human

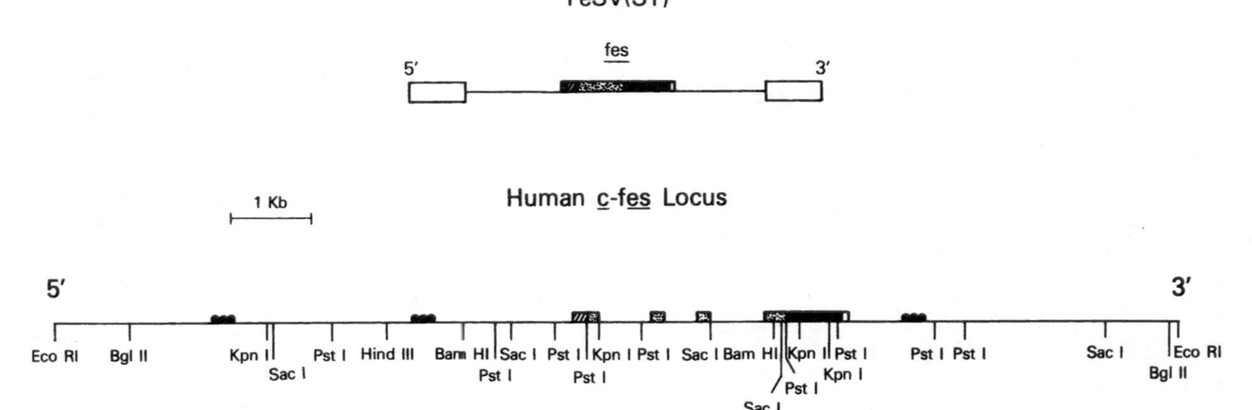

Fig. 5. Restriction enzyme map of the human c-fes locus. The map was derived from results analyzing three overlapping c-fes recombinant clones purified from a human DNA library. The boxes indicate the putative regions of homology to various regions of v-fes as designated. Placement of these regions is only approximate.

leukemia-lymphoma cell lines have been established and these display stable marker characteristics of the original neoplastic cells[23]. Furthermore, a human promyelocytic cell line (HL60) can be induced to differentiate to granulocytes *in vitro*[24,25]. The availability of these cell lines as well as fresh leukemic cells affords one with an ideal system to study gene regulation as a function of differentiation and the mechanism of leukemogenesis. For expression of *onc* gene homologues, we have used probes made from molecularly cloned genomes of avian myeloblastosis virus, AMV; avian myelocytomatosis virus, MC29; Abelson murine leukemia virus, A-MuLV; Harvey murine sarcoma virus, HaMSV; and simian sarcoma virus, SSV. AMV causes myeloid leukemias in chickens and transforms cells of the myeloid lineage[26]. MC29 induces myelocytomatosis and less frequently sarcomas and carcinomas *in vivo*, and transforms macrophage-like cells *in vitro*[26]. A-MuLV causes lymphosarcomas in mice and transforms fibroblasts and cells of the B-lymphocyte series[27]. HaMSV induces erythroleukemias as well as fibrosarcomas in mice[28]. Finally, SSV induces sarcomas in newborn primates and seems to specifically transforms fibroblasts[29]. Nick translated probes were hybridized to poly(A)-containing RNAs using RNA gel blotting techniques[30].

The types of cells used and the results with different probes are summarized in Table 1. The designations for different *onc* genes are as follows: HaMSV, Ha-*ras*; Abelson-MuLV, *abl*; MC29, *myc*; AMV, *amv*; SSV, *sis*; FeSV, *fes*[31]. The *abl* and Ha-*ras* genes are detectably expressed (1-5 copies per cell) in all hematopoietic cells examined. Furthermore, multiple mRNA species were transcribed from these genes. The *myc* and *amv* genes code for single size transcripts of 2.7 Kb and 4.5 Kb respectively. However, the expression of *amv* is more restricted than *myc*. The *myc* gene is transcribed in all hematopoietic cells examined including normal peripheral blood lymphocytes prior to or after stimulation with PHA. The only exception is terminally differentiated HL60 cells where *myc* transcription is turned off. The *amv* gene is expressed in the early precursor cells of lymphoid, myeloid and erythroid lineages, but there is little or no expression relatively early in B-lymphoid cells differentiation, and late in T-cell or myeloid cell differentiation. Like *myc*, *amv* is transcribed in undifferentiated HL60 cells but not in HL60 cells induced to differentiate with either DMSO or retinoic acid. The *sis* and *fes* genes are not commonly transcribed in hematopoietic cells and may be genes specifically involved in fibroblast differentiation.

SUMMARY

Acutely transforming retroviruses have been isolated from avian, feline and primate species. These viruses contain transformation-specific sequences derived from normal cellular genes that are

Table 1

Expression of onc Genes in Human Hematopoietic Cells

Cell Type	Source	Stage of Differentiation	mRNA Species Detected with v-abl (Kb) 7.2, 6.4, 3.8 & 2.0	v-myc (Kb) 2.7	v-amv (Kb) 4.5	v-Ha-ras (Kb) 6.5, 5.8	v-sis (Kb) 4.3	v-fes ?
Myeloid	KG-1	Myeloblast	+ +	+ +	+ +	+	–	–
	HL60	Promyelocyte	+ +	+ + + +	+ +	+	–	–
	HL60 + DMSO, RA	Granulocyte	+ +	±	–	+	–	–
	Fresh AML Cells (4 Pnts)	Myeloblast	+ +	+ +	+ +	+	–	–
Erythroid	K562	(Immature Erythroid Precursor)	+ +	+ +	+ +	+	–	–
Lymphoid								
T-Cells:	CEM	Immature T Cell	+ +	+ +	+ + +	+	–	–
	MOLT 4	Immature T Cell	+ +	+ +	+ + +	+	–	–
	HUT78	Mature T Cell	+ +	+ +	–	+	–	–
	HUT102	Mature T Cell	+ +	+ +	–	+	+	–
B-Cells:	Raji	Burkitt Lymphoma Line	+ +	+ +	–	+	–	–
	Daudi	Burkitt Lymphoma Line	+ +	+ +	–	+	–	–
	NC37	EBV Transformed Normal B-Cell Line	+ +	+ +	–	+	–	–
Normal Peripheral Lymphocytes			NT	+ +	–	NT	NT	NT
Normal Peripheral Lymphocytes + PHA			NT	+ +	–	NT	NT	NT

NT = Not Tested.

conserved among vertebrates. We have cloned the human genes homologous to the transforming (onc) genes of simian sarcoma virus (SSV) and feline sarcoma virus (Snyder-Theilen strain). Simian sarcoma virus (SSV) is the only defective, transforming virus of primate origin. We had previously obtained molecular clones of SSV and the helper associated virus SSAV. Comparison of the SSV and SSAV genomes by restriction enzyme mapping and heteroduplex analysis revealed the 5.6 Kb SSV genome to contain deletions in all three replicative genes derived from SSAV and a substitution of 1.2 Kb cell-derived sequences (sis) near the 3' end. V-sis was shown to have been derived from woolly monkey DNA and genetic loci homologous to v-sis were detected in all vertebrate DNA. Further, v-sis was shown to be distinct from all other viral onc genes. Using cloned SSV as a probe, a human DNA fragment comprising the entire v-sis homologous region was cloned. The v-sis related sequences were found to be dispersed over a stretch of 12 Kb with at least four intervening sequences. The human gene related to FeSV (fes) was likewise obtained from a human DNA library. The human c-fes gene spans 4.5 Kb and contains three discernible intervening sequences.

We also examined the level of expression of six different onc genes in human hematopoietic cells using probes from molecularly cloned virus genomes. The results indicated that the sis and fes genes were not expressed in human hematopoietic cells, the genes related to Harvey MSV and Abelson MuLV were transcribed at low levels as multiple RNA species in all cells, the gene related to MC29 is widely transcribed as a single mRNA species of 2.7 Kb, and expression of the gene related to AMV is limited to early precursor cells of lymphoid, myeloid and erythroid lineages and blocked at different stages of differentiation in different pathways. Our data are insufficient to warrant a general role for onc genes in human neoplasias.

REFERENCES

1. J. M. Bishop, Retroviruses, Ann. Rev. Biochem. 47:35 (1978).
2. P. H. Duesberg, Transforming genes of retroviruses, in: "Cold Spring Harbor Symposium Quant. Biol.", 44:13 (1979).
3. E. P. Gelmann, F. Wong-Staal, R. A. Kramer and R. C. Gallo, Molecular cloning and comparative analyses of the genomes of simian sarcoma virus (SSV) and its helper associated virus (SSAV), Proc. Natl. Acad. Sci. USA 78:3373 (1981).
4. C. J. Sherr, L. A. Fedele, M. Oskarsson, J. Maizel and G. Vande Woude, Molecular cloning of Snyder-Theilen feline leukemia and sarcoma viruses: Comparative studies of feline sarcoma virus with its natural helper virus and with Moloney murine sarcoma virus, J. Virol. 34:200 (1980).

5. E. P. Gelmann, L. Petri, A. Cetta and F. Wong-Staal, Specific regions of the genome of simian sarcoma associated virus are deleted in defective genomes and in SSV, J. Virol. (in press).
6. K. C. Robbins, S. G. Devare and S. A. Aaronson, Molecular cloning of integrated simian sarcoma virus: Genome organization of infectious DNA clones, Proc. Natl. Acad. Sci. USA 78: 2918 (1981).
7. F. Wong-Staal, R. Dalla Favera, E. Gelmann, V. Manzari, S. Szala, S. Josephs and R. C. Gallo, The transforming gene of simian sarcoma virus (sis): A new onc gene of primate origin, Nature (in press).
8. R. C. Gallo and F. Wong-Staal, Molecular biology of primate retroviruses, in: "Viral Oncology", G. Klein, ed., Raven Press, N.Y. (1980).
9. F. Wong-Staal, R. Dalla Favera, G. Franchini, E. P. Gelmann and R. C. Gallo, Three distinct genes in human DNA related to the transforming genes of mammalian sarcoma retroviruses, Science 213:226 (1981).
10. R. Dalla Favera, E. P. Gelmann, R. C. Gallo and F. Wong-Staal, A human onc gene homologous to the transforming gene (v-sis) of simian sarcoma virus, Nature 292:31 (1981).
11. M. Oskarsson, W. L. McClements, D. G. Blair, J. V. Maizel and G. F. Vande Woude, Properties of a normal mouse cell DNA sequence (sarc) homologous to the src sequence of Moloney sarcoma virus, Science 207:1222 (1980).
12. M. S. Collett, J. S. Brugge and R. L. Erikson, Characterization of a normal avian cell protein related to the avian sarcoma virus transforming gene product, Cell 15:1363 (1978).
13. D. Shalloway, A. D. Zelenetz and G. M. Cooper, Molecular cloning and characterization of the chicken gene homologous to the transforming gene of Rous sarcoma virus, Cell 24:531 (1981).
14. S. P. Snyder and G. H. Theilen, Transmissable feline fibrosarcoma, Nature 221:1074 (1969).
15. M. B. Gardner, R. W. Rongey, P. Arnstein, J. D. Estes, P. Sarma, R. J. Huebner and G. G. Rickard, Experimental transmission of feline fibrosarcoma to cats and dogs, Nature 226:807 (1970).
16. S. K. McDonough, S. Larsen, R. S. Brodey, N. D. Stock and W. D. Hardy Jr., A transmissible feline fibrosarcoma of viral origin, Cancer Res. 31:953 (1971).
17. A. G. Frankel, J. H. Gilbert, K. J. Porzig, E. M. Scolnick and S. A. Aaronson, Nature and distribution of feline sarcoma virus nucleotide sequences, J. Virol. 30:821 (1979).
18. M. Shibuya, T. Hanafusa, H. Hanafusa and J. R. Stephenson, Homology exists among the transforming sequences of avian and feline sarcoma viruses. Proc. Natl. Acad. Sci. USA 77:6536 (1980).

19. G. Franchini, J. Even, C. J. Sherr and F. Wong-Staal, Onc sequences (v-fes) of Snyder-Theilen feline sarcoma virus are derived from discontiguous regions of a cat cellular gene (c-fes), Nature 290:154 (1981).
20. D. G. Bergman, L. M. Souza and M. A. Baluda, Vertebrate DNA contains nucleotide sequences related to the putative transforming gene of avian myeloblastosis virus, J. Virol. (in press).
21. S. P. Goff, E. Gilboa, O. N. Witte and D. Baltimore, Structure of the Abelson murine leukemia virus genome and the homologous cellular gene: Studies with cloned viral DNA, Cell 22:777 (1980).
22. A. Eva, K. C. Robbins, P. R. Andersen, A. Srinivasan, S. R. Tronick, E. P. Reddy, N. W. Ellmore, A. T. Galen, J. A. Lautenberger, T. Papas, E. H. Westin, F. Wong-Staal, R. C. Gallo and S. A. Aaronson, Cellular genes analogous to retroviral onc genes are transcribed in human tumor cells, (submitted).
23. J. Minowada, K. Sagawa, M. S. Lok, I. Kubonishi, S. Nakazawa, E. Tatsumi, T. Ohnuma and N. Goldblum, A model of lymphoid-myeloid cell differentiation based on the study of marker profiles of 50 human leukemia-lymphoma cell lines, in: "Int. Sym. on New Trends in Human Immunology and Cancer Immunotherapy", B. Serrou and C. Rosenfeld, eds., Doin Publisher, Paris (1980).
24. S. J. Collins, R. C. Gallo and R. E. Gallagher, Continuous growth and differentiation of human myeloid leukemic cells in suspension culture, Nature 270:347 (1977).
25. S. J. Collins, F. W. Ruscetti, R. E. Gallagher and R. C. Gallo, Terminal differentiation of human promyelocytic leukemia cells induced by dimethylsulfoxide and other polar compounds, Proc. Natl. Acad. Sci. USA 75:2458 (1978).
26. T. Graf and H. Beug, Avian leukemia viruses: Interaction with their target cells in vivo and in vitro, Biochim. Biophys. Acta 516:269 (1978).
27. H. T. Abelson and L. S. Rabstein, Lymphosarcoma: Virus-induced thymic independent disease in mice, Cancer Res. 30:2208 (1970).
28. J. J. Harvey, An unidentified virus which causes the rapid production of tumors in mice, Nature 204:1104 (1964).
29. L. Wolfe, F. Deinhardt, G. Theilden, T. Kawakami and L. Bustad, Induction of tumors in marmoset monkeys by simian sarcoma virus type 1 (Largothrix): A preliminary report, J. Natl. Cancer Inst. 47:1115 (1971).
30. P. S. Thomas, Hybridization of denatured RNA and small DNA fragments transferred to nitrocellulose, Proc. Natl. Acad. Sci. USA 77:5201 (1980).

31. J. M. Coffin, H. E. Varmus, H. E. Bishop, J. M. Bishop, M. Essex, W. D. Hardy, Jr., G. T. Martin, N. E. Rosenberg, E. M. Scolnick, R. A. Weinberg and P. K. Vogt, A proposal for naming host cell-derived inserts in retrovirus genomes. J. Virol. (in press).

PROPERTIES OF AN RNA RETROVIRUS CONTINUOUSLY PRODUCED AT LOW LEVELS BY A HUMAN HISTIOCYTIC LYMPHOMA CELL LINE

Henry S. Kaplan

Cancer Biology Research Laboratory
Department of Radiology
Stanford University School of Medicine
Stanford, CA 94305

Type-C RNA viruses (retroviruses) are capable of inducing leukemias, lymphomas, and sarcomas in several mammalian species, including mice, cats, guinea pigs, cattle, and subhuman primates (Kaplan 1974), suggesting that they may also be involved in the etiology of these types of neoplasms in man. Retroviruses have been detected in culture fluids after short-term culture of hematopoietic cells of patients with leukemias (Gallagher and Gallo, 1975; Nooter et al.,1975; Teich et al., 1975) and of diploid human embryonic cell strains (Panem et al., 1975). These viral isolates have been shown to have typical type-C ultrastructural morphology (Hall and Schidlovsky 1976) and to contain RNA genomes with extensive sequence homologies to the genomes of simian sarcoma virus, SSV-1, as well as to an endogenous virus of baboons, BaEV (Chan et al., 1976, Okabe et al., 1976). The structural proteins and RNA-dependent DNA polymerases (reverse transcriptases) of these isolates also appear to be immunologically related to the corresponding proteins of SSV-1 and BaEV (Chan et al., 1976; Okabe et al., 1976). Incomplete type-C viruses have been detected in the cytoplasm of human leukemia and lymphoma cells, and genetic sequences homologous to those of viral cDNA in the DNA of a leukemic individual, but not in that of his normal identical twin, have been described (Baxt et al., 1973). Intracytoplasmic particles possessing reverse transcriptase activity and a density characteristic of type-C viruses have been reported in leukemic cells (Baxt et al, 1972; Gallo et al., 1972; Gallagher and Gallo, 1975), as well as in the spleens of patients with leukemias and Hodgkin's disease (Witkin et al., 1975; Chezzi et al., 1976). However, the sporadic availability and inconsistent viral yield of primary human neoplastic tissue specimens have presented serious obstacles to the further study of such type-C viruses.

Our studies in this laboratory began with an attempt to establish the human malignant lymphomas in continuous culture. Permanent cell lines have now been established from more than 15 patients with diffuse histiocytic lymphomas (designated SU-DHL-1 through -5), three with North American Burkitt's lymphomas (SU-AmB-1 through -3), and one with acute lymphoblastic leukemia (SU-ALL-1). The growth behavior, surface marker and functional characteristics, neoplastic properties, and cytogenetic abnormalities of these cell lines have been detailed previously (Epstein et al., 1976a,b; Epstein and Kaplan, 1974; Kaiser-McCaw et al., 1977; Kaplan et al., 1979). The isolation and partial characterization of a type-C RNA virus from SU-DHL-1, the first of the diffuse histiocytic lymphoma lines, have also been reported (Kaplan et al., 1977; Kaplan, 1978; Kaplan et al., 1979; Goodenow et al., 1980; Goodenow and Kaplan, 1979; Goodenow et al., 1981.) In this paper, the properties of this virus are briefly summarized. Recently, Gallo and his colleagues have used the same approach successfully to obtain permanently established cell lines of human cutaneous T-cell lymphomas and leukemias which can be induced to release a retrovirus with distinctly different biological, biochemical, and immunological properties (Poiesz et al., 1980 a,b; Reitz et al., 1981; Rho et al., 1981; Kalyanaraman et al., 1981).

RESULTS

Continuous Production and Release of Virus by SU-DHL-1 Cells

Microsomal pellets prepared from SU-DHL-1 cells and banded to equilibrium in sucrose density gradients revealed reverse transcriptase activity in a particulate fraction which banded in the 1.13-1.15 g/ml density region of the gradient (Kaplan et al., 1977.) Reverse transcriptase activity was also detected in culture fluids of serially passaged SU-DHL-1 cells, as well as in cultures reinitiated from cells stored in liquid nitrogen after being frozen at -1° C/minute on day 40 after their initiation in culture. Levels of reverse transcriptase activity fluctuated somewhat with variations in viability of the cells under different culture conditions - but persisted at significantly elevated levels in all cultures tested. Activity was also detected at comparable levels in supernatant fluids of several subclones of SU-DHL-1 cells.

The culture fluids from several of the other histiocytic lymphoma cell lines, as well as from a number of nonneoplastic lymphoblastoid cell lines concurrently established and propagated in this laboratory, have been harvested, concentrated by centrifugation, and assayed for reverse transcriptaselike activity by previously described procedures (Kaplan et al., 1977). Of the nine histiocytic lymphoma cell lines established after SU-DHL-1, seven were tested at least once, and three of these (SU-DHL-4, -8, and -10) revealed significantly elevated levels of transcriptaselike activity. Two other cell lines, SU-DHL-2 and SU-AmB-3, which had borderline levels at the time of the first assay, were strongly positive in additional tests performed after further subcultivation. Thus, six of the cell lines (including SU-DHL-1) tested to date appeared to be releasing particles with reverse transcriptaselike activity into their supernatant culture fluids. In striking contrast to the results with the lymphoma cell lines, the culture fluids of all of the nonneoplastic lymphoblastoid cell lines tested to date have been consistently negative for the corresponding enzymatic activity, in keeping with the report by Klucis et al. (1976) that only one of 13 lymphoblastoid cell lines yielded sporadic, unsustained enzyme activity.

Electron microscopy of pelleted SU-DHL-1 cell culture fluids revealed irregularly shaped spherical particles with a diameter of about 100 μm, composed of a moderately dense central nucleoid surrounded by an envelope, the inner layer of which was electron-dense (Kaplan et al., 1977). The size, morphology, and detergent-sensitivity of these particles were very similar to those of type-C RNA viruses isolated from other mammalian species (Dalton, 1972.)

Syncytium Formation

Co-cultivation of SU-DHL-1 cells with rat XC cells induced typical syncytium formation. The fact that the virus exhibited the capacity to induce syncytia in rat XC cells, a biological attribute characteristic of type-C RNA viruses isolated from mice (Klement et al., 1969), cats (Rangan et al., 1972b), and subhuman primates (Rangan et al., 1974; Rangan et al., 1972b), provides further support for its identification as a type-C virus. Interestingly, another of our reverse transcriptase-positive lymphoma cell lines, SU-AmB-3, induced syncytia of somewhat different morphology in another indicator cell line, the fu-7 rat line (Kaplan et al., 1979.)

Reverse Transcriptase

The enzyme was able to use endogenous RNA, (rA)n · $(dT)_{12-18}$, and (rC)n · $(dG)_{12-18}$ as primer-templates in the presence of all four deoxynucleoside triphosphates. Appropriate control assays excluded termincal deoxynucleotidyl transferase as the source of the observed enzymatic activity.

Studies of the susceptibility of the human lymphoma cell viral reverse transcriptase to inhibition by antibodies to other known type-C viral reverse transcriptases were carried out with antibodies to the polymerases of RD-114, an endogenous feline virus; SSV-1/SSAV, a type-C virus complex isolated from a woolly monkey tumor; and the NZB mouse xenotropic virus, as previously described (Kaplan et al., 1977). Antibody to the murine viral enzyme had no inhibitory activity against the SU-DHL-1 cell viral reverse transcriptase, even at

IgG concentrations that were > 90% inhibitory for the homologous murine viral enzyme. In contrast, partial inhibition was observed with antibodies to the polymerases of RD-114 and SSV-1/SSAV, although the extent of inhibition was very significantly less than that observed in each instance with the homologous viral enzyme. Thus, it appears that the human lymphoma viral polymerase carries antigenic determinants cross-reactive with those of the enzymes both of RD-114 and SSV-1/SSAV. Under the same conditions, no inhibition of cellular DNA polymerases α, β, or γ was observed with the SSV-1/SSAV antibody.

The human lymphoma cell viral reverse transcriptase was partially purified by DEAE-cellulose chromatography (Kaplan et al., 1979) and further purified by poly(rC)-agarose chromatography (Goodenow and Kaplan, 1979.) The eluted enzyme was inoculated into rabbits, and the resultant heterologous antibody was used in enzyme inhibition studies similar to those described above. At the lowest IgG concentrations that exhibited $\geq$75% inhibition of the homologous human lymphoma cell viral enzyme, the gibbon ape leukemia virus (GaLV) enzyme was inhibited only about 35%, those of BaEV, SSV-1/SSAV, RD-114, and NZB-MuLV about 10-15%, and there was little or no inhibition of feline leukemia virus (FeLV) or avian myeloblastosis virus (AMV). Tryptic digest peptide maps of the purified enzyme revealed several peptides common to the subhuman primate viral enzymes, and 1-2 unique peptides in addition (Goodenow and Kaplan, 1979). Subsequently, these observations were confirmed with a mouse monoclonal antibody, 3H4, which strongly inhibited the SU-DHL-1, SSV-1, and GaLV enzymes and immunoprecipitated polypeptides of ∿ 75-80 kilodaltons from the SU-DHL-1 and SSV-1 viruses (Goodenow et al., 1980.) Thus, the antigenic cross-reactivity patterns suggest that the reverse transcriptase of the human lymphoma cell virus may be distantly related to the viral reverse transcriptases of subhuman primates.

Virion Structural Proteins

Heterologous antibody prepared against the envelope glycoprotein (gp 70) of GaLV (grown on bat cells to avoid cross-reaction with human cellular antigens) yielded positive membrane immunofluorescence reactions on SU-DHL-1 cells (Kaplan et al., 1979), and immunoprecipitated a 70 kilodalton polypeptide from preparations of SU-DHL-1 virus. Evidence that the major internal core protein of SU-DHL-1 virus is also antigenically related to the corresponding protein of subhuman primate type-C retroviruses was obtained with a mouse monoclonal antibody, 3D7(2B9), which immunoprecipitated polypeptides of 28 kilodaltons (p28) from both SU-DHL-1 and SSV-1/SSAV. This antibody also immunoprecipitated 28 and 65 kilodalton polypeptides from ^{35}S-methionine labelled SU-DHL-1 cell lysates (Goodenow et al., 1980). Moreover, in a solid phase radioimmunoassay using the 3D7(2B9) monoclonal antibody, competition was observed between lysates of SSV-1/SSAV and radiolabelled lysates of SU-DHL-1 virus (Goodenow et al., 1981). The 3D7 (2B9) antibody also precipitated a 65 kilodalton product (presumed to be the *gag* gene precursor product, $Pr65^{gag}$) following *in vitro* translation of SSV-1/SSAV viral RNA (Goodenow et al., 1981).

Viral Nucleic Acid

All attempts to isolate high molecular weight, undegraded RNA from large-scale preparations of SU-DHL-1 virus have been unsuccessful. It has thus not been possible to prepare cDNA or to undertake molecular cloning of the viral genome. Instead, an alternative approach has been taken, based on the above-cited evidence of the antigenic cross-relatedness of three SU-DHL-1 viral proteins (the reverse transcriptase, gp 70, and p28) to the corresponding proteins of the subhuman primate type-C retroviruses. Unintegrated linear proviral DNA of the San Francisco isolate of the gibbon ape leukemia virus, $GaLV_{Sf}$, has been cloned in a lambda phage vector and mapped with several restriction endonucleases (Scott et al., 1981). The cloned $GaLV_{Sf}$ proviral genome has been radiolabeled by nick translation for use as a probe to detect related sequences in the DNA of SU-DHL-1 and other lymphoma cells; these studies are currently in progress.

Infectivity Studies

In an extensive series of co-cultivation experiments, SU-DHL-1 cells failed to infect a spectrum of human and other mammalian normal and neoplastic cells of nonhematopoietic origin (Kaplan et al., 1977). However, when SU-DHL-1 virus was added to cultures of human lymphoblastoid cells (SU-LB-7 and SU-LB-8), the supernatant culture fluids revealed elevated levels of reverse transcriptaselike activity in a particulate fraction which eluted from DEAE-cellulose in 0.2M KCl (Kaplan et al., 1979). Thus, the virus may have limited infectivity for selected target cells, though amplification of the levels of virus production has not been achieved.

An extensive series of experiments has been performed to ascertain whether normal human peripheral blood mononuclear (PBM) and bone marrow cells would reveal any significant changes in growth behavior following infection with free SU-DHL-1 virus or cocultivation with SU-DHL-1 cells. Limited studies have also been carried out with SU-AmB-3 cells. Control cultures of adherent human and PBM cells, composed almost entirely of monocytes, have exhibited little or no evidence of proliferative activity. Instead, the cells have enlarged, spread out on the culture well surface and assumed the varied shapes typical of macrophages, and persisted with slow attrition over a period of several weeks. When such adherent cells were infected with SU-DHL-1 virus within 24 hours after the initiation of the cultures, changes in their growth behavior were observed in several experiments (Kaplan et al., 1979), whereas cultures infected after an interval of several days, at a time when the cultures comprised mature macrophages, were apparently unaffected by the virus. Infected monocytes failed to convert to mature macrophages; instead, they remained irregularly spherical in shape and appeared to proliferate slowly, often becoming detached from the culture well surface and forming clusters of large, ovoid or spherical cells of increasing size lying in a plane immediately above the culture vessel surface. Other monocytes remained adherent to the culture vessel surface, but retained the size and shape of monocytes, and only a small minority converted to mature macrophages. After 2-3 weeks, the nonadherent cell population of these cultures could be subpassaged. After subpassage, some of the cells became adherent, but most of these failed to convert to the morphology of mature macrophages. Instead, they again detached from the culture vessel surface and appeared to replicate, forming elongated, irregular clusters of large cells. In some experiments, the infected monocytes were carried through three or four serial subpassages with evidence of continued proliferative activity. Adherent human PBM co-cultivated with mytomycin-C-killed SU-DHL-1 cells in the presence of inactivated Sendai virus exhibited similarly altered in vitro growth behavior. Cultures of the nonadherent PBM cell population, composed almost entirely of lymphocytes, when infected with SU-DHL-1 virus in the absence of concomitant or sequential mitogenic stimulation, failed to proliferate.

Stained cytocentrifuge preparations of the virus-infected adherent cell cultures revealed large monocytoid cells, many of which had binucleate or bilobate, large, hyperchromatic nuclei. In some instances, abnormal, occasionally tripolar mitotic figures were observed in these preparations. Cells with these atypical cytologic features were not seen in similar preparations from the control cultures. Even more striking morphologic abnormalities, in some instances closely resembling the changes observed in acute myelomonocytic leukemias, have been observed when normal human bone marrow cells were cocultivated with our virus-producing human lymphoma cell lines. Demecolcine-treated cells from the virus-infected cultures revealed obviously aneuploid cells (Kaplan et al., 1979).

DISCUSSION

The virus spontaneously and continuously released by the SU-DHL-1 human histiocytic lymphoma cell line appears to have many of the characteristics of the other mammalian type-C RNA viruses, including endogenous reverse transcriptase activity inhibitable by antibodies to subhuman primate viral reverse transcriptases, an envelope glycoprotein and an internal core protein antigenically related to the corresponding proteins of the subhuman primate type C retroviruses, the capacity to induce syncytia in rat XC cells, and appropriate electron microscopic morphology. That the production and release of this virus during continuous in vitro cultivation of the SU-DHL-1 cell line is not an isolated and sporadic occurence is suggested by the presence of elevated levels of reverse transcriptaselike activity in the supernatant culture fluids of several other human lymphoma cell lines established in this laboratory. One of these cell lines, SU-DHL-2, has also been shown to induce typical syncytia when co-cultivated with rat XC cells, and SU-AmB-3 induces syncytia on another rat indicator cell line, fu-7.

The virus is apparently noninfectious for a spectrum of human and other mammalian normal and neoplastic cells of nonhematopoietic origin, but appeared to have at least limited infectivity for selected human lymphoblastoid cell lines.

Human monocytes and bone marrow cells exhibited unusual changes in their growth behavior *in vitro* following infection with the virus produced by SU-DHL-1 cells. Morphologic and cytogenetic abnormalities, including the presence of frankly aneuploid chromosome numbers, have been observed in these virus-infected normal human hematopoietic cells. The biological significance of these responses cannot be assessed at this time.

Acknowledgements

This work was supported by research contract NO1-CP-43228 (later NO1-CP-91044) from the National Cancer Institute, National Institutes of Health, Bethesda, Maryland; and by gifts to the Joseph Edward Luetje Memorial Fund for Lymphoma Research. The author is grateful to Suzanne Gartner for invaluable assistance with the infectivity and cell-culture studies, to Robert Goodenow for the viral protein and reverse transcriptase studies and for assays on the other lymphoma and lymphoblastoid cell lines, to Dr. Marcia Bieber for the preparation of heterologous antibody and the indirect immunofluorescence tests for the detection of cytoplasmic viral antigens, and to Barbara Dieckmann for the XC cell assays of the SU-DHL-2 and infected SU-AmB-3 cell lines. Dr. Jack Griffith, Department of Biochemistry, Stanford University, performed the electron microscopy of pelleted SU-DHL-1 cell culture fluids. Dr. Werner Henle and Dr. Gertrude Henle performed the Epstein-Barr nuclear antigen tests of the North American Burkitt's lymphoma cell line.

REFERENCES

Baxt, W., Hehlmann, R., and Spiegelman, S., 1972, Human leukaemic cells contain reverse transcriptase associated with a high molecular weight virus-related RNA, *Nat. New Biol.* 240:72.

Baxt, W., Yates, J.W., Wallace, H.J. Jr., Holland, J.F., and Spiegelman, S., 1973, Leukemia-specific DNA sequences in leukocytes of the leukemic member of identical twins, *Proc. Nat. Acad. Sci. USA,* 70:2629.

Chan, E., Peters, W.P., Sweet, R.W., Ohno, T., Kufe, D.W., Spiegelman, S., Gallo, R.C., and Gallagher, R.E., 1976, Characterisation of a virus (HL23V) isolated from cultured acute myelogenous leukaemic cells, Nature, 2:266.

Chezzi, C., Dettori, G., Manzari, V., Agliano, A.M., and Sanna, A., 1976, Simultaneous detection of reverse transcriptase and high molecular weight RNA in tissue of patients with Hodgkin's disease and patients with leukemia. Proc. Natl. Acad. Sci. USA, 73:4649.

Dalton, A.J., 1972, RNA tumor viruses: Terminology and ultrastructural aspects of virion morphology and replication, J. Natl. Cancer Inst. 49:323.

Epstein, A.L., and Kaplan, H.S., 1974, Biology of the human malignant lymphomas. I. Establishment in continuous cell culture and hetero-transplantation of diffuse histiocytic lymphomas, Cancer 34:1851.

Epstein, A.L., Henle, W., Henle, G., Hewetson, J.F., and Kaplan, H. S., 1976a, Surface marker characteristics and Epstein-Barr virus studies of two established North American Burkitt's lymphoma cell lines, Proc. Nat. Acad. Sci. USA, 73:228.

Epstein, A.L ., Herman, M.M., Kim, H., Dorfman, R.F., and Kaplan, H.S., 1976b, Biology of the human malignant lymphomas. III. Intracranial transplantation in the nude, athymic mouse, Cancer, 37:2158.

Gallagher, R.E., and Gallo, R.C., 1975, Type C RNA virus tumor isolated from cultured human acute myelogenous leukemia cells, Science, 187:350.

Gallo, R.C., Miller, N.R., Saxinger, W.C., and Gillespie, D., 1972, Primate RNA tumor virus-like DNA synthesized endogenously by RNA - dependent DNA polymerase in virus-like particles from fresh human acute leukemic blood cells, Proc. Nat. Acad. Sci. USA, 70:3219.

Goodenow, R,S,, Brown, S., Levy, R., and Kaplan, H.S., 1980, Partial characterization of the virus proteins of a type-C RNA virus produced by a human histiocytic lymphoma cell line, in: "Viruses in Naturally Occuring Cancers," M. Essex, G. Todaro, and H. zur Hausen, eds., Cold Spring Harbor Laboratory, New York, pp. 737-752.

Goodenow, R.S., and Kaplan, H.S., 1979, Characterization of the reverse transcriptase of a type C RNA virus produced by a human lymphoma cell line, Proc. Nat. Acad. Sci. USA, 76:4971-4975.

Goodenow, R.S., Liu, S, Fry, K., Levy, R., and Kaplan, H.S., 1981, Expression of C-type RNA viral proteins by a human lymphoma cell line, in:"Recent Advances in Malignant Lymphomas: Etiology, Immunology, Pathology, Treatment," S.A. Rosenberg and H.S. Kaplan, eds., Academic Press, New York, 1981, in press.

Hall, W.T., and Schidlovsky, 1976, Typical Type-C virus in human leukemia, J. Natl. Cancer Inst., 56:639.

Kaiser-McCaw, B., Epstein, A.L., Kaplan, H.S., and Hecht, F., 1977, Chromosome 14 translocation in African and North American Burkitt's lymphoma, Int. J. Cancer, 19:482.

Kalyanaraman, V.S., Sarngadharan, M.G. Poiesz, B., Ruscetti, F., and Gallo, R., 1981, Immunological properties of a type C retrovirus isolated from cultured human T-lymphoma cells and comparison to other mammalian retroviruses, J. Virol., 38:906.

Kaplan, H.S., 1974, Leukemia and lymphoma in experimental and domestic animals, Ser. Haematol., 7:94.

Kaplan, H.S., 1978, Studies of an RNA virus isolated from a human histiocytic lymphoma cell line, in: "Differentiation of Normal and Neoplastic Hematopoietic Cells," B. Clarkson, P.A. Marks, and J.E. Till, eds., Cold Spring Harbor Laboratory, New York.

Kaplan, H.S., Goodenow, R.S., Epst n, A.L., Gartner, S., Decleve, A., and Rosenthal, P.N., 1977, Isolation of a C RNA virus from an established human histiocytic lymphoma cell line, Proc. Nat. Acad. Sci. USA, 74:2564.

Kaplan, H.S., Goodenow, R.S., Gartner, S., and Bieber, M., 1979, Biology and virology of the human malignant lymphomas: 1st Milford D. Schulz lecture, Cancer 43:1.

Klement, V., Rowe, P., Hartley, J.W., and Pugh, W.E., 1969, Mixed culture cytopathogenicity: A new test for growth of murine leukemia viruses in tissue culture, Proc. Nat. Acad. Sci. USA, 63:753.

Klucis, E., Jackson, L., and Parsons, P.G., 1976, Survey of human lymphoblastoid cell lines and primary cultures of normal and leukemic leukocytes for oncornavirus production, Int. J. Cancer, 18:413.

Nooter, K., Aarssen, A.M., Bentvelzen, P., De Groot, F.G., and van Pelt, F.G., 1975, Isolation of infectious C-type oncornavirus from human leukaemic bone marrow cells, Nature, 256:595.

Okabe, H., Gilden, R.V., Hatanaka, M., Stephenson, J.R., Gallagher, R.E., Gallo, R.C., Tronick, S.R., and Aaronson, S.A., 1976, Immunological and biochemical characterization of the type C viruses isolated from cultured human AML cells, Nature, 260:264.

Panem, S., Prochownik, E.V., Reale, F.R., and Kirsten, W.H., 1975, Isolation of type C virions from a normal human fibroblast strain, Science, 189:297.

Poiesz, B.J., Ruscetti, F.W., Gazdar, A.F., Bunn, P.A., Minna, J.D., and Gallo, R.C., 1980, Detection and isolation of type C retrovirus particles from fresh and cultured lymphocytes of a patient with cutaneous T-cell lymphoma, Proc. Nat. Acad. Sci. USA, 77:7415.

Poiesz, B.J, Ruscetti, F.W., Mier, J.W., Woods, A.M., and Gallo, R. C., 1980, T-cell lines established from human T-lympocytic neoplasias by direct response to T-cell growth factor, Proc. Nat. Acad. Sci. USA, 77:6815.

Rangan, S.R.S., 1974, C-type oncogenic viruses of nonhuman primates, Lab. Anim. Sci., 24:193.

Rangan, S.R.S., Moyer, P.P., Cheong, M.P., and Jensen, E.M., 1972a, Detection and assay of feline leukemia virus (FeLV) by a mixed-culture cytopathogenicity method, Virology, 47:247.

Rangan, S.R.S., Wong, M.C., Ueberhorst, P.J., and Ablashi, D.V., 1972b, Mixed culture cytopathogenicity induced by virus preparations derived from cultures infected by simian sarcoma virus, J. Natl. Cancer Inst., 49:571.

Reitz, M.S., Poiesz, B., Ruscetti, F.W., and Gallo, R.,1981, Characterization and distribution of nucleic acid sequences of a novel retrovirus isolated from neoplastic human T-lymphocytes, Proc. Nat. Acad. Sci. USA, 78:1887.

Rho, H.M., Poiesz, B., Ruscetti, F.W., and Gallo, R.C., 1981,Characterization of the reverse transcriptase from a new retrovirus (HTLV) produced by a human cutaneous T-cell lymphoma cell line, Virology, 112:355.

Scott, M.L., McKereghan, K., Kaplan, H.S., and Fry, K.E., Molecular cloning and partial characterization of unintegrated gibbon ape leukemia virus DNA, Proc. Nat. Acad. Sci. USA, 78:4213.

Teich, N.M., Weiss, R.A., Salahuddin, S.Z., Gallagher, R.E., Gillespie, D.H., and Gallo, R.C., Infective transmission and characterisation of a C-type virus released by cultured human myeloid leukaemia cells, Nature, 256:551.

Witkin, S.S., Ohno, T., and Spiegelman, S., 1975, Purification of RNA-instructed DNA polymerase from human leukemic spleens, Proc. Nat. Acad. Sci. USA, 72:4133.

HUMAN T-CELL RETROVIRUS AND ADULT T-CELL LYMPHOMA AND LEUKEMIA: POSSIBLE FACTORS ON VIRAL INCIDENCE

R. C. Gallo[1], M. Robert-Guroff[2], V. S. Kalyanaraman[2], L. Ceccherini Nelli[1], F. W. Ruscetti[1], S. Broder[3], M. G. Sarngadharan[2], Y. Ito[4], M. Maeda[1], M. Wainberg[1] and M. S. Reitz, Jr.[1]

[1]Laboratory of Tumor Cell Biology
Division of Cancer Treatment
National Cancer Institute
Bethesda, Maryland 20205

[2]Department of Cell Biology
Litton Bionetics, Inc.
Kensington, Maryland 20895

[3]Metabolism Branch
National Cancer Institute
Bethesda, Maryland 20205

[4]Department of Microbiology
Kyoto University
Faculty of Medicine
Kyoto, Japan

SUMMARY

We have recently described the isolation of several related or identical retroviruses (called HTLV) from cultured human neoplastic T-lymphocytes. HTLV has so far been detected only in patients with adult T-cell leukemia-lymphoma. The presence of virus seems to be specific for subsets of T-cells in these patients, and those T-cells are relatively mature (e.g., terminal transferase-negative). Cultured T-cells from two of the more carefully studied patients express abundant intracellular viral components but produce relatively low levels of virions, suggesting that these cells are restricting the virus life cycle at some stage later than synthesis of viral proteins and RNA. We have attempted to transmit HTLV to heterologous cells. So far, transmission of the virus (as judged by prolonged expression by the exposed target cells but not the unexposed cells of viral protein and RNA) has been successful only with T-lymphocytes from blood relatives of patients with T-cell disease for whom there was evidence for the presence of HTLV. These data suggest a genetic basis for T-cell susceptibility to HTLV infection. We have tested numerous sera for the presence of antibodies to HTLV. Among those positive for antibodies to p24 and p19 were almost all serum samples from Japanese patients with adult T-cell leukemia. This disease is endemic to a restricted area in southern Japan, suggesting that geographical factors may also play a role in the incidence of HTLV.

INTRODUCTION

Since T-cell leukemias are often caused by retroviruses in animals (1), we have long been interested in looking for similar viruses in humans. Some years ago we reported (2,3) the discovery of a factor released into media by subsets of T-cells after stimulation with phytohemagglutinum (PHA), which supported the long-term growth of T-lymphocytes. This factor, called TCGF for T-cell growth factor, has been recently purified (4). T-cells from normal individuals respond to TCGF after activation with antigen or lectin, such as PHA (4). In contrast, neoplastic T-cells from patients with leukemias and lymphomas involving mature T-cells grow directly with TCGF without prior stimulation (5). This has made it possible to obtain and maintain lines of neoplastic T-cells and examine them for evidence of the presence of retrovirus.

The first of these lines (HUT-102) in which a virus was discovered (6) was from a lymph node biopsy of a patient with an aggressive variant of mycosis fungoides, a cutaneous form of T-cell lymphoma. The virus, called HTLV (strain CR) was unrelated to any of a wide spectrum of previously described animal retroviruses (including those from nonhuman primates) by homology of its high molecular weight RNA (7) and by antigenic determinants of its reverse transcriptase (RT) (6,8) and structural core protein p24 (9). Virus also was observed in cultured T-cells from independently obtained clinical specimens of peripheral blood from the same patient. These lines are called CTCL-3 (6) and CTC-16 (see below). The second cell line found to be producing a virus was a T-cell line from a patient with Sezary leukemia-lymphoma, another T-cell neoplasia with cutaneous manifestations (10). This virus (called HTLV strain MB) was closely related or identical to the first isolate. These data are summarized in Table 1. We have recently obtained preliminary evidence, based on p19 and p24 assays, for the presence of HTLV in two other cell lines; one a T-cell line from a putative hairy cell leukemia and the other from an adult T-cell leukemia of Japanese origin (unpublished results).

Table 1: Lack of Relatedness of HTLV to Other Retroviruses

Virus[1]	Related to $HTLV_{CR}$ by[2]			
	Nucleic Acid	p24	p19	RT
$HTLV_{MB}$	+++	+++	+++	+++
SSAV-GaLV	-	-	-	-
SMRV	-	-	ND[3]	-
MPMV	-	-	ND[3]	-
BLV	-	-	-	-
BaEV	-	-	-	-
MuLV	-	-	-	-
FeLV	-	-	-	-

[1]Abbreviations used include: SSAV, simian sarcoma associate or woolly monkey virus; GaLV, gibbon ape leukemia virus; SMRV, squirrel monkey retrovirus; MPMV, Mason-Pfizer monkey virus; BLV bovine leukemia virus; BaEV, baboon endogenous virus; MuLV, murine leukemia virus; and FeLV, feline leukemia virus. Other viruses tested for one or more of the above relationships include endogenous retroviruses from colobus, langur, owl monkey, deer, pig, guinea pig, hamster, viper and asian mouse, as well as equine infectious animal virus and various avian retroviruses.

[2]Nucleic acid homology was measured by liquid hybridization using ^{3}H-cDNA as described (7); protein homology was measured by radioimmunoassays and indirect immune flourescent assays using a monoclonal antibody as described elsewhere (6,8,9,11). +++, highly related or identical; -, not detectably related.

[3]ND = not done.

HTLV is not an Endogenous Virus

We had previously shown that HTLV was not an endogenous virus of humans, i.e., HTLV is not a genetic element in the germline of all humans (7). It was possible, however, that the proviral DNA for HTLV could be transmitted genetically only within certain rare families. To test this possibility, DNA was purified from a B-cell line, established in Dr. T. Waldmann's laboratory, from the same patient (CR) from whom the HTLV-producing line HUT-102 was established. The identity of these lines as originating from patient CR has been confirmed by HLA typing (D. Mann, personal communication). This DNA as well as DNA from HUT-102 were assayed for HTLV proviral DNA sequences. As shown in Figure 1, HUT-102 DNA contains readily detectable provirus (about 3-4 copies per haploid genome). DNA from the B-cell line (CR-B), in contrast, has no detectable proviral sequences, indicating

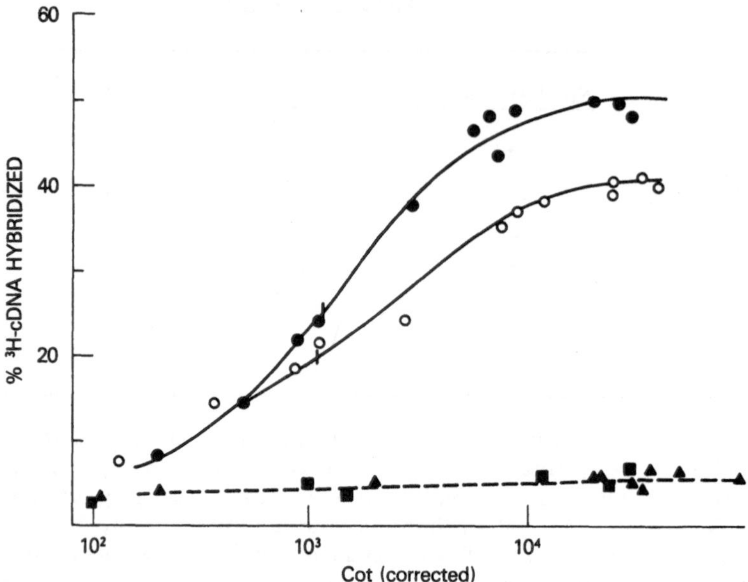

Fig. 1. Presence of HTLV provirus in T- but not B-cells of patient CR. Percent hybridization is assay as described for Table 1 and plotted as a function of corrected Cot expressed in mol sec per l. O, T-lymphocytes (HUT-102) from CR; ●, T-lymphocytes from clone B2; ▲, DNA from B-lymphocytes from CR (CR-B); ■, normal human liver.

that there is substantially less than one proviral equivalent per haploid genome in these cells. This means that HTLV is unlikely to have been inherited in the germ line of this patient, and that it was most likely acquired by infection, either congenital or otherwise.

We further investigated the distribution of HTLV in the hematopoietic cells of this patient (as judged by the presence of viral proteins) in HUT-102 and CR-B cells, as well as in a T-cell line (CTCL-3) established with TCGF from the same patient in our laboratory (6) and CTC-16, T-cell line established from this patient in Dr. T. Waldmann's laboratory (12) with TCGF plus PHA. CTC-16 would thus likely be a mixture of normal and malignant T-cells. As shown in Table 2, all three T-cell lines express p19 and p24, while CR-B cells, in agreement with the hybridization data do not express detectable p19 or p24.

Table 2: Distribution of HTLV in Hematopoietic Cells from Patient CR

Cell Line[1]	p24[2]	p19[2]	Viral Nucleic Acids[2]
T-Cells			
HUT-102	+++	+++	+++
CTCL-3	+++	+++	ND[3]
CTC-16	+++	+++	ND[3]
G5	+	-	-
C6	+	-	-
B-Cells			
CR-B	-	-	-

[1]HUT-102 was established from the neoplastic lymph node of patient CR with purified TCGF (16). CTCL-3 was established from peripheral blood of the same patient with purified TCGF (6). CTC-16 was from CR peripheral blood and was established with partially purified TCGF containing PHA (12). G5 and C6 are partially cloned derivatives of CTC-16. CR-B was established from B-lymphocytes of peripheral blood of patient CR by *in vitro* infection with Epstein-Barr virus strain B-95.

[2]Protein and nucleic acid assays were performed as described for Table 1. +++, highly positive; +, very slightly positive; and -, not detectable.

[3]ND = not done.

Several cultures were derived by end-point dilution from CTC-16. Two of these, G5 and C6, were also tested for protein and provirus. As shown in Table 2, proviral DNA and p19 assays were negative, although low levels (~1% of that with HUT-102) of p24 were detected. Since proviral DNA would have been detected at levels of one copy per haploid, it is likely that these lines are not completely cloned, and contain predominantly virus negative cells. It would thus appear that not all T-cells are infected with HTLV.

Susceptibility of T-Cells to HTLV Infection

In view of the association of HTLV with T-cells, we attempted to transmit the virus *in vitro* to a variety of human T-cells. Initial attempts were unsuccessful, but it seemed possible that there could be genetic factors which might make some T-cells permissive for HTLV infection. To test this we used T-cells from relatives of patients with T-cell neoplasms for which there was some evidence of HTLV infection (actual virus, proviral sequences, or antibodies) since these relatives might have such factors. Infection was monitored by the persistent expression of p19 and HTLV RNA (detected by *in situ* hybridization assays) and was confirmed by electron microscopy. These selected T cells, in contrast to those from the general population, were indeed productively infected with HTLV (Table 3). B-cells from the same individuals could not be infected. Various animal fibroblast lines were also not permissive for HTLV. Production remained stable for many weeks. An unusual feature of this infection is that viral reverse transcriptase (measured by enzymatic assays) was present in the media, but generally diminished to non-detectable levels (even though p19 and viral RNA were still expressed), indicating that although these cells are able to be infected they can also restrict late stages in the viral life cycle. In the absence of added HTLV, proteins and RNA of HTLV (shown for RNA in Fig. 2) were not expressed, indicating that such expression is not due to endogenous HTLV already present.

Distribution of Serum Antibodies to HTLV

Serum samples were obtained from numerous patients from the United States with T-cell neoplasias, as well as from normal donors and tested for the presence of antibodies to HTLV antigens, including homogeneously purified HTLV p24 (13,14). Of over 200 patient sera examined, two possessed HTLV-specific antibodies (Table 4). In contrast, over 100 sera of random normal donors were negative for HTLV-specific antibodies. Among healthy relatives of cutaneous T-cell lymphoma-leukemia patients, two were positive for antibodies to HTLV. One positive relative was the wife of patient C.R. (from whom $HTLV_{CR}$ was isolated) and the second positive relative was the daughter of patient M.B. (from whom $HTLV_{MB}$ was isolated). The presence of viral-specific antibodies in sera of relatives is suggestive of some type of infectious event.

Sera were also obtained from patients from Japan with T-cell neoplasias and normal donors. An adult form of T-cell leukemia (ATL) with frequent cutaneous involvement has been reported to be endemic to southern Japan (15). Sera from patients with ATL were almost uniformly positive for antibodies to HTLV (Table 4). At least one of the negative ATL patients was in remission after undergoing chemotherapy. In addition, 50% of sera from patients with other T-cell lymphomas possessed HTLV-specific antibodies. Presumably some of these other T-cell lymphoma patients had diseases closely related to ATL. In contrast to patient sera, sera from 106 normal donors (including 59 from the endemic region) were all negative.

Table 3: Specificity of *In Vitro* Transmission of HTLV

Types of Cells[1]	Express Cellular p19[2]	Express Extracellular RT[2]	Express Extracellular Particles by EM
Normal Donors			
B-cells	No	No	No
T-cells	No	No	No
Relatives of HTLV-positive patients with T-cell neoplasias			
B-cells	No	No	No
T-cells	Yes	Yes[3]	Yes

[1]Cells were from peripheral blood and grown as EBV-immortalized lines or maintained with TCGF. Normal thymocytes were obtained as a result of surgery on congenital heart disease patients. T- and B-cells were either from random normal donors or from relatives of patients with mycosis fungoides who were positive for HTLV or antibodies to p19 and p24.

[2]p19 was assayed as described for Table 1. RT and virus particles were assayed by enzymatic activity with synthetic primer-template or electron microscopy as described (6).

[3]RT was detectable for about one month but not thereafter, even though the media remained positive for virus particles.

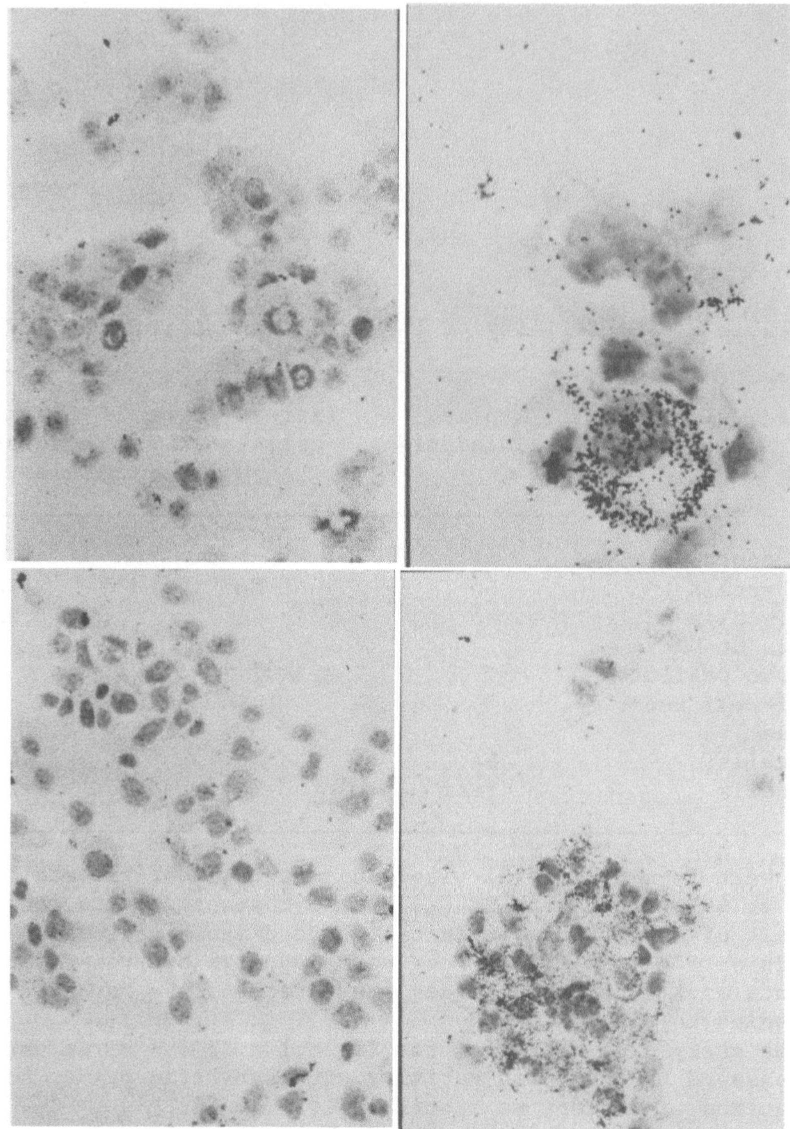

Fig. 2. Transmission of HTLV to peripheral blood T-lymphocytes of relatives of patients with T-cell lymphoma-leukemia. Infection was assayed by hybridization _in situ_ of HTLV ^{3}H-cDNA to slides of fixed cells. Top left, T-lymphocytes of relative, cultured but not exposed to HTLV; bottom left, normal T-lymphocytes, cultured but not exposed to HTLV; bottom right, T-lymphocytes of relative exposed to HTLV and cultured for two months; top right, higher magnification of bottom right showing positive and negative cells in same culture.

Table 4: Distribution of Antibody to HTLV by Clinical and Geographical Status

Sera	Origin	Antibody to HTLV[1] by Solid Phase RIA	Antibody to HTLV[1] by RIP HTLVp24
		#Positive/#Tested-	#Positive/#Tested
Normal donors	USA	0/141	0/114
Cutaneous T-cell lymphoma-leukemia[2]	USA	2/218	2/94
Healthy Relatives of CTCL patients[3]	USA	2/18	2/18
Normal donors[4]	Japan	0/106	0/106
Adult T-cell leukemia	Japan	18/20	19/21
Other T-cell lymphomas	Japan	6/12	5/12

[1]HTLV-specific antibodies were assayed as described elsewhere (13,14).

[2]Sezary syndrome or mycosis fungoides. Two positives include one of each disease.

[3]Positive sera were from the wife of patient C.R. and the daughter of patient M.B.

[4]59 sera were from region of Japan where ATL is endemic, and 47 sera were from random healthy donors.

CONCLUSION

HTLV, a virus recently isolated from several patients with T-cell leukemia-lymphoma is not transmitted vertically (i.e., genetically) in patients who are infected. This is shown by the fact that cultured B-cells from patient CR are negative for HTLV provirus, which is present and expressed in cultured T-cells from the same patient. Some of the T-cells from the same patient are also negative. HTLV must therefore be transmitted by some type of infectious route, as is indeed further suggested by the presence of serum antibody to p24 in a non-blood relative (spouse) of patient CR.

We have attempted, through several lines of investigation, to gain some understanding of how HTLV might be transmitted. One way has been to attempt to transmit the virus *in vitro*. Productive infection seems to be limited both to specific cell types (i.e., T- vs B-cells) and to cells from specific individuals (relatives of

T-cell leukemia-lymphoma patients with evidence of natural HTLV infection). Thus both tissue specificity and genetic susceptibility may well be factors in infection by HTLV.

A second approach to help understand how HTLV might be transmitted has been to look at sera from different geographical areas for antibodies to HTLV. One area of interest is the southern part of Japan where a type of adult T-cell leukemia is endemic (15). Sera from these patients were almost uniformly (90%) positive for antibodies to HTLV, whereas sera from geographically matched normal donors were not. This may mean that there is a geographical distribution of reservoirs of HTLV or, less likely (since ATL is not endemic to other parts of Japan), that the local population is genetically more susceptable to infection. Alternatively, HTLV could be transmitted primarily by congenital infection, which might tend to restrict it to certain areas to a greater degree than if it were contagiously transmissible.

In order to ascertain more clearly the molecular biology of this virus, its routes of infection, and its pathogenic properties, much further work will be needed. Because virus titer is low, it is difficult to obtain nucleic acid probes for HTLV and molecular cloning of the provirus is necessary to solve this problem. The only antibodies that are available are to core proteins, whereas it is likely that the highest titer of natural antibodies would probably be to the virus envelope. It is thus important to obtain serologic reagents for purified envelope protein. These approaches are currently being pursued in our laboratories.

REFERENCES

1. R. C. Gallo, Leukemia, environmental factors and viruses, in: "Viruses and Environment", Academic Press, N.Y., p. 43 (1978).
2. D. A. Morgan, F. W. Ruscetti and R. C. Gallo, Selective in vitro growth of T-lymphocytes from normal human bone marrows, Science 193:1007 (1976).
3. F. W. Ruscetti, D. A. Morgan and R. C. Gallo, Functional and morphologic characterization of human T cells continuously grown in vitro, J. Immunol. 119:131 (1977).
4. J. W. Mier and R. C. Gallo, Purification and some characteristics of human T-cell growth factor from phytohemagglutinin-stimulated lymphocyte-conditioned media, Proc. Natl. Acad. Sci. USA 77:6134 (1980).
5. B. J. Poiesz, F. W. Ruscetti, J. W. Mier, A. M. Woods and R. C. Gallo, T-cell lines established from human T-lymphocytic neoplasias by direct response to T-cell growth factor, Proc. Natl. Acad. Sci. USA 77:6815 (1980).

6. B. J. Poiesz, F. W. Ruscetti, A. F. Gazdar, P. A. Bunn, J. D. Minna and R. C. Gallo, Detection and isolation of type C retrovirus particles from fresh and cultured lymphocytes of a patient with cutaneous T-cell lymphoma, Proc. Natl. Acad. Sci. USA 77:7415 (1980).
7. M. S. Reitz, Jr., B. J. Poiesz, F. W. Ruscetti and R. C. Gallo, Characterization and distribution of nucleic acid sequences of a novel type C retrovirus isolated from neoplastic human T lymphocytes, Proc. Natl. Acad. Sci. USA 78:1887 (1981).
8. H. M. Rho, B. Poiesz, F. W. Ruscetti and R. C. Gallo, Characterization of the reverse transcriptase from a new retrovirus (HTLV) produced by a human cutaneous T-cell lymphoma cell line, Virology 112:355 (1981).
9. V. S. Kalyanaraman, M. G. Sarngadharan, B. Poiesz, F. W. Ruscetti and R. C. Gallo, Immunological properties of a type C retrovirus isolated from cultured human T-lymphoma cells and comparison to other mammalian retroviruses, J. Virol. 38:906 (1981).
10. B. J. Poiesz, F. W. Ruscetti, M. S. Reitz, V. S. Kalyanaraman and R. C. Gallo, Isolation of a new type C retrovirus (HTLV) in primary uncultured cells of a patient with Sezary T-cell leukaemia, Nature, in press (1981).
11. M. Robert-Guroff, F. W. Ruscetti, L. E. Posner, B. J. Poiesz and R. C. Gallo, Detection of the human T-cell lymphoma virus p19 in cells of some patients with cutaneous T-cell lymphoma and leukemia using a monoclonal antibody, J. Exp. Med., in press (1981).
12. T. Uchiyama, S. Broder, G. Bonnard and T. Waldmann, Immunoregulatory functions of cultured human T-lymphocytes, Trans. Amer. Assoc. Physic. 93:251 (1980).
13. L. E. Posner, M. Robert-Guroff, V. S. Kalyanaraman, B. J. Poiesz, F. W. Ruscetti, B. Fossieck, P. A. Bunn, Jr., J. D. Minna and R. C. Gallo, Natural antibodies to the human T cell lymphoma virus in patients with cutaneous T cell lymphomas, J. Exp. Med. 154:333 (1981).
14. V. S. Kalyanaraman, M. G. Sarngadharan, P. A. Bunn, J. D. Minna and R. C. Gallo, Antibodies in human sera reactive against an internal structural protein of human T-cell lymphoma virus, Nature, in press (1981).
15. T. Uchiyama, J. Yodoi, K. Sagawa, K. Takatsuki and H. Uchino, Adult T-cell leukemia: Clinial and hematological features of 16 cases, Blood 50:481 (1977).

REVERSE TRANSCRIPTASE: A MOLECULAR PROBE FOR THE EXPRESSION OF RETROVIRAL INFORMATION IN HUMAN TUMORS

Prakash Chandra

Center of Biological Chemistry
Laboratory of Molecular Biology
University Medical School
Theodore-Stern-Kai 7,
D-6 Frankfurt, West Germany

The discovery of reverse transcriptase in RNA tumor viruses (Temin and Mizutani, 1970; Baltimore, 1970), implicated by the provirus hypothesis of Temin (Temin, 1964a, 1964b), was a new dimension in the understanding of how such viruses multiply and moreover, how do they catalyze neoplastic transformation. Soon after the discovery of reverse transcriptase in retroviruses, Gallo and his associates detected a similar enzymic activity in lymphoblasts of patients with leukemia (Gallo et al., 1970). This activity was absent in mitotically stimulated normal human peripheral blood leucocytes. The first detection of intracellular reverse transcriptase in leukemic cells (Gallo et al., 1970) provided incentive to examine this activity in other human tumor tissues (Chandra, 1979; Chandra and Steel, 1977, 1980; Chandra et al., 1975, 1978, 1980, 1981; Ebener et al.,

1979; Ohno et al., 1977; Poiesz et al., 1980; Welte et al., 1979; and Witkin et al., 1975). The main objective of these studies was: 1) to understand whether this is a new enzyme or is simply a modified form of one of the constitutive cellular DNA polymerases α, β and γ, 2) to develop new procedures of purifying this viral-like reverse transcriptase from neoplastic tissues, and 3) to characterize the purified reverse transcriptase from human tumor tissues to determine whether the detection of this enzyme indicates an expression of retrovirus information in these tissues.

Reverse transcriptases purified from extracellular virions exhibit some unique biochemical properties, such as template-primer specificity, divalent-ion requirements and their ability to transcribe heteropolymeric regions of genomic RNA (reviewed by Wu and Gallo, 1975). These biochemical properties have been very useful in establishing the purification procedures for reverse transcriptase from tumor cells and tissues. However, these properties have a limited use in telling us whether the purified enzyme is a new entity or related to one of the cellular polymerases. A clear understanding of the cross-reactivity between human reverse transcriptase and the reverse transcriptases purified from the retroviruses of primates and sub-human primates will contribute towards the nature and origin of this retroviral information. Furthermore, the immunological procedures are very useful in detecting low levels of reverse transcriptase activity, as in the case of human tumor cells and tissues. For this reason, the characterization of intracellular reverse transcriptase from human tumor tissues based entirely on biochemical criteria is not safe.

Using antisera to different mammalian viral polymerases, Gallo et al. (1975) reported that the reverse transcriptase purified from human leukemic cells is antigenically similar to the reverse transcriptase of the SSV-GaLV group of viruses. This was the first report constituting the fact that the reverse transcriptase from human neoplastic cells cross reacts with antisera to purified viral polymerases from primates. Soon after this, Chandra et al. (1975) reported high levels of reverse transcriptase activity in spleen of a child with myelofibrotic syndrome, a preleukemic disorder. The purified reverse transcriptase exhibited all the biochemical characteristics of a mammalian retroviral reverse transcriptase, including chromatographic behavior on ion exchange columns, primer-template preferences, pH optima and molecular weight (Chandra and Steel, 1977; Chandra et al., 1975, 1980). This purified enzyme was reported to be specifically inhibited by antibodies to the reverse transcriptases of the two primate viruses, SSV and GaLV (Chandra and Steel, 1977; Chandra et al., 1980). These authors succeeded in obtaining antisera, though with low titer, against the reverse transcriptase from human myelofibrotic spleen. These antisera were able to inhibit the homologous enzyme and also the reverse transcriptase of SSV. Cellular polymerases (α, β and γ) purified from the same spleen specimen were not inhibited by the antisera. These results constituted the first indication of a possible retroviral etiology from myelofibrotic syndrome.

The findings of Chandra et al. (Chandra and Steel, 1977; Chandra et al., 1975, 1979, 1980) have a considerable relevance to results obtained by Gallo and his associates (1975) regarding retroviral components in human leukemic cells. A development that added to the significance of the correlation between the two independent findings was that the child developed AML 3 weeks after the surgical removal of spleen. Since SSV and GaLV are exogenous (oncogenic) to all primates(see Gillespie et al., 1975) these results indicated that this type of reverse transcriptase was acquired, and the results can best be interpreted as an indication that SSV-GaLV virus-related information is expressed in man.Using the technique of isoelectric focusing, Chandra et al. (1980) succeeded in obtaining a highly purified reverse transcriptase from human AML cells. This enzyme was inhibited by antisera against reverse transcriptases from SSV and myelofibrotic spleen. However, to our surprise we found (Chandra et al., 1980) that antisera against reverse transcriptase from FeLV, an exogenous virus of cat, were quite affective in neutralizing the enzyme activity. We cannot interpret the significance of this finding, but Gallo et al. (Dr. R.C. Gallo, pers. comm.) have also found a similar type of cross reactivity between reverse transcriptases from AML cells of some patients and FeLV.

Followed by our observations with the serological specificity of reverse transcriptase from myelofibrotic spleen, we got interested to look for reverse transcriptase in other leukemia-associated diseases.

Chloroma like ocular lesions in Turkish children with AMML were reported by Cavdar et al. (1971); these lesions were associated with approximately one third of the cases of AMML. The eye lesions consisted of orbital tumor, exophthalmos, proptosis, chemosis, corneal opacity, and swollen eyelids. The authors did not believe that the term chloroma was suited to these tumors because (1) the tumors were devoid of the characteristic green color, (2) no periosteal or other bone changes were present, and (3) no grayish color was observed in the bone marrow. For this reason, Cavdar et al. (1971) suggested the term granulocytic sarcoma. In most cases, the ocular lesions were noted before the diagnosis of leukemia; varying from 20 days to 8 months before the onset of leukemic disease. A viral probe carried out on one orbital tumor revealed the presence of a group specific (gs-3) antigen (Cavdar et al., 1978) previously observed in murine cells as an expression of type-C viruses. This motivated us to look for the presence of an oncornavirus-related reverse transcriptase activity in the orbital tumor associated with AMML .

The purified reverse transcriptase from orbital tumors showed a strong preference for the template-primer $(rA)_n.(dT)_{12}$, followed by $(rC).(dG)_{12}$ and $(rC.Ome)_n.(dG)$ (Chandra 1979, 1979a; Chandra and Steel, 1980; Chandra et al., 1979, 1979a, 1979b, 1980); whereas, $(dA)_n.(dT)_{12}$ was clearly ineffective. These results agree with those reported for other mammalian type-C oncornaviral DNA Polymerases. Transcription of heteropolymeric regions of a 70S RNA from MuLV and stimulation of its utilization by the addition of oligo dT further supported the oncogenic nature of the orbital enzyme and its similarity to other known RNA tumor virus reverse transcriptases. Using antisera against reverse transcriptases from various mammalian retroviruses, we could show that the orbital enzyme is specifically inhibited by antisera against GaLV-polymerase.

The studies reported on reverse transcriptases associated to leukemia and leukemia-associated diseases indicate a type of group-specific antigen on the reverse transcriptase molecule. To verify this, we undertook studies to purify and characterize immunologically the reverse transcriptases from other human tumor tissues.

The molecular probing experiments in Spiegelman's laboratory (Balda et al., 1975) indicated the presence of a C-type viral information in human malignant melanomas. This motivated us to purify the reverse transcriptase from human melanoma tissue and characterize it immunologically (Chandra 1979; Chandra et al., 1978, 1979a, 1980, 1981). Unlike the group-specific cross-reactivity between myelofibrotic and orbital enzymes and the enzymes from SSV-GaLV group, no serological similarity between the melanoma enzyme and the enzymes from SSV-GaLV group was found. The activity of purified reverse transcriptase from the melanoma tissue, challenged with antibodies to DNA polymerases from baboon endogenous virus (BEV) and RD-114 (originally isolated from human rhabdomyosarcoma cells that had been cultivated in a fetal cat; see McAllister et al., 1971, 1972) was found to be strongly inhibited , almost as effectively as they inhibited their homologous enzymes. The fact that the purified polymerases from BEV and RD114 are antigenically closely related to each other (Sarin et al., 1977) suggests the presence of a common antigen in the human melanoma tissue.

The antigenic similarity between the human melanoma enzyme and both the endogenous viral polymerases raises several interesting questions for future investigations, including, for example, Is the viral information in human melanoma cells derived from endogenous genetic information or from exogenous sources? The presence of proviral sequences of BEV in DNA of leukemic tissues (Wong-Staal et al., 1976) suggests that, in analogy to the cat virus, a similar transmission to humans could have occured in the evolutionary history. Another question, and perhapse a more important one, is: Can we prove whether this viral information is causally related to to the malignancy or is any exogenous viral information involved? In this connection, the induction of malignant melanomas in gnotobiotic cats inoculated with Gardner-Feline-fibrosarcoma-virus (G-FeSV) is very interesting (McCullough et al., 1972).

A very similar type of cross reactivity was observed between the reverse transcriptases from human osteosarcoma tissue and from RD-114 and BEV (Welte et al., 1979; Chandra et al., 1980). Reverse transcriptase purified from human Osteosarcoma tissue was not inhibited when challenged with antibodies against reverse transcriptases from AMV, RLV, SSV, GaLV and myelofibrotic spleen. A strong inhibition of the enzymatic activity was obtained when antibodies to DNA polymerase from BEV and RD-114 were preincubated with the enzyme.

The spectrum of antigenic specificity of human osteosarcoma reverse transcriptase follows some well-established facts concerning the nature of antigenic determinants in reverse transcriptases from endogenous or exogenous classes of viruses (see Gallo et al, 1975). For example, the human osteosarcoma enzyme cross reacts with the DNA polymerase from BEV but has no antigenic relationship to DNA polymerases from the exogenous primate viruses SiSV and GaLV. This is analogous to the fact that BEV DNA polymerase does not cross react with DNA polymerases from SSV and GaLV. The inhibition of osteosarcoma reverse transcriptase by antibodies to DNA polymerase from RD114 again points to a common determinant.

The search for a reverse transcriptase activity in normal human cells and tissues was implicated by the "Protovirus" hypothesis of Temin (Temin, 1971a, 1971b, 1972, 1974a, 1974b). He proposed that normal cells have a system of DNA to RNA, and RNA to DNA information transfer which probably functions during embryogenesis and perhaps in adult tissues. The process of reverse information flow may generate new nucleic acid sequences containing the oncogenic potential. This led to the presumption that normal human cells may contain reverse transcriptase, and the embryonic tissues might contain more activity than the adult tissues. Attempts to purify such a reverse transcriptase from normal human cells or tissues have been repeatedly unsuccessful (Chandra et al., unpublished results; Vogel and Chandra 1981). Nelson et al. (1978) reported a retroviral-like reverse transcriptase activity in normal human placenta. This activity was established in crude homogenates which contained endogenous nucleic acids. Ateempts in our laboratory to purify this enzyme from normal human placenta were repeatedly unsuccessful (Vogel and Chandra, 1981). Though in crude homogenates we observed a small $(rC)_n.(dG)_{12}$ -catalyzed activity, but after the removal of endogenous nucleic acids on DEAE-cellulose columns, no $(rC)_n.(dG)_{12}$-catalyzed activity was detectable. We believe that the activity reported by Nelson etal. (1978) is due to some non-specific transcription of oligomeric nucleic acids present in the crude fractions, and does not constitute any proof for the presence of reverse transcriptase in normal human placenta.

In another study on human placenta, we have recently succeeded in purifying two biochemically distinct reverse transcriptases from the placenta of a patient with advanced mammary cancer (Vogel and Chandra, 1981). The two activities were separated on a preparative isoelectric focusing columns at pH 5.5 and 7.2. The pH 5.5 enzyme (molec. wt. 68,000) transcribes $(rC)_n.(dG)_{12}$ preferably, whereas the pH 7.2 enzyme (molec. wt. approx. 98,000) exhibits a strong preference for $(rA)_n.(dT)_{12}$. The immunological characterization of pH 7.2 enzyme has been carried out by us. No inhibition of the placental reverse transcriptase was observed when challenged with anti-polymerase IgG from AMV, SSV, RD114 and RLV. interestingly, this enzyme was inhibited strongly by the anti-polymerase IgG fraction of a human spleen from a patient with myelofibrosis (Chandra and Steel, 1977).

The serological relatedness of placental reverse transcriptase to the reverse transcriptase from human myelofibrotic spleen is strange for two reasons: Firstly, the purified reverse transcriptase from human myelofibrotic spleen cross-reacts immunologically with the enzyme from SSV (Chandra and Steel, 1977, Chandra et al. 1980). Since the placental enzyme does not cross-react with the enzyme from SSV, its cross-reaction with reverse transcriptase from myelofibrotic spleen was unexpected. This result would indicate that immunological relatedness between polymerases from placenta and myelofibrotic spleen is due to a common determinant between the two enzymes, different from the determinant recognized by SSV polymerase. Secondly, the source of this placenta was a patient with metastatic breast cancer, and by analogy with similar neoplasia in animals, one would expect the involvement of B-type virus particles. Thus one would expect a strong immunological cross-reactivity between the placental enzyme and the enzymes from human breast-cancer tissue, or from MPMV.Spiegelman and his associates have purified reverse transcriptase from human breast tumor tissue (Ohno et al., 1977). This enzyme was shown to be immunologically cross reactive with the reverse transcriptase from M-PMV (Ohno and Spiegelman, 1977). For this reason, further characterization of the plcental enzyme will be of interest to understand the origin of these two reverse transcriptase activities.

It is hoped that the increasing use of advanced biochemical techniques will lead to the purification and characterization of reverse transcriptases from other human tumors. Immunological probing of these enzymes would find a wide and increasing application to study the nature of viral information involved in human malignancies; and moreover, for the preparation of specific antibodies in looking for virus related information in other related cancers.

ACKNOWLEDGMENTS

I am grateful to Dr. Jack Gruber (Chief, Office of

Program Resources and Logistics, National Cancer Institute, U.S.A.) and Dr. Robert C. Gallo (National Cancer Institute, Tumor Cell Biology, U.S.A.) for providing antisera against DNA polymerases from RNA tumor viruses. I also thank Dr.H. Sonneborn (Biotest, Frankfurt) for his help in preparing antibodies against the splenic reverse transcriptase. I am indebted to Professor A.O. Cavdar, University of Ankara, for making available the orbital tumor specimen. The collaboration of my coworkers, cited in the literature, is gratefully acknowledged.

These studies were supported by grants from German Research Organization (Deutsche Forschungsgemeinschaft), Kind-Philipp-Stiftung, and Stiftung Volkswagenwerk (No. 14 0305).

REFERENCES

Balda, B.R., Hehlmann,R., Cho,J.R. and Spiegelman,S. (1975): Proc. Natl. Acad. Sci. U.S.A. 72, 3697-3700.

Baltimore, D.(1970): Nature, Lond. 226, 1209-1211.

Cavdar,A.O., Arcasoy,A., Gözdasoglu,S. and Demirag,B. (1971): Lancet I, 680-682.

Cavdar,A.O., Arcasoy,A., Babcan,E., Gözdasoglu,S., U. Topuz and Fraumeni,J.F., Jr. (1978): Cancer 41,1606-1609.

Chandra,P. (1979):In, Proceedings of the XIIth Intern. Cancer Congress, Buenos Aires (Edt. G.P. Margison), Vol. 1 , 29-41, Pergamon Press, Oxford and New York.

Chandra,P. (1979a): In, Proceedings of the Federal of European Biochemical Societies Meeting, Dubrovnik, (Edt. P. Mildner) Vol 61, 215-227, Pergamon Press.

Chandra,P. and Steel,L.K.(1977): Biochem. J. (Lond.) 167, 513-524.

Chandra,P. and Steel, L.K.(1980): Cancer Lett. 9, 67-74.

Chandra,P., Steel,L.K., Laube,H. and Kornhuber, B. (1975) : FEBS- Lett. 58, 71-75.

Chandra,P., Balikcioglu,S., and Mildner,B.(1978): Cancer Lett. 5, 299-310.

Chandra,P., Steel,L.K., Laube,H. and Kornhuber,B.(1979): In, Antiviral Mechanisms in the Control of Neoplasia, (Edt. P. Chandra) pp 177-198, Plenum Press, N.Y.

Chandra,P., Steel,L.K., Laube,H., Balikcioglu,S., Ebener, U. and Welte,K. (1979a): In, Proceedings of the 8th Internatl. Symposium on the Biological Characterization of Human Tumors, Athens, Excerpta Medica, Amsterdam, pp.

Chandra,P., Steel,L.K. and Cavdar,A.O. (1979b):In, Modern Trends in Human Leukemia, Vol. III, (Edts. R. Neth et al.), pp 497-500, Springer-Verlag, Berlin-N.Y.

Chandra,P., Steel,L.K., Laube,H., Balikcioglu,S., Mildner B., Ebener,U., Welte,K. and Vogel,A.(1980): Cold Spring Harbor Conf. Cell Proliferation 7, 775-791.

Chandra,P., Balikcioglu,S. and Mildner,B.(1981): Cell. and Molec. Biol.(Pergamon Press) 27, 239-251.

Ebener,U., Welte,K. and Chandra,P.(1979): Cancer Lett. 7, 179-188.

Gallo,R.C., Yang,S.S. and Ting,R.C.(1970): Nature(Lond.) 228, 927-929.

Gallo,R.C., Gallaghar,R.E., Miller,N.R., Mondal,H., Saxinger,W.C., Mayer,R.T., Smith,R.G. and Gillespie, D.H.(1975): Cold Spring Harbor Symp. Quant. Biol. 39, 933-961.

Gillespie,D., Saxinger,W.C. and Gallo,R.C.(1975): Progr. Nucl.Acid Res. Mol. Biol. 15, 1-108.

McAllister,R.M., Nelson,Res$W.A., Johnson,E.Y., Rongey, R.W. and Gardner,M.B.(1971): J. Natl. Cancer Inst. 47, 603-611.

McAllister,R.M., Nicolson,M., Gardner,M.B., Rongey,R.W., Rasheed,S., Sarma,P.S., Huebner,R.J., Hatanaka, M., Oroszlan,S., Gilden,R.V.,Kabigiting,A. and Vernon,L. (1972): Nature New Biol. 235, 3-6.

Nelson,J., Leong,J.A. and Levy,J.A. (1978): Proc.Natl. Acad.Sci. U.S.A. 75, 6263-6267.

Ohno,T. and Spiegelman,S. (1977): Proc.Natl.Acad.Sci. U.S.A. 74, 2144-2148.

Ohno,T., Sweet,R.W., Hu,R., Dejak,D. and Spiegelman,S. (1977): Proc.Natl.Acad.Sci. U.S.A. 74, 764-768.

Poiesz,B.J., Ruscetti,E.W., Gazdar,A.F., Bunn,P.A., Minna,J.D. and Gallo,R.C.(1980): Proc.Natl.Acad.Sci. U.S.A. 77, 7415-7419.

Sarin,P., Friedman,B. and Gallo,R.C.(1977): Biochem. Biophys.Acta 479, 198-206.

Temin,H.(1964a): Natl.Cancer Inst. Monograph 17,557-570.

Temin,H.(1964b): J. Virol. 23, 486-494.

Temin,H.(1971a): J. Nazl.Cancer Inst. 42, 11

Temin,H.(1971b): J. Natl.Cancer Inst. 46, III-VIII

Temin,H.(1972): Proc.Natl.Acad.Sci. U.S.A. 69, 1016-1020

Temin,H.(1974a): Cancer Res. 34, 2835-2841.

Temin,H.(1974b): Annual Rev. Genetics 8, 155-176.

Temin,H. and Mizutani,S.(1970): Nature(Lond.)226,1211-1214.

Vogel,A. and Chandra,P.(1981): Biochem.J. (Lond) 197, 553-563.

Welte,K., Ebener,U. and Chandra,P.(1979): Cancer Lett. 7, 189-195.

Witkin,S.S., Ohno,T. and Spiegelman,S.(1975): Proc. Natl. Acad.Sci. U.S.A. 72, 4133-4136.

Wong-Staal,F., Gillespie,D.H. and Gallo,R.C.(1976): Nature(Lond.) 262, 190-195.

Wu,A. and Gallo,R.C.(1975): CRC Crit.Rev.Biochem. 1975 289-347.

CLINICAL TRIALS WITH A PARTIALLY THIOLATED POLYCYTIDYLIC ACID, A POTENT INHIBITOR OF RETROVIRAL REVERSE TRANSCRIPTASE, IN THE TREATMENT OF ACUTE CHILDHOOD LEUKEMIA

Bernhard Kornhuber and Prakash Chandra

Zentrum der Kinderheilkunde und Zentrum der Biolog.-Chemie, Klinikum der Universität
Theodor-Stern-Kai 7,
D-6 Frankfurt/M.70, BRD.

A partially thiolated polycytidylic acid, containing 5-mercapto substituted cytosine bases, abbreviated as MPC was found to inhibit the oncornaviral DNA polymerase in a very specific manner (Chandra and Bardos, 1972; Chandra, 1974; Chandra et al., 1975,1977,1980). Compared to the cellular DNA polymerases α, β and γ, the activity of viral DNA polymerase was, by far, more sensitive to MPC (Chandra et al., 1979, 1980). The antileukemic effect of MPC was studied on FLV-induced leukemia (Chandra et al., 1977, 1979), and under vitro-vivo conditions it was found to be very effective. Toxicological studies have shown that the compound is not mutagenic in Ames test, and does not produce any chromosomal abnormalities in mice and rats (unpublished results).

The fact that reverse transcriptase is present in human leukemic cells, has led to the clinical trials of MPC in the treatment of childhood acute leukemia. This report describes a pilot clinical trial on 23 patients.

The results are encouraging, and have led to a second pilot study including other clinical centers on the clinical efficacy of MPC in the treatment of childhood leukemia.

TABLE 1

Clinical data of patients submitted to MPC trials.

No.	Init.	Age	Sex	Diag.	Stage	Results
1	D. M.	8	♀	ALL	3rd rel.	PR, WBC ↓
2	T. I.	6 6/12	♀	ALL	2nd rel.	?
3	B. M.	6 7/12	♂	ALL	2nd rel.	?
4	M. M.	8 4/12	♂	ALL	2nd rel.	CR
5	J. O.	3 6/12	♂	ALL	2nd rel.	?
6	N. A.	10	♀	ALL	2nd rel.	CR
7	B. C.	5 8/12	♀	ALL	3rd rel.	CR
8	N. N.	7 11/12	♂	ALL	3rd rel.	?
9	M. A	12	♂	ALL	1st rel.	?
10	M. I.	8 6/12	♂	ALL	1st rel.	?
11	K. C.	11 3/12	♀	ALL	4th rel.	?
12	K. K.	10 9/12	♀	ALL	5th rel.	PR
13	L. J.	7	♂	ALL	3rd rel.	WBC ↓
14	F. D.	2 3/12	♂	ALL	init. ph.	WBC ↓
15	S. B.	7 11/12	♀	ALL	init. ph.	WBC ↓
16	H. B.	12 5/12	♀	ALL	init. ph.	WBC ↓
17	S. N.	4	♀	AML	init. ph.	WBC ↓
18	W. H.	5 6/12	♂	AML	init. ph.	WBC ↓
19	S. K.	4 9/12	♂	ALL	init. ph.	WBC ↓
20	F. A.	7 10/12	♀	AML	init. pH.	WBC ↓
21	L. F.	3 4/12	♂	ALL	init. ph.	WBC ↓
22	F. T.	4 9/12	♂	ALL	init. ph.	WBC ↓
23	C. E.	13 4/12	♂	ALL	init. ph.	WBC ↓

Clinical data of patients submitted to MPC trials are shown in Table 1. Of the 23 cases treated with MPC, 13 were in the terminal phase of the disease. These patients were resistent to all previous chemotherapeutic regimes which involved drugs, such as prednisone, vincristin, daunorubicin, L-asparaginase, Ara-C, 6-mercaptopurine, methotrexate, cyclophosphamide and actinomycin-D.

MPC used in our clinical trials contained 12-15% of thiolated cytosine bases. The lyophilized product (MPC) was dissolved in 0.1 M Tris/HCl buffer, pH 7.6 and diluted with 0.9% NaCl before use. This solution was sterilized by passing through a membrane filter (Millipore, Neu Isenburg, Germany). It was kept at 4 ^{o}C and used immediately, or within the next five days; solutions older than 5 days were reprecipitated, purified on the column and resterilized. In our clinical trials, MPC (sterile) was given intravenously at a dose 0.5 mg/kg body weight. The injections were given once a week.

Of the 13 terminal cases, complete remission was achieved in 3, and a partial remission achieved in 2 other cases. Fever, occasionally accompanied by shivering was frequently observed under MPC treatment in the first hour after injection. However, these symptoms never lasted more than the first hour, and no other side-effects could be observed.

On the basis of our experience with MPC on terminal cases, we were motivated to give MPC a clinical trial in the beginning of leukemia. A monotherapy with MPC, as devised for terminal cases, is, however, not possible. We therefore decided to introduce MPC (0.5 mg/kg body weight) therapy in the beginning of treatment of cases which at the time of diagnosis had leucocytosis. This initial treatment, a single dose of MPC, was then followed up by a polychemotherapeutic protocol, developed by Rhiem et al. (1974). As shown in Fig. 1, 24 hrs. after MPC injection, there was a significant reduction of leukemic cells in all the cases.

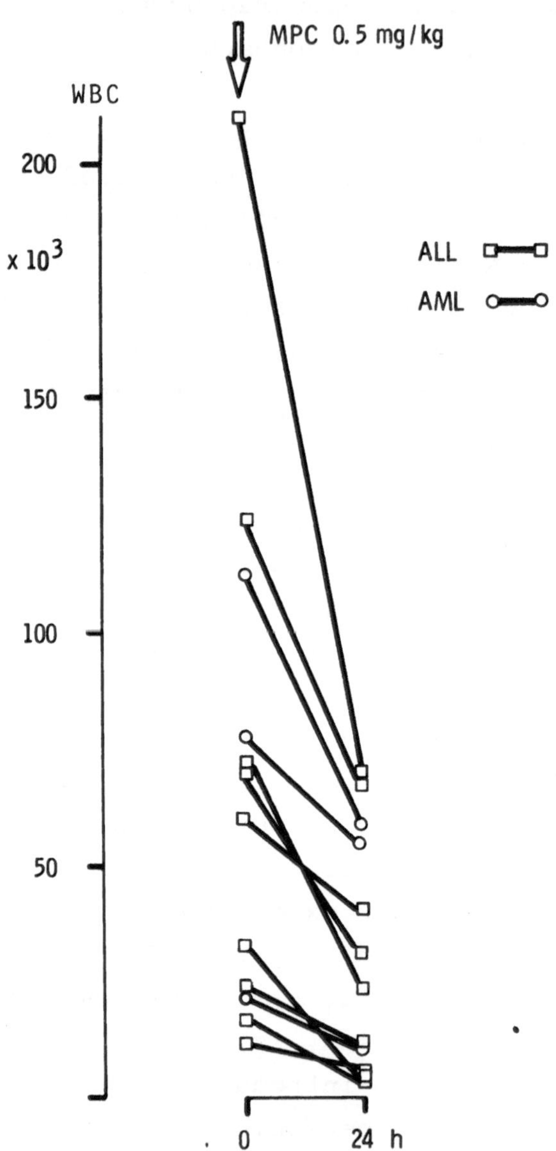

Fig.1. MPC trials in the freshly diagnosed leukemic cases.

The status of this drug in the chemotherapy of fresh leukemic cases is not known, since monotherapy with MPC in such cases has not been done. The fact that under the present polychemotherapeutic protocols one can frequently achieve longterm remissions, hinders one ethically to use MPC as a monotherapeuticum in fresh cases. However, its use in the initial phase of the acute disease, and its use as a monotherapeutic agent in terminal cases are quite encouraging. On the basis of our to-date experience with MPC we could summarize by saying: a)MPC is useful to initiate the therapy in freshly diagnosed acute leukemic cases, b) it has shown promise as an effective drug in the treatment of leukemic cases in the terminal phase, and c) it could be used in the remission maintenance therapy. This, latter aspect, is yet to be investigated. This, as a matter of fact, is the rationale for its therapeutic application, since it is a potent inhibitor of reverse transcriptase.

Our experimental and clinical studies have led us to postulate the following working hypothesis: The selective cytolysis of leukemic cells by MPC indicates that this compound binds to some "novel" component present in these cells, which may be the virus-related reverse transcriptase. The reason for this selectivity may be the concentration range, which does not effect the normal cellular polymerases, as indicated by our enzymatic data (Chandra et al., 1979, 1980). Once the compound is bound to this "target", the cytotoxicity could then be a secondary event, probably by release of lysosomal enzymes. This is supported by two facts: (1) The lysosomal membranes of tumor cells are less stable than those of normal liver cells (Taniguchi et al., 1976), and 2) after treatment with antitumor agents the lysosomal enzyme activities in tumor cells are higher than in normal cells (Haddow, 1947). Taniguchi et al. (1976) have shown that lysosomal labilizers, plasmin and lipoprotein lipase are capable of enhancing the cytocidal effect of mitomycin-C on tumor cells. The advantage of our compound is that unlike other lysosomal labilizers which cannot distinguish between normal and the tumor cells, this compound could act more specifically on cells which bear its specific receptor, the reverse transcriptase.

ACKNOWLEDGMENT

This work was financially supported by grants from Stiftung Volkswagenwerk (grant No. 14 0305).

REFERENCES

CHANDRA,P. (1974): In, Topics in Current Chem., Springer Verlag Berlin-Heidelberg-New York, Vol 52, 99-139.

CHANDRA,P. and BARDOS,T.J. (1972): Res. Commun. Chem. Pathol. and Pharmacol. 4, 615-622.

CHANDRA,P., EBENER,U and GÖTZ,A. (1975): FEBS-Lett. 53, 10-14.

CHANDRA,P. , EBENER,U., BARDOS,T.J., GERICKE,D., KORNHUBER,B., GÖTZ,A. (1977): Fogarty International Center Proceedings (USA) 28, 169-186.

CHANDRA,P., EBENER,U. and GERICKE,D. (1979): In, Antiviral Mechanisms in the Control of Neoplasia (Ed. P. Chandra), Plenum Press, N.Y. 523-538.

CHANDRA,P., STEEL,L.K., EBENER,U., WOLTERSDORF,M., LAUBE,H., KORNHUBER,B., MILDNER,B., GÖTZ,A. (1980): Intnl.Encycl.Pharm.Ther. Section 103, 47-90

HADDOW,A. (1947): Growth 11, 339.

RHIEM, H., GARDNER,H., JESSENBERGER,K. and TARIVERDIAN, G. (1974): Proc.Amer.Assoc.Cancer Res. 15, 58.

TANIGUCHI, T., NIITANI,H., SUZUKI,A., SAIJO,N., KAWASE, I. and KIMURA,K. (1976): In, Chemotherapy (Edts. K. HELLMANN and T.A. CONNORS), vol. 8, 175, Plenum Press New York.

SUBCELLULAR LOCALIZATION OF SIMIAN VIRUS 40 T-ANTIGEN

Wolfgang Deppert and Matthias Staufenbiel

Dept. of Biochemistry
University of Ulm
D-7900 Ulm, Federal Republic of Germany

INTRODUCTION

SV40 T-antigen - a polyfunctional regulator protein

The outcome of an infection of tissue culture cells with the small DNA tumor virus simian virus 40 (SV40) depends on the type of cell used in such an experiment: (i) infection of permissive cells (epithelial monkey cells) will lead to production of progeny virus, resulting in cell lysis; (ii) infection of nonpermissive or semipermissive cells (e.g. mouse or human cells), however, may lead to integration of at least part of the viral genome and an alteration of these cells from a "normal" to a "transformed" phenotype. Both events, viral replication and viral transformation, are initiated, maintained and regulated by the expression of an early SV40 gene, the SV40 A gene. The product of the SV40 A-gene, SV40 T-antigen (T-Ag), is a phosphoprotein with a polypeptide molecular weight of about 96 K on SDS polyacrylamide gels (reviewed in ref.[1]). So far, quite a number of different functions and activities have been attributed to T-Ag[1] (Table 1). SV40 T-Ag, therefore, seems to be a multifunctional (pleiotropic) molecule[2]. Modulation of T-Ag to perform these diverse functions might occur via two known posttranslational modifications, phosphorylation of different serine and threonine residues [3-5] and poly ADP-ribosylation[6], and/or via an association with a host cell protein, the 53 K nonviral tumor antigen (pp 53, NVT)[7,8]. Consequently, heterogeneity of T-Ag with regard to monomeric and polymeric forms, degree of phosphorylation and association with NVT has been demonstrated by analysis of total T-Ag extracts from SV40-infected and transformed cells[9-11].

Table 1. Postulated functions and activities of SV40 T-Ag

(1) initiation of each round of viral DNA replication
(2) stimulation of cellular DNA synthesis
(3) stimulation of cellular RNA synthesis
(4) stimulation of cellular protein synthesis
(5) stimulation of synthesis of cellular enzymes involved in DNA synthesis (e.g. thymidine kinase)
(6) initiation of transcription of late SV40 genes
(7) regulation of its own biosynthesis (autoregulation)
(8) induction and/or stabilisation of a nonviral tumor antigen (NVT, 53 K protein, pp 53)
(9) "helper" function for the growth of human adenovirus in monkey cells
(10) mediation of SV40 tumor transplantation immunity (TSTA-activity)
(11) establishment and maintenance of cellular transformation
(12) DNA binding activity
  a) nonspecific (to the polyphosphate backbone of nucleic acids)
  b) specific (to the origin of viral DNA replication)
(13) ATPase activity
(14) protein kinase activity (perhaps in association with a cellular kinase)

Functions (1) - (7) can be observed in SV40 infected cells, functions (8) - (10) both in SV40 infected and in SV40 transformed cells. Activities (12) - (14) can be measured in T-Ag extracts from SV40 infected or transformed cells or with purified T-Ag (see review in ref.[1]).

Looking at the rather diverse and complex functions of T-Ag, one has to assume that at least some of these functions are exerted in association with different viral and cellular structures. So, for example, it has been shown that the preferred binding of SV40 T-Ag to the regulatory region of the SV40 DNA is an important requirement for regulating viral DNA synthesis, transcription of SV40 early and late genes as well as its own biosynthesis.[12-14] Whereas the interaction of SV40 T-Ag with SV40 DNA[12-17] and SV40 minchromosomes[18,19] has been studied in some detail, very little, so far, is known about its interaction with different cellular structures in SV40 infected and transformed cells. However, such analyses should provide valuable clues for understanding the multiple biological functions of SV40 T-Ag.

RESULTS AND DISCUSSION

SV40 T-Ag is present in the cell nucleus in distinct subclasses

Most of the functions listed in Table 1 are nuclear functions, like stimulation of cellular DNA and RNA synthesis and binding to viral and cellular DNA. Consequently, the majority of T-Ag can be detected in the nuclei of SV40 infected and transformed cells by immunofluorescence microscopy using anti-T-sera (mostly sera from SV40 tumor bearing animals).[20] The nuclei appear to be evenly stained, with only the nucleoli being devoid of stain, indicating a rather homogeneous distribution of T-Ag. However, the multiple functions postulated for nuclear T-Ag (see Table 1) as well as its demonstrated heterogeneity[9-11] suggested to us that there may be different subclasses of T-Ag associated with different nuclear subcompartments. We, therefore, have fractionated nuclei of SV40 transformed cells into nucleoplasm, chromatin and a nuclear residual structure, the nuclear matrix, and analyzed each of these nuclear fractions for the presence of SV40 T-Ag. Fig. 1 reveals that about one third of the total nuclear T-Ag is present in the nucleoplasm, about two thirds are associated with chromatin, and a few percent of the nuclear T-Ag molecules are tightly bound to the nuclear matrix. The possibility to separately isolate nuclear subclasses of T-Ag already indicates differences between these T-Ag molecules. In this line, our preliminary experiments to further characterize the nuclear subclasses of T-Ag suggest that they differ in their degree of phosphorylation, with the chromatin and nuclear matrix associated T-Ag molecules being more highly phosphorylated than T-Ag molecules in the nucleoplasm.

The separate isolation of the different nuclear T-Ag subclasses will allow their biochemical, immunological and biological analysis and will eventually lead to a more functional characterization of T-Ag subclasses present in the different nuclear subcompartments. So, for instance, one may be able to determine whether T-Ag associated with chromatin may bind to origins of cellular DNA replication, an important feature for T-Ag being directly involved in the regulation of cellular DNA synthesis.

TSTA-activity of SV40 T-Ag requires its presence on the cell surface

Although the bulk of SV40 T-Ag is located in the cell nucleus, at least one T-Ag function by definition cannot be exerted by nuclear T-Ag: the mediation of SV40 tumor transplantation immunity. SV40 tumor specific transplantation antigen (TSTA) classically is defined as an antigen present on the surface of SV40 tumor cells. TSTA will be recognized by host-immunocompetent cells when the animals had been preimmunized with either infectious SV40 virus,

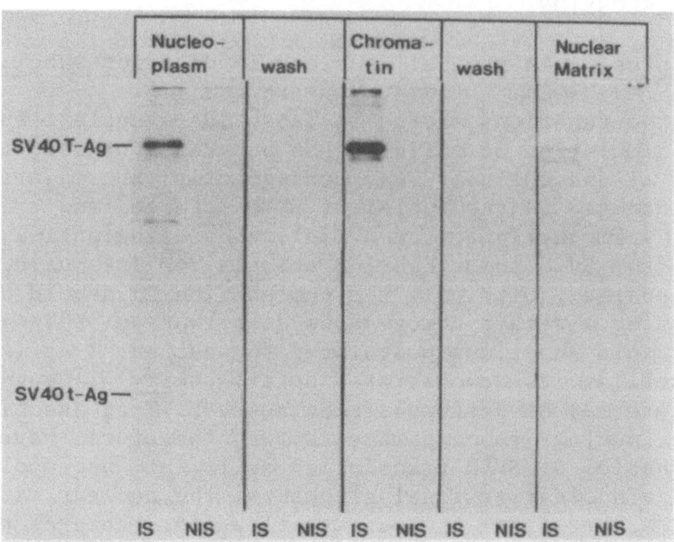

Fig. 1. Demonstration of distinct subclasses of SV40 T-Ag in nuclear subcompartments of SV40 transformed cells. SV40 transformed balb/c mouse cells (mKSA tumor line) grown in suspension culture were labeled with ^{35}S-methionine. Fractions containing the nucleoplasm, the dissolved chromatin and the dissolved nuclear matrix were prepared as described in Experimental Procedures. Fractions were immunoprecipitated with rabbit antiserum against purified, SDS-denatured T-Ag (IS) or rabbit non-immune serum (NIS) by the protein A-sepharose technique[36] and immunoprecipitates analyzed on a 12.5 % SDS-polyacrylamide gel. A fluorogram of radioactive polypeptides is shown. The separate isolation of distinct T-Ag subclasses could be demonstrated by immunoprecipitating second time extractions (wash), see Experimental Procedures, of nucleoplasm or chromatin, respectively. These washes contained only trace amounts of T-Ag.

or SV40 infected or transformed cells (live or dead). Preimmunized animals then are resistant to subsequent challenges with transplantable tumors induced by SV40. The tumor-rejection mechanism is highly specific: no crossreaction between SV40 TSTA and a TSTA of any other transformed cell has been reported so far (reviewed in ref.[1]).

There are several lines of evidence indicating that TSTA and T-Ag are closely related, culminating in the demonstration that immunization with low amounts of T-Ag ($\leq$1 μg) purified to homogeneity is sufficient to protect mice against subsequent SV40 tumor challenges.[21] This implies that T-Ag has to be located on the surface of SV40 transformed cells. However, it was not until recently that the serological demonstration of this postulated "surface T" became feasable. The difficulty to detect surface T on SV40 transformed cells in immunofluorescence analysis can be attributed to the observations that surface T (i) exists in only very small amounts, (ii) is a rather "cryptic" antigen on SV40 transformed monolayer cells and, (iii) most importantly, does not readily react with SV40 tumor sera. These problems, however, can be overcome by pretreatment of the cells with formaldehyde before analysis, which apparently exposes surface T-determinants and performing the immunofluorescence assay with special antisera, like antisera directed against purified SDS-denatured T-antigen.[22] These sera recognize antigenic determinants on the T-Ag polypeptide not recognized by antisera directed against native T-Ag like SV40 tumor sera (W. Deppert and G. Walter, manuscript in preparation). Under these conditions a positive, T-Ag specific surface staining can be observed on SV40 transformed cells. Fig. 2 a shows the surface staining on formaldehyde treated SV40 transformed mouse 3T3 monolayer cells, stained with rabbit anti-SDS T-serum and fluorescein-conjugated goat anti-rabbit-immunoglobulins. In Fig. 2 b the same cells were permeabilized after the cell surface staining by additional treatment with methanol/acetone and stained in a second reaction with mouse SV40 tumor serum followed by rhodamine-conjugated swine-anti-mouse immunoglobulins. Using this double staining immunofluorescence technique it has become possible to analyze the same cells simultaneously for both cell surface and nuclear T-Ag. This approach allowed the demonstration of a correlated expression of surface T and nuclear T-Ag in cells transformed by a temperature-sensitive SV40 A-gene mutant (tsA 28.3 cells):[23] At permissive growth temperature, the cells are phenotypically transformed and express nuclear T-Ag as well as surface T. After shift up to nonpermissive growth temperature, the cells shut off the synthesis of T-Ag and revert to a non-transformed phenotype. Concomitantly, they loose nuclear T-Ag and surface T. These studies not only provided genetic evidence that nuclear T-Ag and surface T are derived from the same gene but also showed that expression of nuclear T-Ag <u>and</u> of surface T correlate with the transformed phenotype of these cells.

A very important question in the analysis of surface T is how one can explain the presence of T-Ag both in the nucleus and on the cell surface. In principle, two rather different pathways for the association of T-Ag with the cell surface can be envisioned:

(i) After its synthesis, T-Ag will only and specifically accumulate in the nuclei of transformed cells. From the nuclei of dead or permeable cells, T-Ag may leak out and adsorb to the surface of surrounding cells. In this model surface T actual-

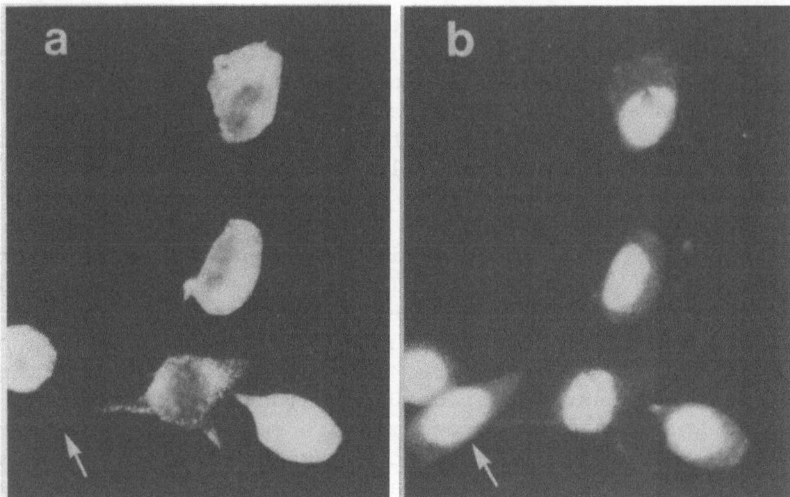

Fig. 2. Double-staining of SV3T3 cells for surface T (a) and nuclear T-Ag (b). Cells grown on coverslips were fixed with formaldehyde and first stained for cell surface fluorescence with rabbit anti-SDS T-serum followed by fluoresceine-coupled sheep anti-rabbit-immunoglobulin (a). The same cells were then permeabilized with methanol and acetone and stained additionally for nuclear T-Ag with mouse SV40 tumor serum followed by rhodamine-coupled rabbit-anti-mouse-immunoglobulin (b). Arrows point to a cell which is surface T negative (a) but brilliantly stained for nuclear T-Ag (b), indicating heterogeneous expression of surface T (reprinted from Deppert et al.[22]).

ly would be exogeneous T-Ag adsorbed to the cell surface but originating from nuclei of other, dead cells.

(ii) Alternatively, the existence of nuclear and of surface T might be the result of a programmed compartmentalization of this molecule both into the nucleus and onto the cell surface, i.e., after its synthesis T-Ag will reach both the nucleus and the cell surface of the same cell. In this case T-Ag would be a nuclear protein and, simultaneously, a membrane protein.

To decide between these alternatives, i.e. to answer the question of the origin of surface T as detected in the immunofluorescence assay, is difficult using SV40 transformed cells. Most of the functions of T-Ag in transformed cells supposedly are nuclear functions. In accordance, the majority of T-Ag (an estimated

>95 %), is located in the nuclei. Therefore, all studies on surface T using SV40 transformed cells are confronted with the problem of a possible "contamination" by nuclear T-Ag. This problem, however, can be circumvented by using a system in which most of the nuclear T-Ag functions are not expressed, but where TSTA still can be demonstrated. Fortunately, such a system is provided by the Ad2$^+$SV40 hybrid viruses.

Ad2$^+$SV40 hybrid viruses - a tool for analyzing functional domains of SV40 T-Ag

The Ad2$^+$SV40 hybrid viruses Ad2$^+$ND1, Ad2$^+$ND2 and Ad2$^+$ND4 are nondefective type 2 adenoviruses (Ad2), which contain overlapping portions of the SV40 genome inserted into the Ad2 genome. The integrated SV40 DNA has a common endpoint at 0.11 SV40 m.u. in the late region of the SV40 genome and extends for various length into the early region of the SV40 genome (Fig. 3). At the integration sites, 4 to 6 % of the Ad2 genome to the left of Ad2 map position 85.5 is deleted. During productive infection, the SV40 information in the hybrid viruses is expressed and codes for overlapping fragments of SV40 T-Ag (reviewed in ref.[1]). All fragments contain the COOH-terminus of T-Ag[26] and extends for various length towards its NH_2-terminus.[27] These COOH-terminal T-Ag fragments all are biologically active and, depending on their sizes, they exert different T-Ag functions (reviewed in ref.[1]). This demonstrates that the T-Ag molecule harbors different functional domains. Some of these domains can be analyzed independently from one another in Ad2$^+$SV40 hybrid virus infected cells.

Ad2$^+$ND1 as a model system to study the origin of surface T

To study the origin of surface T in the absence of interfering nuclear T-Ag, we have chosen to work with Ad2$^+$ND1, the hybrid virus with the smallest SV40 information (see Fig. 3) Ad2$^+$ND1 is able to induce SV40 TSTA;[28] the only other T-Ag function known to be mediated by Ad2$^+$ND1 is the helper function.[29] This function allows human adenoviruses to grow in monkey cells, which otherwise are nonpermissive for these viruses.[30] The block to Ad2 multiplication in monkey cells most likely is due to incorrect processing (splicing) of some late Ad2 mRNAs.[31] The SV40 helper function, therefore, must act on the level of RNA processing (splicing), which is a nuclear event.[32]

Ad2$^+$ND1 codes for a SV40-specific 28 K protein,[33] corresponding to the COOH-terminal third of T-Ag. Its subcellular locations as determined by biochemical cell fractionation correlate well with its postulated functions: it is associated with the nuclear matrix,[34] where it may exert the helper function, and with the plasma membrane,[24,28] where it may be part of TSTA. The direct involvement of the 28 K protein in the formation of TSTA

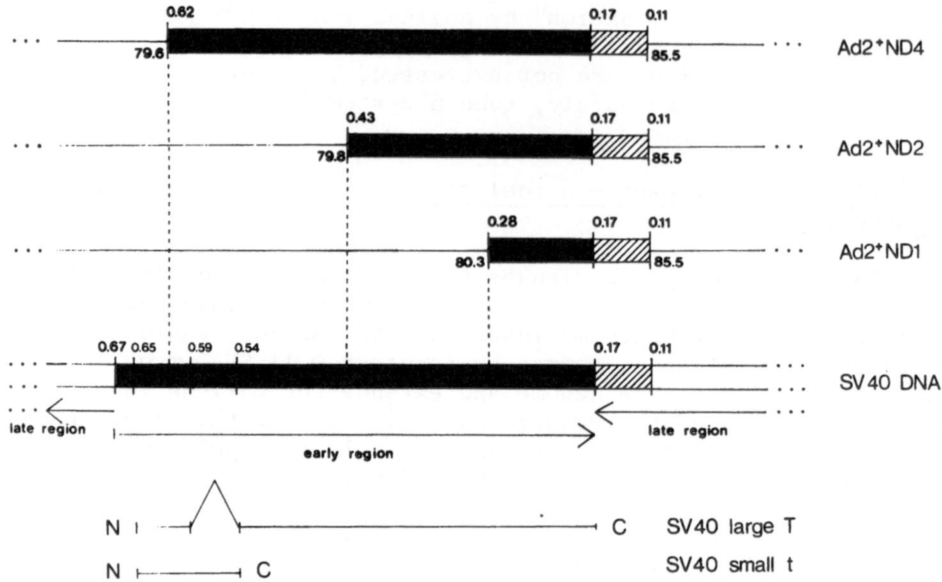

Fig. 3. SV40 DNA segments in the nondefective Ad2$^+$SV40 hybrid viruses Ad2$^+$ND1, Ad2$^+$ND2 and Ad2$^+$ND4. Mapping of SV40 DNA fragments and their integration sites on the Ad2 genome is according to Henry et al.[37] and Morrow et al.[38] for Ad2$^+$ND1 and Ad2$^+$ND2 and according to Westphal et al.[39] for Ad2$^+$ND4. The positions of the early and late regions on the SV40 DNA maps are derived from Reddy et al.[40] Arrows indicate the direction of transcription. Coding regions for SV40 large T-Ag and small t are according to Crawford et al.[25] The Ad2$^+$SV40 hybrid viruses direct the synthesis of the following SV40-specific proteins: 28 K (Ad2$^+$ND1);[33] 42 K, 56 K (Ad2$^+$ND2);[24,41] 42 K, 56 K, 60 K, 64 K, 72 K, 74 K, 94 K (Ad2$^+$ND4).[24] These proteins contain the COOH-terminus of T-Ag,[26] and extend for various length towards its NH_2-terminus.[27]

has been further substantiated by demonstration of its cell surface location by immunofluorescence microscopy.[35] Since the only subnuclear location of the 28 K protein is the nuclear matrix, to which it is tightly bound,[34] leakage of this protein from the nuclei should not occur. One, therefore, can assume that the surface membrane location of the Ad2$^+$ND1 28 K protein most likely is not due to adsorption of released nuclear 28 K protein. These data instead strongly argue for a programmed compartmentalization of the Ad2$^+$ND1 28 K protein both into the nuclear matrix and onto the cell surface.

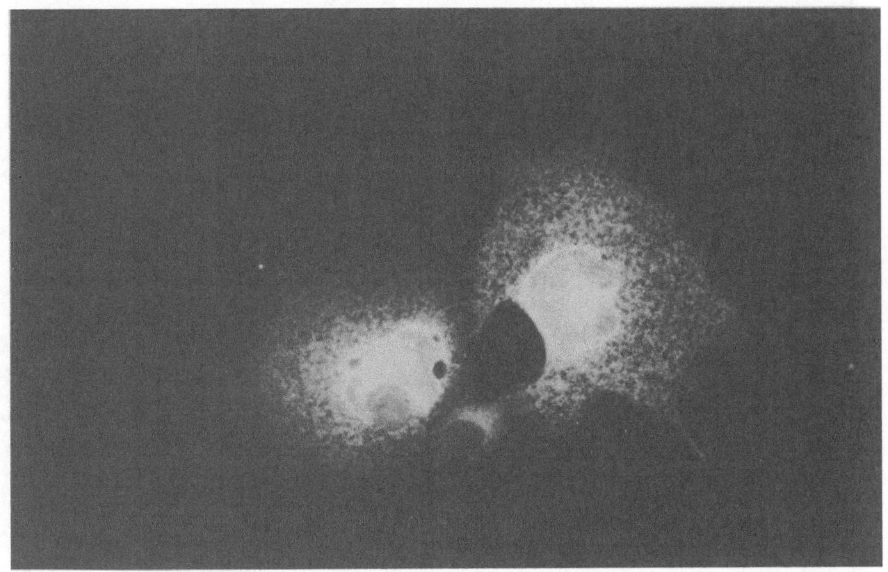

Fig. 4. SV40-specific fluorescence in Ad2+ND1 infected monkey cells. Ad2+ND1 infected monkey (TC-7) cells grown on coverslips were fixed with methanol and acetone at 30 h postinfection and stained with rabbit anti-SDS T-serum followed by fluoresceine coupled sheep anti-rabbit-immunoglobulin.

Association of the SV40 28 K protein with intracellular membranes

A rather astonishing observation, however, is made, when the intracellular location of the SV40 28 K protein is analyzed in Ad2+ND1 infected cells by immunofluorescence microscopy. Fig. 4 demonstrates that,in addition to a nuclear staining, a predominant cytoplasmic fluorescence is obtained, extending from a perinuclear area into the ectoplasm.

In an attempt to resolve the seeming discrepancy in the subcellular locations of the 28 K protein as determined by biochemical cell fractionation and by immunofluorescence, we have characterized the cytoplasmic 28 K fluorescence. A typical feature of this fluorescence is that it is not evenly distributed within the cell, but clearly has a structured appearance. This strongly suggests that the Ad2+ND1 28 K protein does not exist in the cytoplasm in a free (unbound) form, but instead is associated with one of the major cytoplasmic structural systems: (i) the cytoskeletal filament systems or (ii) the intracellular membrane systems (mostly membranes of the endoplasmic reticulum).

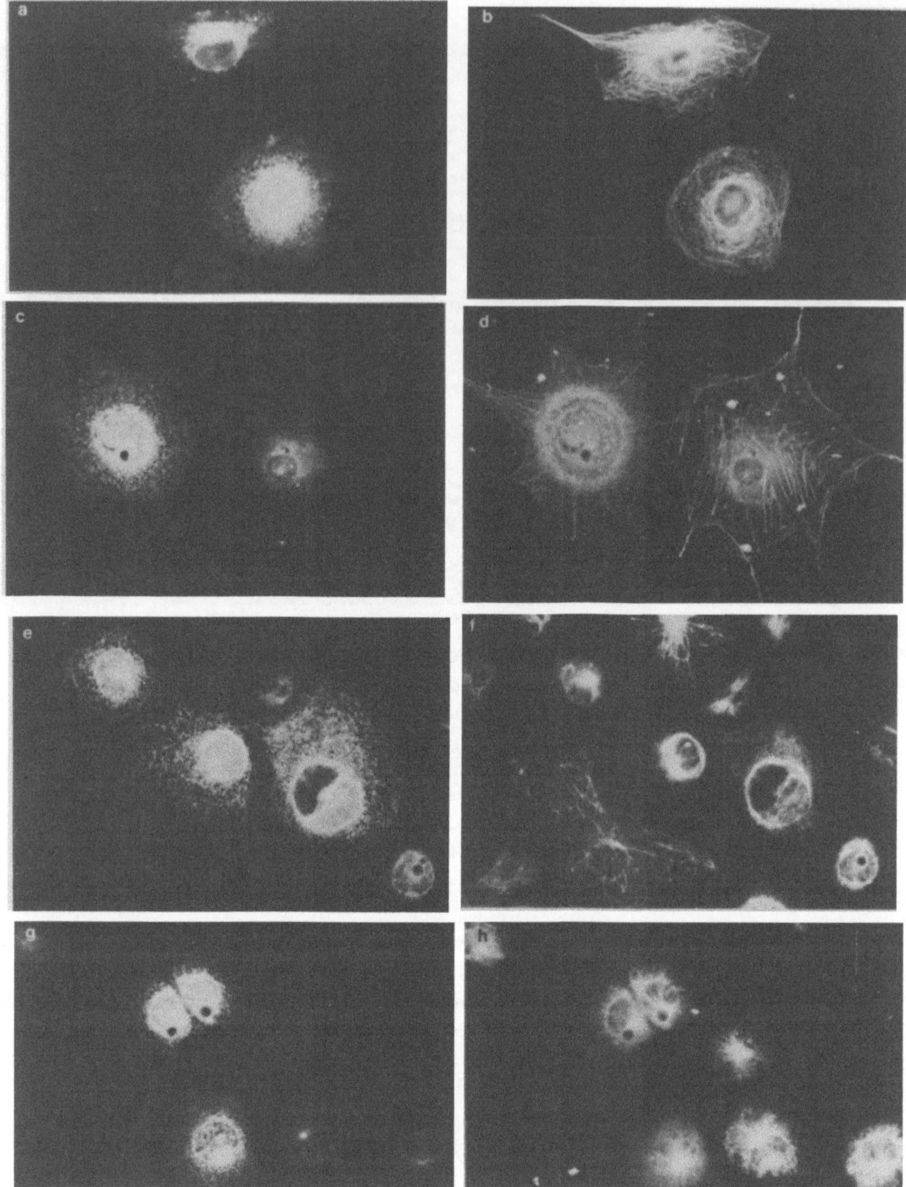

Fig. 5. Double-staining of Ad2^{+}ND1 infected TC-7 cells for the SV40-specific 28 K protein and cytoplasmic filament systems. Ad2^{+}ND1 infected TC-7 cells grown on coverslips were fixed with methanol and acetone at 30 h postinfection and first stained with guinea pig anti-SDS T-serum[35] followed by fluoresceine coupled goat anti-guinea pig immunoglobulin (panels a, c, e, g). The same cells were (continued)

then stained with monospecific rabbit antibodies prepared against cytoplasmic filament subunit proteins tubulins (b), actin (d), vimentin (f) and cytokeratin (h), followed by rhodamine coupled anti-rabbit immunoglobulin.

Fig. 5 demonstrates that by double staining immunofluorescence analysis an association of the 28 K protein with one of the cytoskeletal filament systems can be ruled out, since the SV40-specific cytoplasmic fluorescence in $Ad2^{+}ND1$ infected cells is different from the staining patterns for actin, tubulin and intermediate filament proteins. This already strongly argues for an association of the 28 K protein with intracellular membrane systems. This association can be demonstrated more directly, when cytoskeletal preparationsof $Ad2^{+}ND1$ infected cells are analyzed: treatment of unfixed living cells with appropriate buffers containing nonionic detergents removes most constituents of the cellular membranes, but leaves an intact cytoskeletal framework.[43] After such a treatment the 28 K protein no longer can be visualized in the cytoplasm (Fig. 6). This is in accordance with an association of the 28 K protein with intracellular (cytoplasmic) membranes. This association of the 28 K protein with intracellular membranes might explain why during biochemical cell fractionation only very little 28 K protein is found in the cytoplasmic fraction, but, instead, is enriched in the plasma membrane fraction: it has been observed that part of the intracellular membranes copurify with plasma membranes in such experiments.[42] Fig. 6, in addition, shows the nuclear location of the 28 K protein in cytoskeletal preparations of $Ad2^{+}ND1$ infected cells. This immunofluorescence result confirms the nuclear matrix location of the 28 K protein as determined by biochemical cell fractionation.[34]

The association of the 28 K protein with intracellular membranes most likely does not only reflect newly synthesized 28 K molecules at membrane-bound polysomes of the rough endoplasmic reticulum: when initiation of protein synthesis is blocked by addition of the drug verrucarin[43] (50 μg/ml for 1 h) the cytoplasmic 28 K protein fluorescence persists. This demonstrates that this protein is directly associated with intracellular membrane systems. At present, we do not know whether this association reflects an affinity of the 28 K protein for membranes in general. In this case the 28 K protein not only will associate with intracellular membranes, but also with plasma membranes. Alternatively, as part of its pathway to the plasma membrane the 28 K protein might selectively associate with certain intracellular membranes, like the membranes of the endoplasmic reticulum. For both possibilities the surface membrane association of the 28 K protein is not fortuitous, but is programmed by a domain(s) or a site(s) on the polypeptide backbone of this molecule.

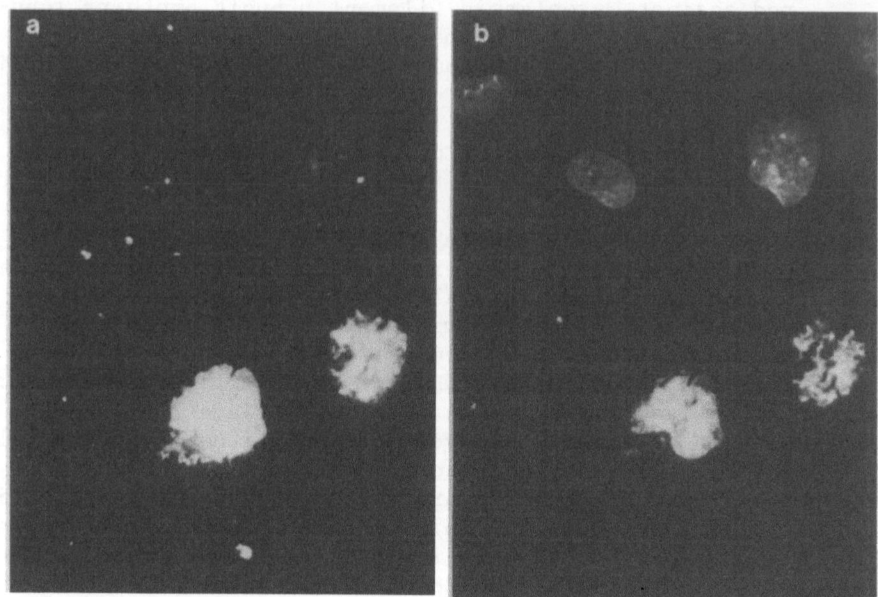

Fig. 6. Double staining of cytoskeletal preparations of Ad2+ND1 infected TC-7 cells with anti-SDS T-serum (a) and with antiserum against a structural protein of the nuclear matrix (b). Cytoskeletal preparations of Ad2+ND1 infected TC-7 cells grown on coverslips were prepared as described in Experimental Procedures and fixed with methanol and acetone. Cytoskeletons were first stained with rabbit anti-SDS T-serum followed by fluoresceine-conjugated sheep anti-rabbit immunoglobulin (a) and then with guinea pig antiserum directed against nuklin B, one of the major structural proteins of the nuclear matrix, followed by rhodamine-conjugated goat anti-guinea pig immunoglobulin (b).

Model for a programmed compartmentalization of T-Ag

The Ad2+ND1 28 K protein is predominantly associated with membranes (both intracellular and plasma membranes) as its main subcellular location. A fraction of this protein also is found in the nucleus, but there it is exclusively associated with the nuclear matrix. In SV40 transformed cells, in contrast, only a minor fraction of T-Ag is plasma membrane-associated. The main subcellular location of full length T-Ag instead is the cell nucleus, with the majority of nuclear T-Ag found in the nucleoplasm and associated with chromatin. Only a few percent of the nuclear T-Ag molecules are associated with the nuclear matrix (see Fig. 1). This indicates that full length T-Ag must contain domains, which direct its accumulation in the nucleoplasm and its association with chromatin. These domains must be located in the NH_2-terminal half of T-Ag, since in $Ad2^+SV40$ hybrid virus infected cells the COOH-terminal T-Ag fragments up to the 56 K protein exclusively associate with membranes and with the nuclear matrix.[24,34]

According to this concept we suggest a model for the programmed compartmentalization of T-Ag: The distribution of full length T-Ag into its different subcellular locations is mediated by specific domains on the T-Ag molecule. Domains in the COOH-terminal region of T-Ag will mediate the membrane location of T-Ag and its association with the nuclear matrix. Domains in its NH_2-terminal half, on the other hand, will cause the accumulation of T-Ag in the nucleoplasm and its association with chromatin. SV40 T-Ag, therefore, contains several distinct domains, each determining a different subcellular location. These domains are located in different parts of the T-Ag molecule and can act independently from one another, as can be demonstrated by analyzing the subcellular distribution of T-Ag fragments. The compartmentalization of T-Ag, therefore, is the result of influences of several domains on this molecule, directing it into different subcellular locations. In addition to domains determined by the amino acid sequence of T-Ag, posttranslational events, like phosphorylation at specific sites, might modulate the distribution of T-Ag into certain subcellular compartments.

The majority of T-Ag in SV40 transformed cells is found in the nucleoplasm and associated with chromatin (see Fig. 1). This indicates that the domains directing T-Ag into these nuclear subcompartments have a strong influence on its overall location. In contrast, domains on full length T-Ag directing its association with membranes or the nuclear matrix have a relatively weak influence. The existence of such domains, however, can be demonstrated, when fragments of T-Ag are analyzed, which lack the domains responsible for the nucleoplasm and chromatin location.

ACKNOWLEDGEMENT

We thank Ms. Petra Epple for skilful technical assistance. This study was supported by the Stiftung Volkswagenwerk (VW I/37084) and the Deutsche Forschungsgemeinschaft (De 212/3).

EXPERIMENTAL PROCEDURES

Cells and viruses. Monkey TC-7 cells and mouse SV3T3 cells were grown in Dulbecco's modified Eagle medium (DMEM, Gibco) supplemented with 5 % fetal calf serum. SV40 transformed balb mouse cells (mKSA tumor line) were grown in suspension culture using Eagle minimum essential medium (MEM) without calcium for suspension culture (Gibco, F13) supplemented with 5 % calf serum.

Ad2$^+$SV40 hybrid virus Ad2$^+$ND1 (kindly provided by Dr. A. Lewis, Jr., NIH) was used for infection of TC-7 cells.

Cell fractionation. A detailed fractionation procedure for the preparation of nuclear subfractions from mKSA cells will be described elsewhere (M. Staufenbiel and W. Deppert, manuscript in preparation). Briefly, 5 x 10^8 mKSA cells were labeled for 2 h with ^{35}S-methionine (2 mCi) and lysed at 4°C in hypotonic buffer (RSB),[44] pH 7.2, containing 0.5 % NP40. Under these conditions demembranated nuclear structures are obtained, from which soluble nuclear proteins (nucleoplasm) are extracted. The nuclear structures were then reextracted with the same hypotonic buffer ("wash") and treated with 2 M NaCl and DNase (0.1 mg/ml) for 30 min to remove chromatin. After additional treatment with RNase (0.1 mg/ml) and DNase (0.1 mg/ml) ("wash") a residual nuclear structure, the nuclear matrix, is obtained.[45]

For immunoprecipitation analysis extracted chromatin was diluted 5-fold with RSB and cleared by centrifugation at 100,000 x g for 30 min at 4°C. The nuclear matrix fraction was dissolved in TNM buffer containing 1 % sodium dodecyl sulfate (SDS) and the lysate diluted 10-fold with the same buffer containing 0.5 % NP40.[46] The nucleoplasm fraction and the washes were immunoprecipitated directly.

Antisera. Antisera against denatured T-Ag and mouse SV40 tumor sera described in ref.[35] were used. Antisera against vimentin and cytokeratin were described in ref.[46] Antisera against tubulin and actin were prepared as described by Weber et al.[47] and by Lazarides and Weber.[48] Fluorochrome labeled second antibodies were purchased from Byk Mallinckrodt, Dietzenbach, Federal Republic of Germany.

Immunofluorescence microscopy. Cells were grown on glass coverslips and prepared for cell surface staining as described.[22] Intracellular antigens were visualized as described.[22,23] Cytoskeletal preparation of Ad2$^+$ND1 infected TC-7 cells were prepared according to Fulton et al.[43] and processed for immunofluorescence microscopy as described by these authors.
Cells and cytoskeletal preparations were viewed with a Zeiss photomicroscope III and pictures taken with Neofluar x25 and x40 lenses.

REFERENCES

1. J. Tooze, "Molecular Biology of Tumor viruses, Part 2, DNA Tumor Viruses", Cold Spring Harbor Laboratory, Cold Spring Harbor, N.Y. (1980).
2. R. Weil, Viral "tumor antigens". A novel type of mammalian regulator proteins. Biochim. Biophys. Acta 516:301 (1978).

3. P. Tegtmeyer, K. Rundell and J. K. Collins. Modification of simian virus 40 protein A. J. Virol. 21:647 (1977).
4. K. Rundell, J. K. Collins, P. Tegtmeyer, H. L. Ozer, C. J. Lai and D. Nathans. Identification of simian virus 40 protein A. J. Virol. 21:636 (1977).
5. G. Walter and P. J. Flory, Jr. Phosphorylation of SV40 large T-antigen, in "Cold Spring Harbor Symposia on Quantitative Biology" Vol. XLIV, Cold Spring Harbor Laboratory, Cold Spring Harbor, N.Y. (1980).
6. N. Goldman, M. Brown and G. Khoury. Modification of SV40 T-antigen by poly ADP-ribosylation. Cell 24:567 (1981).
7. D. P. Lane and L. V. Crawford. T-antigen is bound to a host protein in SV40 transformed cells. Nature 278:261 (1979).
8. D. I. H. Linzer and A. J. Levine. Characterization of a 54 K cellular SV40 tumor antigen present in SV40 transformed cells and uninfected embryonal carcinoma cells. Cell 17: 43 (1979).
9. M. Montenarh and R. Henning. Simian virus 40 T-antigen phosphorylation is variable. FEBS Lett. 114:107 (1980).
10. E. Fanning, B. Nowak and C. Burger. Detection and characterization of multiple forms of simian virus 40 large T-antigen. J. Virol. 37:92 (1981).
11. D. S. Greenspan and R. B. Carroll. Complex of simian virus 40 large tumor antigen and 48 000-dalton host tumor antigen. Proc. Natl. Acad. Sci. U.S.A. 78:105 (1981).
12. G. Khoury and E. May. Regulation of early and late simian virus 40 transcription: overproduction of early viral RNA in the absence of a functional T-antigen. J. Virol. 23:167 (1977).
13. D. Rio, A. Robbins, R. Myers and R. Tijan. Regulation of simian virus 40 early transcription in vitro by a purified tumor antigen. Proc. Natl. Acad. Sci. U.S.A. 77:5706 (1980).
14. R. Mc Kay and D. Di Maio. Binding of an SV40 T-antigen-related protein to the DNA of SV40 regulatory mutants. Nature 289:810 (1981).
15. D. Jessel, T. Landau, J. Hudson, T. Lalor, D. Tenen and D. M. Livingston. Identification of regions of the SV40 genome which contain preferred SV40 T-antigen-binding sites. Cell 8:535 (1976).
16. R. Tijan. The binding site on SV40 DNA for a T-antigen related protein. Cell 13:165 (1978).
17. C. Prives, Y. Beck, D. Gidoni, M. Oren and H. Shure. DNA binding and sedimentation properties of SV40 T-antigens synthesized in vivo and in vitro, in "Cold Spring Harbor Symposia on Quantitative Biology XLIV", Cold Spring Harbor Laboratory, Cold Spring Harbor, N.Y. (1980).
18. M. Persico-di Lauro, R. G. Martin and D. M. Livingston. Interaction of simian virus 40 chromatin with simian virus 40 T-antigen. J. Virol. 27:451 (1977).

19. J. Reiser, J. Renart, L. V. Crawford and G. E. Stark. Specific association of simian virus 40 tumor antigen with simian virus 40 chromatin. J. Virol. 33:78 (1980).
20. J. W. Pope and W. P. Rowe. Detection of specific antigens in SV40 transformed cells by immunofluorescence. J. Exp. Med. 120:121 (1964).
21. C. Chang, R. G. Martin, D. M. Livingston, S. W. Luborsky, C. P. Hu and P. T. Mora. Relationship between T-antigen and tumor-specific transplantation antigen in simian virus 40 transformed cells. J. Virol. 29:69 (1979).
22. W. Deppert, K. Hanke and R. Henning. Simian virus 40 T-antigen-related cell surface antigen: serological demonstration on simian virus 40-transformed monolayer cells in situ. J. Virol. 35:505 (1980).
23. W. Deppert. SV40 T-antigen-related surface antigen: correlated expression in cells transformed by an SV40 A-gene mutant. Virology 104:497 (1980).
24. W. Deppert, G. Walter and H. Linke. Simian virus 40 tumor-specific proteins: subcellular distribution and metabolic stability in HeLa cells infected with nondefective adenovirus type 2 - simian virus 40 hybrid viruses. J. Virol. 21:1170 (1977).
25. L. V. Crawford, C. N. Cole, A. E. Smith, E. Paucha, P. Tegtmeyer, K. Rundell and P. Berg. Organization and expression of early genes of simian virus 40. Proc. Natl. Acad. Sci. U.S.A. 75:117 (1978).
26. G. Walter, K. H. Scheidtmann, A. Carbonne, A. P. Laudano and R. F. Doolittle. Antibodies specific for the carboxy- and amino-terminal region of simian virus 40 large tumor antigen. Proc. Natl. Acad. Sci. U.S.A. 77:5197 (1980).
27. K. Mann, T. Hunter, G. Walter and H. Linke. Evidence for simian virus 40 (SV40) coding of SV40 T-antigen and the SV40-specific proteins in HeLa cells infected with nondefective adenovirus type 2 - SV40 hybrid viruses. J. Virol. 24:151 (1977).
28. G. Jay, F. T. Jay, C. Chang, R. M. Friedman and A. S. Levine. Tumor-specific transplantation antigen: use of the $Ad2^{+}ND1$ hybrid virus to identify the protein responsible for simian virus 40 tumor rejection and its genetic origin. Proc. Natl. Acad. Sci. U.S.A. 75:3055 (1978).
29. A. M. Lewis, Jr. and W. P. Rowe. Studies on nondefective adenovirus - simian virus 40 hybrid viruses. I. A newly characterized simian virus 40 antigen induced by the $Ad2^{+}ND1$ virus. J. Virol. 7:189 (1971).
30. S. G. Baum, M. S. Horwitz and J. V. Maizel, Jr. Studies on the mechanism of enhancement of human adenovirus infection in monkey cells by simian virus 40. J. Virol. 10: 211 (1972).

31. D. F. Klessig and L. T. Chow. Incomplete splicing and deficient accumulation of the fiber messenger RNA in monkey cells infected by human adenovirus type 2. J. Mol. Biol. 139:221 (1980).
32. Y. Aloni. Splicing of mRNAs. Progr. Nucl. Ac. Res. 25:1 (1981).
33. R. Lopez-Revilla and G. Walter. Polypeptide specific for cells infected with adenovirus 2-SV40 hybrid Ad2$^+$ND1. Nature (London) New Biol. 244:165 (1973).
34. W. Deppert. Simian virus 40 (SV40)-specific proteins associated with the nuclear matrix isolated from adenovirus type 2 - SV40 hybrid virus-infected HeLa cells carry SV40 U-antigen determinants. J. Virol. 26:165 (1978).
35. W. Deppert and R. Pates. Cell surface location of simian virus 40-specific proteins on HeLa cells infected with adenovirus type 2 - simian virus 40 hybrid viruses Ad2$^+$ND1 and Ad2$^+$ND2. J. Virol. 31:522 (1979).
36. M. Schwyzer. Purification of SV40 T-antigen by immuno-affinity chromatography on staphylococcal protein A-sepharose. Colloq. INSERM 69:63 (1977).
37. P. H. Henry, L. E. Schnipper, R. J. Samaha, C. S. Crumpacker, A. M. Lewis, Jr. and A. S. Levine. Studies of nondefective adenovirus 2 - simian virus 40 hybrids. VI. Characterization of the DNA from five nondefective hybrid viruses. J. Virol. 11:665 (1973).
38. J. F. Morrow, P. Berg, T. J. Kelly, Jr. and A. M. Lewis, Jr. Mapping of simian virus 40 early functions on the viral chromosome. J. Virol. 12:653 (1973).
39. H. Westphal, S. P. Lai, C. Lawrence, T. Hunter and G. Walter. Mosaic adenovirus 2 - SV40 DNA specified by the nondefective hybrid virus Ad2$^+$ND4. J. Mol. Biol. 130:337 (1979).
40. V. B. Reddy, P. K. Ghosh, P. Lebowitz, M. Piatak and S. M. Weissman. Simian virus 40 early mRNA's. I. Genomic localization of 3' and 5' termini and two major splices in mRNA from transformed and lytically infected cells. J. Virol. 30:279 (1979).
41. G. Walter and H. Martin. Simian virus 40-specific proteins in HeLa cells infected with nondefective adenovirus 2 - simian virus 40 hybrid viruses.
42. J. Avruch and D. F. Hoelzl Wallach. Preparation and properties of plasma membrane and endoplasmic reticulum fragments isolated from rat fat cells. Biochim. Biophys. Acta 233:334 (1971).
43. A. B. Fulton, K. M. Wan and S. Penman. The spatial distribution of polyribosomes in 3T3 cells and the associated assembly of proteins into the skeletal framework. Cell 20:849 (1980).
44. S. Penman. RNA metabolism in the HeLa cell nucleus. J. Mol. Biol. 17:117 (1966).

45. R. Berezney and D. S. Coffey. The nuclear protein matrix: isolation, structure and functions. Adv. Enzyme Regul. 14: 63 (1976).
46. M. Staufenbiel and W. Deppert. Intermediate filament systems are collapsed onto the nuclear surface after isolation of nuclei from tissue culture cells. Exp. Cell Res., in press (1982).
47. K. Weber, R. Pollack and T. Bibring. Antibody against tumulin: the specific visualization of cytoplasmic microtubules in tissue culture cells. Proc. Natl. Acad. Sci. U.S.A. 72:459 (1975).
48. E. Lazarides and K. Weber. Actin antibody: the specific visualization of actin filaments in non-muscle cells. Proc. Natl. Acad. Sci. U.S.A. 71:2268 (1974).

INVOLVEMENT OF HERPES SIMPLEX VIRUS IN CERVICAL CARCINOMA

Fred Rapp and Mary K. Howett

Department of Microbiology and Specialized Cancer Research Center
The Pennsylvania State University College of Medicine
Hershey, Pennsylvania, 17033

INTRODUCTION

Cancer of the uterine cervix is the most common cancer of the reproductive system. Richart (1975) has stated that cervical intraepithelial neoplasia begins as a single-cell lesion at the squamocolumnar junction, more commonly on the anterior than on the posterior lip of the cervix, and only infrequently at the lateral angles. The progression in growth of lesions exhibiting irreversible malignant cellular changes advances from dysplasia to carcinoma *in situ* to invasive carcinoma. Massive screening of semi-annual cervical smears via Papanicolaou staining has resulted in increased detection of this cancer prior to invasion, and the cure rate of cancers detected at early stages (dysplasia or carcinoma *in situ*) is close to 100% (Gellman, 1976). However, women in lower socioeconomic groups and in Third World countries lack access to regular gynecologic examination and present with an increased incidence of invasive tumors. In the United States, the American Cancer Society estimates 16,000 new cases of invasive cervical cancer during 1981 and predicts 7,200 deaths (Table 1).

Many studies support the hypothesis that infection plays a major role in a production of cervical neoplasia. Factors associated with increased risk for cervical cancer include early age of first intercourse, early pregnancy, promiscuity, low socioeconomic status and poor personal hygiene (Nahmias and Norrild, 1980). The evidence that a venereally transmitted microorganism is involved can be supported by epidemiologic, serologic and direct studies of cervical cancer patients. The leading contender

Table 1. American Cancer Society Predictions for Genital Cancers-1981[a]

Site	New Cases	Deaths
Genital cancers, total	151,600	46,400
Invasive cancer of uterine cervix	16,000	7,200
Corpus, endometrium (uterus)	38,000	3,100
Ovary	18,000	11,400
Prostate	70,000	22,700
Testes, other male genital	5,200	1,000
Other and unspecified genital, female	4,400	1,000

[a] Reprinted with permission from "Cancer Facts and Figures - 1981" American Cancer Society

among suspected etiologic agents of this tumor is herpes simplex virus type 2 (HSV-2); however, human cytomegalovirus (HCMV) and human papilloma virus have been implicated to a lesser degree (de-Thé, 1977). The evidence linking HSV-2 to cervical cancer is summarized in Table 2. The suspicion that DNA-containing viruses may cause cancer in humans is supported by the fact that many of these viruses can transform mammalian cells in culture to a malignant state. HSV-2 was first reported capable of transforming rodent cells *in vitro* in 1971 (Duff and Rapp, 1971a,b) and there are sporadic reports that this virus can induce cervical carcinoma in mice (Muñoz, 1973; Wentz et al., 1975, 1981). In this review we will discuss in detail the evidence for the involvement of HSV-2 in this important cancer. Additional reviews on this subject have been written by Nahmias and Norrild (1980) and by Rapp and Jenkins (1981).

Table 2. Evidence for the Association of HSV-2 with Cervical Cancer

1. Epidemiology of virus isolation and anti-HSV-2 antibodies
2. Detection of virus proteins in cultured transformed and tumor cells and neoplastic cervical biopsies
3. Detection of HSV-2 DNA in one cervical carcinoma biopsy
4. Detection of HSV-2 messenger RNA in neoplastic biopsies of the cervix
5. Transformation in culture of normal animal cells to malignancy by HSV-2 and HSV-2 DNA fragments
6. Induction of cervical carcinoma in mice by intravaginal administration of inactivated HSV-2

PATHOLOGY OF CERVICAL DYSPLASIA AND NEOPLASIAS

The current classification of cervical dysplasias and neoplasias was endorsed by the International Federation of Gynecology and Obstetrics in 1971 and represents a widely-adopted system for staging (Table 3) these abnormalities (Novak and Woodruff, 1979). However, it should be stated for the purpose of this review that to date, virus-related studies on cervical biopsies have largely confined themselves to assigning specimens to one of only four categories: normal cervix, dysplasia, carcinoma *in situ* or invasive carcinoma. Because it is important for virologists to consider the cell types involved in tumor formation, a brief description of these categories follows.

The surface of the normal cervix can be differentiated into two regions: the exocervix, covered by stratified epithelium and the endocervix, lined with high columnar, mucus-secreting elements. The junction of the two types of epithelia usually occurs at the cervical os, although this may vary in individuals. A thorough review of the histology of normal and abnormal cervices can be found in Novak and Woodruff (1979). The junction between the epithelium and the underlying stroma is usually referred to as the basement membrane or the basal lamina.

Table 3. Stage-grouping in Cervical Neoplasias

Stage 0	Carcinoma *in situ*, intraepithelial carcinoma.
Stage I	Carcinoma strictly confined to the cervix (extension to the corpus should be disregarded).
Ia	The cancer cannot be diagnosed by clinical examination. 1) early stomal invasion 2) occult cancer
Ib	All other cases of Stage I.
Stage II	The carcinoma extends beyond the cervix but not to pelvic wall. The carcinoma involves the vagina but not the lower third.
IIa	No obvious parametrial involvement.
IIb	Obvious parametrial involvement.
Stage III	The carcinoma has extended onto the pelvic wall. On rectal exam there is no tumor-free space between the tumor and the pelvic wall. There is involvement of the lower third of the vagina.
IIIa	No extension onto the pelvic wall.
IIIb	Extension onto the pelvic wall.
Stage IV	The carcinoma has extended beyond the true pelvis or has involved the mucosa of the bladder or rectum.

Carcinoma *in situ* and dysplasia were defined by an International Committee on Histological Definitions in Vienna in 1961 as follows:

> "Carcinoma *in situ*. Only these cases should be classified as carcinoma *in situ* which, in the absence of invasion, show as surface lining, an epithelium in which, throughout its whole thickness, no differentiation takes place. It is recognized that the cells of the uppermost layers may show some flattening. The very rare case of an otherwise characteristic carcinoma *in situ* that shows a greater degree of differentiation belongs to the exception for which no classification can provide.
>
> All other disturbances of differentiation of the squamous epithelial lining of surface and glands are to be classified as dysplasia. They may be characterized as of high or low degree, terms which are preferable to suspicious and nonsuspicious, as the proposed terms describe the histological appearances and do not express an opinion" (Gore and Hertig, 1964).

The distinction described above is based on the concept that many dysplasias are capable of spontaneous regression. However, increasingly rigid diagnostic guidelines for the determination of dysplasia have yielded several studies that suggest that true dysplasia is not reversible but progresses to carcinoma *in situ* and further to invasive carcinoma (Lerch, et al., 1963; Richart, 1963; Richart and Barron, 1969). The diagnosis of dysplasia undoubtedly often includes benign atypias associated with cervicitis and immature squamous metaplasia. Benson and Norris (1978) proposed adoption of a classification that does not distinguish dysplasia from carcinoma *in situ* but lists the lesions as "mild," "moderate" or "severe" intraepithelial neoplasia.

Martzloff (1945) recognized that 94.5% of cervical cancer was of the "epidermoid variety" arising at the squamocolumnar junction, although occasional pure squamous cell carcinomas can appear. In addition, adenoepidermoid tumors and true adenocarcinomas can arise in the gland-like areas of the endocervix. However, most studies undoubtedly deal with the epidermoid variety of tumor. A carcinoma is classified as "invasive" if penetration of the basement membrane is detected in biopsy sections of the lesions. Without defining the nuances of the pathology, it should be stated that penetration of less than 5 mm is considered microinvasion.

EPIDEMIOLOGIC AND SEROLOGIC EVIDENCE FOR THE ROLE OF HERPES SIMPLEX VIRUS IN ONCOLOGIC DISEASES OF THE CERVIX

As early as 1842, Rigoni-Stern postulated that cervical cancer was related to the frequency of sexual activity. He reported that women in cloistered religious orders were at low risk for the development of this neoplasm compared with married women. This observation was confirmed by Towne in 1955 and again by Fraumeni and coworkers in 1969. In two reports (Pereyra, 1961; Keighly, 1968) prostitutes demonstrated a 4- to 6-fold increased risk of developing cervical cancer. In contrast, Kessler (1976) pointed out that Jewish women have a very low incidence of this tumor and postulated that sexual abstinence during menstruation, reduced promiscuity and circumcision of male partners may be contributing factors.

These observations enhanced suspicion that a venereally transmitted organism plays a role in development of this tumor. In 1960, Ayre suggested a virus etiology based on cytological examination of premalignant lesions. Naib and coworkers (1966) detected increased cervical anaplasia in women with cytologically detectable genital herpesvirus infections. Since biopsies from vesicular herpetic lesions of the cervix frequently show intra-epithelial or subepithelial blebs, which may contain classic virus inclusions, and since virus disease often causes degenerative multinucleation and vacuolization of the nuclei (Novak and Woodruff, 1979), it is not surprising that similarities between such biopsies and neoplasia can be observed. Naib and coworkers (1969) extended their previous observations and found active herpetic infections in 23.7% of biopsies from women with reported cervical anaplasia. Venereal transmission of HSV-2 was further substantiated when Rawls and coworkers (1968) isolated HSV-2 from four smegma samples obtained at a venereal disease clinic. Martinez (1969) reported eight cases of cervical cancer in wives of 889 men with penile cancer; this represents an 8-fold increase over the expected incidence. Additionally, Kessler (1977) reported a 2.7% incidence of cervical cancer in wives of men whose previous mates had contracted this tumor. The control group, reported by Martinez, had no incidence of tumor. Unfortunately, many early studies are flawed due to small sample numbers and some of the data, while suggestive, proved somewhat anecdotal.

With the advent of modern virologic and serologic technology scientists began to examine retrospectively the incidence of HSV-2 infection in cervical cancer patients. Extensive reviews of the studies performed have been written by Nahmias and coworkers (1974), Kessler (1974) and Rawls and Adam (1977). Nahmias et al. (1974) stressed the importance of using control patients who are carefully matched for age, promiscuity, socio-economic background and age at first coitus. He reported (1974) a dramatic increase (65% versus 15%) in the incidence of HSV-2 antibodies in patients with carcinoma *in situ*. Aurelian (1975) extended this observation by reporting that 100% of patients with invasive carcinoma were seropositive for HSV-2 compared with 65% of the controls. Similar studies and results were also reported by Rawls and Adam (1977) and Thiry and coworkers (1977).

Nahmias and his coworkers (1970) carried out a prospective follow-up study of 870 women with genital herpes infection and 562 matched controls for 1-6 years. They reported a 2-fold increase in dysplasia and an 8-fold increase in carcinoma *in situ* in the virus-infected women. Based on these observations, they believe that virus infection precedes neoplasia. Continuing prospective studies have led to the hypothesis that females infected with HSV-2 prior to acquisition of herpes simplex virus type 1 (HSV-1) antibody may represent a population at extremely high risk for developing cervical cancer. Additional patients in this category are needed to confirm this observation.

ONCOGENIC PROPERTIES OF HERPES SIMPLEX VIRUS

In Vitro

The seroepidemiologic data presented in the last section are strengthened by experiments demonstrating the ability of HSV-2 to transform cells *in vitro*. Of the five human herpesviruses, four have been shown to be capable of *in vitro* transformation of cells to an oncogenic state. HSV-1 and HSV-2 are normally lytic viruses with broad host ranges, but by limiting the ability of these viruses to replicate, via ultraviolet irradiation, Duff and Rapp (1971a,b, 1973) transformed normal hamster cells into tumorigenic cells. Animals bearing tumors induced by these cells produce HSV-2 neutralizing antibodies although these cells do not synthesize infectious virus. Since that time, cells of various species have been transformed with virus inactivated by photodynamic methods (Rapp et al., 1973) or elevated temperature (Darai and Munk, 1973; Geder et al., 1973). Macnab (1974) has reported extensive studies illustrating that cells infected with temperature-sensitive mutants of HSV-2 and maintained at nonpermissive temperature can also transform. Takahashi and Yamanishi (1974) also transformed cells using this approach.

Initial demonstrations that the transformed cells contain virus nucleic acids involved detection of messenger RNA (mRNA) (Collard et al., 1973) and extensive studies by Frenkel and her coworkers (1976) showed that these cells contain virus-specific DNA in varying amounts, usually less than one virus genome per cell.

Because of the large protein coding capacity of the virus genome and because studies by Frenkel and her coworkers (1976) clearly showed that tumorigenic HSV-2-transformed hamster cells need not contain the entire virus genome, numerous investigators are attempting to isolate the fragment of virus DNA capable of oncogenic transformation (see Table 4). This approach is supported by the fact that some DNA tumor viruses utilize as little as 1.5×10^6 daltons of DNA to effect transformation. Galloway and McDougall (1981) recently reported that a fragment of HSV-2 DNA of 4.8×10^6 daltons and mapping between 0.58 and 0.63 on the virus genome is capable of morphologically transforming rodent cells. These investigators used focus formation in low serum and anchorage independence as criteria of transformation and derived cell lines capable of forming tumors. Dr. Galloway's work extended observations by Reyes and coworkers (1979) that this region of the genome may have morphologic transforming potential. The latter group, however, has not reported tumor formation by their transformants. Southern blotting of the cell DNA from tumorigenic transformants using the putative region of the virus genome as a probe, however, demonstrated that less than one genome copy of this region per cell was present in the transformed cell cultures (Galloway and McDougall, 1981). The exact role of this region in the maintenance or initiation of transformation is not clear. Jariwalla and his coworkers (1980) have reported oncogenic tranformation of hamster embryo cells with a DNA fragment of 16.5×10^6 daltons and mapping on the genome between 0.42 and 0.58. These same investigators recently narrowed their map coordinates for the transforming region to a fragment of DNA of 6.7×10^6 daltons and mapping between 0.41 and 0.52 on the virus genome (Jariwalla, 1981). This region is to the left of the transforming region reported by Galloway and McDougall (1981) and Reyes et al. (1979) and reasons for this disparity are unclear. One possibility advanced is differences in methods of

Table 4. DNA Fragments of Herpes Simplex Viruses That Can Effect Transformation

Virus	Region of Genome	Reference
HSV-1	0.31-0.42	Camacho and Spear, 1978
HSV-1	0.31-0.42	Reyes et al., 1979
HSV-2	0.42-0.58	Jariwalla et al., 1980
HSV-2	0.41-0.52	Jariwalla, 1981
HSV-2	0.58-0.63	Reyes et al., 1979
HSV-2	0.58-0.63	Galloway and McDougall, 1981

selecting transformants (Jariwalla and coworkers passed transfected cell cultures in growth medium and looked for outgrowth of transformed foci). It seems clear, however, that additional studies are necessary to elucidate the minimal transforming region and its gene products.

Certain virus gene products have been found in the transformants. Virus thymidine kinase can be detected in some of the cell lines, although its presence does not seem to be a prerequisite for transformation (Rapp and Westmoreland, 1976). Flannery and coworkers (1977) have reported the presence of a 143,000 molecular weight polypeptide in HSV-2-transformed lines. It is also clear that one or more of the HSV-specific glycoproteins can be detected in the transformed cells. Evidence of glycoproteins D (Reed et al., 1975), A and C has been presented and it is known that sera from animals bearing tumors induced by HSV-transformed cells can neutralize HSV, indicating that the sera contain antibody to virion glycoproteins (Duff and Rapp, 1971a,b, 1973). There are no clear data, however, identifying the protein products of the putative transforming regions of HSV DNA.

In line with the *in vitro* transformation experiments described, limited attempts have been made to induce tumors *in vivo* by infection with HSV-2. The lack of experimental animal models is due mainly to the ability of the virus to cause disseminated disease unless very low doses are administered (Rapp and Falk, 1964; Trentin et al., 1969). However, Nahmias and his coworkers (1975) were able to induce low-level tumor development (less than 4%) in baby hamsters inoculated with less than 10^3 TCD_{50} of live or inactivated HSV-2. The resulting tumors failed to demonstrate HSV-2 markers. More interestingly, investigators in several laboratories have now reported the effects of intravaginal administration of HSV-2. Muñoz (1973) reported induction of a few cervical cancers in mice preimmunized with inactivated HSV-2 and subsequently exposed vaginally to HSV-2 and hormones. This study has some problems, however, because the number of animals used was small and because a few controls also developed cancer. More thorough studies (Wentz et al., 1975) examined intravaginal infection of mice with formalin-inactivated HSV-2. After insertion for extended time periods (5 times per week for 88 weeks) of tampons soaked with inactivated virus they noted neoplastic disease of the cervix. Seventy-seven percent of the animals developed preinvasive lesions as early as 15 weeks after the first administration of virus and a large number of animals developed carcinoma of the cervix and adenocarcinoma of the uterine horn.

This team of investigators (Wentz et al., 1981) has also published studies using HSV-1 and HSV-2 inactivated with formalin or ultraviolet light. After insertion of virus-soaked tampons five times per week for 20-90 weeks, premalignant and malignant cervical lesions were detected in 78-91% of the animals and dysplasia and invasive carcinoma were detected in 18-66% and 24-60% of the animals, respectively. The controls remained normal. Furthermore, the frequency of invasive cancer was twice as high after exposure to ultraviolet-inactivated virus. In addition to rodent studies, one group of investigators detected an increased frequency of dysplasia in Cebus monkeys exposed to virus (Palmer et al., 1976).

It is clear that HSV-2 possesses oncogenic potential. However, the *in vivo* data at this time, though suggestive, require independent confirmation.

HERPES SIMPLEX VIRUS MARKERS IN CERVICAL CANCER

Antigens

Aurelian and her coworkers (1973a) reported the detection of an HSV-2 antigen, AG-4, in 90% of cervical tumor biopsies compared with 10% of controls. In addition, 91% of patients with invasive carcinoma and 68% of patients with carcinoma *in situ* had positive antibody titers against AG-4 compared with 5-9% of control patients (Aurelian et al., 1975). Kawana and his coworkers (1978) also examined the frequency of AG-4 antibodies in patients with uterine cervical cancer and found a 47% incidence of antibody compared with 14% in controls. Aurelian's group also claims that antibody to AG-4 is prognostic in cervical cancer patients since in one study AG-4 antibody became undetectable in women successfully treated for cervical carcinoma (Aurelian et al., 1973b).

Levels of AG-4 antibody have been shown to correlate positively with antibody against another HSV tumor-associated antigen (HSV-TAA) (Notter and Docherty, 1976). Similar results were also reported by Hollinshead and coworkers in 1976. Melnick and his coworkers (1976) reported an increased incidence of reactivity against an early nonstructural protein (VP134) found in HSV-2-infected cells; 93% of cervical cancer patients demonstrated antibody to VP134 compared with 30% of breast cancer patients and 40% of controls. A similar antigen has also recently been detected in material from carcinoma *in situ* of the vulva (Kaufman et al., 1981).

More recently, Gilman and coworkers (1981) demonstrated antibodies to two HSV-2-associated proteins with molecular weights of 38,000 and 118,000 in cervical cancer patients. The 38,000 protein may be coded by a region of HSV-2 DNA capable of *in vitro* transformation. Cabral and coworkers (1980), by use of immunoperoxidase staining, found an HSV-2-specific polypeptide (VP143) in a significant number of dysplasias (31%) carcinomas *in situ* (29%) and invasive squamous cell carcinomas (41%) but not in normal biopsies.

Nucleic Acids

The direct demonstration of HSV-2 nucleic acids in cervical biopsies has been difficult for valid technological reasons and consequently the issue remains unresolved despite positive reports by a few investigators. Frenkel and her coworkers (1972) reported the presence of HSV-2 DNA in one cervical carcinoma by DNA-DNA reassociation between labeled HSV-2 DNA and unlabeled biopsy DNA. The results indicated that approximately 39% of the genome was present at a frequency of 1 to 3.5 copies per cell genome. Hybridization kinetics between HSV-2 DNA and unlabeled biopsy RNA indicated that as little as 5% of the virus genome might be transcribed in the tumor cells (Frenkel et al., 1972).

Because attempts to confirm the DNA-DNA reassociation results of Frenkel and others failed to yield useful data, investigators turned to *in situ* hybridization techniques in an effort to demonstrate that labeled HSV-2 DNA probes could hybridize to HSV-2-specific mRNA in the cytoplasm of transformed and/or tumor cells. The rationale of this approach was based on the supposition that DNA sequences in the tumors may be amplified as multiple copies in the form of mRNA. Using this approach Jones et al. (1978) detected HSV-2 mRNA in 5 of 8 cervical cancer biopsies; 3 of 3 controls were negative. Since that time, data accumulated from 160 biopsy specimens indicate that 42% of biopsies from dysplasia or carcinoma *in situ*, 30% of biopsies from invasive cancer but only 13% of biopsies from normal cervix will hybridize to a labeled virus DNA probe representative of the whole HSV-2

Table 5. *In situ* Hybridization of HSV-2 DNA Probes to Cervical Biopsy Specimens

Specimen	Number Positive	(%)
Cin I-III[b]	28/67	42
Invasive Cancer	4/13	30
Normal Cervix	10/80	13

[a]Summarized from data of McDougall et al., 1981.
[b]Cin I, mild to moderate dysplasia; Cin II, moderate and severe dysplasia; Cin III, severe dysplasia and carcinoma *in situ*.

genome (McDougall et al., 1981) (Table 5). When these same investigators used cloned DNA fragments to probe cervical samples, DNA representing map positions 0.07-0.32 and 0.84-0.86 (a repeated sequence also was found at 0.96-0.98) hybridized most frequently, whereas sequences representing one putative transforming region (BglII N; 0.58-0.62) hybridized infrequently. The meaning of these results is unclear and more work is necessary before we can define with certainty exact regions of the genome that may be conserved in the transformants.

ETIOLOGIC ROLE OF OTHER MICROORGANISMS IN CERVICAL CANCER

As previously mentioned, HCMV and papilloma viruses have also been implicated in the etiology of cervical cancers. Although not the subject of this review, evidence for the involvement of these two viruses should be mentioned briefly.

HCMV can cause persistent infection of the genitourinary tract (for review, see Geder, 1980) and has been isolated from cervical smears and from semen (Lang and Kummer, 1972; Montegomery et al., 1972; Amstey, 1975). This virus, like the other herpesviruses, can establish latent infections even in the presence of antibody (Lang et al., 1974) and frequently can be isolated from female patients at venereal disease clinics (Willmott, 1975). We have already noted that HCMV can transform cells *in vitro* (Albrecht and Rapp, 1973; Geder et al., 1976); however, evidence linking HCMV to cervical cancer is sparse. Pacsa and coworkers (1975) found that 61% of patients with cervical atypia were positive for anti-HCMV antibodies compared with 33% of matched controls. Melnick and his collaborators (1978) found infectious HCMV in two invasive carcinoma biopsies.

The evidence for papilloma virus involvement in cervical cancer is also only suggestive. The papilloma viruses are known to induce three different benign epithelial tumors: common warts, juvenile warts and genital warts. At least eight different human papilloma viruses have been identified (for review see Orth et al., 1977; zur Hausen, 1977), but the etiology of warts is confusing because there have been instances when more than one virus has been isolated from a single lesion. It is known, however, that genital warts (condylomata acuminata) contain papilloma viruses that are distinct from common wart virus (zur Hausen, 1976). Genital warts are predominantly found in promiscuous individuals and are undoubtedly transmitted sexually. There have been rare reports that condylomata acuminata can progress to carcinomas of the anus (Siegel, 1962) penis (Dawson et al., 1965) and vulva (Sims and Garb, 1951) and zur Hausen (1977) has postulated that this virus may cause up to 5% of carcinomas of the vulva. Woodruff and Peterson (1958) noted cervical atypia and occasional carcinomas *in situ* in patients with condylomata acuminata of the cervix, but to date there are no reports that associate papilloma viruses with invasive carcinomas.

EVIDENCE REQUIRED FOR PROOF OF HERPES SIMPLEX VIRUS INVOLVEMENT IN CERVICAL CANCER

Even considering the data presented in this review, conclusive proof that HSV-2 is the etiologic agent of cervical carcinoma remains elusive. Two other possibilities must also be considered: (1) HSV-2 may require other cofactors for tumorigenesis in the human; and (2) separate etiologic agents not related to the virus may also produce identical pathology in the uterine cervix. Neither of these alternatives, however, diminishes the importance of HSV-2 in cancer of the cervix if it can be shown that this virus plays a causative role in some cases. It should be remembered that investigation of HSV-2 will also lead to important findings regarding the natural history of genital HSV-2 infection.

In order to more clearly ascertain the precise role of HSV-2 in cervical cancer, continued studies are required to show beyond doubt which virus antigens and nucleic acids are present in cervical biopsies. At the same time the question of whether neoplastic cervical cells are predisposed to virus infection needs to be answered, and it is imperative to determine whether cervical biopsies that are positive for virus nucleic acids contain complete or partial genomes and/or virus. While these studies are ongoing, researchers studying the molecular biology of *in vitro* transformation should define the transforming region and its gene products so that comparisons can be made between animal and human tumorigenesis.

Studies demonstrating tumor formation *in vivo* by herpesviruses have not been nearly as extensive as they should be. The initial results demonstrating induction of cervical neoplasia in mice are promising, but this work needs independent confirmation. If cervical carcinogenesis in the mouse can be accomplished by vaginal administration of HSV-2, this will be an important animal model and similar experiments should be attempted in higher animals, such as nonhuman primates.

Finally, conclusive proof that HSV-2 causes a significant portion of cervical carcinomas will be obtained if an effective and safe vaccine that prevents HSV-2 infection is developed and tumor incidence is subsequently reduced. Kessler (1974) pointed out that this would involve a long-range prospective study of women vaccinated against HSV-2 and monitored for the incidence of genital herpetic infection and subsequent incidence of cervical neoplasia. The topic of the development and efficacy of herpesvirus vaccines is a subject apart from the scope of this chapter, but suffice it to say that many years will probably elapse before such direct proof could possibly become available. In the meantime, virologists must concentrate their efforts on strengthening what now appears to be a convincing indirect association of this virus with cancer in the human female.

ACKNOWLEDGMENTS

This work was supported by grants CA 27503, CA 25305 and CA 18450 from the National Cancer Institute.

REFERENCES

Albrecht, T., and Rapp, F., 1973, Malignant transformation of hamster embryo fibroblasts following exposure to ultraviolet-irradiated human cytomegalovirus, Virology, 55:53.

Amstey, M.S., 1975, Genital herpesvirus infection, Clin. Obstet. Gynecol. 18:89.

Aurelian, L., Cornish, J.D., and Smith, M.F., 1975, Herpesvirus type 2-induced tumor specific antigen (AG-4) and specific antibodies in patients with cervical cancer and controls, IARC Sci. Publ. No. 11:79.

Aurelian, L., Davis, H.J., and Julian, C.G., 1973a, Herpesvirus type 2 induced tumor specific antigen in cervical carcinoma, Am. J. Epidemiol., 98:1.

Aurelian, L., Schumann, B., Marcus, R.L., and Davis, H.J., 1973b, Antibodies to HSV-2 induced tumor specific antigens in serums from patients with cervical carcinoma, Science, 181:161.

Ayre, J.E., 1960, Role of the halo cell in cervical cancerigenesis. A virus manifestation in premalignancy?, Obstet. Gynecol., 15:481.

Benson, W.L. and Norris, H.J., 1978, Current problems in gynecologic pathology, in: "Gynecologic Oncology" L. McGowen, ed., Appleton-Century Crofts, New York.

Cabral, G.A., Courtney, R.J., Schaffer, P.A., and Marciano-Cabral, F., 1980, Ultrastructural characterization of an early, nonstructural polypeptide of herpes simplex virus type 1, J. Virol., 33:1192.

Camacho, A. and Spear, P.G., 1978, Transformation of hamster embryo fibroblasts by a specific fragment of the herpes simplex virus genome, Cell, 15:993.

Cancer Facts and Figures, 1980, American Cancer Society.

Collard, W., Thornton, H., and Green, M., 1973, Cells transformed by HSV-2 transcribe virus specific RNA sequences shared by herpes virus types 1 and 2, Nature 243:264.

Darai, G., and Munk, K., 1973, Human embryonic lung cells abortively infected with herpesvirus hominis type 2 show some properties of cell transformation, Nature 241:268.

Dawson, D.F., Duckworth, J.K., Bernhardt, H., and Young, J.M., 1965, Giant condyloma and verrucous carcinoma of the genital area, Arch. Pathol., 79:225.

de-The, G., 1977, Viruses as causes of some human tumors? Results and prospectives of the epidemiologic approach, in: "Origins of Human Cancer," H.H Hiatt, J.D. Watson, and J.A. Winsten, eds. Cold Spring Harbor Laboratory, New York.

Duff, R., and Rapp, F., 1971a, Oncogenic transformation of hamster cells after exposure to herpes simplex virus type 2, Nature, 233:48.

Duff, R., and Rapp, F., 1971b, Properties of hamster embryo fibroblasts transformed in vitro after exposure to ultra-violet-irradiated herpes simplex virus type 2, J. Virol., 8:469.

Duff, R., and Rapp, F., 1973, Oncogenic transformation of hamster embryo cells after exposure to inactivated herpes simplex virus type 1, J. Virol., 12:209.

Flannery, V.L., Courtney, R.J., and Schaffer, P.A., 1977, Expression of an early, nonstructural antigen of herpes simplex virus in cells transformed in vitro by herpes simplex virus, J. Virol., 21:284.

Fraumeni, J.F., Lloyd, J.M., Smith, E.M., and Wagoner, J.K., 1969, Cancer mortality among nuns: role of marital status in etiology of neoplastic disease in women, J. Natl. Cancer Inst., 42:455.

Frenkel, M., Locker, H., Cox, B., Roizman, B., and Rapp, F., 1976, Herpes simplex virus DNA in transformed cells: sequence complexity in five hamster cell lines and one derived hamster tumor, J. Virol., 18:885.

Frenkel, M., Roizman, B., Cassai, E., and Nahmias, A., 1972, A DNA fragment of herpes simplex 2 and its transcription in human cervical cancer tissue, Proc. Natl. Acad. Sci. USA, 69:3784.

Galloway, D.A., and McDougall, J.K., 1981, Transformation of rodent cells by a cloned DNA fragment of herpes simplex virus type 2, J. Virol., 38:749.

Geder, L., 1980, Oncogenic properties of human cytomegalovirus, in: "Oncogenic Herpesviruses," F. Rapp, ed., CRC Press, Boca Raton, Florida.

Geder, L., Lausch, R., O'Neill, F., and Rapp, F., 1976, Oncogenic transformation of human embryo lung cells by human cytomegalovirus, Science, 192:1134.

Geder, L., Vaczi, L., and Boldough, I., 1973, Development of cell lines after exposure of chicken embryonic fibroblasts to herpes simplex virus type 2 at supraoptimal temperature, Acta Microbiol. Acad. Sci. Hung., 20:119.

Gellman, D.D., 1976, Cervical cancer screening programs. Can. Med. Assoc. J., 114:1003.

Gilman, S.C., Docherty, J.J., Clark, A., and Rawls, W.E., 1981, The reaction patterns of HSV type 1 and type 2 proteins with the sera of patients with uterine cervical carcinoma and matched controls, Cancer Res., in press.

Gore, H., and Hertig, A.T., 1964, Definitions, in: "Dysplasia Carcinoma In Situ and Microinvasive Carcinoma of the Cervix Uteri," L.A. Gray, ed., Charles C. Thomas, Springfield.

Hollinshead, A.C., Chretien, P.B., Lee, O., Tarpley, J.L., Kerney, S.E., Silverman, N.A., and Alexander, J.C., 1976, In vivo and in vitro measurements of the relationship of human squamous carcinomas to herpes simplex virus tumor-associated antigens, Cancer Res., 36:821.

Jariwalla, R.J., Aurelian, L., and Ts'o, P.O.P., 1980, Tumorigenic transformation induced by a specific fragment of DNA from herpes simplex virus type 2, Proc. Natl. Acad. Sci. USA, 77:2279.

Jariwalla, R.J., 1981, Neoplastic transforming activity of defined HSV-2 DNA fragments representing 6.5-13% of the Viral Genome, in: "Proceedings of the International Workshop on Herpesviruses," July 29-31, Bologna, Italy.

Jones, K.W., Fenoglio, C.M., Shevchuk-Chaban, M., Maitland, W.J., and McDougall, J.K., 1978, Detection of herpes simplex virus type 2 mRNA in human cervical biopsies by in situ cytological hybridization, IARC Sci. Publ. No. 24:917.

Kaufman, R.H., Dreesman, G.R., Burek, J., Korhonen, M.O., Matson, D.O., Melnick, J.L., Powell, K.L., Purifoy, D.J.M., Courtney, R.J., and Adam, E., 1981, Herpesvirus-induced antigens in squamous-cell carcinoma in situ of the vulva, N. Engl. J. Med., 305:483.

Kawana, T., Sakamoto, S., Kasamatsu, T., and Aurelian, L., 1978, Frequency of anti-AG-4 antibodies in patients with uterine cervical cancer and controls, Gann, 69:389.

Keighley, E., 1968, Carcinoma of the cervix among prostitutes in a women's prison, Br. J. Vener. Dis., 44:254.

Kessler, I.I., 1974, Perspectives on the epidemiology of cervical cancer with special reference to the herpesvirus hypothesis, Cancer Res., 34:1091.

Kessler, I., 1976, Human cervical cancer as a venereal disease, Cancer Res., 36:783.

Kessler, I.I., 1977, Venereal factors in human cervical cancer, Cancer, 39:1912.

Lang, P.J., and Kummer, J.F., 1972, Demonstration of cytomegalovirus in semen, N. Engl. J. Med., 287:756.

Lang, D.J., Kummer, J.F., and Hartley, D.P., 1974, Cytomegalovirus in semen: persistence and demonstration in extracellular fluids, N. Engl. J. Med., 291:121.

Lerch, V., Okagaki, T., Austin, J.H., Kevorkian, A.Y., Youge, P.A., 1963, Cytologic findings in progression of anaplasia (dysplasia) to carcinoma in situ: a progress report. Acta. Cytol., 7:183.

Macnab, J.C.M., 1974, Transformation of rat embryo cells by temperature-sensitive mutants of herpes simplex virus, J. Gen. Virol., 24:143.

Martinez, I., 1969, Relationship of squamous cell carcinoma of the cervix uteri to squamous cell carcinoma of the penis among Puerto Rican women married to men with penile cancer, Cancer, 24:777.

Martzloff, K.H., 1945, Cancer of the cervix. Some fundamental considerations, West. J. Surg., 53:255.

McDougall, J.K., Crum, C.P., Levine, R.U., Richart, R.M., and Fenoglio, C.M., 1981, Expression of limited regions of the HSV-2 genome in cervical neoplasia, in: "Proceedings of the International Workshop on Herpesviruses," July 27-31, Bologna, Italy.

Melnick, J.L., Courtney, R.J., Powell, K.L., Schaffer, P.A., Benyesh-Melnick, M., Dreesman, G.R., Anzai, T., and Adam, E., 1976, Studies on herpes simplex virus and cancer, Cancer Res., 36:845.

Melnick, J.L., Lewis, R., Wimberly, I., Kaufman, R.H., and Adam, E., 1978, Association of cytomegalovirus (CMV) infection with cervical cancer: isolation of CMV from cell cultures derived from cervical biopsy, Intervirology, 10:115.

Montgomery, R., Youngblood, L., and Medearis, D.N., Jr., 1972, Recovery of cytomegalovirus from the cervix in pregnancy, Pediatrics, 49:524.

Muñoz, M., 1973, Effect of herpes simplex virus type 2 and hormonal imbalance on the uterine cervix of the mouse, Cancer Res., 33:1504.

Nahmias, A.J., Del Buono, I., and Ibrahim, I., 1975, Antigenic relationship between herpes simplex viruses, human cervical cancer and HSV-associated hamster tumours, IARC Sci. Publ. No. 11:309.

Nahmias, A.J., Josey, W.E., Naib, Z.M., Luce, C.F., and Guest, B., 1970. Antibodies to herpesvirus hominis types 1 and 2. II. Women with cervical cancer, Am. J. Epidemiol., 91:547.

Nahmias, H.J., Naib, Z.M., and Josey, W., 1974, Epidemiological studies relating genital herpes to cervical cancer, Cancer Res., 34:1111.

Nahmias, A.J., and Norrild, B., 1980, Oncogenic potential of herpes simplex viruses and their association with cervical neoplasia, in: "Oncogenic Herpesviruses," F. Rapp, ed., CRC Press, Boca Raton, Florida.

Naib, Z.M., Nahmias, A.J., and Josey, W.E., 1966, Cytology and histopathology of cervical herpes simplex infection, Cancer, 19:1026.

Naib, Z.M., Nahmias, A.J., Josey, W.E., and Kramer, J.H., 1969, Genital herpetic infection: association with cervical dysplasia and carcinoma, Cancer, 23:940.

Notter, M.F.D., and Docherty, J.J., 1976, Comparative diagnostic aspects of herpes simplex virus tumor-associated antigens, J. Natl. Cancer Inst., 57:483.

Novak, E.R., and Woodruff, J.D., 1979, "Gynecologic and Obstetric Pathology," W.B. Saunders Co., Philadelphia.

Orth, G., Breitburd, F., Favre, M., and Croissant, O., 1977, Papillomaviruses: possible role in human cancer, in: "Origins of Human Cancer," H.H. Hiatt, J.D. Watson and J.A. Winston eds., Cold Spring Harbor Laboratory, New York.

Palmer, A.E., London, W.T., Nahmias, A.J., Naib, Z.M., Tunca, J., Fuccillo, D.A., Ellenberg, J.H. and Sever, J.L., 1976, A preliminary report on investigation of oncogenic potential of herpes simplex virus type 2 in Cebus monkeys, Cancer Res., 36:807.

Pasca, A.S., Kummerländer, L., Pejtsik, B., and Pali, K., 1975, Herpesvirus antibodies and antigens in patients with cervical anaplasia and in controls, J. Natl. Cancer Inst., 55:775.

Pereyra, A.J., 1961, The relationship of sexual activity to cervical cancer. Cancer of the cervix in a prison population, Obstet. Gynecol., 17:154.

Rapp, F., and Falk, L.A., 1964, Study of virulence and tumorigenicity of variants of herpes simplex virus, Proc. Soc. Exp. Biol. Med., 110:361.

Rapp, F., and Jenkins, F., 1981, Genital cancer and viruses, Gynecol. Oncol., in press.

Rapp, F., Li, J.L.H., and Jerkofsky, M., 1973, Transformation of mammalian cells by DNA-containing viruses following photodynamic inactivation, Virology, 55:339.

Rapp, F., and Westmoreland, D., 1976, Cell transformation by DNA-containing viruses, Biochim. Biophys. Acta, 458:167.

Rawls, W.E. and Adam, E., 1977, Herpes Simplex viruses and human malignancies, in: "Origins of Human Cancer," H.H. Hiatt, J.D. Watson, and J.A. Winston, eds., Cold Spring Harbor Laboratory, New York.

Rawls, W.E., Laurel, D., Melnick, J.L., Glicksman, J.M., and Kaufman, R.H., 1968, A search for viruses in smegma, premalignant and early malignant cervical tissues. The isolation of herpesviruses with distinct antigenic propterties, Am. J. Epidemiol., 87:647.

Reed, C.L., Cohen, G.H., and Rapp, F., 1975, Detection of a virus specific antigen on the surface of herpes simplex virus transformed cells, Virology, 15:668.

Reyes, G.R., Lafemina, R., Hayward, S.D. and Hayward, G.S., 1979, Morphological transformation by DNA fragments of human herpes viruses: evidence for two distinct tranforming regions in HSV-1 and HSV-2 and lack of correlation with biochemical transfer of the thymidine kinase gene. Cold Spring Harbor Symp. Quant. Biol., 44:629.

Richart, R.M., 1963, Cervical neoplasmia in pregnancy. A series of pregnant and postpartum patients followed without biopsy or therapy. Am. J. Obstet. Gynecol., 87:474.

Richart, R.M., 1975, Cervical intraepithelial neoplasia, in: "Genital and Mammary Pathology, Decennial 1966-1975," S.C. Sommers, ed., Appleton, New York.

Richart, R.M., and Barron, B.A., 1969, A follow-up of patients with cervical dysplasia, Am. J. Obstet. Gynecol., 105:386.

Rigoni-Stern, D., 1842, Fatti statistia relativi alle malatie cancerose, Giorn Dervire Progr. Pathol. Terap., 2:507.

Siegel, A., 1962, Malignant transformation of condyloma acuminatum, Am. J. Surg., 103:613.

Sims, C.F., and Garb, J., 1951, Giant condylomata acuminata of the penis associated with metastatic carcinoma of the right inguinal lymph node, Arch. Dermatol. Syphil., 68:383.

Takahashi, M., and Yamanishi, K., 1974, Transformation of hamster embryo and human embryo cells by temperature sensitive mutants of herpes simplex virus type 2, Virology, 61:306.

Thiry, L., Sprecher-Goldberger, S., Hannecart-Pokorni, E., Gould, I., and Bossens, M., 1977, Specific non-immunoglobulin G antibodies and cell mediated responses to herpes simplex virus antigens in women with cervical carcinoma, Cancer Res., 37:1301.

Towne, J.E., 1955, Carcinoma of the cervix in nulliparous and celibate women, Am. J. Obstet. Gynecol., 69:606.

Trentin, J.J., Van Hoosier, G.L., Boren, D., Ferguson, P.G., and Kitamuru, T., 1969, Oncogenic evaluation in hamsters of human picorna-paramyxo-and herpesviruses, Proc. Soc. Exp. Biol. Med. 132:912.

Wentz, W.B., Reagan, J.W., and Heggie, A.D., 1975, Cervical carcinogenesis with herpes simplex virus type 2, Obstet. Gynecol., 46:117.

Wentz, W.B., Reagan, J.W., Heggie, A.D., Fu, Y., and Anthony, D.D., 1981, Induction of uterine cancer with inactivated herpes simplex virus, types 1 and 2, Cancer, in press.

Willmott, F.E., 1975, Cytomegalovirus in female patients attending a VD clinic, Br. J. Vener. Dis., 51:278.

Woodruff, J.D., and Peterson, W.F., 1958, Condylomata acuminata of the cervix, Am. J. Obstet. Gynecol., 75:1359.

zur Hausen, H., 1976, Condylomata acuminata and human genital cancer, Cancer Res., 36:794.

zur Hausen, H., 1977, Human papillomaviruses and their possible role in squamous cell carcinomas, Curr. Top. Microbial. Immun., 78:1.

THE ROLE OF EPSTEIN-BARR VIRUS IN BURKITT'S LYMPHOMA

H. zur Hausen

Institut für Virologie, Zentrum für Hygiene
Universität Freiburg, Hermann-Herder-Str. 11
7800 Freiburg, West Germany

The pathogenesis of Epstein-Barr virus (EBV) infections has been analyzed extensively in recent years. In particular, the virus latency of this agent can be approached experimentally: it is the only one of the human herpes viruses that permits the establishment of latent infections under tissue culture conditions, thus facilitating studies of the mode of regulation of viral genome persistence.

The virus has been identified as the causative agent of infectious mononucleosis (Henle et al., 1968). The age of primary infection depends on socioeconomic conditions (Henle and Henle, 1972). In our region clinically overt disease occurs most frequently in the 6-18 year age group. The disease apparently results from transformation of B-lymphocytes and the subsequent interaction of cell-mediated immune functions with these cells (zur Hausen, 1976). Cells of the B lineage appear to be the primary target of EBV infection. Their rapid and efficient transformation into B-lymphoblasts was an early hint for a possible role of this virus as an oncogenic agent (Henle et al., 1967; Pope et al., 1968).

EBV particles were first demonstrated by Epstein et al. (1964), in all lymphoblastoid lines derived from Burkitt's lymphoma. This, in combination with seroepidemiological studies which demonstrated significantly elevated EBV antibody titers in patients with Burkitt's lymphoma when compared to appropriately matched controls (Henle et al., 1971), raised the suspicion that EBV may play a role in the etiology of this peculiar tumor.

The subsequent years have added substantially more information supporting the possible involvement of EBV in Burkitt's lymphoma.

This may be summarized as follows:

1. Burkitt's lymphoma patients reveal high antibody titers against structural and nonstructural antigens of EBV, exceeding significantly those of controls (Henle et al., 1971); some of them appear to have prognostic value (Henle et al., 1973).
2. Burkitt's lymphoma cells contain EBV DNA in multiple copies, (zur Hausen et al., 1970), (Nonoyama and Pagano, 1971), (usually between 20 and 40) which persists predominantly in an episomal state, being not integrated into host cell DNA (Lindahl et al., 1976).
3. Persisting EBV DNA is partially expressed resulting in EBNA synthesis (Reedman and Klein, 1973), which represents a useful marker for the presence of the EBV genome.
4. The virus transforms efficiently cells of the B lineage; the efficiency of this process is remarkably high.
5. EBV induces lymphoproliferative disease resembling malignant lymphoma in marmosets, thus revealing its oncogenic properties even under in vivo conditions (Shope et al., 1973; Epstein et al., 1973).
6. A prospective seroepidemiological study performed in Uganda (de Thé et al., 1978) clearly underlined an elevated risk for the development of Burkitt's lymphoma in children suffering from severe EBV infections, with the development of high antibody titers if infection occurs early in life.

Taken together, these results clearly demonstrate the EBV possesses oncogenic properties and strongly indicate a role of this virus in Burkitt's lymphoma.

There are, however, some additional data available which unfortunately do not support the assumption that EBV causes all cases of Burkitt's lymphoma. Here the reasoning goes as follows:

1. In the African tumor belt approximately 5% of histologically typical Burkitt's lymphomas are devoid of detectable EBV genomes and do not express the EBNA antigen. The rarely occurring European and U.S. cases of Burkitt's lymphoma reveal only in 15-20% evidence for EBV genome persistence. Approximately 80% of these tumors are negative (review zur Hausen, 1980a).
2. The reasoning that negative and positive cases represent distinct etiological entities can be challenged on the basis of similar chromosomal changes, commonly affecting chromosomes 8 and 14 showing reciprocal translocations and occasionally also other chromosomes (Manolov and Manolova, 1972).
3. The polyclonal lymphoblastoid lines obtained from cells transformed in vitro by EBV or from patients with infectious mononucleosis or other conditions clearly differ in their biological behavior from the monoclonal malignant Burkitt's lymphoma cells. This is evidenced by certain morphological differences, by high cloning efficiencies of the latter under soft agar conditions,

and by their ability of form tumors upon subcutaneous inoculation into nude mice (Diehl et al., 1977; Nilsson et al., 1977). These studies stress that Burkitt's lymphoma cells are different from merely EBV-transformed lymphoblasts.

4. Severe immunosuppression rarely results in the development of Burkitt's lymphoma. EBV infections under those conditions may emerge in a polyclonal lymphoproliferative syndrome which may be fatal for the patient (Purtillo et al., 1977).

In analyzing these pros and cons we are left with a dilemma: there is clearly reason to claim that EBV must play some role in the induction of EBV-positive Burkitt's lymphomas. But there is also justification for the statement that EBV-negative cases of Burkitt's lymphoma have no relationship to EBV infections. This despite a similar histology, the same specific chromosomal changes, and similar biological behavior.

The available data appear to be most easily reconciled by the assumption that transformation of cells by EBV puts this population at higher risk for a yet undefined subsequent event resulting in the outgrowth of a malignant cell clone. We call this effect "target cell conditioning," implying that target cells conditioned by EBV have a higher chance of malignant conversion than their non-EBV-DNA-containing counterparts (zur Hausen, 1980b).

If this assumption is correct the postulated second event could be a genetic as well as an epigenetic change, and some experiments have been performed to test this hypothesis: experiments performed by Dr. Böcker in my laboratory to test whether ultimate carcinogens like N-acetylaminofluorene or 4-nitroquinoline-1-oxide would lead to mutations in EBV-transformed lymphoblastoid lines affecting their cloning efficiency as well as their oncogenicity after subcutaneous implantation into nude mice failed to provide positive data. An alternative possibility that the conditioned cells would become susceptible to specific - possibly supertransforming - virus infections proved to be somewhat more successful. It has at least been possible to show that a virus exists, belonging to a group of oncogenic agents which preferentially or even exclusively infect transformed B-lymphoblasts. This agent, originally obtained from a lymphoblastoid line derived from an African green monkey, is a typical papovavirus, belonging into the polyoma subgroup (zur Hausen and Gissmann, 1979). Serological tests as well as analysis of its DNA revealed that it is distinct from all other characterized members of this group. Its in vitro host range is most interesting: it infects only lymphoblastoid cells of B-cell origin of its original host and cannot be propagated in monolayer cells. In addition, however, some human cell lines of B-cell origin proved to be susceptible to this infection (Brade et al., 1981). All 14 lines tested originating from Burkitt's lymphoma or other B-cell malignant lymphomas supported replication of this virus (Brade and zur Hausen, unpublished data). Susceptibility did not depend on the presence or absence of EBV DNA within these cells, since several EBV-negative B-lymphoma lines were highly susceptible. In marked contrast, only 2 out of 18 lymphoblastoid lines obtained either by in vitro transformation with EBV or representing spontaneously established lines from healthy donors revealed a low degree of susceptibility.

This virus, which we tentatively designated as B-lymphotropic papovavirus, or LPV, thus shows a remarkable lymphoma-specificity and discriminates to a substantial extent between malignant B-lymphoma cells and nonmalignantly EBV-transformed B-lymphoblasts. The molecular basis for this discrimination is presently under investigation. Similar agents could be suitable candidates for cells conditioned by EBV or other factors. In this respect it is of interest to note that approximately 30% of human sera above 30 years of age contain antibodies to the major capsid proteins of this virus, some at quite high titers, which can be demonstrated by immunofluorescence or immunoprecipitation, and also by their capacity to neutralize LPV-infectivity (Brade et al., 1980).

We can therefore assume that a similar agent, closely related to AGM-LPV, infects the human population. Although serological testing of human tumor sera, including those from Burkitt's lymphoma patients, did not provide a clue to its involvement in human malignant tumors, we believe that the demonstration of a host range of exclusively transformed B-lymphoblasts for a papovavirus should be of some interest for a postulated role of additional agents in the target cell conditioning model. Moreover, the lymphoma tropism of this virus in human cells may open a new approach to study differences between malignantly transformed and EBV-transformed cells at the molecular level.

Another approach to understanding the role of EBV in human malignant tumors could originate from studies on the regulation of virus latency in lymphoblastoid cells. In a variety of these lines EBV replication and particle synthesis can be induced by various inducing chemicals, such as pyrimidine analogues, n-butyric acid, intercalating drugs, and tumor promoters. We studied the latter group in particular, and demonstrated that tumor promoters of the diterpene ester type efficiently induced EBV synthesis (zur Hausen et al., 1978). This induction can be inhibited by antipromoting chemicals such as retinoic acid (Yamamoto and zur Hausen, 1979), indomethacine (Daniel et al., unpublished), and L-canavanine (Yamamoto et al., 1980a). All these substances do not affect EBV replication following infection of susceptible cells. Thus, they are specific inhibitors of induction but not of infection, indicating that the regulation of persisting genomes must be different from that of infecting EBV DNA.

Experiments performed with inhibitors of protein or RNA synthesis first suggested the existence of a cellular control for EBV persistence (Yamamoto et al., 1980b). Inhibition of protein synthesis by cycloheximide immediately after induction with TPA followed 2 days later by removal of the inhibitor and the addition of actinomycin D in order to permit the translation of preexisting mRNA by stopping further transcriptional events did not lead to measurable induction of virus-specific proteins. Infection inhibition, followed by the same regimen of actinomycin D treatment, clearly resulted in viral antigen synthesis.

These data were interpreted to mean that the synthesis of a cellular protein is induced by the promoter, which in turn activates EBV early antigens. This interpretation was further underlined by results showing that treatment of EBV genome-free cells by TPA followed 2 days later by removal of the drug and fusion of these cells with genome-positive cells resulted in significant induction of the latter (Takaki and zur Hausen, unpublished data). Obviously a factor accumulated in the negative cells which upon fusion induced the positive ones. Moreover, TPA induction of early EBV mRNA was almost completely blocked by cycloheximide treatment, emphasizing again the existence of an inducible cellular mediator-protein which depresses the persisting viral DNA and the regulation of viral genome latency by cellular genes (Freese and Bornkamm, unpublished data).

The failing control of such genes, particularly if involved in modifying the function of early viral gene products, could contribute to the development of malignant tumors. This failure could originate from mutational events affecting the genome-harboring cell.

Tumor promoters at very low doses enhance substantially transformation of B-lymphatic cells by EBV (Yamamoto and zur Hausen, 1978). The increase is six- to eightfold under appropriate conditions.

Basing himself partly on these data, Prof. Ito of Kyoto has proposed that the consumption of plant extracts containing inducing compounds in endemic areas for Burkitt's lymphoma and nasopharyngeal carcinoma may represent a significant risk factor for the development of these malignant tumors (Ito, 1980).

The data presented clearly indicate that EBV is indeed somehow etiologically involved in the development of many - but not all - Burkitt's lymphomas, as well as in nasopharyngeal cancer, which has not been discussed in this presentation. This "somehow," however, leaves many question marks. We have to admit that, at least at present, the role of EBV as an oncogenic agent is at best poorly understood.

REFERENCES

Brade, L., Müller-Lantzsch, N., and zur Hausen, H. B-lymphotropic papovavirus and possibility of infections in humans. J. Med. Virol. 6, 301-308, 1980.

Brade, L., Vogl, W., Gissmann, L., and zur Hausen, H. Propagation of B-lymphotropic papovavirus (LPV) in human B-lymphoma cells and characterization of its DNA. Virology 114, 228-235, 1981.

de Thé, G., Geser, A., Day, N.E., Tukei, P.M., Williams, E.H., Beri, D.P., Smith, P.G., Dean, A.G., Bornkamm, G.W., Feorino, P., and Henle, W. Epidemiological evidence for causal relationship between Epstein-Barr virus and Burkitt's lymphoma from Ugandan prospective study. Nature 274, 756-761, 1978.

Diehl, V., Krause, P., Hellriegel, K.P., Busche, M., Schedel, I., and Laskewitz, E. In: Immunological Diagnosis of Leukemias and Lymphomas. Thierfelder, T., Rodt, H., and Thiel, E. (eds.). Springer-Verlag, Berlin and New York, pp. 289-296, 1977.

Epstein, M.A., Achong, B.G., and Barr, Y.M. Virus particles in cultured lymphoblasts from Burkitt's lymphoma. Lancet I, 702-703, 1964.

Epstein, M.A., Hunt, R.D., and Rabin, H. Pilot experiments with EB virus in owl monkeys (*Aotus trivirgatus*). I. Reticuloproliferative disease in an inoculated animal. Int. J. Cancer 12, 309-318, 1973.

Henle, W., and Henle, G. Epstein-Barr virus: the cause of infectious mononucleosis - a review. In: Oncogenesis and Herpesviruses. Biggs, P.M., de Thé, G., and Payne, L.N. (eds.), IARC, Lyon, pp. 269-274, 1972.

Henle, W., Diehl, V., Kohn, G., zur Hausen, H., and Henle, G. Herpes-type virus and chromosome markers in normal leukocytes after growth with irradiated Burkitt cells. Science 157, 1064, 1065, 1967.

Henle, G., Henle, W., and Diehl, V. Relation of Burkitt's tumor associated herpes-type virus to infectious mononucleosis. Proc. Nat. Acad. Sci. U.S. 59, 94-101, 1968.

Henle, G., Henle, W., Klein, G., Gunvén, P., Clifford, P., Morrow, R.H., and Ziegler, J.L. Antibodies to early Epstein-Barr virus-induced antigens in Burkitt's lymphoma. J. Natl. Cancer Inst. 46, 861-871, 1971.

Henle, W., Henle, G., Gunvén, P., Klein, G., Clifford, P., and Singh, S. Patterns of antibodies to Epstein-Barr virus-induced early antigens in Burkitt's lymphoma. Comparison of dying patients with long-term survivors. J. Natl. Cancer Inst. 50, 1163-1173, 1973.

Ito, Y., Kishihita, M., Morigaki, T., Yanase, S., and Hirayama, T. Induction and intervention of Epstein-Barr virus antigens in human lymphoblastoid cell lines: a simulation model for study of cause and prevention of nasopharyngeal carcinoma and Burkitt's lymphoma. In: Nasopharyngeal Carcinoma: Basic Research as Applied to Diagnosis and Therapy, Gustav-Fischer Verlag, Stuttgart, New York, 1980.

Lindahl, T., and Adam, A. Covalently closed circular duplex DNA of Epstein-Barr virus in a human lymphoid cell line. J. Mol. Biol. 102, 511-530, 1976.

Manolov, G., and Manolova, Y. Marker band in one chromosome 14 from Burkitt lymphomas. Nature 237, 33-34, 1972.

Nilsson, K., Giovanella, B.C., Stehlin, J.S., and Klein, G. Tumorigenicity of human hematopoietic cell lines in athymic nude mice. Int. J. Cancer 19, 337-344, 1977.

Nonoyama, M., and Pagano, J.S. Detection of Epstein-Barr viral genome in non-productive cells. Nature (New Biol.) 233, 103-106, 1971.

Pope, J.H., Horne, M.K., and Scott, W. Transformation of foetal human leukocytes in vitro by filtrates of a human leukemic cell line containing herpes-like virus. Int. J. Cancer 3, 857-866, 1969.

Purtillo, D.T., DeFlorio, D., Hutt, L.M., Bhawan, J., Yang, J.P.S., Otto, R., and Edwards, W. Variable phenotypic expression of an X-linked recessive lymphoproliferative syndrome. New Engl. J. Med. 297, 1077-1081, 1977.

Reedman, B.M., and Klein, G. Cellular localization of an Epstein-Barr virus (EBV)-associated complement-fixing antigen in producer and non-producer lymphoblastoid cell lines. Int. J. Cancer 11, 499-520, 1973.

Shope, T., Dechario, D., and Miller, G. Malignant lymphoma in cottontop marmosets after inoculation with Epstein-Barr virus. Proc. Natl. Acad. Sci. U.S. 70, 2487-2491, 1973.

Yamamoto, N., Bister, K., and zur Hausen, H. Retinoic acid inhibition of Epstein-Barr virus induction. Nature 278, 553-554, 1979.

Yamamoto, N., Müller-Lantzsch, N., and zur Hausen, H. Differential inhibition of Epstein-Barr virus induction by the amino acid analogue, L-canavanine. Int. J. Cancer 25, 439-443, 1980a.

Yamamoto, N., Müller-Lantzsch, N., and zur Hausen, H. Effect of actinomycin D and cycloheximide on Epstein-Barr virus induction in lymphoblastoid cells. J. Gen. Virol. 51, 235-261, 1980b.

zur Hausen, H. DNA viruses in human cancer: biochemical approaches. Cancer Res. 36, 414-416, 1976.

zur Hausen, H. The role of viruses in human tumors. Adv. Cancer Res. 33, 77-107, 1980a.

zur Hausen, H. The role of Epstein-Barr virus in Burkitt's lymphoma and nasopharyngeal carcinoma. In: Oncogenic Herpesviruses. Rapp, F. (ed.), CRC Press, Boca Raton, Florida. Vol. II, pp. 13-24, 1980b.

zur Hausen, H., and Gissmann, L. Lymphotropic papovavirus isolated from African green monkey and human cells. Med. Microbiol. Immunol. 167, 137-153, 1979.

zur Hausen, H., Schulte-Holthausen, H., Klein, G., Henle, W., Henle, G., Clifford, P., and Santesson, L. EBV-DNA in biopsies of Burkitt tumours and anaplastic carcinomas of the nasopharynx. Nature 228, 1056-1058, 1970.

zur Hausen, H., O'Neill, F.J., Freese, U.K., and Hecker, E. Persisting oncogenic herpesvirus induced by the tumor promoter TPA. Nature 272, 373-375, 1978.

IMMUNOLOGICAL CONTROL OF EBV INFECTED B CELLS

Eva Klein and George Klein

Department of Tumor Biology
Karolinska Institutet
S 104 01 Stockholm, Sweden

INTRODUCTION

The majority of individuals harbors EB-virus infected B cells (B_{EBV}). (For characteristics of the virus-cell and virus host relationship see references 1 and 2). When blood lymphocytes or tissues from donors who experienced virus infection are explanted, the B_{EBV} cells grow and can be easily established as permanent lines - LCL_{EBV}, lymphoblastoid cell lines. Limiting dilution experiments estimated 500-5,000 B cells with proliferative potential in the blood. In acute IM (infectious mononucleosis) the number is higher and cells that carry the viral genome and consequently express a nuclear antigen, EBNA, can be visually detected in the isolated B subset. Robinson et al. showed that these cells divide.[3] In fatal IM cases EBNA positive cells were detected also in the spleen, liver and thymus.

Generally the B_{EBV} cells are kept under control and proliferative "accidents" are very rare. These occur either because the cells escape from the control mechanisms or the control is impaired. The former event may occur in Burkitt's lymphoma and the latter in fatal mononucleosis, in X-linked combined variable progressive immunoproliferative syndrome (XLP) and in renal transplant patients.[4]

In vitro only B lymphocytes can be infected with virus. The viral receptor is either identical or associated with the C3d receptor. The virus-cell interaction and the response of the host has several aspects which are of great interest to the immunologist. Such are 1) the autonomous (T-cell and macrophage independent) polyclonal activation of the B-cells. 2) the proliferative capacity of B_{EBV} cells can be exploited to establish antibody producing monoclonal human cell lines. 3) the complexity of the surveillance mechanisms that prevent the in vivo growth of the B_{EBV} cells. 4) the regulatory mechanism that limits the activation and proliferation of T cells during the acute phase of IM.

With the help of plasma membrane transplantation which equipped cells with the EBV receptor, it was shown that a variety of cells could be infected with the virus. In contrast to the natural host cell, the human B lymphocyte, in this "artifical" situation virus infection proceeded to the lytic cycle, and did not establish the latent virus carrier state.[5] This suggests that the present EBV may be a host range mutant of an originally lytic herpesvirus; - its survival may have been favoured by the moderate, transforming interaction with the B-lymphocyte. The factors which determine the differences in the virus-cell interaction are not known.

In the evolution leading to the present balanced EBV-human host relationship, selection of virus with low virulence, and selection of hosts with efficient control mechanisms had to occur.

Independent of the clinical picture the primary infection has two life lasting consequences: 1. persisting antibody titers against at least 3 virus determined antigens (VCA - viral capsid antigen, MA - cell membrane antigen, viral envelope antigen, EBNA - nuclear antigen); 2. persistence of the virus in a non-lytic, latent form in a proportion of B-lymphocytes.

The cells which are the source of infectious virus production have not been identified with certainty.

Immune response against viral antigens

During the acute phase of IM antibodies to VCA are the first to appear. This is followed by antibodies to the early (EA) and to the viral envelope antigen. Antibodies directed against the nuclear antigen, EBNA, emerge usually after several weeks, sometimes after a couple of months.

In T-cell suppressed conditions the anti-EA and VCA titers increase and the anti-EBNA antibodies decrease or disappear.[6] Since EBNA is only present in the transformed, latently infected cell that has not entered the lytic cycle whereas EA and VCA appear during the viral cycle, this would suggest that either the activation of viral cycle occurs in the cells and/or the elimination of the cells are impaired in these cases. Experimental evidence suggests that cells that are on the way in the lytic cycle are sensitive to the cytotoxic effect of natural killer effect, a function of the T lymphocyte subset.[7]

The antibody patterns (specificities and titers) have a prognostic significance and the three major EBV-associated diseases, infectious mononucleosis, IM, Burkitt's lymphoma, BL, and nasopharyngeal carcinoma, NPC, each have their own characteristics.

Cellular memory to EBV determined antigens has also been demonstrated. T cells of healthy seropositive individuals were shown to enter DNA synthesis when exposed to UV irradiated EBV or to EBV absorbed on the surface of autologous B cells.[8,9,10] Extracts and membranes from EBV-genome carrying cells were shown to elicit the production of leukocyte migration inhibitory factor (LIF) in T cells.[11]

Immune response against B_{EBV} cells

Uncontrolled proliferative capacity of B_{EBV} cells in vivo as it is exhibited in vitro, would lead to malignant lymphocytosis. Consequently, either the behaviour of the B_{EBV} cell change upon explantation, or in the culture they are released from an in vivo control. It seems that the immunological control operates through two essentially different mechanisms. One is the homeostasis within the immune system which regulates the T and B lymphocyte compartments the other is the specific recognition of EBV determined antigen(s) on the surface of B_{EBV} cells.

The cellular response against B_{EBV} cells has been studied by assaying blood lymphocytes of seropositive healthy individuals and of patients during the acute phase of IM. It is assumed that the T cell proliferation in IM represents an immune response against viral and/or virally determined antigens. The reaction towards B_{EBV} cells was tested by direct cytotoxicity or inhibition of the EBV induced activation and transformation of the B cells.

The atypical lymphocyte populations in the acute phase of IM share certain properties with T cells activated in vitro by exposure to antigens. Compared to freshly harvested lymphocytes of healthy persons, they 1. have fewer cells with high avidity Fcγ receptors.[12,13] 2. have lower proportion of cells defined with OKM1 monoclonal reagent (the subset that contributes in the natural killer effect of blood lymphocyte populations)[14] and 3. are cytotoxic for a broader target panel.[15,16,17,18]

The initial cytotoxic experiments with IM patients were interpreted to show recognition of EBV determined antigen(s) on the target.[15] However, in view of the accumulated knowledge concerning lymphocyte mediated cytotoxicity the lytic effects of the IM blood lymphocytes had to be reevaluated.[17,18,19]

In the interpretation of the direct lymphocyte mediated cytotoxicity experiments the characteristics of the effector cells and the specificity of the effects are decisive. In order to designate an effect as natural killer the presence of Fc receptor and the absence of high avidity E receptors is considered as an important characteristic. However in activated lymphocyte populations further effector subsets are recruited, most importantly FcR negative cells. Thus not all the rules governing the natural cytotoxicity assays - when unmanipulated lymphocytes are used - are valid with activated populations. In the acute phase of mononucleosis various proportions of T cells are activated as indicated by their morphology.

The lysis of cell lines by the natural and the activated killer lymphocytes seem to reflect polyclonal effects and the majority of the effector-target cell interactions do not occur due to recognition of antigens. However recognition of the species is involved.[20]

We compared the effect of IM lymphocytes and in vitro activated T cells, because the latter were shown to acquire lytic potential against certain cell lines, even such which are not damaged by unmanipulated lymphocytes in short term assays.[18]

We assumed that a substantial part of the lytic function of the T lymphocytes in the acute phase of IM is a consequence of their activated state because 1. the lytic effect does not show the histocompatibility restriction which is a rule for T cell mediated lysis based on recognition of surface antigens. 2. EBV genome negative B cell lines are also affected. 3. the lytic potential in short term assays follows closely the proportion of atypical cells.[21]

In a subsequent series of experiments the cytotoxic potential of IM patients lymphocytes was again assayed in our laboratory with special emphasis on the question of HLA-restriction and selectivity for EBV positive targets.[22] The pooled results showed that the effect against K562 and Molt-4, the prototypes of NK sensitive targets, was lower in patients than in the controls, and it was further lowered when FcR positive cells were depleted. The results were similar when EBV negative B cell lines were the targets (BJAB, U698 and U715). However the results with EBV positive targets were different and suggested an EBV related component. Cytotoxicities were higher with IM patients' than with the controls' lymphocytes and they were not lowered with FcR-cell depletion. While the comparison of the normal and patient derived effectors may not be justified, due to the difference in the composition of the effector cell populations, the comparison of the IM lymphocytes acting on EBV negative and EBV positive B targets is meaningful. The EBV-specific effect did not show restriction to autologous targets.

These results were substantiated in a study of a patient who was identified to aquire a primary infection.[23] Several lymphocyte samples were tested for cytotoxicity prior, during and after the acute phase of IM. Cytotoxicity against autologous LCL_{EBV} and the EBV positive Kaplan line was elevated during the disease. The effect against K562 decreased and the effects against the former 2 lines were not depleted when FcR positive cells were eliminated.

It may be argued that the recognition of EBV determined antigens is unique in that it does not follow the rule of MHC restriction. However, the T cell memory of healthy seropositive donors revealed by activation in vitro, shows a clear preference for the lysis of autologous target.[24]

Thus there is suggestive evidence for the EBV related cytotoxic function of the acute phase IM lymphocytes.

In the initial stage of IM - during the first week - the EBV induced polyclonal B cell activation is reflected in hypergammaglobulinemia. T-cells collected during the second week of the disease were found to suppress in vitro B cell activation assayed by immunoglobulin production. This inhibition affected both EBV and pokeweed mitogen-induced activation of autologous and allogeneic B cells.[25]

The dissection of an EBV specific immunological effector function of the IM T lymphocytes requires that it can be distinguished from effects due to their activated state. Such studies may be complicated by the fact that results with patients who are in different phases of the disease may differ considerably due to the evolution of EBV related humoral and cellular immunity as well as the properties of the lymphocyte population.

It has been shown that the evolution of the virus specific

T cell memory - assayed by blastogeneic response - is slower than the antibody response.[10]

However, for the clinical course of the disease the activated state of T cells is important. In vitro activated T cells have been shown to be cytotoxic to autologous LCL_{EBV}.[18] Activated T cells may be instrumental in the elimination or control of the B_{EBV} cells even if they do not act on the basis of EBV induced plasma membrane antigen recognition.

The cellular response against newly transformed B_{EBV} cells has been also shown in vitro. T cells of IM patients inhibited the outgrowth of B cells in primary in vitro infection system.[26]

Usually EBV infection of blood lymphocytes in vitro leads only to temporary growth of B_{EBV} cells and the proliferation "regresses".[27] The role of the T cells in this phenomenon was shown by the uninhibited growth of purified B cell populations and the abrogation of the regression phenomenon after treatment with the T cell suppressive drug cyclosporin A.[28] In the majority of such experiments there was a difference between seropositive and seronegative individuals.

Shope and Kaplan interpreted the results of various systems to show that the inhibition of the outgrowth of B_{EBV} cells in vitro is brought about by T cells of different functional characteristics.[29] We can assume at least three types of T cell responses: One category acts promptly and occurs irrespective of the serological status of the donor, another is affected by T cells activated and induced to proliferate by encounter with B blasts, and finally there is an antigen specific component which is represented by the emergence of the specific clone, an EBV related memory.

Evidence for the first mechanism was provided by several experiments. T cells were shown to suppress the EBV induced proliferation of B cells even if they were allowed to present only for 3 hours. Their presence for longer time (2 days) was not necessary and did not improve the growth inhibitory effect.[30] T cells attaching to the B cells were detected in EBV infected lymphocyte populations[18] and the outgrowth of B cells were shown to be inhibited by the Fcγ receptor carrying T cell subset which is responsible for natural killer function.[29] Neither of these phenomena differed when the assays were performed with seropositive and seronegative individuals.

The EBV specific memory has been demonstrated in experiments in which cytotoxic T lymphocytes were generated in the "regressing" cultures and in mixed cultures of T cells and autologous LCL_{EBV}.[24,31] In these systems two phenomena may appear. T cells activated by B blasts, without the involvement of EBV memory, develop cytotoxic potential that is polyclonal and therefore is not restricted to autologous target but can also act on it.

It has been shown in vitro that contact with autologous B cells triggers T cells to proliferate and to acquire cytotoxic potential.[32]

Stimulation of the T cells by autologous B cells was found to be considerably stronger when the B cells have been blast transformed by mitogen or EBV.[18] Since this was true also for cells derived from EBV-seronegative donors, the phenomenon was unrelated to EBV determined antigens.[18,33] The second component is the outgrowth of the clone representing the EBV specific memory. In cultures of seropositive donors the latter mechanism would act as the third line of response. Detected as cytotoxic potential this response follows the rule of MHC restriction in that autologous LCL_{EBV} were found to be preferentially killed and allogeneic LCL_{EBV} only when they showed a certain degree of HLA matching.[24]

Dissection of the two effects is highly dependent on the conditions of the experiments. It has been shown for the generation of alloreactive response that the responder stimulation ratio, the density of the culture influence the events in the cultures.[34,35] Repeated encounter with the autologous LCL_{EBV} increased the specific effect of the culture and decreased the cytotoxicity against lines sensitive to the effect of natural and activated lymphocytes.[36]

We did not discuss the involvement of antibodies in these cell mediated phenomena. As mentioned before antibodies which react with cell surface of EBV infected cells are present in the serum of seropositive individuals. These antibodies react with viral envelope antigens. After superinfection of EBV genome carrying non-producer lines with the P3HR type virus or induction of the viral cycle the cells reacted with EBV positive sera.[7]

In vivo and also in vitro, particularly in those experiments which involve long term cultures, antibodies may contribute to the events. The explanted lymphocyte populations of seropositive individuals may contain cells which produce EBV specific antibodies. The influence of antibodies in the cellular immunity experiments has not been demonstrated yet.

It may be possible that the EBV specific cell surface antigens recognized by T cells, LYDMA, do not elicit antibodies or have extremely low titers in the natural EBV infection.

In vivo proliferation of B_{EBV} cells

The in vitro experiments show that EBV is the most powerful transforming virus known. Consequently EBV must be a potentially dangerous agent in immunologically unprepared hosts.

EBV was found to induce progressive lymphoproliferative disease in the marmoset monkey. While often fatal because of its polyclonal nature it is more akin to infectious mononucleosis in man. In all likelihood, the B_{EBV} grows progressively in the immunologically naive host species. Because this species did not encounter the EBV virus naturally relevant host defense mechanisms may not have evolved.

It can be therefore expected that EBV-carrying lymphoproliferative disease in man can occur when the immune defenses break down. There is indeed evidence for such events. A recently described disease complex, designated as the X-linked lymphoproliferative syndrome (XLP) was postulated to be due to the unrestrained proliferation of EBV-transformed B-cells.[4] This assumption was substantiated by the demonstration of multiple EBV-genomes in the cells of 6 cases tested. In addition 7 "lymphomas" of renal transplant patients and one lymphoma that developed in a child with ataxia telangiectasia also were found to contain EBV genome.

Patients with EBV-carrying lymphoproliferative disease (two with chronic mononucleosis and four XLP) were found to have seriously impaired immune effector mechanisms (NK, even after interferon activation, ADCC, antibodies and cellular response against various EBV antigens).[37]

These polyclonal EBV-carrying lymphoproliferative diseases must be distinguished from the Burkitt's lymphoma (BL). BL arises in immunocompetent patients, and is always monoclonal. Its evolution is apparently due to a specific genetic change in a single cell and not to the impairment of the host to handle the B_{EBV} cells.

Burkitt lymphoma biopsies and derived lines contain a marker, which is the result of a reciprocal translocation between chromosomes 8 and 14. For discussion of the significance of this marker and the implications for oncogenesis, see ref. 38. The breaking points on these chromosomes were found to be the same in different tumors, 8q24 and 14q32, respectively. While other types of human hemo- and lymphopoetic neoplasias were found to carry 14q+ markers, these were different from the translocation characteristic for BL: the donor chromosome of the extra band was variable, and so was the breaking point on chromosome 14. B-cell derived ALL was the only exception: it contains the BL type 8;14 reciprocal translocation. This leukemia is believed to originate from the same target cell as BL.

The chromosomal "variant" cases of BL had also translocation of the distal segment of chromosome 8 however the recipient chromosome was not no. 14 but no. 22 and no. 2. This regularity has become meaningful when it was recently discovered that in humans, the heavy chain immunoglobulin locus is localized to chromosome 14, the kappa gene on 2 and the lambda gene on 22.

The cytogenetical properties of these lymphomas suggest that the distal segment of human chromosome 8 may contain a presumptive oncogene that is switched on if translocated to another chromosomal region that is active in that particular cell.

EBV is not involved in the generation of the 8;14-22-2 translocation. First, because it is present also in EBV-negative Burkitt lymphomas and B-ALL and secondly, it is absent in the LCL_{EBV}-s which are diploid, as a rule, for several months, up to a year, after their establishment. The translocation is likely to be a random event, occurring by chance, in persistent, EBV-carrying, latently infected B-cells that are urged to divide, but cannot differentiate properly. Division may be stimulated by the environmental co-factor, possibly chronic holo- or hyperendemic malaria.

BL cells are closely similar to LCL_{EBV} with regard to the number of viral genome copies, the occurrence of integrated vs. free episomal viral DNA, the antigenic specificity of EBNA, and the latency and inducibility of the viral cycle. Attempts to detect DNA sequence differences between BL associated EBV, on the one hand, and normal LCL or IM associated virus, on the other, failed to reveal any disease related difference; the variations appear to reflect the properties of the individual isolates. BL cells are less differentiated than LCL_{EBV} of normal origin.

The marked phenotypic differences between LCL and BL cells and the characteristic cytogenetical change in the latter suggest that the malignant behaviour of the BL cell is not determined at the level of the virus-cell interaction, but at the level of the cell-host relationship.

Because BL patients do not show immunological impairment and the tumor is monoclonal it could be postulated that one of its important phenotypic characteristics is a relative immunoresistance. In vitro experiments with BL derived culture lines and LCL_{EBV} do not suggest that this is the case.[39] However freshly harvested BL cells and newly transformed B_{EBV} were not compared for this property. The localized appearance of the tumor with dissemination only in the terminal stages may suggest that the growth is compartmentalized because of a host response. Such situation has been shown with transplanted antigenic lymphomas. In unimmunized animals the cells grew disseminated, in animals with developed immunity the cells were rejected, as incomplete immunity allowed growth of the tumor at the inoculation site.[40]

ACKNOWLEDGEMENTS

This project has been funded with Federal Funds from the Department of Health, Education and Welfare under Grant No. 5R01 CA 25250-03 and Grant No. 1 R01 CA 30264-01 awarded by the National Cancer Institute, DHEW and by the Swedish Cancer Society.

REFERENCES

1. The Epstein-Barr virus, M.G. Epstein and B.G. Achong, eds., Springer Verlag, Berlin, Heidelberg, New York (1979).
2. Viral Oncology, G. Klein, ed., Raven Press, New York (1980).
3. J. Robinson, D. Smith, and J. Niederman, Mitotic EBNA-positive lymphocytes in peripheral blood during infectious mononucleosis. Nature 287: 334 (1980).
4. D.T. Purtilo, D. De Floria, L.M. Hutt, J. Bhawan, J.P.S. Yang, R. Otto, and W. Edwards, Variable phenotypic expression of an X-linked recessive lymphoproliferative syndrome. New Engl. J. Med. 297: 1077 (1977).
5. D. Volsky, G. Klein, B. Volsky, and I. Shapiro, Production of infectious Epstein-Barr virus in mouse lymphocytes. Nature 293: 399 (1981).
6. W. Henle, and G. Henle, Epstein-Barr specific serology in immunologically compromised individuals. Cancer Res., in press.
7. M. Patarroyo, B. Blazar, G., Pearson, E. Klein, and G. Klein, Induction of the EBV cycle in B-lymphocyte-derived lines is accompanied by increased natural killer (NK) sensitivity and the expression of EBV-related antigen(s) detected by the ADCC reaction. Int. J.

Cancer 26: 365 (1980).

8. P. Gergely, I. Ernberg, G. Klein, and M. Steinitz, Blastogenic responses of purified human T-lymphocyte populations to Epstein-Barr virus (EBV). Clin. Exp. Immunol. 30: 347 (1977).

9. L.M. Hutt-Fletcher, and C.J. Gilbert, Cytotoxic cells induced in vitro by Epstein-Barr virus antigens. Cell. Immunol. 63: 91 (1981).

10. C.H.H. Ten Napel, and T.H. The, Lymphocyte reactivity in infectious mononucleosis. J. Inf. Dis. 141: 716 (1980).

11. R. Szigeti, L. Timár, and T. Révész, Leukocyte migration inhibition with Epstein-Barr virus negative and positive cell extracts. Allergy 35: 97 (1980).

12. A. Poros, and E. Klein, Cultivation with K562 cells leads to blastogenesis and increased cytotoxicity with changed properties of the active cells when compared to fresh lymphocytes. Cell. Immunol. 41: 240 (1978).

13. B.F. Haynes, R.T. Schooley, J.E. Grouse, C.R. Payling-Wright, R. Dolin, and A.S. Fauci, Characterization of thymus-derived lymphocyte subsets in acute Epstein-Barr virus-induced infectious mononucleosis. J. Immunol. 122: 699 (1979).

14. I.M. Zarling, and P.C. Kung, Monoclonal antibodies which distinguish between human NK cells and cytotoxic T lymphocytes. Nature 288: 394 (1980).

15. E. Svedmyr, and M. Jondal, Cytotoxic effector cells specific for B-cell lines transformed by Epstein-Barr virus are present in patients with infectious mononucleosis. Proc. Natl. Acad. Sci. USA 72: 1622 (1975).

16. M.G. Masucci, E. Klein, and S. Argov, Disappearance of the NK effect after explantation of lymphocytes and generation of similar non-specific cytotoxicity correlated to the level of blastogenesis in activated cultures. J. Immunol. 124: 2458 (1980).

17. M.G. Masucci, G. Masucci, E. Klein, and W. Berthold, Target selectivity of interferon-induced human killer lymphocyte related to their Fc receptor expression. Proc. Natl. Acad. Sci. 77: 3620 (1980).

18. E. Klein, I. Ernberg, M.G. Masucci, R. Szigeti, Y.T. Wu, G. Masucci, and E. Svedmyr, The role of T cell response to B cells and EBV antigens in control of EBV infected B cells in infectious mononucleosis. Cancer Res., in press.

19. E. Klein, M.G. Masucci, W. Berthold, and B.A. Blazar, Lymphocyte-mediated cytotoxicity toward virus-induced tumor cells; natural and activated killer lymphocytes in man. Cold Spring Harbor Conferences on Cell Proliferation, Cold Spring Harbor, USA 7: 1187 (1980).

20. E. Klein, Natural and activated cytotoxic T lymphocytes. Immunology Today 1: IV (1980).

21. E. Svedmyr, M. Jondal, W. Henle, O. Weiland, L. Rombo, and G. Klein, EBV specific killer cells and serological responses after onset of infectious mononucleosis. J. Clin. Lab. Immunol. 1: 225 (1978).
22. L. van der Waal, Epstein-Barr virus selective T cells in infectious mononucleosis are not restricted to HLA-A and B antigens. J. Immunol. 127: 293 (1981).
23. E. Svedmyr, I. Ernberg, J. Seeley, O. Weiland, G. Masucci, K. Tsukuda, R. Szigeti, M.G. Masucci, H. Blomgren, W. Henle, and G. Klein, Virologic, immunologic and clinical observations on a patient during the incubation acute and convalescent phases of infectious mononucleosis. To be published.
24. I.S. Misko, D.J. Moss, and J.H. Pope, HLA-antigen-related restriction of T-lymphocyte cytotoxicity to Epstein-Barr virus. Proc. Natl. Acad. Sci. USA 77: 1247 (1980).
25. G. Tosato, I. Magrath, I. Koski, M. Dooley, and M. Blease, Activation of suppressor T cells during Epstein-Barr virus induced infectious mononucleosis. New Engl. J. Med. 301: 1133 (1979).
26. A.B. Rickinson, D. Crawford, and M.A. Epstein, Inhibition of the in vitro outgrowth of Epstein-Barr virus-transformed lymphocytes by thymus-dependent lymphocytes from infectious mononucleosis patients. Clin. Exp. Immunol. 28: 72 (1977).
27. D.A. Thorley-Lawson, L. Chess, and J. Strominger, Suppression of in vitro Epstein-Barr virus infection. J. Exp. Med. 146: 495 (1977).
28. A.G. Bird, S.M. McLachlan, and S. Britton, Cyclosporin promotes the spontaneous outgrowth in vitro of Epstein-Barr virus induced B-cell lines. Nature 289: 300 (1981).
29. T.C. Shope, and J. Kaplan, Inhibition of the in vitro outgrowth of Epstein-Barr virus-infected lymphocytes by T_G lymphocytes. J. Immunol. 123: 2150 (1979).
30. D.A. Thorley-Lawson, The suppression of Epstein-Barr virus infection in vitro occurs after infection but before transformation of the cell. J. Immunol. 124: 745 (1980).
31. D.J. Moss, A.B. Rickinson, and J.H. Pope, Long term T cell mediated immunity to Epstein Barr virus in man. III. Activation of cytotoxic cells in virus infected leukocyte cultures. Int. J. Cancer 23: 618 (1979).
32. M.E. Weksler, and R. Kozak, Lymphocyte transformation induced by autologous cells. V. Generation of immunologic memory and specificity during the autologous mixed lymphocyte reaction. J. Exp. Med. 146: 1833 (1977).
33. R.S. Chang, and Y.Y. Chang, Activation of lymphocytes from Epstein-Barr virus seronegative donors by autologous Epstein-Barr virus transformed cells. J. Infect. Dis. 142: 156 (1980).
34. H. Spits, J.E. de Vries, and C. Terhorst, A permanent

human cytotoxic T-cell line with high killing capacity against a lymphoblastoid B-cell line shows preference for HLA A, B target antigens and lacks spontaneous cytotoxic activity. Cell. Immunol. 59: 435 (1981).

35. L.E. Wallace, A.B. Rickinson, M. Rowe, D.J. Moss, D.J. Allen, and M.A. Epstein, Stimulation of human lymphocytes with irradiated cells of the autologous Epstein-Barr virus transformed cell line. I. Virus specific and non-specific components of the cytotoxic response. In press.

36. Y. Tanaka, K. Sugamura, Y. Hinuma, H. Sato, and K. Okochi, Memory of Epstein-Barr virus specific cytotoxic T cells in normal seropositive adults as revealed by an in vitro restimulation method. J. Immunol. 125: 1426 (1980).

37. M.G. Masucci, R. Szigeti, I. Ernberg, G. Masucci, G. Klein, J. Pritchard, C. Sieff, S. Lie, A. Glomstein, L. Businco, W. Henle, G. Henle, G. Pearson, K. Sakamoto, and D. Purtilo, Cellular immune defects to Epstein-Barr virus determined antigens in males with chronic mononucleosis. Cancer Res., in press.

38. G. Klein, Changes in gene dosage and gene expression: a common denominator in the tumorigenic action of viral oncogenes and non-random chromosomal changes? Nature, in press.

39. M. Jondal, C. Spina, and S. Targan, Human spontaneous killer cells selective for tumor-derived target cells. Nature 72: 62 (1978).

40. E. Klein, Effect of immunization on the spread of transplanted lymphoma cells. J. Natl. Cancer Inst. 51: 1991 (1973).

PAPILLOMAVIRUSES

H. zur Hausen, L. Gissmann, and E.M. de Villiers

Institut für Virologie, Zentrum für Hygiene
Universität Freiburg, Hermann-Herder-Str. 11
7800 Freiburg, West Germany

Papillomaviruses are found in a wide variety of species (see review zur Hausen, 1977). They induce epithelial or fibroepithelial proliferations, usually for limited periods of time. In most cases spontaneous regression occurs within a period of several weeks or months after initial appearance. The following report summarizes the present knowledge on papillomavirus infections in man, their mode of maturation, and their possible involvement in the etiology of malignant tumors.

1. Maturation of Papillomaviruses

Papillomaviruses successfully infect cells of the basal cell layer of skin and mucosa, preferentially after traumatization of the intact skin. The uptake of viral DNA seems to stimulate the respective cell to repeated rounds of replication without essentially changing the capacity of these cells to differentiate. Enhanced growth and resulting thickened layers of epithelial cells are typical features of papillomavirus infection. Viral DNA replication starts in the cell layer on top of the basal cells as demonstrated by <u>in situ</u> hybridizations (Orth et al. 1971; Grussendorf and zur Hausen, 1980). Further keratinization of the cells is accompanied by synthesis of larger quantities of viral particles, which are preferentially found in the most superficial layers, being easily released for new rounds of infections.

Nothing is known of the mode of restriction of viral DNA replication within actively growing cells. As a working hypothesis we favor the speculation that proliferating cells of the basal cell layer actively control papillomavirus gene expression, i.e., that cellular control mechanisms inhibit late viral functions. In the course of differentiation a regulated switch-off of cellular control functions should take place, lifting eventually the block for viral replication and permitting the accumulation of mature particles. Of course, alternative speculations are possible and further work is needed to elucidate the underlying mechanism.

It is interesting to note that the quantities of viral particles produced depend on the type of virus infecting skin or mucosa. As a rule, papillomas induced, e.g., by human papillomavirus (HPV) type 1 produce large quantities of virus, whereas HPV-6-induced papillomas (genital warts) contain only minute quantities of virus particles.

2. Characterization of Human Papillomaviruses

Typing of papillomaviruses started in 1976 (Gissmann and zur Hausen, 1976; Orth et al., 1977; Gissmann et al., 1977) based on restriction endonuclease cleavage and nucleic acid hybridizations (Coggin and zur Hausen, 1978). So far eight distinct types of human papillomaviruses have been identified, causing common warts (verrucae vulgares), flat warts (verrucae planae), genital warts (condylomata acuminata), or warts of epidermodysplasia verruciformis. The characterized types are summarized in Table 1. With the exception of types 5 and 8 (both found in epidermodysplasia verruciformis), all others appear to be only distantly related with each other, as evidenced by the lack of cross-hybridization under conditions of high stringency. It can be expected that additional types of papillomaviruses will be found in the near future since viral DNA found in laryngeal papillomas and in some atypical condylomata of the cervix

Table 1. Papillomavirus types in human papillomas

Type of papilloma	Virus type
verruca vulgaris (common wart)	HPV 1 "plantar" wart virus HPV 2 "hand" wart virus HPV 4 small common warts HPV 7 "butcher" wart virus
verruca plana (flat wart)	HPV 3
condyloma acuminatum (genital wart)	HPV 6 (HPV 1 and HPV 2 occasionally)
epidermodysplasia verruciformis	HPV 3 HPV 5) warts with increased) tendency for malignant HPV 8) conversion

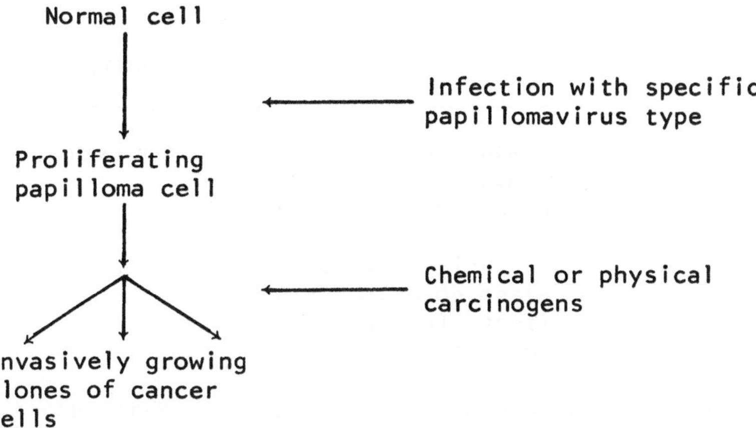

Fig. 1. Schematic outline of malignant cell transformation by papillomaviruses.

is different from that of the characterized papillomaviruses but has not yet been extensively analyzed.

3. Papillomaviruses and Cancer

Papillomavirus infections appear to be involved in specific cancers occurring under natural conditions (see reviews zur Hausen, 1977, 1980; Orth et al., 1980) in man and animals. In many instances infection with a specific papillomavirus type requires the additional application of chemical or physical carcinogens before malignant conversion takes place. This is outlined schematically in Fig. 1.

Papillomavirus types specifically involved in the development of malignant tumors are listed in Table 2.

Epidermodysplasia verruciformis is a disease of special interest (Jablonska et al., 1966; Orth et al., 1980): it is a hereditary disease resulting in a disseminated verrucosis, probably on the basis of impaired cellular immune functions. Malignant conversion appears to take place exclusively within light-exposed papillomas, pointing to a synergistic function of the ultraviolet part of the sunlight. Malignant conversion, in addition, seems to depend on the presence of HPV-5 or HPV-8 DNA, which can also be demonstrated within the malignant tumors (Orth et al., 1980). Viral DNA persists within the malignant cells in an episomal state without any evidence for integration into host cell DNA. This seems to be a common feature of papillomavirus DNA persistence within infected

Table 2. Papillomavirus types involved in malignant tumors

Virus type	Malignant tumor	Synergistic factor
HPV-5)) HPV-8)	Squamous cell carcinomas in epidermodysplasia verruciformis	Sunlight
HPV-6	Invasively growing giant condylomata acuminata (Buschke-Löwenstein tumors)	?
HPV of laryngeal papillomas	Squamous cell carcinomas arising in laryngeal papillomas after X-irradiation	X-irradiation
HPV of atypical genital condylomata	Squamous cell carcinomas of cervix, penis, and vulva?	?
BPV-4	Alimentary tract carcinomas of cattle	Carcinogens in bracken fern
Shope papillomavirus	Squamous cell carcinomas of the skin in cottontail and domestic rabbits	Chemical carcinogens
Papillomavirus of mastomys natalensis	Invasively growing acanthomas of the skin	?

tissues (Lancaster 1981; Law et al., 1981; Pfister et al., 1981; Moar et al., 1981). The rarity of epidermodysplasia verruciformis renders it difficult to assess the risk malignant conversion in patients infected with HPV-5 or HPV-8 accurately. It seems, however, that the majority of these patients acquire malignant tumors within a period of 5 to 10 years. Pigmented skin appears to provide some protection against malignant conversion (zur Hausen, 1977; Pfister et al., 1981).

The possible involvement of papillomaviruses in human genital cancer deserves some attention (zur Hausen, 1975, 1976). Human genital warts are frequent infections of the genital tract (Editorial, 1979) and anecdotal reports of malignant conversion are often encountered in the literature (see review zur Hausen, 1977). Only very recently has the viral DNA of the responsible agent, HPV-6, been identified (Gissmann and zur Hausen, 1980) and characterized after molecular cloning in bacterial vectors (de Villiers et al., 1981). HPV-6 DNA is regularly demonstrated in genital warts and invasively growing Buschke-Löwenstein tumors, although it has not been detected in typical squamous cell carcinomas of cervix, vulva, and penis (Gissmann et al., 1982). The available data disclosed that atypical condylomata of the cervix were occasionally only due to HPV-6 infection. A number of them contain distinct yet noncharacterized papillomaviruses which are at most distantly related to HPV-6. It has not been possible until now to obtain sufficient DNA from these lesions for further analysis.

It is interesting to note that typical papillomavirus particles have been demonstrated electron-microscopically within the superficial cell layers of atypical condylomata (Meisels et al., 1981). In addition, papillomavirus group-specific antigens were demonstrated in more than 50% of cervical dysplastic lesions by immunoperoxidase staining procedures (Shah et al., 1980). These data suggest that besides HPV-6 infections additional papillomavirus infections occur within the human genital tract. Their association with premalignant lesions of the cervix clearly deserves study. They may represent important candidates for an etiological role in human genital cancer.

Laryngeal papillomas occurring in young children are frequently multifocal and reveal a high tendency for recurrence after surgical removal. They show a peculiar age distribution, being found preferentially in children of 2 - 5 years of age (zur Hausen et al., 1975). Spontaneous malignant conversion of such tumors appears to be a very rare event. Therapeutic X-irradiation, however, led in many patients to the development of malignant tumors 5 to 40 years later (see review zur Hausen, 1977). The virus in laryngeal papillomas is produced in minute quantities only, which has prevented its characterization thus far. Only very recently has molecular cloning of its DNA been achieved (Gissmann and zur Hausen, unpublished), which should make possible its further characterization and an analysis of its possible role in malignant tumors.

This review does not cover the involvement of animal papillomaviruses in cancer development. The presently available data show, however, that this virus group deserves special attention: it represents an obviously heterogeneous group of agents with wide variations in the oncogenic potential of individual types. The characterization of these viruses should be important for a further assessment of their possible role in the induction of specific types of human cancer.

REFERENCES

Coggin, J.R., Jr., and zur Hausen, H., Workshop on papillomaviruses and cancer. Cancer Res. 39, 545-546, 1979.

de Villiers, E.M., Gissmann, L., and zur Hausen, H., Molecular cloning of viral DNA from human genital warts. J. Virol., Dec. issue, 1981.

Editorial: Sexually transmitted disease surveillance 1978. Brit. Med. J. 2, 1375-1376, 1979.

Gissmann, L., and zur Hausen, H. Human papilloma viruses: physical mapping and genetic heterogeneity. Proc. Natl. Acad. Sci. U.S. 73, 1310-1313, 1976.

Gissmann, L., and zur Hausen, H. Partial characterization of viral DNA from human genital warts (condylomata acuminata). Int. J. Cancer 25, 605-609, 1980.

Gissmann, L., Pfister, H., and zur Hausen, H. Human papilloma viruses (HPV): characterization of 4 different isolates. Virology 76, 569-580, 1977.

Gissmann, L., de Villiers, E.M., and zur Hausen, H. Analysis of human genital warts (condylomata acuminata) and other genital tumors for human papillomavirus type 6 DNA. Int. J. Cancer, Febr. issue, 1982.

Grussendorf, E., and zur Hausen, H. Localization of viral DNA replication in sections of human warts by nucleic acid hybridization with complementary RNA of human papilloma virus type 1. Arch. Derm. Res. 264, 55-63, 1979.

Jablonska, S., Fabjanska, L., and Formas, I. On the viral aetiology of epidermodysplasia verruciformis. Dermatologica 132, 369-385, 1966.

Lancaster, W.D. Apparent lack of integration of bovine papillomavirus DNA in virus-induced equine and bovine tumor cells and virus-transformed mouse cells. Virology 108, 251-255, 1981.

Law, M.F., Lowy, D.R., Dvoretzky, I., and Howley, P.M. Mouse cells transformed by bovine papillomavirus contain only extra-chromosomal viral DNA sequences. Proc. Natl. Acad. Sci. U.S. 78, 2727-2731, 1981.

Meisels, A., Roy, M., Fortier, M., Morin, C., Casas-Cordero, M., Shah, K.V., and Turgeon, H. Human papillomavirus infection of the cervix. The atypical condyloma. Act. Cytol. 25, 7-16, 1981.

Moar, M.H., Campo, M.S., Laird, H.M., and Jarrett, W.F.H. Unintegrated viral DNA sequences in a hamster tumor induced by bovine papilloma virus. J. Virol. 39, 945-949, 1981.

Orth, G., Jeanteur, P., and Croissant, O. Evidence for and localization of vegetative viral DNA replication by autoradiographic detection of RNA-DNA hybrids in sections of tumors induced by Shope papilloma virus. Proc. Natl. Acad. Sci. U.S. 68, 1876-1889, 1971.

Orth, G., Favre, M., and Croissant, O. Characterization of a new type of human papillomaviruses that causes skin warts. J. Virol. 24, 108-120, 1977.

Orth, G., Favre, M., Breitburd, F., Croissant, O., Jablonska, S., Obalek, S., Jarzybek-Chorzelska, M., and Rzesa, G. Epidermodysplasia verruciformis: a model for the role of papilloma viruses in human cancer. In: Viruses in Naturally Occurring Cancer, Essex, M., Todaro, G., and zur Hausen, H. (eds.). Cold Spring Harbor Lab. Press, Vol. A, 259-282, 1980.

Pfister, H., Nürnberger, F., Gissmann, L., and zur Hausen, H. Characterization of a human papillomavirus from epidermodysplasia verruciformis lesions of a patient from Upper Volta. Int. J. Cancer 27, 645-650, 1981.

Pfister, H., Fink, B., and Thomas, C. Extrachromosomal bovine papillomavirus type 1 DNA in hamster fibromas and fibrosarcomas. Virology 115, 414-418, 1981.

Shah, K.H., Lewis, M.G., Jenson, A.B., Kurman, R.J., and Lancaster, W.D. Papillomaviruses and cervical dysplasia. Lancet II, 1190, 1980.

zur Hausen, H. Oncogenic herpesviruses. Biophys. Biochim. Acta 417, 25-53, 1975.

zur Hausen, H. Condylomata acuminata and human genital cancer. Cancer Res. 36, 530, 1976.

zur Hausen, H. Human papillomaviruses and their possible role in squamous cell carcinomas. Curr. Top. Microbiol. Immunol. 78, 1-30, 1977.

zur Hausen, H. The role of viruses in human tumors. Adv. Cancer Res. 33, 77-107, 1980.

zur Hausen, H., Gissmann, L., Steiner, W., Dippold, W., and Dreger, J. Human papillomaviruses and cancer. Bibl. Haematol. 43, 569-571, 1975.

HEPATITIS B VIRUS: MOLECULAR BIOLOGY STUDIES ON ITS PRESUMPTIVE ROLE IN HUMAN PRIMARY LIVER CANCER

Rajen Koshy and Peter Hans Hofschneider

Abt. Virusforschung
Max-Planck-Institut für Biochemie
8033 Martinsried, FRG

INTRODUCTION

Hepatitis B virus (HBV) is the aetiological agent for the most widespread and virulent form of hepatitis in man. In some countries of the world, particularly tropical Africa and the Far East, the virus is endemic and the incidence of hepatitis is extremely high. It has long been observed that the frequency of primary liver cancer (PLC) is also unusually high among peoples of those particular countries (1). It has further been extensively documented that this frequency of liver cancer is correlated to previous infection with HBV and indeed, that PLC is commonly associated with the presence in the serum of HBs Ag, the major surface antigen of HBV and/or other HBV antigens and their corresponding antibodies (2). These observations lead to the obvious question of an aetiological role for HBV in malignant liver disease.

The hepatitis B virus is a small DNA virus belonging to a unique class of viruses along with the more recently discovered woodchuck hepatitis virus (3), the beechy ground squirrel hepatitis virus (4) and the Pekin duck hepatitis virus (5). The genome is a circular, partially double stranded DNA molecule with a nick in one strand and a variable gap in the other. The DNA is associated with viral core proteins and a DNA polymerase which can 'fill in' the gap in the short strand of the DNA, in an endogenous reaction (6).

The DNA of this virus has been cloned by several groups (7,8,9,10) and is available in large quantities for further study.

In an attempt to understand the role of HBV in liver cancer we are studying the relationship of the HBV genome with that of the malignant cell, at a molecular level. In an initial survey we have analysed DNA from liver tissue of patients with PLC and of those with cirrhosis.

By means of the blot-hybridization technique of Southern (11), using $\alpha^{32}P$ labelled cloned HBV DNA as probe we have shown that HBV specific sequences are present integrated in the DNA of PLC cells as well as of cirrhotic cells: Sequences were also found in DNA of cultured cells of lines established from primary liver cancer tissues.

MATERIALS AND METHODS

Liver specimens were obtained post surgically or at autopsy and frozen immediately at -80°C until they were used for DNA extractions. Samples were received coded and studies were done blind. In table I are presented relevant information about the samples analysed. Seven PLC, 5 cirrhosis and 1 angiosarcoma cases were studied.

Six cell lines derived from primary liver carcinoma tissues were also studied. These were PLC/PRF/5 and Mahlavu (12), Hep 3B (13), BEL 7402, BEL 7404 and BEL 7405 (14). PLC/PRF/5 and Hep 3B are cell lines which are continuously producing the HBs Ag whereas the others do not. Cells were grown in plastic flasks (Falcon) in Dulbecco's minimal essential medium supplemented with 10% fetal calf serum.

DNA extraction

High molecular weight DNA was extracted from the tissues as follows: Tissues were minced finely and then homogenized in a Sorvall omnimix homogeniser in 3 vols. of 50 mM tris, pH 9.0, and 10 mM EDTA. The homogenates were then adjusted to 1% sodium dodecyl sulfate (SDS) and incubated at 60°C for 20 minutes, in order to lyse the nucleic. The mixture was then gently shaken with one volume of a buffered solution (PCI-9) of phenol, chloroform and isoamyl alcohol (5:5:1, pH 9.0) for 2 hours. One volume of chloroform was then added and the mixture shaken for another 2 hours. The aqueous phase containing the DNA was recovered by centrifugation at 12,000 x g in a Sorvall RC 2-B centrifuge. This solution was shaken overnight with one volume of chloroform and recovered by centrifugation as before, and precipitated with 2 vol. of ethanol at -20°C. The precipitated DNA was air dried and dissolved in 50 mM tris (pH 9.0) and 0.5% SDS and incubated with 300 μg/ml of proteinase K for 4 hours at 37°C. This was followed by a second cycle of extractions with PCI-9 and chloroform as described and the purified DNA was ethanol precipitated and the DNA was dissolved in distilled water and stored at 4°C.

Cultured cells were harvested by treatment with 0.25% trypsin and washed with phosphate buffered saline. The cells were recovered by centrifugation at 800 rpm for 10 min. Cell pellets were gently resuspended in 3 vol. of 50 mM tris (pH 9.0) and 10 mM EDTA. They were then lysed in 1% SDS and further processed as described for the tissue homogenates.

Restriction endonuclease digestions

Restriction endonucleases Hind III and Hha I were used in this study. Hind III digestions were done in 20 mM tris-HCl (pH 7.4), 7 mM $MgCl_2$ and 60 mM NaCl for 6-8 hrs at 37°C. Hha I digestions were done in 50 mM tris-HCl (pH 7.4), 5 mM $MgCl_2$ and 0.5 mM dithiothreitol, at 37°C for 6-8 hrs.

Gel electrophoresis and DNA transfer

DNA samples containing bromophenol blue tracking dye were applied to slots in horizontal agarose slab gels (0.8% in 40 mM tris pH 8.0, 18 MM NaCl, 20 mM sodium acetate and 2 mM EDTA). 40 μg of DNA were loaded per slot. Electrophoresis was done in the same solution as was used to prepare the gel, at 100 V and 80 mA. Hind III digested bacteriophage lambda DNA was used as molecular weight markers in an adjacent slot of the gel.

Following electrophoresis, the gels were prepared for DNA transfer to nitrocellular filters, as follows: 20 minutes in 0.5 N NaOH and 1-5 M NaCl, to denature the DNA in situ, 60 minutes in 0.5 M tris-HCl (pH 7.5) and 3.0 M NaCl in order to neutralize the alkali and then they were placed in 6 x SSC (SSC = 0.15 M NaCl and 0.015 M sodium citrate) for 10 minutes prior to transfer. Between each change of solution, the gels were well rinsed in water.

A modification (15) of the procedure of Southern was made to transfer the DNA from the gels to nitrocellulose membranes (Schleicher and Schüll). Following transfer, the membranes were baked in a vacuum oven at 80°C for 4 hrs.

Radio-labelling of HBV DNA and hybridization

The HBV used in the experiments had been cloned in E. coli using plasmid pAO1 (pAO1-HBV). The recombinant plasmid was kindly supplied by Dr. J. Summers (10). The DNA was labelled in vitro with $\alpha^{32}P$ dATP (400 Ci/mMol) (New England Nuclear), by the nick translation reaction described by Rigby et al. (16). Specific activities of 1×10^8 cpm per microgram were obtained. Unreacted triphosphates were removed from the preparation by passing it through a Sephadex G-50 (Pharmacia) column. The DNA was eluted with 50 mM tris (pH 7.5) and 10 mM EDTA. Before hybridization, the DNA was denatured by boiling for 5 minutes and quickly added to the hybridization solution. The nitrocellulose filters were prehybridized in a solution containing 5 x SSC, 50% formamide, 50 mM sodium phosphate (pH 6.5), 200 µg/ml of sonicated, denatured, salmon sperm DNA and 5 x Denhardt's reagent. (1 x Denhardt's reagent = 0.02% bovine serum albunin, 0.02% polyvinylpyrollidone and 0.02% ficoll.) Prehybridization was done at 37°C for at least 6 hours.

The filters were then transferred to the hybridization mixture containing 5 x SSC, 50% formamide, 50 mM sodium phosphate, 200 µg/ml of sonicated, denatured salmon sperm DNA, 5 x Denhardt's reagent, 10% sodium dextran sulfate (Pharmacia), and 10-20 nanograms of the $\alpha^{32}P$ labelled HBV DNA probe per filter. Incubation was at 37°C for about 20 hours. The filters were washed extensively in 1 x SSC and 0.5% SDS at 60°C until no radioactivity could be detected in the wash solution. SDS was removed by a final rinse in 1 x SSC. The filters were dried and autoradiographed on Kodak X'omat R X-ray films in the presence of Quanta III (Dupont) intensifying screens.

RESULTS AND DISCUSSION

Two restriction endonucleases were used to analyse the DNA samples. Hind III was selected because no HBV DNA has so far been shown to have a cleavage site for this enzyme. Bands seen at positions indicating sizes greater than unit length of the HBV genome (3.2 kilobase pairs) could be taken as evidence of integration of HBV

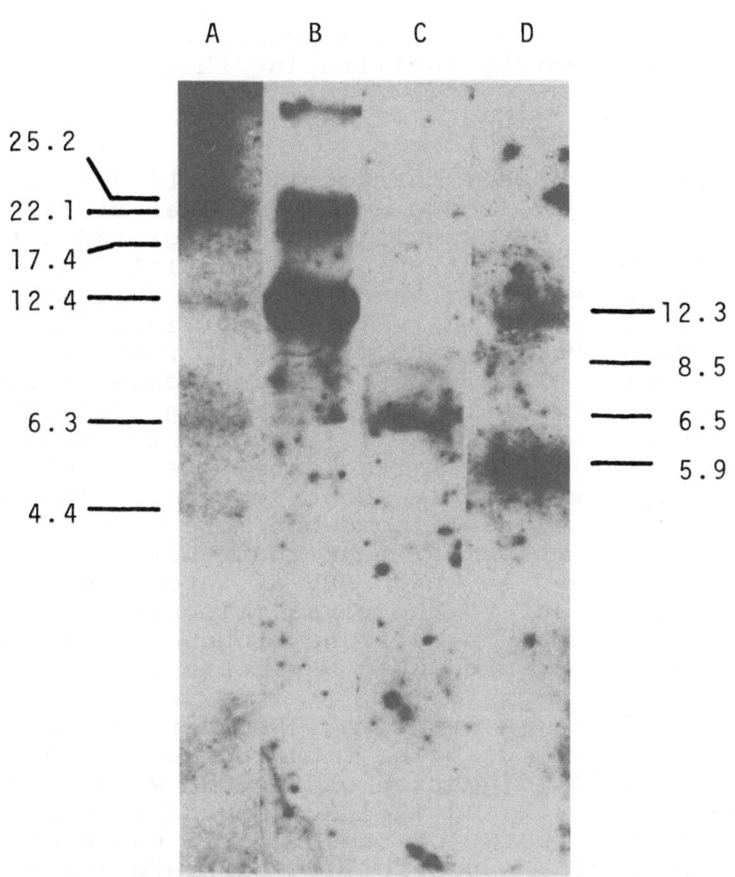

Fig. 1. Hind III analyses of DNA from cell lines and tissues.
a) PLC/PRF/5; b) Hep 3B; c) D-9 tumour; d) D-2 tumour.

in the host genome; additional information about the number of integrated copies and a comparison of integration patterns in different samples can be obtained.

Hha I was used because it cuts within the HBV genome more than once. If HBV is present rarely or in a variety of sites, they may not be detectable by the use of Hind III. In such a case the use of an internal cutting enzyme would concentrate purely viral fragments of the same size thereby facilitating their detection.

Analyses of cell lines

PLC/PRF/5, a well characterized PLC derived cell line, continuously producing HBs Ag, has previously been shown to contain HBV specific DNA integrated within the genome (17-20). Hybridization of Hind III digested DNA revealed 6 HBV specific bands of 25.2, 22.1, 17.4, 12.4, 6.3 and 4.4 kilobase pairs (Kb) (Fig. 1a). This result is in close agreement with previously reported results. Hha I digestion of this DNA produced bands of 7.0, 5.6, 4.2, 3.0, 2.4, 1.7, 1.6 and 1.3 Kb (Fig. 2a). The greater intensities of the 1.7 and 1.6 Kb bands suggest that they are 'internal' fragments, i.e. purely viral fragments without flanking host sequences. The sum of these two bands alone is somewhat more than the size of the HBV genome. This must mean that there are at least two different ways in which HBV is integrated in these cells, with respect to the viral molecule. A further possibility is that there is heterogeneity of restriction sites in the integrated HBV copies.

A second HBs Ag producing, PLC derived cell line designated Hep 3B, was also found to contain HBV sequences. Hind III digestion of the DNA gave 2 HBV specific bands at 24.2 and 12.1 Kb (Fig. 1b). These are similar in size to two of the Hind III bands of PLC/PRF/5, which suggest the possibility of preferential integration of HBV in some sites.

Four other PLC-derived cell lines, viz., Mahlavu, BEL 7402, BEL 7404 and BEL 7405 were all negative for HBV sequences. These cells did not produce any HBs Ag as assayed by radioimmunoassay (AUSRIA-ABBOTT laboratories). The results are summarized in table 1.

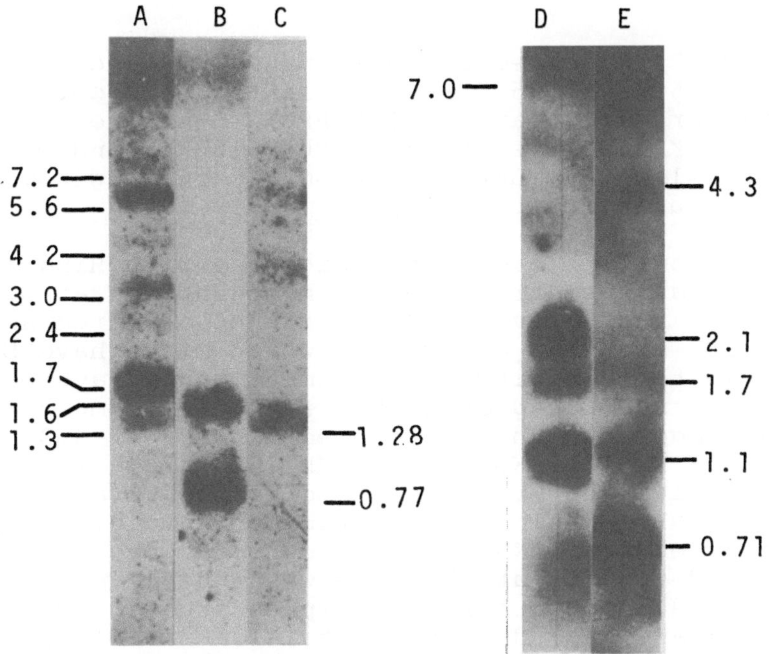

Fig. 2. Hha I analyses of cell line and liver tissue DNA.
a) PLC/PRF/5; b) D-18 peritumour; c) F-1 cirrhotic liver; d) G-1 cirrhotic liver; e) G-5 PHC.

Analyses of tissues

a) Hepatocellular carcinomas

Three out of seven DNA samples from PLC tissue were seen to contain HBV sequences (table 1).

Hind III digested DNA in the case of sample D-2 revealed 2 bands of 12.3 Kb and 5.9 Kb. The 12.3 Kb band corresponds to similar sized bands in PLC/PRF/5 and Hep 3B DNAs.

Hind III digested DNA of sample D-9 generated two bands of 8.5 Kb and 6.5 Kb. The latter band corresponded to a similar sized band in PLC/PRF/5 DNA. The faintness of the 8.5 Kb band relative to the 6.5 Kb band probably reflects a lower amount of HBV DNA (partial genome) in that particular site.

DNA from sample G-5 when digested with Hind III produced a dark smear which was most intense at 3.2 Kb (not shown) indicative of the presence of large amounts of free HBV DNA. Any other band which might have been present in the vicinity of the smear would have been obscured by the background. DNA from this same sample, upon cleavage with Hha I produced 5 bands of 4.3 Kb, 2.1 Kb, 1.7 Kb, 1.1 Kb and 0.7 Kb (Fig. 2e). These results indicate the presence of both free viral DNA and integrated HBV DNA in this particular sample. The presence of free viral DNA is to be expected because the patient had viremia at the time when the liver sample was obtained. Four other PLC samples studied (D-10, D-12, D-21 and G-3) did not contain detectable HBV sequences, irrespective of whether Hind III or Hha I was used to cleave the DNA.

Analyses of cirrhosis samples

Two out of five samples of liver cirrhosis also were seen to contain integrated HBV sequences. The DNA of sample F-1 showed no bands after digestion with Hind III. However, digestion with Hha I showed the presence of a single band of 1.28 Kb (Fig. 2c). This suggests that in this case only a small portion of the HBV genome might have been inserted into the host DNA and possibly only in a proportion of the cells, or alternatively at different sites.

Table 1

Relation between HBs Ag expression and integrated HBV sequences in PHC tissues, cirrhosis and hepatoma cell lines.

	CODE	ORIGIN	HBs Ag Serum	HBs Ag Tissue	Integrated HBV Sequences
P.H.C. TISSUE	D 2	Senegal	+	+	+
	D 9	"	+	+	+
	D12	"	+	+	-
	G 5	F.R.Germany	+	NT**	+
	D10	Senegal	-	-	-
	D21	"	-	-	-
	G 3	F.R.Germany	NT	NT	-
CIRRHOSIS TISSUE	F 1	France	+	+	+
	G 1	F.R.Germany	+	NT	+
	D18PTA*	Senegal	+	Necrotic	+
	F 2	France	-	NT	-
	G 2	F.R.Germany	-	NT	-
	G 4	"	-	NT	-
			Culture medium		
CELL LINES	PLC/PRF/5	S.Africa	+		+
	HEP.3B	USA	+		+
	MAHLAVU	S.Africa	-		-
	BEL.7402	China	-		-
	BEL.7404	"	-		-
	BEL.7405	"	-		-

** NT: Not tested.
* PTA: Peritumoral area.

DNA of sample G-1 digested with Hind III produced a smear on the autoradiogram, similar to that in the case of G-5. Again a vast amount of free viral DNA was present as a contaminant due to the viremic state of the patient. In this case too, Hha I digestion of the DNA eliminated the smear, revealing bands of 7.0 Kb, 2.0 Kb, 1.5 Kb, 1.1 Kb and 0.9 Kb (Fig. 2d). The presence of several intense bands which could only be 'internal' viral fragments suggest that in addition to the population of episomal viral DNA, there is also integrated HBV DNA. Samples F-2, G-2 and G-4 did not contain detectable amounts of HBV DNA.

In most of the carcinoma and cirrhosis samples studied above, normal healthy tissue around the diseased areas were also similarly analysed. In each case the DNA was negative for HBV sequences.

One last case, D-18 merits mention. This was an angiosarcoma in the liver. The DNA of the tumor tissue contained no HBV specific DNA. However the DNA from the surrounding normal areas (peritumoral area) produced two HBV specific bands of 1.4 Kb and 0.8 Kb (Fig. 2b), when cleaved with Hha I. Hind III digestion had failed to produce any bands. In this particular case, insertion of subgenomic fragments of HBV may have occurred. The tissue was rather necrotic precluding an accurate histological identification of the cells in the peritumoral area.

The presence of HBV sequences in the tumour cells lends strength to the hypothesis that HBV plays a role in hepatocarcinogenesis.

The presence of HBV specific DNA in cirrhosis is significant considering the fact that cirrhosis frequently progresses to hepatocellular carcinoma and an important question would be whether there is a correlation between cirrhosis which ultimately result in carcinoma, and the presence of the HBV genome in those cases.

Another observation in this study is that in every case where integrated HBV was detected, both in fresh tissues as well as cultured cells, HBs Ag was expressed, suggesting that the site of integration may be fairly specific with respect to the viral molecule. The gene coding for the HBs Ag is not interrupted as a consequence of integration. On the other hand, except in the viremic cases, there was no expression of HBc Ag (HBV core antigen) suggesting that the gene coding for it might be interrupted.

The integration of viral information (either complete or relevant subgenomic fragments) in most known tumour virus systems appear to be a prerequisite to malignant transformation of the host cell concerned (21a,b). It is tempting to draw comparisons with such known systems in speculating on the significance of HBV integration in liver carcinoma cells. Some important questions we are now trying to answer are:

a) Are there specific differences in the association of HBV genome with respect to the HBV related diseases such as acute and chronic hepatitis, cirrhosis and carcinoma? We are continuing our survey of larger numbers of samples of the different diseases to elucidate the question.

b) What is the nature of the integration sites in the host DNA? Are there some specific preferred sites as might be indicated by the similarities in the sizes of some of the bands in different samples described earlier? What are the host sequences flanking the HBV inserts? In order to address these questions, we have attempted to clone integrated HBV fragments from the samples described above. At this time we have obtained the 6.3 Kb integrated fragment from PLC/PRF/5 cell DNA (R. Koshy, A. Freytag von Loringhoven, P. H. Hofschneider and K. Murray), by cloning it in a λ-vector. This fragment should contains the entire HBV genome and approximately 3 Kb of flanking host sequences. These are now being sequenced and we are attempting to obtain more cloned inserts for comparative studies.

c) Does HBV DNA in its integrated state have transforming potential? We are presently studying the ability of the isolated cloned HBV fragment referred to above, to transform mouse 3T3 cells. Preliminary evidence indicates the development of transformed foci in the treated cells. The stability of these foci (transformed phenotype) is to be still to be established and further studies on the genetic content of these cells, are being done. If there are true transformants, further proof will be required to attribute the transformation to the DNA used in the transfection experiments. These and other experiments to study the transforming genes in liver carcinoma are currently being pursued.

ACKNOWLEDGMENTS

We are grateful to Dr. Ph. Maupas and R. Müller for providing us with tissue samples, to Dr. S. Plotkin for the Hep 3B cell line, to Dr. Chen Ruiming for the BEL lines and to Drs. J. Summers and W. Salzer for the plasmid pAO1-HBV.

REFERENCES

1. W. Szmuness, E. J. Harley, H. Ikram and C. E. Stevens. Sociodemographic aspects of the epidemiology of hepatitis B: In Viral Hepatitis, G. N. Vyas, S. N. Cohen and R. Schmid, ed., Franklin Institute Press, Philadelphia, 297-320 (1978).
2. A. Goudeau, Ph. Maupas, P. Coursaget, J. Drucker, J. P. Chiron, F. Denis and I. Diopmar. Hepatitis B virus antigens in human primary hepatocellular carcinoma tissues. International Journal of Cancer 24, 421-429 (1979).
3. J. Summers, J. M. Smolec and R. Snyder. A virus similar to human hepatitis B virus associated with hepatitis and hepatoma in woodchucks. Proceedings of the National Academy of Sciences, U.S.A. 75, 4533-4537 (1978).
4. P. L. Marion, L. Oshira, D. C. Regnery, G. H. Scullard and W. S. Robinson. A virus in Beechy ground Squirrels that is related to hepatitis B virus of humans. Proceedings of the National Academy of Sciences, U.S.A. 77, 2941-2945 (1980).
5. W. S. Mason, G. Seal and J. Summers. Virus of Pekin ducks with structural and biological relatedness to human hepatitis B virus. J. Virology 36, 829-836 (1980).
6. W. S. Robinson. The genome of hepatitis B virus. Annual Review of Microbiology 31, 357-377 (1977).
7. C. J. Burrell, P. Mackay, P. J. Greenaway, P. H. Hofschneider and K. Murray. Expression in Escherichia coli of hepatitis B virus DNA sequences cloned in plasmid pBR 322. Nature 279, 43-47 (1979).
8. P. Charnay, C. Pourcel, A. Louise, A. Fritsch and P. Tiollais. Cloning in Escherichia coli and physical structure of hepatitis B virion DNA. Proceedings of the National Academy of Sciences, U.S.A. 76, 2222-2226 (1979).
9. J. J. Sninsky, A. Siddiqui, W. S. Robinson and S. N. Cohen. Cloning and endonuclease mapping of the hepatitis B viral genome. Nature 279, 346-348 (1979).
10. I. W. Cummings, J. K. Browne, W. A. Salser, G. V. Tyler, R. L. Snyder, J. M. Smolec and J. Summers. Isolation, characterization, and comparison of recombinant DNAs derived from genomes of human hepatitis B virus and woodchuck hepatitis virus. Proceedings of the National Academy of Sciences, U.S.A. 77, 1842-1846 (1980).

11. E. M. Southern. Detection of specific sequences among DNA fragments separated by gel electrophoresis. Journal of Molecular Biology 98, 503-517 (1975).
12. G. M. Macnab, J. J. Alexander, G. Lecatsas, E. M. Bey and J. M. Urbanowicz. Hepatitis B surface antigen produced by a human hepatoma cell line. British Journal of Cancer 34, 509-515 (1976).
13. D. P. Aden, A. Fogel, S. Plotkin, I. Damjanov and B. B. Knowles. Controlled synthesis of HBs Ag in a differentiated human liver carcinoma-derived cell line. Nature 282, 615-616 (1979).
14. C. Ruiming, Z. Dehou, Y. Xiuzheu, S. Dingwu and L. Ronghua. Establishment of three human liver carcinoma cell lines and some of their biological characteristics in vitro. Scientia Sinica 23, 236-247 (1980).
15. R. Koshy, R. Gallo and F. Wong-Staal. Characterization of the endogenous feline leukemia virus related DNA sequences in cats and attempts to identify exogenous viral sequences in tissues of virus negative leukemic animals. Virology 103, 434-445 (1980).
16. P. W. Rigby, M. A. Dieckmann, C. Rhodes and P. J. Berg. Labelling of deoxyribonucleic acid to high specific activity in vitro by nick translation with DNA polymerase I. Journal of Molecular biology 113, 237-251 (1977).
17. P. L. Marion, F. H. Salazar, J. J. Alexander and W. S. Robinson. State of hepatitis B viral DNA in a human hepatoma cell line. Journal of Virology 33, 795-806 (1980).
18. C. Brechot, C. Pourcel, A. Louise, B. Rain and P. Tiollais. Presence of integrated hepatitis B virus DNA sequences in cellular DNA of human hepatocellular carcinoma. Nature 286, 533-535 (1980).
19. P. R. Chakraborty, N. Ruiz-Opazo, D. Shouval and D. A. Shafritz. Identification of integrated hepatitis B viral DNA and expression of viral RNA in an HBs Ag-producing human hepatocellular carcinoma cell line. Nature 286, 531-533 (1980).
20. J. C. Edman, P. Gray, P. Valenzuela, L. B. Rall and W. J. Rutter. Integration of hepatitis B virus sequences and their expression in a human hepatoma cell. Nature 286, 535-538 (1980).
21a. J. Tooze. The molecular biology of tumour viruses. Cold Spring Harbor publications. (1973).
21b. J. Tooze. DNA tumour viruses. In Molecular Biology of Tumour Viruses. 2nd Edn. (1980).

VIRAL MARKERS IN HUMAN MALIGNANCIES - EPIDEMIOLOGICAL STUDIES

Guy de-Thé

CNRS Laboratory, Faculty of Medicine Alexis Carrel

69372 LYON CEDEX 2 and I.R.S.C./CNRS, Villejuif, France

On these shores, twenty four centuries ago, in the treaty on the "Origin of Diseases", Hippocrates urged to investigate "first, the effect of seasons, winds and rains, second, the way in which the inhabitants live, drink, eat, were indolent or active".

Twenty four centuries later, we have accumulated an immense knowledge on the origins of diseases, but still have a long way to go to control two main killers associated with our ways of life, namely cardio-vascular and malignant diseases. With most powerful tools, from computers to nucleic acid hybridization, our aim in epidemiology remains the same, i.e. to study first the distribution and second the determinants of diseases. "Determinants" mean causes and cause does not mean the same for the epidemiologists and the experimental researchers. The epidemiological concept of cause relates to _risk factors_ to be eliminated to prevent the diseases, whereas the aim of experimentalists is to uncover the _origins_ and the molecular mechanism of disease process.

Viral oncology is involved in both these complementary aspects of cancer research: some viruses do represent "_critical risk factors_" for naturally occurring tumors in animal and man, and oncogenic viruses represent unique tools to uncover the molecular mechanism leading to a cancer cell.

Cancer epidemiologists have made two important discoveries in the last decades. The first refers to the role of the environmental factor in tumor development: geographical comparisons of the distribution of cancers around the world showed that cancer on specific sites in specific organs varies widely between geographical areas and cultural groups (Berg, 1977). Each cultural pattern carries specific risk factors for certain diseases including cancers. Thus, research on the origins and determinants of human tumors necessitates a multidisciplinary approach involving epidemiology, life-science technology up to social sciences. The second contribution of cancer epidemiologists was to suggest that malignant diseases may result from a multistep process, each step having independent cause(s) (Peto, 1977). Viruses may act as the cause of one step, without representing the necessary nor sufficient agent for oncogenesis (de-Thé, 1980). A critical question is to know whether antiviral intervention could succeed in preventing virus associated cancers.

Three types of viruses are candidates to play a role in human malignancies:
1) the hepatitis B virus (HBV) in primary liver carcinoma (PLC),
2) the herpes viruses, mainly the Epstein-Barr virus (EBV) , closely associated with endemic Burkitt's lymphoma (BL) and endemic and non-endemic nasopharyngeal carcinoma (NPC) ; the association between cytomegalovirus and Kaposi sarcomas and that of herpes simplex virus type 2 as a risk factor for cervical carcinoma are still a matter for debate,
3) oncorna viruses, associated with many naturally recurring leukemias and sarcomas in animals, might exist in man as presented by Gallo at this meeting.

I. COMPARATIVE CHARACTERISTICS OF HBV-EBV AND FeLV IN ONCOGENESIS

The relationship between hepatitis B virus (HBV) and primary liver carcinoma (PLC) is being covered in this meeting by Hofschneider (for review, see also Maupas and Melnick, 1981). The association between feline leukemia virus with spontaneous leukemia and lymphoma in cats is being presented by Dr. Essex. I would like to compare here the epidemiological properties of the EBV and the associated malignancies versus the two above mentioned systems (Essex and Gutensohn, 1981 ; Francis et al, 1981).

Table 1 compares the three systems: HBV-EBV-FeLV and associated malignancies. It is worth noting that in spite of the widely different type of viruses involved in tumor development, certain epidemiological properties are in common:

a) the viruses involved are ubiquitous,
b) the early age at viral infection is critical for the risk of developing the associated malignancies,
c) tumor development is a late manifestation of viral infection,
d) the transmission of virus is horizontal, through saliva, leading to a long latent period elapsing between primary infection and tumor development.

Table 1
COMPARATIVE EPIDEMIOLOGICAL PROPERTIES OF THREE VIRUS-ASSOCIATED NATURALLY OCCURRING MALIGNANCIES

virus involved	DNA-Hepatovirus Hepatitis B virus (HBV)	DNA-Herpesvirus Epstein-Barr virus (EBV)		RNA-Retrovirus Feline Leukemia Virus (FeLV)
associated tumor	liver carcinoma	Burkitt's lymph. (BL)	nasopharyng. carcinoma (NPC)	cat leukemias lymphomas
age of primary infection	very early (0–1 y)	very early (0–2 y)	early (0–5y)	early or late
ubiquitous virus	+/–	+	+	+
virus transmission-horizontal	saliva	saliva	saliva	saliva
relationship dose/effect	yes	yes	?	yes
pick of age incidence of Tm	early adults	infants	early adults	yound and old
viral markers in tumor cells in endemic areas in non endemic areas	70 % 5 to 30 %	95 % 5 to 10 %	98 %	early infection: virus positive Tm late infection: virus free tumor
pre-cancerous states	chronic carrier → cirrhosis → liver carcinoma	very early EBV infection → latency → BL	early infect. latency → reactiv. → NPC	early infection → chronic viremia → leukemia
transformation in vitro of target cells by the virus	?	immortalization of human B lymphocytes		transformation of cat cells
need for other environmental factors	aflatoxin	malaria	nitrosamines	??
preventive intervention	vaccine trial (heat inactiv. HBV)	vaccine development		successful vaccination
Final proof of causality	on going in Senegal	theoretically possible		achieved

The long"incubating period" represents 30 to 70 % of the life span for the primary liver carcinoma and nasopharyngeal carcinoma, whereas it represents only 10 to 20 % of life span for both Burkitt's lymphoma and cat leukemia. The sequence of events elapsing between primary infection and tumor development is that of a chronic carrier state in the hepatitis B virus in relation to liver carcinoma and of active viremia in the case of feline leukemia. The situation is somewhat different in the EBV-associated malignancies where there is a long and quiet latency between primary infection and the development of Burkitt's lymphoma and nasopharyngeal carcinoma. For nasopharyngeal carcinoma however, a reactivation preceeds clinical onset of the disease ; the critical question in the later system being to know whether reactivation can "precipitate" tumor development and represent the immediate cause of NPC or only a passive parasitism accompanying sublicinal NPC (de-Thé and Zeng, in press),

e) the association between these 3 types of viruses and the corresponding malignancies is based on the regular presence of viral markers in tumor cells. Such an association may vary from endemic to non-endemic situations. In the endemic areas for primary liver carcinoma, 70 % at least or more of the tumors have HBV markers including viral DNA, whereas in non-endemic areas, only 5 to 30 % of primary liver carcinomas have viral markers detectable at the tumor cell level (Hofschneider, this meeting ; Maupas and Melnick, 1981 ; Francis et al, 1981). In the EBV-associated malignancies, the situation is similar for Burkitt's lymphoma: 95 % of Burkitt's lymphoma in the endemic areas of Equatorial Africa have viral markers in tumor cells (Olweny et al, 1977), versus 5 to 10 % in the non-endemic Burkitt type lymphomas observed in temperate climates (Ziegler et al, 1976). In sharp contrast, undifferentiated nasopharyngeal carcinomas exhibit the same level of association (95 to 98 % of the cases exhibit viral markers in the tumor cells) in endemic or non-endemic areas. The situation of cat leukemias and lymphomas might enlighten the human situation in BL: early and massive FeLV infection favors virus positive cat leukemias and lymphomas, whereas late or mild infection by FeLV induces viral free leukemias (Essex et al, 1977). As Dr. Essex will present here, the Focma assay involving virally induced antigens represented a critical tool to uncover this fact. Unfortunately, we do not have a similar test for EBV.

Needs for Other Environmental Factors

In the case of hepatitis B virus and Epstein-Barr virus, there is an obvious need for other environmental factors in the development of the associated malignancies. In the case of HBV, aflatoxin plays a critical role (Linsell and Peers, 1977). In the case of Burkitt's lymphoma, malaria and chromosome anomalies are epidemiologically and experimentally involved (Klein, 1979 ; de-Thé, 1980). In the case of nasopharyngeal carcinoma, nitrosamines have been implied to play an important role (Ho, 1978). Finally, the comparison between these three systems should help to assess the possibilities of prevention of these tumors by anti-viral measures. HBV vaccine trials involving human subjects are going on in Senegal, using inactivated HBV virus (Maupas et al, 1981). Vaccines for the Epstein-Barr virus are not yet available but a certain number of laboratories in the United States and Europe are involved in such a research.

II. RECENT EPIDEMIOLOGICAL DEVELOPMENT IN EBV-ASSOCIATED MALIGNANCIES

A - Burkitt's Lymphoma (BL)

Two new cases have been detected (Geser et al, submitted for publication) since the last results in the Ugandan prospective study (de-Thé et al, 1978). Both have much higher VCA antibody titer than controls, 5 and 6 years prior to clinical onset of BL. A girl is of particular interest since she was bled at three months of age and bled again at one year of age. On both occasions, she had a high VCA titer (1280) with presence of antibodies to EA but no IgM antibodies. Her mother had low antibody titer which indicated that this girl had been infected very early after birth by EB virus. This further supported the hypothesis put forward (de-Thé, 1977) that early and massive infection represented a major risk factor for developping EBV-associated and endemic Burkitt's lymphoma. In fact, the increased relative risk of developping Burkitt's lymphoma is now 5, 25 and 125 times higher for a child having VCA antibody titers 1, 2 or 3 dilutions above the mean of age, sex and locality matched control group (Geser et al, submitted for publication).

However, there is in Africa a small proportion of EBV-free lymphomas (from 1 to 4 %) and in temperate climates, 90 to 95 % of Burkitt type lymphomas are EBV-free. It is therefore clear that EBV is not the necessary nor sufficient factor for all Burkitt's lymphomas but represents a critical risk factor for the endemic tumor. A situation which calls to mind the relationship between cigarette smoking and bronchial carcinoma. Other factors in BL pathogenesis appear to be malaria in the endemic zone and for both endemic and non-endemic Burkitt type lymphomas, chromosomal anomalies involving chromosome 8 and translocations 8;14 but also 8;22 and 8;2 (Berger et al, 1979). In conclusion, early EBV infection in Equatorial Africa seems to act as an initiating factor followed by further oncogenic steps involving malaria and chromosomal changes, leading to Burkitt's lymphoma (Klein, 1979 ; de-Thé, 1980).

B – Nasopharyngeal Carcinoma (NPC)

Recent development involved early detection of the tumor and the characterization of pre-cancerous conditions. In utilizing the regular presence of IgA antibodies specific to EBV-VCA, Zeng et al (1979, 1980) implemented large serological surveys in high NPC risk areas in the Guang-Xi Autonomous Region, involving 150,000 individuals and leading to the finding of 1,200 IgA-VCA positive individuals and 47 nasopharyngeal carcinomas. Among the IgA positive individuals followed-up twelve months later, 12 new NPC were detected, representing an incidence of 1 % (1,000 for 100,000) per year. Thus, the detection of IgA antibody is an excellent tool for early detection of this cancer and for the characterization of the group at immediate risk for NPC.

Pre-NPC conditions

The nasopharyngeal biopsy of 56 individuals IgA positive since 18 months, allowed us to uncover 4 nasopharyngeal carcinomas plus 14 individuals (IgA positive but symptomless) with viral markers in their nasopharyngeal mucosa. EBV/DNA was detected in their biopsies by using a cloned internal repeat of EBV/DNA (Desgranges et al, in press). This technique required 16 micrograms of DNA, which in some cases cannot be obtained. The spot hybridization test as developed by Brandsma and Miller (1980) is now being tested to see if it can be applied to large numbers of specimens as necessary in field studies. A sequence of events preceeds NPC clinical onset: future NPC patients develop IgA antibodies to VCA, possibly in a transient manner first, then chronically with the presence of viral DNA and EBNA in their nasopharyngeal mucosa, these being an immediate risk for NPC. There is a gradient of increasing immunological response to EBV from normal individuals to those carrying EBV in their nasopharyngeal mucosa and to NPC patients (de-Thé and Zeng, in press). From unpublished data from Zeng et al, it appears as if the IgA/VCA marker is present 12 to 18 months prior to clinical onset of the disease. This is the period during which anti-viral interventions should be implemented in order to try and prevent clinical onset of NPC.

III. ADULT T-CELL LEUKEMIA AND LYMPHOMA IN JAPAN

Dr. Gallo will present new data concerning the presence of a retro virus in the culture of T-cell malignancies. We would like to stress the interesting epidemiological characteristics of these T-cell malignancies in Japan. When looking at the proportion of T versus B cell lymphomas in the different regions of Japan, one observes that in the island of Kyushu, T-cell malignancies prevail largely over B-cell malignancies, in sharp contrast with the situation existing in USA and Europe (Tajima et al, 1979).

Furthermore, the birth place of the majority of patients with T-cell leukemia observed in the major specialized cancer centers in Japan, was found to be in the western region of the island of Kyushu and to a lesser extent the western part of the island of Shikoku (Tajima et al, 1979). Are environmental factors associated with such malignancy microbiological agents or chemical agents ? Such epidemiological feature may reflect a particular situation in these islands which existed 10 to 20 years ago but which no longer exists today.

In conclusion, the role of viruses in human malignancies is a matter of both basic and practical interests. Considering that primary liver carcinoma and nasopharyngeal carcinoma represent the main cancer killer for respectively 500 and 250 million inhabitants in Equatorial Africa and in South East Asia, the priorities should lie in developping anti-viral intervention with the aim to find ways of preventing these leading cancers in the developping world and to bring final proof of the causal relationship between the putative virus and the associated tumor.

REFERENCES

- Berg, J.W., 1977, World-Wide Variations in Cancer Incidence as Clues to Cancer Origins, in: Origin of Human Cancer, Hiatt, Watson and Winsten eds. Cold Spring Harbour Conferences on Cell Proliferation vol. 4. pp. 15-20.
- Berger, R., Bernheim, A., Weh, H.J., Flandrin, G., Daniel, M.T., Brouet, J.C. and Colbert, N. 1979b, A new translocation in Burkitt tumor cells, Hum Genet, 53:111-112.
- Brandsma, J. and Miller, G. 1980, Nucleic acid and spot hybridization: rapid quantitative screening of lymphoid cell lines for Epstein-Barr viral DNA. Proc. Natl. Acad. Sci. 77:6581-6855.
- Desgranges, C., Bornkamm, G.W., Zeng, Y., Wang, P.C., Zhu, J.S., Shang, M. and de-Thé, G. Detection of Epstein-Barr viral DNA internal repeats in the nasopharyngeal mucosa of Chinese with IgA/EBV-specific antibodies. Int. J. Cancer, submitted for publication.
- de-Thé, G. Is Burkitt's lymphoma (BL) related to a perinatal infection by Epstein-Barr virus (EBV) ? 1977, Lancet, i, 335-338.
- de-Thé, G., 1979, The epidemiology of Burkitt's lymphoma: evidence for a causal relationship with Epstein-Barr virus, Am. J. Epidemiol. 1:32-54.
- de-Thé, G., 1980, Multistep Carcinogenesis, Epstein-Barr virus and Human Malignancies. in: Viruses in naturally occurring cancers, M. Essex, G. Todaro and H. zur Hausen eds. Cold Spring Harbor Conferences on Cell Proliferation, vol. 7, pp. 11-21.
- de-Thé, G. and Zeng, Y. Epidemiology of Epstein-Barr virus: recent results on endemic versus non-endemic Burkitt's lymphoma and EBV-related pre-nasopharyngeal carcinoma conditions. In: Comparative

Research on Leukemia and Related Diseases, Yohn, ed. Elsevier North Holland, New-York, in press.

- de-Thé, G., Geser, A., Day, N.E., Tukei, P.M., Williams, E.H., Beri, D.P., Smith, P.G., Dean, A.G., Bornkamm, G.W., Feorino, P; and Henle, W. 1978, Epidemiological evidence for causal relationship between Epstein-Barr virus and Burkitt's lymphoma: results of the Ugandan prospective study, Nature, 274, 756-761.
- Essex, M. and Gutensohn, N. 1981, A Comparison of the Pathobiology and Epidemiology of Cancers Associated with Viruses in Humans and Animals, Prog. med. Virol.vol 27, pp. 114-126.
- Essex, M., Cotter, S.M., Stephenson, J.R., Aaronson, S.A. and Hardy, R., W.D., 1977 Leukemia, Lymphoma and Fibrosarcoma of Cats as Models for Similar Diseases of Man, in: Origins of Human Cancer, Hiatt, Watson and Winsten, eds. Cold Spring Harbor Conferences on Cell Proliferation, vol. 4, pp. 1197-1214.
- Francis, D.P., Essex, M. and Maynard J.E., 1981, Feline Leukemia Virus and Hepatitis B Virus: a Comparison of Late Manifestations, Prog. med. Virol, vol 27, pp. 127-132.
- Geser, A., de-Thé, G., Day, N.E. and Williams, E.H., Final case reporting from the Ugandan prospective study of the relationship between EBV and Burkitt's lymphoma. Int. J. Cancer, submitted for publication.
- Ho, J.H.C., 1978, An epidemiologic and clinical study of nasopharyngeal carcinoma, Int. J. Radiol. Oncol. Biol. Phys, 4:181-198.
- Klein, G., 1979, Lymphoma development in mice and humans: Diversity of initiation is followed by a convergent cytogenetic evolution, Proc. Natl. Acad. Sci. 76:2442.
- Linsell, C.A. and Peers, F.G., 1977, Field Studies on Liver Cell Cancer, in: Origins of Human Cancer, Hiatt, Watson, Winsten, eds, Cold Spring Harbor Conferences on Cell Proliferation, vol. 4, pp. 549-556.
- Maupas, P. and Melnick, J.L., 1981, Hepatitis B Infection and Primary Liver Cancer, Prog. med. Virol, vol 27. pp. 1-5
- Maupas, P., Coursaget, P., Chiron, J.P., Goudeau, A., Barin, F., Perrin, J., Denis, F. and Diop Mar, I, 1981, Active Immunization against Hepatitis B in an Area of High Endemicity, part I. Prog. med. Virol. vol 27, pp. 168-184.
- Maupas, P., Chiron, J.P., Goudeau, A., Coursaget, P., Perrin, J., Barin, F. and Diop Mar, I., 1981, Active Immunization against Hepatitis B in an Area of High Endemicity, part II. Prog. med. Virol. vol. 27, pp. 185-201.
- Olweny, C.L.M., Atine, I., Kaddu-Mukasa, A., Owor, R., Anderson, M., Klein, G., Henle, W. and de-Thé, G., 1977, Epstein-Barr virus genome studies in Burkitt and non-Burkitt lymphomas in Uganda. J. Natl. Cancer Inst. 58, 1191-1196.
- Peto, R., 1977, Epidemiology, Multistage Models and Short-Term Mutagenicity Tests, in: Origins of Human Cancer, Hiatt, Watson and Winsten, eds. Cold Spring Harbor Conferences on Cell Proliferation vol. 4, pp. 1403-1428.

- Tajima,K., Tominaga, S., Kuroishi, T., Shimizu, H. and Suchi, T. 1979, Geographical Features and Epidemiological Approach to Endemic T-Cell Leukemia/Lymphoma in Japan, Jpn. J. Clin. Oncol. 9, 495-504.
- Zeng, Y., Liu, Y.X., Wei, J.N., Zhu, J.S., Cai, S.L., Wang, P.Z., Zhong, J.M., Li, R.C., Pan, W.J., Li, E.J. and Tan, B.F. 1979, Serological mass survey of nasopharyngeal carcinoma, Acta. Acad. Med. Sin. 1, 123-126.
- Zeng, Y., Liu, Y.X., Liu, Z.R., Zhen, S.W., Wei, J.N., Zhu, J.S. and Zai, H.S., 1980, Application of an immunoenzymatic method and an immunoautoradiographic method for a mass survey of nasopharyngeal carcinoma, Intervirology, 13, 162-168.
- Ziegler, J.L., Andersson, M., Klein, G. and Henle, W., 1976, Detection of Epstein-Barr virus DNA in American Burkitt's lymphoma, Int. J. Cancer, 17: 701-706.

PARTICIPANTS

BACH, FRITZ: Immunobiology Center, Mayo Building, Box 724, University of Minnesota, Minneapolis 55455, U.S.A.

BALDWIN, R.W.: Cancer Research Compaign Labs., University of Nottingham, University Park, Nottingham NG7 2RD, England.

BARLATI, SERGIO: Laboratorio di Genetica Biochimica ed Evoluzionistica, via S. Epifanio, 27100 Pavia, Italy.

BAUER, HEINZ: Institut für Virologie, Frankfurter Str. 107, D-63 Giessen/Lahn, W. Germany.

BAUTZ, E.K.F.: Institute of Molecular Genetics, University of Heidelberg, im Neuenheimer Feld 230, 6900 Heidelberg, W. Germany.

BAUTZ, F.A.: Institute of Molecular Genetics, University of Heidelberg, Im Neuenheimer Feld 230, D-6900 Heidelberg, W. Germany.

BECKER, YECHIEL: Department of Molecular Virology, Hebrew University-Hadassah Medical School, 91000 Jerusalem, Israel.

BENZ, CHRISTOPH C.: Medical Oncology, NS 290, Yale University School of Medicine, 333 Cedar Street, New Haven, Conn. 06510, U.S.A.

BERTAZZONI, UMBERTO: Laboratorio di Genetica Biochemica ed Evoluzionitica, via S. Epifanio 14, 27100 Pavia, Italy.

SEGAL, SHERAGA: Department of Cell Biology, The Weizmann Institute of Science, Rehovot, Israel.

SEKERIS, CONSTANTINE: National Hellenic Research Foundation, 48 Vassileos Constantinou Avenue, Athens, Greece.

BOUMSELL, LAURENCE: Hopital Saint Louis, 2 Place du Dr. Fournier, 75475 Paris Cedex 10, France.

CHANDRA, PRAKASH: Center of Biological Chemistry, Laboratory of Molecular Biology, University Medical School, Theodor-Stern-Kai 7, D-6000 Frankfurt, W. Germany.

CHIECO-BIANCHI, LUIGI: Universita Degli Studi, Instituto di Anatomia Patologica, Cattedra di Oncologica, via A. Gabelli 61, 35100 Padova, Italy.

DAUSSET, JEAN: Hopital Saint Louis, 2 Place du Dr. Fournier, 75475 Paris Cedex 10, France.

DEMIRHAN, ILHAN: Center of Biological Chemistry, Laboratory of Molecular Biology, University Medical School, Tjeodor-Stern-Kai 7, D-6000 Frankfurt, W. Germany.

DESCHAMPS, J.: Department of Molecular Biology, University of Brussels, 67 rue des Chavaux, 1640 Rhode St.-Genese, Belgium.

DEPPERT, WOLFGANG: Abteilung Biochemie, Universität Ulm, Oberer Eselberg M23 Niv3, D-79 ULM,W.Germany.

DUESBERG, PETER: University of California, Department of Molecular Biology, Virus Laboratory, Berkeley California 94720, U.S.A.

EKMAN, PETER: Brady Urological Institute, The John Hopkins Hospital, Baltimore, Md. 21205, U.S.A.

ESSEX, MYRON: Harvard School of Public Health, Department of Viruses in Natorally Occuring Cancer, 665 Huntington Avenue, Boston,Ma.02115, U.S.A.

FELDMAN, MICHAEL: Department of Cell Biology, The Weizmann Institute of Science, P.O.B. 26, Rehovot 76000, Israel.

BOLLUM FRED: Department of Biochemistry, University of Health Services, 4301 Jones Bridge, Bethesda, Maryland 20014, U.S.A.

GALLO, ROBERT C.: Tumor Cell Biology Department, National Cancer Institute, Building 37, 6B 04, Bethesda, Maryland 20014, U.S.A.

GISSMANN, L.: Institut für Virologie, Zentrum der Hygiene, Hermann-Herder-Str. 11, 7800 Freiburg, W. Germany.

HILZ, HELMUTH: Institute of Physiological Chemistry, University of Hamburg, Martini Str. 52, D-2000 Hamburg 20, W. Germany.

HNILICA, LUBOMIR S.: Biochemistry Department, School of Medicine, Vanderbilt University, Nashville, Tennessee 37232, U.S.A.

HOFSCHNEIDER, PETER HANS: Max-Planck Institut für Biochemie, 8033 Martinsried/Munich,W.Germany.

HOMO, FRANCOISE: Physiologie-Pharmacologie, INSERM U.7, Hopital Necker, 161 Rue de Sevres, 75015 Paris, France.

JANIAUD, PAUL: Service de Chimi Cancerologique,Lab. Curie, Institut Curie, 11 Rue Pierre et Marie Curie, 75231 Paris Cedex 05, France.

KAPLAN, HENRY S.: Stanford University, Department of Radiology, Medical Center, Stanford, CA 94305, U.S.A.

KATZAV, SHULAMIT: Department of Cell Biology, The Weizmann Institute of Science, Rehovot 76100, Israel.

KING, R.J.B. : Imperial Cancer Research Fund Labs., Hormonal Biochemistry Department, P.O. Box 123, Lincoln's Inn Fields,London WC2A 3PX, England.

KLEIN EVA: Karolinska Institute, Department of Tumor Biology, S 10401 Stockholm 60, Sweden.

KNUDSON, ALFRED G., Jr.: The Fox Chase Cancer Center, 7701 Burholme Avenue, Philadelphia, Pennsylvania 19111, U.S.A.

KORNHUBER, BERNHARD: Department of Pediatric Oncology, University Medical School, Theodor-Stern-Kai 7, D.6000 Frankfurt 70, W. Germany.

KOTLAR, HANS KR. : Laboratory of Environmental and Occupational Cancer, Norsk Hydro´s Institute of Cancer, Montebello, Oslo 3, Norway.

LARIZZA, LIDIA: Istituto di Biologia Generale, Facolta di Medicina, via Viotti 3/5, 20133 Milano, Italy.

LENNOX, EDWIN S.: Laboratory of Molecular Biology, University Medical School, Hills Road, Cambridge CB 2QH, England.

LIEHL, E. : Sandoz Forschungsinstitut, Brunner Str. 59, A-1235 Wien, Austria.

MARA, HAREUVENI : Tel-Aviv University, George S. Wise Faculty of Life Sciences, Department of Microbiology, Tel-Aviv, Israel.

MEADOWS, ANNA: Children´s Hospital of Philadelphia, Philadelphia, U.S.A.

MITCHISON, N.A.: Department of Zoology, University College of London, Gower Street, London WC 1E 6 BT, England.

NETH, ROLF: Kinderklinik der Universität Hamburg, Martini Str. 52, D-2000 Hamburg 20, W.Germany.

PAFFENHOLZ, VOLKER: Center of Biological Chemistry, Laboratory of Molecular Biology, University Medical School, Theodor-Stern-Kai 7, D-6000 Frankfurt 70, W. Germany.

PAPAMICHAEL, MICHAEL: Immunology Department, Hellenic Anticancer Institute, 171 Alexandra Avenue, Athens 603, Greece.

PAPAS, TAKI: National Cancer Institute, National Institutes of Health, Building 37, 1B 19, Bethsda, Maryland 20205, U.S.A.

PARMIANI, GIORGIO: Istituto Nazionale per Lo Studio e La Cura Dei Tumori, Via Venezian 1, 20133 Milano, Italy.

RAPP,FRED:Department of Microbiology, Hershey Medical Center, Hershey, PA 17033 , U.S.A.

(Due to sudden illness could not come to the meeting but, served as member of the Scientific Committee).

RIBONI, LAURA: Laboratorio Prof.B. Berra, Istituto di Chimica Biologica, Via Salsini 50, 20100 Milano, Italy.

RING,KLAUS: Center of Biological Chemistry, Laboratory of Microbiological Chemistry, University Medical School, Theodor-Stern-Kai 7, D-6000 Frankfurt 70, W. Germany.

SACCHI, NICOLLETA: Istituto di Biologia Generale, Facolta di Medicina, Via Viotti 3/5, 20133 Milano, Italy.

SARIN, PREM: Tumor Cell Biology Laboratory, National Cancer Institute, Building 37, 6C 19, Bethesda, Maryland 20205, U.S.A.

SCHIRRMACHER, VOLKER: DKFZ, Postfach 101949, D-69 Heidelberg 1, W. Germany.

SCHLOSSMAN, STUART F.: Sidney Farber Cancer Institute, Division of Tumor Immunology, 44 Binney Street, Boston,MA 02115, U.S.A.

SCHWARTZ, ELISABETH: Istitut für Virologie, Zentrum der Hygiene, Hermann-Herder-Str. 11, D-78 Freiburg, W. Germany.

FORCHHAMMER, JES: The Fibiger Laboratory, NDR. Frihavensgade 70, DK 2100 Copenhagen, Denmark.

SELIGMANN, M: Laboratoire D'Immunochemie et D'Immunopathologie, Hopital Saint-Louis, 2 Place du Dr. Fournier, 75475 Paris Cedex 10, France.

SHAPIRO, JOSEPH: Department of Cell Biology, The Weizmann Institute of Sciences, Rehovot 76100, Israel.

SHOYAB, MOHAMMED: Laboratory of Viral Carcinogenesis National Cancer Institute, Frederick, Maryland 21701, U.S.A.

TARTAKOVSKY, B.: Department of Cell Biology, The Weizmann Institute of Science, Rehovot 76100, Israel.

THE', GUY BLAUDIN DE: Centre National de la Recherche Scientifique, Rue G. Paradin, 69372 Lyon Cedex 2, France.

TREWYN, RONALD W.: Department of Physiological Chem., 5170 Graves Hall, 333 West 10th Avenue, Columbus, Ohio 43210, U.S.A.

TSAWDAROǴLOU, NIKOS: National Hellenic Research Foundation, 48 Vassileos Const.Avenue,Athens, Greece.

IOANNIDIS, CONSTANTINE: See Tsawdaroglou, Nikos.

TSIAPALIS, C.M.: Biochemistry Department, Hellenic Anticancer Institute, 171 Alexandras Avenue, Athens 603, Greece.

VILLANUEVA, V.R.: C.N.R.S., Charge de Recherche, Institut de Chimie des Substances Naturelles, 91190 Gif-sur Yvette, France.

VILLIERS, E.M. DE: Institut für Virologie, Zentrum der Hygiene, Hermann-Herder-Str. 11, D-7800 Freiburg, W. Germany.

VOGEL, ANGELIKA: Center of Biological Chemistry, Laboratory of Molecular Biology, University Med. Schhol, Theodor-Stern-Kai 7, D-6000 Frankfurt/M., W. Germany.

WAELE, PETER DE: Laboratorium voor Moleculaire Bio-Logie, Rijksuniversiteit Gent, Ledeganckstraat 35, B-9000 Gent, Belgium.

WITZ, ISSAC P.: Faculty of Life Sciences, Tel-Aviv University, Tel Aviv, Israel.

WONG-STAAL, FLOSSIE: Tumor Cell Biology Laboratory, National Cancer Institute, Building 37, 6B 04, Bethesda, Maryland 20205 ,U.S.A.

ZUR HAUSEN, HARALD: Institut für Virologie, Zentrum der Hygiene, Hermann-Herder-Str. 11, D-7800 Freiburg, W. Germany.

INDEX